T0255129

Lecture Notes in Computer Science　　　14287

The series Lecture Notes in Computer Science (LNCS), including its subseries Lecture Notes in Artificial Intelligence (LNAI) and Lecture Notes in Bioinformatics (LNBI), has established itself as a medium for the publication of new developments in computer science and information technology research, teaching, and education.

LNCS enjoys close cooperation with the computer science R & D community, the series counts many renowned academics among its volume editors and paper authors, and collaborates with prestigious societies. Its mission is to serve this international community by providing an invaluable service, mainly focused on the publication of conference and workshop proceedings and postproceedings. LNCS commenced publication in 1973.

Nils Jansen · Mirco Tribastone
Editors

Quantitative Evaluation of Systems

20th International Conference, QEST 2023
Antwerp, Belgium, September 20–22, 2023
Proceedings

Springer

Editors
Nils Jansen ⓘ
Radboud University Nijmegen
Nijmegen, The Netherlands

Mirco Tribastone
IMT School for Advanced Studies Lucca
Lucca, Italy

ISSN 0302-9743   ISSN 1611-3349 (electronic)
Lecture Notes in Computer Science
ISBN 978-3-031-43834-9   ISBN 978-3-031-43835-6 (eBook)
https://doi.org/10.1007/978-3-031-43835-6

This Springer imprint is published by the registered company Springer Nature Switzerland AG
The registered company address is: Gewerbestrasse 11, 6330 Cham, Switzerland

Paper in this product is recyclable.

Preface

This book proudly presents the proceedings of the 20th International Conference on Quantitative Evaluation of SysTems (QEST 2023). This edition of QEST is a special one. The conference has now been running for 20 years. It has, despite strong competition, always been an outstanding venue that has maintained a standard of high-quality publications, reviews, talks, speakers, and discussions. This year has been no exception. QEST 2023 was co-located with CONCUR 2023, FMICS 2023, and FORMATS 2023, as part of CONFEST 2023 in Antwerp.

We are delighted to present a strong and diverse QEST 2023 program with a great and reliable program committee (PC). The 38 PC members, with help from their 27 external reviewers, decided to accept 23 papers out of 44 submissions to QEST. Each submission received at least 3 single-blind reviews. In many cases, 4 or even 5 reviews were delivered. All accepted papers were very strong, and most decisions were unanimous. Artifact evaluation is a well-established part of QEST, and we can report that 12 papers submitted an artifact, out of which 11 were accepted. We plan to edit special issues in the journals International Journal on Software Tools for Technology Transfer (STTT) and Performance Evaluation (PEVA).

As usual, the purpose of QEST was to promote and foster progress in the quantitative evaluation and verification of computer systems and networks. The particular focus was to stay on the cutting edge of research by welcoming contributions that tap into data-driven and machine-learning systems, case studies, and tool papers. Moreover, this year we also experimented with two new types of submissions: work-in-progress and journal-first, to allow the community to share early-stage research or relevant results previously published in journals, respectively.

A great highlight was having two excellent invited speakers, David Parker from the University of Oxford and Frans Oliehoek from Delft University of Technology. Both are known worldwide for their strong research on quantitative aspects of systems. We were particularly happy to reflect on the growing push towards joint efforts in verification and artificial intelligence exemplified by these speakers. David Parker's background is in the verification community, and Frans Oliehoek's is in artificial intelligence.

Finally, we have to thank several people sincerely. First, we want to thank all authors who submitted their work to QEST 2023. Likewise, we thank all PC members and all external reviewers for their hard work and timely reviews. The process of evaluating artifacts is now well established for QEST, and we want to thank Carlos E. Budde and Tim Quatmann for their excellent work in chairing the artifact evaluation. Special thanks go to David Safranek for serving as QEST publicity chair and the local organizing committee of CONFEST. Thanks to the QEST steering committee, particularly its chair Enrico Vicario, for trusting us with the program of this year's special edition. Finally, we thank Joost-Pieter Katoen for stepping in to ensure that these proceedings could be published in time. QEST 2023 was a memorable event, and we want to thank everybody who contributed in any way.

September 2023

Nils Jansen
Mirco Tribastone

Organization

Program Committee

Alessandro Abate	University of Oxford, UK
Ezio Bartocci	TU Wien, Austria
Luca Bortolussi	University of Trieste, Italy
Peter Buchholz	TU Dortmund, Germany
Carlos E. Budde	University of Trento, Italy
Laura Carnevali	University of Florence, Italy
Giuliano Casale	Imperial College London, UK
Milan Ceska Brno	University of Technology, Czech Republic
Pedro R. D'Argenio	Universidad Nacional de Córdoba – CONICET, Argentina
Rayna Dimitrova	CISPA Helmholtz Center for Information Security, Germany
Clemens Dubslaff	Eindhoven University of Technology, The Netherlands
Nathanaël Fijalkow	CNRS, University of Bordeaux, France
Marco Gribaudo	Politecnico di Milano, Italy
Shibashis Guha	Tata Institute of Fundamental Research, India
Sofie Haesaert	Eindhoven University of Technology, The Netherlands
Arnd Hartmanns	University of Twente, The Netherlands
Holger Hermanns	Saarland University, Germany
Nils Jansen	Radboud University, The Netherlands
Sebastian Junges	Radboud University, The Netherlands
Benjamin Lucien	Kaminski Saarland University, Saarland Informatics Campus, Germany
Michaela Klauck	Bosch Research, Germany
Jan Kretinsky	Technical University of Munich, Germany
Diwakar Krishnamurthy	University of Calgary, Canada
Luca Laurenti	TU Delft, The Netherlands
Andrea Marin	Università Ca' Foscari Venezia, Italy
Daniel Neider	TU Dortmund, Germany
Gethin Norman	University of Glasgow, UK
Marco Paolieri	University of Southern California, USA
Dave Parker	University of Oxford, UK
Pavithra Prabhakar	Kansas State University, USA

Tim Quatmann	RWTH Aachen University, Germany
Anne Remke	WWU Münster, Germany
Matteo Sereno	Università di Torino, Italy
Markus Siegle	Universität der Bundeswehr München, Germany
Mirco Tribastone	IMT School for Advanced Studies Lucca, Italy
Max Tschaikowski	Aalborg University, Denmark
Benny Van Houdt	University of Antwerp, The Netherlands
Andrea Vandin	Sant'Anna School of Advanced Studies, Italy
Matthias Volk	University of Twente, The Netherlands
Verena Wolf	Saarland University, Germany
Katinka Wolter	Freie Universität zu Berlin, Germany
David Šafránek	Masaryk University, Czech Republic

Artifact Evaluation Committee

Oyendrila Dobe	Michigan State University, USA
Joshua Jeppson	Utah State University, USA
Bram Kohlen	University of Twente, The Netherlands
Riccardo Reali	University of Florence, Italy
Raphaël Reynouard	Reykjavík University, Iceland
Robert Rubbens	University of Twente, The Netherlands
Andreas Schmidt	Saarland Informatics Campus, Germany
Philipp Schröer	RWTH Aachen University, Germany
Landon Taylor	Utah State University, USA
Djurre van der Wal	University of Twente, The Netherlands
Marck Vandervegt	Radboud University, The Netherlands
Gabriele Venturato	KU Leuven, Belgium
Lulai Zhu	Politecnico di Milano, Italy

Additional Reviewers

Amparore, Elvio Gilberto	Gouberman, Alexander
Andriushchenko, Roman	Grobelna, Marta
Banse, Adrien	Gros, Timo P.
Becker, Lena	Grover, Kush
Benes, Nikola	Großmann, Gerrit
Cairoli, Francesca	Horváth, András
Camerota Verdù, Federico Julian	Lopez-Miguel, Ignacio D.
Edwards, Alec	Mohr, Stefanie
Engelaar, Maico	Padoan, Tommaso

Pontiggia, Francesco
Qi, Shuhao
Romao, Licio
Salvato, Erica
Schupp, Stefan

Stankovic, Miroslav
Stübbe, Jonas
Verscht, Lena
Weininger, Maximilian

Contents

Multi-agent Verification and Control with Probabilistic Model Checking

David Parker$^{(\boxtimes)}$ ⓘ

Department of Computer Science, University of Oxford, Oxford, UK
`david.parker@cs.ox.ac.uk`

Abstract. Probabilistic model checking is a technique for formal automated reasoning about software or hardware systems that operate in the context of uncertainty or stochasticity. It builds upon ideas and techniques from a diverse range of fields, from logic, automata and graph theory, to optimisation, numerical methods and control. In recent years, probabilistic model checking has also been extended to integrate ideas from game theory, notably using models such as stochastic games and solution concepts such as equilibria, to formally verify the interaction of multiple rational agents with distinct objectives. This provides a means to reason flexibly about agents acting in either an adversarial or a collaborative fashion, and opens up opportunities to tackle new problems within, for example, artificial intelligence, robotics and autonomous systems. In this paper, we summarise some of the advances in this area, and highlight applications for which they have already been used. We discuss how the strengths of probabilistic model checking apply, or have the potential to apply, to the multi-agent setting and outline some of the key challenges required to make further progress in this field.

1 Introduction

Probabilistic model checking is a fully automated approach for formal reasoning about systems exhibiting uncertain behaviour, arising for example due to faulty hardware, unpredictable operating environments or the use of randomisation. Probabilistic models, such as Markov chains, Markov decision processes (MDPs) or their variants, are systematically explored and analysed in order to establish whether formal specifications given in temporal logic are satisfied. For models such as MDPs, controllers (policies) can be automatically generated to ensure that such specifications are met, or are optimised for.

Probabilistic model checking builds on concepts and techniques from a wide array of other fields. Its roots lie in formal verification, and it relies heavily on the use of logic and automata. Since models typically need to be solved or optimised numerically, it also adopts methods from Markov chains, control theory, optimisation and, increasingly, from various areas of artificial intelligence.

In the this paper, we discuss the integration of probabilistic model checking with ideas from *game theory*, facilitating the verification or control of multi-agent systems in the context of uncertainty. We focus on *stochastic games*, which

ⓒ The Author(s), under exclusive license to Springer Nature Switzerland AG 2023
N. Jansen and M. Tribastone (Eds.): QEST 2023, LNCS 14287, pp. 1–9, 2023.
https://doi.org/10.1007/978-3-031-43835-6_1

model the interaction between multiple agents (players) operating in a dynamic, stochastic environment. They were introduced in the 1950 ss by Shapley [26], generalising the classic model of Markov decision processes [4] to the case of multiple players, and techniques for their solution have been well studied [12].

In the context of formal verification, game-theoretic modelling has a number of natural applications, in particular when an agent interacts with an *adversary* that has opposing goals, for example a defender and an attacker in the context of a computer security protocol or honest and malicious participants in a distributed consensus algorithm. For verification or control problems in stochastic environments, game-based models also underlie methods for *robust* verification, using worst-case assumptions of epistemic model uncertainty. Furthermore, game theory provides tools for controller synthesis in a more *cooperative* setting, for example via the use of *equilibria* representing strategies for collections of players with differing, but not strictly opposing objectives.

Verification problems for stochastic games have been quite extensively studied (see, e.g., [5,6,27]) and in recent years, probabilistic model checking frameworks and tools have been developed (e.g., [8,18,19]) and applied to a variety of problem domains. In the next section, we summarise some of these advances. We then go on to discuss the particular strengths of probabilistic model checking, the ways in which these are applicable to multi-agent models and some of the remaining challenges that exist in the field.

2   Model Checking for Stochastic Games

2.1   Turn-based Stochastic Games

Stochastic games comprise a set of n players making a sequence of decisions that have stochastic outcomes. The way in which players interact can be modelled in various ways. The simplest is with a *turn-based stochastic game* (TSG). The state space S of the game is partitioned into disjoint sets $S_1 \uplus \cdots \uplus S_n = S$, where states in S_i are controlled by player i. Players choose between actions from a set A (for simplicity, let us assume that all actions are available to be taken in each state) and the dynamics of the model is captured, like for an MDP, by a probabilistic transition function $P : S \times A \times S \rightarrow [0,1]$, with $P(s,a,s')$ giving the probability to move to state s' when action a is taken in state s.

A probabilistic model checking framework for TSGs is presented in [8], which proposes the logic rPATL (and its generalisation rPATL*) for specifying *zero-sum* properties of stochastic games, adding probabilistic and reward operators to the well known game logic ATL (alternating temporal logic) [2]. A simple (numerical) query is $\langle\!\langle ag_1, ag_2 \rangle\!\rangle P_{\max=?}[\,\mathsf{F}\ \mathsf{goal}\,]$, which asks for the maximum probability of reaching a set of states $\mathsf{goal} \subseteq S$ that is achievable by a coalition of the players ag_1 and ag_2, assuming that any other players in the game have the directly opposing objective of minimising the probability of this event.

Despite a worse time complexity than MDPs for the core underlying problems (e.g., computing optimal reachability probabilities, for the query above, is in NP $\cap$ co-NP, rather than PTIME), [8] shows that value iteration (dynamic programming) is in practice an effective and scalable approach. For maximising probabilistic reachability with two players, the values x_s^k defined below converge, as $k \to \infty$, to the required probability for each state s:

$$
x_s^k = \begin{cases}
1 & s \in \mathsf{goal} \\
0 & s \notin \mathsf{goal} \text{ and } k = 0 \\
\max_{a \in A} \sum_{s' \in S} P(s, a, s') \cdot x_{s'}^{k-1} & s \in S_1 \backslash \mathsf{goal} \text{ and } k > 0 \\
\min_{a \in A} \sum_{s' \in S} P(s, a, s') \cdot x_{s'}^{k-1} & s \in S_2 \backslash \mathsf{goal} \text{ and } k > 0
\end{cases}
$$

The computation yields optimal strategies for players, which are deterministic (i.e., pick a single action in each state) and memoryless (i.e., do so regardless of history). The model checking algorithm [8] extends to many other temporal operators including a variety of reward (or cost) based measures.

Subsequently, various other aspects of TSG model checking have been explored, including the performance of different game solution techniques [17], the use of interval iteration methods to improve accuracy and convergence [10], trade-offs between multiple objectives [3,7] and the development of symbolic (decision diagram based) implementations to improve scalability [21].

Despite the relative simplicity of TSGs as a modelling formalism, they have been shown to be appropriate for various scenarios in which there is natural turn-based alternation between agents; examples include human-robot control systems [11,15] and self-adaptive software systems interacting with their environment [9].

2.2   Concurrent Stochastic Games

To provide a more realistic model of the concurrent execution of agents, we can move to the more classic view of player interaction in stochastic games, where players make their decisions simultaneously and independently. To highlight the distinction with the turn-based model variant discussed above, we call these *concurrent stochastic games* (CSGs); the same model is referred to elsewhere as, for example, Markov games or multi-agent Markov decision processes. In a CSG, each player i has a separate set of actions A_i and the probabilistic transition function $P : S \times (A_1 \times \cdots \times A_n) \times S \to [0, 1]$ now models the resulting stochastic state update that arises for each possible joint player action.

Probabilistic model checking of CSGs against zero-sum objectives, again using the logic rPATL, is proposed in [19]. Crucially, optimal strategies for players are now randomised, i.e., can choose a probability of selecting each action in each state, however, a value iteration approach can again be adopted. Consider again maximal probabilistic reachability for two players: instead of simply picking the highest value action for one player in each state, a one-shot matrix game Z, indexed over action sets A_1 and A_2, is solved at each state:

$$
x_s^k = \begin{cases}
1 & s \in \mathsf{goal} \\
0 & s \notin \mathsf{goal} \text{ and } k = 0 \\
\mathrm{val}(Z) & s \notin \mathsf{goal} \text{ and } k > 0
\end{cases}
\qquad \text{where } Z_{a,b} = \sum_{s' \in S} P(s, (a, b), s') \cdot x_{s'}^{k-1}
$$

The matrix game Z contains payoff values for player 1 and the value val(Z) of Z is the optimal achievable value when player 2 minimises. This is solved via a (small) linear programming problem over variables $\{p_a \mid a \in A_1\}$ which yields the optimal probabilities p_a for player 1 to pick each action a. Despite the increase in solution complexity with respect to TSGs, results in [19] show the feasibility of building and solving large CSGs that model examples taken from robotics, computer security and communication protocols. These also highlight deficiencies when the same examples are modelled with TSGs.

2.3 Equilibria for Stochastic Games

Zero-sum objectives, for example specified in rPATL, allow synthesis of optimal controllers in the context of both stochasticity and adversarial behaviour. But there are many instances where agents do not have directly opposing goals. The CSG probabilistic model checking framework has been extended to incorporate non-zero-sum objectives such as Nash equilibria (NE) [19]. Informally, strategies for a set of players with distinct, individual objectives form an NE when there is no benefit to any of them of unilaterally changing their strategy.

It is shown in [19], that by focusing on *social welfare* NE, which also maximise the sum of all players' objectives, value iteration can again be applied. Extending rPATL, we can write for example $\langle\!\langle ag_1{:}ag_2 \rangle\!\rangle_{\max=?}(\mathrm{P}[\mathrm{F}\ \mathsf{goal}_1]+\mathrm{P}[\mathrm{F}\ \mathsf{goal}_2])$ to ask for the social welfare NE in which two players maximise the probabilities of reaching distinct sets of state goal_1 and goal_2. Value iteration becomes:

$$
x_s^k =
\begin{cases}
(1,1) & s \in \mathsf{goal}_1 \cap \mathsf{goal}_2 \\
(1, P_{s,\mathsf{goal}_2}^{\max}) & s \in \mathsf{goal}_1 \backslash \mathsf{goal}_2 \\
(P_{s,\mathsf{goal}_1}^{\max}, 1) & s \in \mathsf{goal}_2 \backslash \mathsf{goal}_1 \\
(0,0) & s \notin (\mathsf{goal}_1 \cup \mathsf{goal}_2)\ \text{and}\ k = 0 \\
\mathrm{val}(Z^1, Z^2) & s \notin (\mathsf{goal}_1 \cup \mathsf{goal}_2)\ \text{and}\ k > 0
\end{cases}
$$

$$
\text{where } Z_{a,b}^i = \sum_{s' \in S} P(s,(a,b),s') \cdot x_{s'}^{k-1}(i),
$$

val(Z^1, Z^2) is the value of a *bimatrix game* and $P_{s,\mathsf{goal}_i}^{\max}$ is the maximum probability of reaching goal_i from state s, which can be computed independently by treating the stochastic game as an MDP. The value of the (one-shot) game defined by payoff matrices Z^i for player i is a (social welfare) NE, computed in [19] using an approach called labelled polytopes [22] and a reduction to SMT. Optimal strategies (in fact, ϵ-optimal strategies) can be extracted. They are, as for zero-sum CSGs, randomised but also require memory.

The move towards concurrent decision making over distinct objectives opens up a variety of interesting directions for exploration. Equilibria-based model checking of CSG is extended in several ways in [20]. Firstly, *correlated* equilibria allow players to coordinate through the use of a (probabilistic) public signal, which then dictates their individual strategies. These are shown to be cheaper to compute and potentially more equitable in the sense that they improve joint outcomes. Secondly, *social fairness* is presented as an alternative optimality criterion to social welfare, which minimises the *difference* between players' objectives, something that is ignored by the latter criterion.

3   Opportunities and Challenges

Probabilistic model checking is a flexible technique, which already applies to many different types of stochastic models and temporal logic specifications. On the one hand, the thread of research described above represents a further evolution of the approach towards a new class of models and solution methods. On the other, it represents an opportunity to apply the strengths of probabilistic model checking to a variety of problem domains in which multi-agent approaches to modelling are applicable and where guarantees on safety or reliability may be essential; examples include multi-robot coordination, warehouse logistics, autonomous vehicles or robotic swarms. We now discuss some of the key benefits of probabilistic model checking and their relevance to the multi-agent setting. We also summarise some of the challenges that arise as a result.

Temporal Logic. Key to the success of model checking based techniques is the ability to precisely and unambiguously specify desired system properties in a formal language that is expressive enough to be practically useful, but constrained enough that verification or control problems are practical.

Temporal logics such as rPATL and its extensions show that it is feasible to combine quantitative aspects (probability, costs, rewards) with reasoning about the strategies and objectives of multiple agents, for both zero-sum optimality and equilibria of various types. This combines naturally with the specification of temporal behaviour using logics such as LTL, from simple reachability or reach-avoid goals, to more complex sequences of events and long-run specifications. The latter have been of increasing interest in, for example, task specification in robotics [16] or reinforcement learning [14]. Another key benefit here is the continual advances in translations from such logics to automata, facilitating algorithmic analysis. From a multi-agent perspective, specifically, challenges include more expressive reasoning about dependencies between strategies or epistemic aspects, where logic extensions exist but model checking is challenging.

Tool Support and Modelling Languages. The main functionality for model checking of stochastic games described in Sect. 2 is implemented within the PRISM-games tool [18], which has been developed over the past 10 years. However, interest in verification for this class of models is growing and, for the simpler model variant of TSGs, support now exists in multiple actively developed probabilistic model checking tools [13, 17, 24, 25].

Currently, these tools share a common formalism for specifying models, namely PRISM-games's extension to stochastic games of the widely used PRISM modelling language. Although relatively simple from a programming language perspective, this has proved to be expressive enough for modelling across a large range of application domains. Key modelling features include the ability to describe systems as the parallel composition of multiple, interacting (sometimes duplicated) components and the use of parameters for easy reconfiguration of models. It also provides a common language for many different types of probabilistic models through various features that can be combined, e.g., clocks (for *real-time* modelling), observations (for *partially observable* models) and model uncertainty such as transition probability intervals (for *epistemic uncertainty*).

Component-based modelling is of particular benefit for the multi-agent setting, but challenges remain as the modelling language evolves. Examples include dealing with the subtleties that arise regarding how components communicate and synchronise, particularly under partial observability, and the specification of particular strategies for some agents.

Exhaustive Analysis. Traditionally, a strength of model checking based techniques is their focus on an exhaustive model analysis, allowing them to identify (or prove the absence of) corner cases representing erroneous or anomalous behaviour. In the stochastic setting this remains true, in particular for models that combine probabilistic and nondeterministic behaviour. The subtle interaction between these aspects can be difficult to reason about without automated tools, and exhaustive approaches can in these cases be preferable to more approximate model solution methods, such as those based on simulation.

Adding multiple players only strengthens the case for these techniques. For example, [8] identifies weaknesses in a distributed, randomised energy management protocol arising when some participants behave selfishly; a simple incentive scheme is then shown to help alleviate this issue. Multi-agent models allow a combination of *control* and *verification*, for example synthesising a controller (strategy) for one player, or coalition of players, which can be verified to perform robustly in the context of adversarial behaviour by other players.

A natural direction is to then deploy verification to controllers synthesised by more widely applicable and more scalable approaches such as multi-agent reinforcement learning [1]. This brings challenges in terms of, for example, extending probabilistic model checking to continuous state spaces, and tighter integration with machine learning methods. Progress in this direction includes the extension of CSGs to a *neuro-symbolic* setting [28,29], which incorporates neural networks for specific model components, e.g. for perception tasks.

Further Challenges. In addition to those highlighted above, numerous other challenges exist in the field. One perennial issue with model checking approaches is their *scalability* to large systems. Symbolic approaches, e.g., based on decision diagrams, have proved to be valuable for probabilistic model checking, and also extended to TSGs [21]. However it is unclear how to adapt these to CSGs: while value iteration is often amenable to a symbolic implementation, methods such as linear programming or bimatrix game solution are typically not.

The use of modelling formalisms like the PRISM-games language should also encourage the development of *compositional* analysis techniques, such as counter abstraction [23] or assume-guarantee methods, but progress remains limited in these directions within probabilistic model checking. On a related note, while human-created models naturally exhibit high-level structure, strategies synthesised by model checking tools typically do not. This hinders comprehension and explainability, which becomes more important when strategies, as here, are more complex due to randomisation, memory and distribution across agents.

There are also major *algorithmic* challenges which arise as the techniques are applied to new problems. For example, there have been steady advances in verification techniques for *partially observable* MDPs, but much less work on this topic for stochastic games. Finally, there are potential benefits from further exploration of ideas from game theory, e.g., other equilibria, such as Stackelberg (with applications, for instance, to security or automotive settings) or the inclusion of epistemic aspects into logics and model checking algorithms.

Acknowledgements. This work was funded in part by the ERC under the European Union's Horizon 2020 research and innovation programme (FUN2MODEL, grant agreement No. 834115).

References

1. Albrecht, S.V., Christianos, F., Schäfer, L.: Multi-Agent Reinforcement Learning: Foundations and Modern Approaches. MIT Press, Cambridge (2023)
2. Alur, R., Henzinger, T., Kupferman, O.: Alternating-time temporal logic. J. ACM **49**(5), 672–713 (2002)
3. Basset, N., Kwiatkowska, M., Wiltsche, C.: Compositional strategy synthesis for stochastic games with multiple objectives. IC **261**(3), 536–587 (2018)
4. Bellman, R.: Dynamic Programming. Princeton University Press, Princeton (1957)
5. Chatterjee, K.: Stochastic ω-Regular Games. Ph.D. thesis, University of California at Berkeley (2007)
6. Chatterjee, K., Henzinger, T.: A survey of stochastic ω-regular games. J. CSS **78**(2), 394–413 (2012)
7. Chatterjee, K., Katoen, J.-P., Weininger, M., Winkler, T.: Stochastic games with lexicographic reachability-safety objectives. In: Lahiri, S.K., Wang, C. (eds.) CAV 2020. LNCS, vol. 12225, pp. 398–420. Springer, Cham (2020). https://doi.org/10.1007/978-3-030-53291-8_21
8. Chen, T., Forejt, V., Kwiatkowska, M., Parker, D., Simaitis, A.: Automatic verification of competitive stochastic systems. FMSD **43**(1), 61–92 (2013)
9. Cámara, J., Garlan, D., Schmerl, B., Pandey, A.: Optimal planning for architecture-based self-adaptation via model checking of stochastic games. In: Proceedings of SAC 2015, pp. 428–435. ACM (2015)
10. Eisentraut, J., Kelmendi, E., Kretínský, J., Weininger, M.: Value iteration for simple stochastic games: stopping criterion and learning algorithm. Inf. Comput. **285**(Part), 104886 (2022)
11. Feng, L., Wiltsche, C., Humphrey, L., Topcu, U.: Controller synthesis for autonomous systems interacting with human operators. In: Proceedings of ICCPS 2015, pp. 70–79. ACM (2015)

12. Filar, J., Vrieze, K.: Competitive Markov Decision Processes. Springer, New York, NY, USA (1996). https://doi.org/10.1007/978-1-4612-4054-9
13. Fu, C., Hahn, E.M., Li, Y., Schewe, S., Sun, M., Turrini, A., Zhang, L.: EPMC gets knowledge in multi-agent systems. In: Finkbeiner, B., Wies, T. (eds.) VMCAI 2022. LNCS, vol. 13182, pp. 93–107. Springer, Cham (2022). https://doi.org/10.1007/978-3-030-94583-1_5
14. Hammond, L., Abate, A., Gutierrez, J., Wooldridge, M.J.: Multi-agent reinforcement learning with temporal logic specifications. In: Proceedings of 20th International Conference on Autonomous Agents and Multiagent Systems (AAMAS 2021), pp. 583–592. ACM (2021)
15. Junges, S., Jansen, N., Katoen, J.-P., Topcu, U., Zhang, R., Hayhoe, M.: Model checking for safe navigation among humans. In: McIver, A., Horvath, A. (eds.) QEST 2018. LNCS, vol. 11024, pp. 207–222. Springer, Cham (2018). https://doi.org/10.1007/978-3-319-99154-2_13
16. Kress-Gazit, H., Fainekos, G.E., Pappas, G.J.: Temporal logic-based reactive mission and motion planning. IEEE Trans. Robot. **25**(6), 1370–1381 (2009)
17. Kretínský, J., Ramneantu, E., Slivinskiy, A., Weininger, M.: Comparison of algorithms for simple stochastic games. Inf. Comput. **289**(Part), 104885 (2022)
18. Kwiatkowska, M., Norman, G., Parker, D., Santos, G.: PRISM-games 3.0: stochastic game verification with concurrency, equilibria and time. In: Lahiri, S.K., Wang, C. (eds.) CAV 2020. LNCS, vol. 12225, pp. 475–487. Springer, Cham (2020). https://doi.org/10.1007/978-3-030-53291-8_25
19. Kwiatkowska, M., Norman, G., Parker, D., Santos, G.: Automatic verification of concurrent stochastic systems. Formal Methods Syst. Des. **58**, 188–250 (2021)
20. Kwiatkowska, M., Norman, G., Parker, D., Santos, G.: Correlated equilibria and fairness in concurrent stochastic games. In: Fisman, D., Rosu, G. (eds.) Tools and Algorithms for the Construction and Analysis of Systems. TACAS 2022. LNCS, vol .13244, pp. 60–78. Springer, Cham (2022). https://doi.org/10.1007/978-3-030-99527-0_4
21. Kwiatkowska, M., Norman, G., Parker, D., Santos, G.: Symbolic verification and strategy synthesis for turn-based stochastic games. In: Raskin, JF., Chatterjee, K., Doyen, L., Majumdar, R. (eds.) Principles of Systems Design. LNCS, vol. 13660, pp. 388–406. Springer, Cham (2022). https://doi.org/10.1007/978-3-031-22337-2_19
22. Lemke, C., J. Howson, J.: Equilibrium points of bimatrix games. J. Soc. Ind. Appl. Math. **12**(2), 413–423 (1964)
23. Lomuscio, A., Pirovano, E.: A counter abstraction technique for verifying properties of probabilistic swarm systems. Artif. Intell. **305**, 103666 (2022)
24. Meggendorfer, T.: PET – a partial exploration tool for probabilistic verification. In: Bouajjani, A., Holík, L., Wu, Z. (eds.) Automated Technology for Verification and Analysis. ATVA 2022. LNCS, vol. 13505, pp. 320–326. Springer, Cham (2022). https://doi.org/10.1007/978-3-031-19992-9_20
25. Pranger, S., Könighofer, B., Posch, L., Bloem, R.: TEMPEST - synthesis tool for reactive systems and shields in probabilistic environments. In: Hou, Z., Ganesh, V. (eds.) ATVA 2021. LNCS, vol. 12971, pp. 222–228. Springer, Cham (2021). https://doi.org/10.1007/978-3-030-88885-5_15
26. Shapley, L.: Stochastic games. PNAS **39**, 1095–1100 (1953)

27. Ummels, M.: Stochastic Multiplayer Games: Theory and Algorithms. Ph.D. thesis, RWTH Aachen University (2010)
28. Yan, R., Santos, G., Norman, G., Parker, D., Kwiatkowska, M.: Strategy synthesis for zero-sum neuro-symbolic concurrent stochastic games. arXiv2202.06255 (2022)
29. Yan, R., Santos, G., Duan, X., Parker, D., Kwiatkowska, M.: Finite-horizon equilibria for neuro-symbolic concurrent stochastic games. In: Proceedings of 38th Conference on Uncertainty in Artificial Intelligence (UAI 2022), AUAI Press (2022)

Formal Controller Synthesis for Markov Jump Linear Systems with Uncertain Dynamics

Luke Rickard[1]([✉]) (ID), Thom Badings[2] (ID), Licio Romao[1] (ID), and Alessandro Abate[1] (ID)

[1] University of Oxford, Oxford, UK
rickard@robots.ox.ac.uk
[2] Radboud University, Nijmegen, The Netherlands

Abstract. Automated synthesis of provably correct controllers for cyber-physical systems is crucial for deployment in safety-critical scenarios. However, hybrid features and stochastic or unknown behaviours make this problem challenging. We propose a method for synthesising controllers for Markov jump linear systems (MJLSs), a class of discrete-time models for cyber-physical systems, so that they certifiably satisfy probabilistic computation tree logic (PCTL) formulae. An MJLS consists of a finite set of stochastic linear dynamics and discrete jumps between these dynamics that are governed by a Markov decision process (MDP). We consider the cases where the transition probabilities of this MDP are either known up to an interval or completely unknown. Our approach is based on a finite-state abstraction that captures both the discrete (mode-jumping) and continuous (stochastic linear) behaviour of the MJLS. We formalise this abstraction as an interval MDP (iMDP) for which we compute intervals of transition probabilities using sampling techniques from the so-called 'scenario approach', resulting in a probabilistically sound approximation. We apply our method to multiple realistic benchmark problems, in particular, a temperature control and an aerial vehicle delivery problem.

Keywords: Markov Jump Linear Systems · Stochastic Models · Uncertain Models · Robust Control Synthesis · Temporal logic · Safety Guarantees

1 Introduction

In a world where autonomous cyber-physical systems are increasingly deployed in safety-critical settings, it is important to develop methods for certifiable control of these systems [32]. Cyber-physical systems are characterised by the coupling of digital (discrete) computation with physical (continuous) dynamical components. This results in a *hybrid system*, endowed with different discrete modes of

This work was supported by funding from the EPSRC AIMS CDT EP/S024050/1, and by NWO grant NWA.1160.18.238 (PrimaVera).

operation, each of which is characterised by its own continuous dynamics [35]. Ensuring that these hybrid systems meet complex and rich formal specifications when controlled is an important yet challenging goal.

Formal Controller Synthesis. Often, these specifications cannot be expressed as classical control-theoretic objectives, which by and large relate to stability and convergence, or invariance and robustness [8]. Instead, these requirements can be expressed in a temporal logic, which is a rich language for specifying the desired behaviour of dynamical systems [44]. In particular, probabilistic computation tree logic (PCTL, [27]) is widely used to define temporal requirements on the behaviour of probabilistic systems. For example, in a building temperature control problem, a PCTL formula can specify that, with at least 75% probability, the temperature must stay within the range 22–23 °C for 10 min. Leveraging probabilistic verification tools [7], it is of interest to synthesise a controller that ensures the satisfaction of such a PCTL formula for the model under study [26].

Markov Jump Linear Systems. Markov jump linear systems (MJLSs) [20] are a well-known class of stochastic, hybrid models suitable for capturing the behaviour of cyber-physical systems [35]. An MJLS consists of a finite set of linear dynamics (also called *operational modes*), where jumps between these modes are governed by a Markov chain (MC) or, if jumping between the modes can be controlled, by a Markov Decision Process (MDP). Despite each mode having linear (though possibly stochastic) dynamics, the overall dynamics are non-linear due to the jumping between modes. MJLSs have been used to model, among other things, networked control systems, where the different operation modes relate to specific packet losses or to distinct discrete configurations [29,42].

Uncertainty in MJLSs. We consider a rich class of discrete-time MJLSs with two sources of uncertainty. First, the continuous dynamics in each mode are affected by an additive stochastic process noise, e.g., due to inaccurate modelling or wind gusts affecting a drone [10]. We only assume sampling-access to the noise, rather than full knowledge of its probability distribution, allowing us to provide probably approximately correct (PAC) guarantees on the behaviour of the MJLS. Second, similar to [42], we assume that the transition probabilities of the Markov jump process are not precisely known. However, unlike [42], we consider two different semantics for this uncertainty: either (1) transition probabilities between modes are given by intervals; or (2) these probabilities are not known at all [31,36]. More details on the considered model are in Sect. 2.

Problem Statement. Several MJLS control problems have been studied, such as stability [12,56], H_∞-controller design [19,21,22,55], and optimal control [30,54]. However, limited research has been done for more complex tasks expressed in, for example, PCTL. In this paper, we thus solve the following problem. Given an MJLS subject to uncertainty in both its continuous dynamics (via additive noise of an unknown distribution) and its discrete behaviour (uncertain Markov jumps), compute a provably correct controller that satisfies a given PCTL formula.

Abstractions of MJLSs. We develop a new technique for abstracting MJLSs by extending methods introduced for linear non-hybrid systems in [4]. In line with [4], we capture the stochastic noise affecting the continuous dynamics by means of transition probability *intervals* between the discrete states of the abstraction. We compute these intervals using sampling techniques from the *scenario approach* [16] and leverage the tighter theoretical bounds developed in [49]. We thus formalise the resulting abstract model as an interval MDP (iMDP), which is an MDP with transition probabilities given as intervals [24]. Different from [4], we also newly capture the discrete mode jumps in the abstract iMDP.

Controller Synthesis. We use the state-of-the-art verification tool PRISM [33] to synthesise a policy on the abstract iMDP that satisfies a given PCTL specification. Leveraging results from the scenario approach, we refine this policy into a controller for the MJLS with PAC guarantees on the satisfaction of the specification.

Contributions. Our main contribution is a framework to synthesise provably-correct controllers for discrete-time MJLSs given general PCTL specifications, based on iMDP abstractions of the MJLSs. Previous work in this area has been limited to linear time-invariant dynamics, and to simpler reach-avoid specifications [4]. We thus extend earlier techniques by developing new methods for a broader class of hybrid models (MJLSs) and for general PCTL formulae. In line with previous work, we propose a semi-algorithm based on iterative refinements of our model, meaning that a synthesised controller will satisfy the required formula, but the inability to find such a controller does not imply the non-existence of one. Technically, we newly show how to capture both the continuous and discrete dynamics of the MJLS in the abstract iMDP model. In particular, our methods are applicable to MJLSs where the stochastic noise in the continuous dynamics and that in the transition probabilities of the Markov jump process are unknown.

Related Work

Techniques for providing safety guarantees for dynamical systems can largely be split into two approaches, respectively called *abstraction-free* and *abstraction-based* [35].

Abstraction-free methods derive safety guarantees without the need to create simpler abstract models. For example, barrier functions [37,43,48] can be used to certify the existence of control inputs that keep the system within safe states. Another approach is that of (probabilistic) reachability computation [3,41], where the goal is to evaluate if the system will reach a certain state over a given horizon.

Abstraction-based methods [8,51] analyse a simpler model of the system, formally shown to be related to the concrete model, and thus allow to transfer the obtained results (safety guarantees, or synthesised policies) back to the original model. Various approaches exist for creating abstractions of different forms, including the celebrated counterexample-guided abstraction/refinement approach [18] and, relevant for this work, a few involve abstractions as Markov models [1,3,5,6,17,50].

Related to the approaches detailed above is robust control, where the goal is to compute a controller that achieves some task while being robust against disturbances. Robust control techniques for MJLSs have been studied in [9,13,52].

2   Foundations and Problem Statement

2.1   Markov Decision Processes

A Markov decision process (MDP) is a tuple $\mathcal{M} = (\mathcal{S}, \mathcal{A}, s_I, P)$ where $\mathcal{S}$ is a finite set of states, $\mathcal{A}$ is a finite set of actions, $s_I \in \mathcal{S}$ is the initial state, and $P\colon \mathcal{S} \times \mathcal{A} \rightharpoonup Dist(\mathcal{S})$ is a (partial) probabilistic transition function, with $Dist(\mathcal{S})$ the set of all probability distributions over $\mathcal{S}$ [7]. We call a tuple (s, a, s') with probability $P(s, a)(s') > 0$ a *transition*. We write $\mathcal{A}(s) \subseteq \mathcal{A}$ for the actions enabled in state s. A Markov chain (MC) is an MDP such that $|\mathcal{A}(s)| = 1, \forall s \in S$. We consider time-dependent deterministic (or pure) policies, $\pi\colon \mathcal{S} \times \mathbb{N} \to \mathcal{A}$, which map states $s \in \mathcal{S}$ and time steps $k \in \mathbb{N}$, to actions $a \in \mathcal{A}(s)$. The set of all policies for MDP $\mathcal{M}$ is denoted by $\Pi_{\mathcal{M}}$.

Interval Markov decision processes (iMDPs) extend regular MDPs with uncertain transition probabilities [24]. An iMDP is a tuple $\mathcal{M}_{\mathbb{I}} = (\mathcal{S}, \mathcal{A}, s_I, \mathcal{P})$, where the states and actions are defined as for MDPs, and $\mathcal{P}\colon \mathcal{S} \times \mathcal{A} \rightharpoonup 2^{Dist(\mathcal{S})}$ maps states and actions to a set of distributions over successor states. Specifically, each $\mathcal{P}(s, a)(s')$ is an *interval* of the form $[\underline{p}, \overline{p}]$, with $\underline{p}, \overline{p} \in (0, 1], \underline{p} \leq \overline{p}$. Intuitively, an iMDP encompasses a set of MDPs differing only in their transition probabilities: fixing an allowable probability distribution in the set $\mathcal{P}(s, a)$ for every state-action pair (s, a) (denoted $P \in \mathcal{P}$ for brevity) results in an MDP, denoted by $\mathcal{M}_{\mathbb{I}}^P$.

2.2   Markov Jump Linear Systems

Let $\mathcal{Z} = \{z_1, \ldots, z_N\}$ be a finite set of discrete modes. Consider the collections of matrices $A = (A_1, \ldots, A_N), A_i \in \mathbb{R}^{n \times n}$, and $B = (B_1, \ldots, B_N), B_i \in \mathbb{R}^{n \times m}$; and of vectors $q = (q_1, \ldots, q_N), q_i \in \mathbb{R}^n$. A discrete-time MJLS model $\mathfrak{J}$ comprises continuous and discrete dynamics. Each triple (A_i, B_i, q_i) defines a linear dynamical system, with discrete-time dynamics in (1a). The discrete jumps between the N modes in $\mathcal{Z}$ are governed by an MDP $(\mathcal{Z}, \mathcal{B}, z_I, T)$ with *switching actions* $\mathcal{B} = \{1, \ldots, M\}$, and *mode switch* transition function $T\colon \mathcal{Z} \times \mathcal{B} \rightharpoonup Dist(\mathcal{Z})$. At any time $k \in \mathbb{N}$, we denote the continuous state by $x(k) \in \mathcal{X} \subseteq \mathbb{R}^n$ ($\mathcal{X}$ bounded), and the discrete mode by $z(k) \in \mathcal{Z}$. Given initial state $x(0) \in \mathcal{X}, z(0) \in \mathcal{Z}$, the (hybrid) state (x, z) is computed as

$$\mathfrak{J}\colon \begin{cases} x(k+1) = A_{z(k)}x(k) + B_{z(k)}u(k) + q_{z(k)} + w_{z(k)}(k) & (1a) \\ z(k+1) \sim T(z(k), b(k)), & (1b) \end{cases}$$

where $u(k) \in \mathcal{U} \subseteq \mathbb{R}^m$ is the control input to the continuous dynamics, and $b(k) \in \mathcal{B}$ is the discrete (MDP) switching action. Note, for each mode $z \in \mathcal{Z}$, the corresponding continuous dynamics are affected by an additive stochastic process noise w_z, with a (potentially) unknown distribution. The distribution of the noise w_z is not required to be the same across different modes, but $\{w_z(k)\}_{k \in \mathbb{N}}$ must be an i.i.d. stochastic process having density with respect to the Lebesgue measure, and independent across modes. Importantly for our setting, the input u and switch b are *jointly determined* by a feedback controller (namely, a policy for the MJLS) of the following form.

Definition 1. *A time-dependent feedback controller $F: \mathcal{X} \times \mathcal{Z} \times \mathbb{N} \to \mathcal{U} \times \mathcal{B}$ is a function that maps a continuous state $x \in \mathcal{X}$, discrete mode $z \in \mathcal{Z}$, and time step $k \in \mathbb{N}$, to a continuous control input $u \in \mathcal{U}$ and discrete switch $b \in \mathcal{B}$.*

Example 1. Consider a temperature regulation problem inspired by [3], in which a portable fan heater and a portable radiator are used to heat a two-room building. We define two modes $\mathcal{Z} = \{1, 2\}$, relating to the fan heater being in room 1 or 2 respectively (and the radiator in the other room). Swapping the heat sources between rooms is modelled by an MDP with actions $\mathcal{B} = \{0, 1\}$, relating to leaving or switching the heaters. Each of these mode-switching actions fails with some probability. This problem is naturally modelled as an MJLS with the matrices

$$
A_{\{1,2\}} = \begin{bmatrix} 1 - b_1 - a_{12} & a_{12} \\ a_{21} & 1 - b_2 - a_{21} \end{bmatrix},
$$

$$
B_1 = \begin{bmatrix} k_f & 0 \\ 0 & k_r \end{bmatrix}, \quad B_2 = \begin{bmatrix} k_r & 0 \\ 0 & k_f \end{bmatrix}, \quad q_{\{1,2\}} = \begin{bmatrix} b_1 x_a \\ b_2 x_a \end{bmatrix}, \tag{2}
$$

where the state $x = [T_1, T_2]^\top \in \mathbb{R}^2$ models the room temperatures, and the power of both heaters can be adjusted within the range $u \in [0, 1]^2$ (the extrema denoting being fully on and off). In Sect. 6, we perform a numerical experiment with this MJLS. □

2.3 Probabilistic Computation Tree Logic

Probabilistic computation tree logic (PCTL) depends on the following syntax [7]:

$$
\begin{aligned}
\Phi ::&= true \mid p \mid \neg\Phi \mid \Phi \wedge \Phi \mid \mathsf{P}_{\sim\lambda}(\psi) \\
\psi ::&= \Phi\mathsf{U}\Phi \mid \Phi\mathsf{U}^{\leq K}\Phi \mid \mathsf{X}\Phi.
\end{aligned} \tag{3}
$$

Here, $\sim \in \{<, \leq, \geq, >\}$ is a comparison operator and $\lambda \in [0, 1]$ a probability threshold; PCTL formulae Φ are state formulae, which can in particular depend on path formulae ψ. Informally, the syntax consists of state labels $p \in AP$ in a set of atomic propositions AP, propositional operators negation $\neg$ and conjunction $\wedge$, and temporal operators until U, bounded until $\mathsf{U}^{\leq K}$, and next X. The probabilistic operator $\mathsf{P}_{\sim\lambda}(\psi)$ requires that paths generated from the initial conditions satisfy a path formula ψ with total probability exceeding (or below, depending on $\sim$) some given threshold λ.

An MJLS $\mathfrak{J}$ with a controller F induces a stochastic process on the hybrid state space $\mathcal{X} \times \mathcal{Z}$. Let $L_{\mathfrak{J}}: \mathcal{X} \times \mathcal{Z} \to 2^{AP}$ be a labelling from hybrid states to a subset of labels. Recall that the noise affecting the continuous dynamics in ((1)) has density with respect to the Lebesgue measure. We assume for each label that the set $\{x \in \mathcal{X} : p \in L_{\mathfrak{J}}(x, z), z \in \mathcal{Z}\} \subseteq \mathcal{X}$ of continuous states with label p is measurable. We follow the same semantics as used in [34] for stochastic hybrid systems, i.e., the (initial) state $x(0), z(0)$ of an MJLS $\mathfrak{J}$ satisfies a property $\Phi = \mathsf{P}_{\sim\lambda}(\psi)$ if the probability of all paths from $x(0), z(0)$ satisfies $\sim \lambda$. For brevity, we shall write this satisfaction relation as $\mathfrak{J} \models_F \Phi$. All the sets of paths $(x(0), z(0)), (x(1), z(1)), \ldots$ expressed by PCTL under the above assumptions are measurable, see, e.g., [35, 46, 53] for details.

For an iMDP $\mathcal{M}_{\mathbb{I}}$, the satisfaction relation $\mathcal{M}_{\mathbb{I}} \models_\pi \Phi$ defines whether a PCTL formula Φ holds true, when following policy π from the initial state(s). Formal definitions for semantics and model checking are provided in [7, 27]. Recall from ((4)) that for iMDPs $\mathcal{M}_{\mathbb{I}}$, the threshold $\sim \lambda$ must hold under the *worst-case realization* of the probabilities $P \in \mathcal{P}$ in their intervals. That is, we are interested in synthesizing an optimal policy $\pi^\star \in \Pi_{\mathcal{M}_{\mathbb{I}}}$ that maximises the probability of satisfying a path specification ψ for the *worst-case* assignment $P \in \mathcal{P}$ (which is determined by a so-called *adversary*). In other words, we seek to solve the max-min decision problem given by

$$\pi^\star = \arg\max_{\pi \in \Pi_{\mathcal{M}_{\mathbb{I}}}} \min_{P \in \mathcal{P}} \lambda \quad \text{s.t.} \quad \mathsf{P}_{\geq\lambda}(\mathcal{M}_{\mathbb{I}}^P \models_\pi \psi). \tag{4}$$

It is shown by [38, 45], and in a much more general setting by [25], that deterministic policies suffice to obtain optimal values for iMDPs.

2.4 Problem Statement

We consider tasks encoded as a PCTL formula Φ. Our goal is to find a feedback controller F that satisfies Φ. As such, we solve the following problem.

Problem 1. Given an MJLS $\mathfrak{J}$ as in (1) and a PCTL formula Φ, find a control policy $F : \mathcal{X} \times \mathcal{Z} \times \mathbb{N} \to \mathcal{U} \times \mathcal{B}$, such that $\mathfrak{J} \models_F \Phi$.

In this paper, we address Problem 1 through the lens of abstractions [51], under two distinct assumptions on the mode-transition function of the MJLS.

Assumption A (Uncertain Markov jumps). *Each transition probability of the MDP $(\mathcal{Z}, \mathcal{B}, z_I, T)$ driving the jumps across modes in $\mathcal{Z}$ is known up to a certain interval $\mathcal{T} \ni T$, i.e., the Markov jump process is an iMDP $(\mathcal{Z}, \mathcal{B}, z_I, \mathcal{T})$.*

Assumption B (Unknown Markov jumps). *The Markov jumps are driven by a Markov chain (MC) for which we can measure the current mode, but the transition function (and hence its underlying graph structure) is unknown.*

For Assumption A, we will exploit the transition function of the jump process to reason over the joint probability distribution over the state $x(k)$ and mode $z(k)$. By contrast, under Assumption B, we do not know the distribution over successor modes $z(k+1)$, so reasoning over the joint distribution is not possible.

Instead, our goal is to attain *robustness* against any mode changes that may occur. An overview of our abstraction-based approach to solve Problem 1 is presented in Fig. 1. We note that it may occur that the PCTL formula is not satisfiable on the abstract model. To alleviate this issue, we propose an iterative refinement of the abstraction (shown by the dashed line in Fig. 1), which we explain in more detail in Sect. 5.

3   Abstractions of Non-Hybrid Dynamical Systems

Our abstraction procedure expands on the techniques from [4,6] to make them applicable to hybrid (and probabilistic) models. We start by summarising the main contributions of these papers, while referring to [4,6] for proofs and more details.

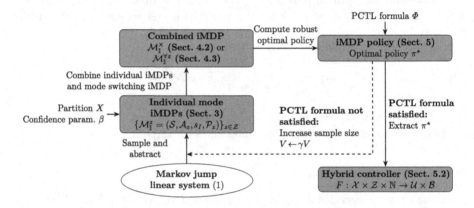

Fig. 1. Approach for synthesising a provably-correct controller for an MJLS.

Consider a discrete-time linear system $\mathfrak{L}$ with additive stochastic noise:

$$\mathfrak{L}\colon x(k+1) = Ax(k) + Bu(k) + q + w(k), \tag{5}$$

where $A \in \mathbb{R}^{n \times n}, B \in \mathbb{R}^{n \times m}, q \in \mathbb{R}^n$, and $w(k)$ defines an i.i.d. stochastic process, and $x(k) \in \mathcal{X} \subseteq \mathbb{R}^n$ and $u(k) \in \mathcal{U} \subseteq \mathbb{R}^m$ are the states and control inputs, respectively. The distribution of the noise $w(k)$ is assumed to be unknown, but instead we have access to a set $\{\delta_1, \ldots, \delta_V\}$ of V i.i.d. samples of $w(k)$. Note that the system in (5) reduces to an MJLS with a single mode. Given such a set of i.i.d. samples, the authors in [6] show how to construct an iMDP which, with a specified confidence level, abstracts the system in (5):

Definition 2 (β-iMDP abstraction). *Choose $\beta \in (0, 1)$ and let $\{\delta_1, \ldots, \delta_V\}$ be a collection of samples from the noise distribution affecting the dynamics in (5). An iMDP $\mathcal{M}_\mathbb{I} = (\mathcal{S}, \mathcal{A}, s_I, \mathcal{P})$ is a β-iMDP abstraction if for every PCTL* *formula Φ and for every policy $\pi \in \Pi_{\mathcal{M}_{\mathbb{I}}}$, there exists a feedback control $F \colon \mathcal{X} \times \mathbb{N} \to \mathcal{U}$ such that, for any initial condition $x(0)$, we have that*

$$\mathbb{P}^V \Big\{ (\mathcal{M}_{\mathbb{I}} \models_{\pi} \Phi) \implies (\mathfrak{L} \models_F \Phi) \Big\} \geq 1 - \beta, \tag{6}$$

where s_I is the initial state of the β-iMDP associated with continuous state $x(0)$, and $\mathbb{P}^V$ is the product probability measure induced by the sample set $\{\delta_1, \ldots, \delta_V\}$.

We remark that $\mathbb{P}^V$ is the product probability measure corresponding with sampling a set $\{\delta_1, \ldots, \delta_V\}$ of $V \in \mathbb{N}$ samples of the noise $w(k)$ in (5) (see, e.g., [14] for details). Definition 2 states that, with a confidence of at least $1 - \beta$, the satisfaction of a formula on the abstract iMDP implies the existence of a feedback controller that allows the satisfaction of the same formula on the concrete model. The confidence bound accounts for the inherent statistical error caused by constructing the iMDP based on a finite set of noise samples only. The iMDP abstraction allows us to synthesise correct-by-design feedback controllers for continuous-state dynamical systems [40], by utilising policies designed for a discrete-state model. Note that Definition 2 applies to general PCTL formulas, while [6] only considers reach-avoid properties (a subset of PCTL).

Definition 3 (Partition) *A partition $X = \{X_1, \ldots, X_p\}$ is an ordered set of subsets of $\mathcal{X}$ such that $\mathcal{X} = \bigcup_{i=1}^{p} X_i$, and $X_i \cap X_j = \emptyset$, $\forall i, j \in \{1, \ldots, p\}$, $i \neq j$.*

Papers [4,6] show how to generate β-iMDP abstractions by combining partitioning of the state space, backward reachability computation, and the scenario approach theory [14]. To this end, these papers create an iMDP abstraction $(\mathcal{S}, \mathcal{A}, s_I, \mathcal{P})$ of the continuous-state dynamics using the following procedure:

- The set of states $\mathcal{S} = \{s_1, \ldots, s_p\} \cup \{s^\star\}$ consists of elements associated with a partition X of the state space. This correspondence is given by the quotient mapping induced by the equivalence relation of the partition (see, e.g., [51]).
- The action space $\mathcal{A} = \{a_1, \ldots, a_q\}$, where each action $a \in \mathcal{A}$ is associated with a target point $d \in \mathcal{X}$ in the continuous state space (a convenient choice is to define each target d as the centre of an element $X_i \in X$ of the partition).
- To decide which actions are enabled at a given state of the abstraction, backward reachable set computations are employed. More specifically, we let

$$\mathcal{R}^{-1}(a) = \{x \in \mathbb{R}^n \mid d = Ax + Bu + q, \, u \in \mathcal{U}\} \tag{7}$$

be the backward reachable set of the target point d associated with the action $a \in \mathcal{A}$. Action a is enabled in state $s \in \mathcal{S}$ if and only if its corresponding element $X_i \in X$ is contained in $\mathcal{R}^{-1}(a)$. Mathematically, we have that

$$\mathcal{A}(s) = \{a \in \mathcal{A} \mid X_i \subseteq \mathcal{R}^{-1}(a)\}. \tag{8}$$

- The initial state s_I of the iMDP is defined by the element of the partition to which the initial state of the continuous dynamics belongs.

- The probability intervals $\mathcal{P}(s, a_j)(s_i)$ of the abstract iMDP can be efficiently computed using the scenario approach [15,49], or using statistical inequalities such as Hoeffding's bound [11].

To show that this procedure indeed yields a β-iMDP abstraction as per Definition 2, we also invoke the following key result from [6]:

Theorem 1 (iMDP abstraction of stochastic linear systems [6]). *Let X be a partition of the state space, then for any $\beta \in (0,1)$ and sample set $\{\delta_1, \ldots, \delta_V\}$, the procedure above yields a β-iMDP abstraction for the dynamics in (5).*

We provide an intuitive proof outline here, while referring to [6] for the full proof. Consider state s, with an associated continuous state $x(k)$; successor state s', with associated partition X_i; action a, with an associated feedback controller F. The true probability of transitioning from s to s', under action a is defined as

$$\mathrm{P}^\star(s, a)(s') = \int_{\mathbb{R}^n} \mathbb{1}_{X_i}\left(Ax(k) + BF(x(k), k) + q + \xi\right) \mathbb{P}_w(d\xi), \tag{9}$$

where $\mathbb{P}_w$ is the (in practice unknown) probability measure induced by the noise distribution, $\mathbb{1}_{X_i}(\cdot)$ is the indicator function (which returns value 1 if its argument belongs to the set X_i). Since transition probability intervals are obtained from the scenario approach theory, they contain probability $\mathrm{P}^\star$ with confidence β:

$$\mathbb{P}^V\left\{\mathrm{P}^\star(s, a)(s') \in \mathcal{P}(s, a)(s'), \forall s \in \mathcal{S}\right\} \geq 1 - \frac{\beta}{|\mathcal{A}| \cdot |\mathcal{S}|}. \tag{10}$$

The generated iMDP has at most $|\mathcal{A}| \cdot |\mathcal{S}|$ unique probability intervals, because $\mathrm{P}^\star(s, a)(s') = \mathrm{P}^\star(s'', a)(s')$ for any $s, s'' \in \mathcal{S}$ in which a is enabled. Thus, using Boole's inequality, we have that for all probabilities $\mathrm{P}^\star(s, a)(s')$

$$\mathbb{P}^V\left\{\mathrm{P}^\star(s, a)(s') \in \mathcal{P}(s, a)(s'), \forall s, s' \in \mathcal{S}, a \in \mathcal{A}\right\} \geq 1 - \beta. \tag{11}$$

Let $\mathcal{M}^{\mathrm{P}^\star}$ denote the MDP under the true transition function $\mathrm{P}^\star$, and let $\pi \in \Pi_{\mathcal{M}^{\mathrm{P}^\star}}$ be any policy for this MDP such that a given PCTL property Φ is satisfied on the iMDP, i.e. $\mathcal{M}^{\mathrm{P}^\star} \models_\pi \Phi$. Using concepts from probabilistic simulation relations [17,28,35], it can be shown that there exists a controller F such that $(\mathcal{M}^{\mathrm{P}^\star} \models_\pi \Phi) \implies (\mathfrak{L} \models_F \Phi)$. Combining this with (11), which states that $\mathbb{P}^V\{\mathcal{M}^{\mathrm{P}^\star} \in \mathcal{M}_\mathbb{I}\} \geq 1 - \beta$, we arrive at the condition for a β-iMDP abstraction in Def 2.

Theorem 1 can be used to synthesise *provably correct* controllers for temporal logic specifications, but is limited to systems *without discrete dynamics*, as for MJLSs. In what follows, we will develop a framework to overcome this limitation.

4   Abstractions of Markov Jump Linear Systems

In this section, we present our main contributions to solving Problem 1. We first explain how we use the results from Sect. 3 to construct an abstraction for the continuous dynamics of an individual mode. Then, we discuss how to "combine" abstractions across discrete modes to obtain a single iMDP abstraction. Finally, we compute an optimal policy $\pi^\star$ on the obtained iMDP and show (using Theorems 2 and 3) that this policy can be refined as a controller for the hybrid dynamics.

4.1   iMDP Abstraction for Individual Modes

We construct an abstraction for each separate mode $z \in \mathcal{Z}$ of the MJLS defined by (1) using the procedure that led to Theorem 1. For simplicity, we consider rectangular partitions, but our methods are applicable for any partition into convex sets satisfying Definition 3, and even to distinct partitions across modes. We then obtain a β-iMDP abstraction $\mathcal{M}_{\mathbb{I}}^z = (\mathcal{S}, \mathcal{A}_z, s_I, \mathcal{P}_z)$ for each mode $z \in \mathcal{Z}$.

In order to reason over the *hybrid* system as a whole, we now need a sound method to "combine" the abstractions $\mathcal{M}_{\mathbb{I}}^z$ for each mode $z \in Z$ into a single abstract model. However, without careful consideration of the enabled discrete actions, the resulting model may fail to soundly abstract the overall MJLS, as different actions may be enabled in the same region of continuous states, and this would lead to spurious trajectories in the abstraction.

To exemplify this issue, consider a specific instance of Example 1, in which the room without the radiator is perfectly insulated from the other. If we naively "combine" single-mode abstractions together, then we might conclude that we will be able to heat either room to any temperature, since the two modes taken individually can heat either room. This is an example of artificial behaviour introduced in the abstraction. In reality, we can only control one room at the same time; any actions which say otherwise will not be realisable on the concrete dynamical model. As our main contribution, we introduce in Sect. 4.2 an approach for combining single-mode iMDPs under Assumption A in a sound manner, and in Sect. 4.3 we discuss the case for Assumption B.

4.2   Abstraction Under Uncertain Markov Jumps (Assumption A)

Under Assumption A, we have access to an iMDP representation $\mathcal{M}_{\mathbb{I}} = (\mathcal{Z}, \mathcal{B}, z_I, \mathcal{T})$ of the discrete-mode Markov jump process, which has modes in $\mathcal{Z}$, switching actions in $\mathcal{B}$, initial mode z_I, and transition probability intervals in $\mathcal{T}$. Let $\{\mathcal{M}_{\mathbb{I}}^z = (\mathcal{S}, \mathcal{A}_z, s_I, \mathcal{P}_z)\}_{z \in \mathcal{Z}}$ be a set of β-iMDPs for each mode $z \in Z$, constructed as described in Sect. 4.1 with a confidence level of $\beta \in (0,1)$. We assume that these β-iMDPs have a common state space $\mathcal{S}$, and an overall action space $\mathcal{A}$. We also allow for a mode-dependent set of enabled actions; and use the notation $\mathcal{A}_z(s)$ to define actions enabled at a state s, in mode z.

To combine these modes, we use a product construction, similar to methods for constructing product automata [23]. We define our product construction among $\mathcal{M}_{\mathbb{I}}$ and $\{\mathcal{M}_{\mathbb{I}}^z\}_{z \in \mathcal{Z}}$. The *joint state/action* space of the product are the sets $\mathcal{Z} \times \mathcal{S}$ and $\mathcal{B} \times \mathcal{A}$. At a particular joint state (z, s), we define the set of enabled actions $\mathcal{A}(z, s) = \mathcal{B}(z) \times \mathcal{A}_z(s)$ as the product between the actions enabled at a particular mode, and the switches allowed in the corresponding state of the discrete iMDP. Thus, an action in the product iMDP corresponds with executing both an action in $\mathcal{A}_z$ (for the current mode z) and a discrete mode switching action in $\mathcal{B}$. The overall product iMDP under Assumption A is defined as follows:

Definition 4 (Product iMDP with mode switch control). *Let* $\{\mathcal{M}_{\mathbb{I}}^z = (\mathcal{S}, \mathcal{A}_z, s_I, \mathcal{P}_z)\}_{z \in \mathcal{Z}}$ *be a set of β-iMDP abstractions for each mode $z \in Z$, and let $\mathcal{M}_{\mathbb{I}} = (\mathcal{Z}, \mathcal{B}, z_I, \mathcal{T})$ be an iMDP for the Markov jump process. Then, the product iMDP $\mathcal{M}_{\mathbb{I}}^{\times} = (\mathcal{S}_{\times}, A_{\times}, s_{\times}^I, \mathcal{P}_{\times})$ is defined with*

- *Joint state space $\mathcal{S}_{\times} = \mathcal{Z} \times \mathcal{S}$;*
- *Joint action space $\mathcal{A}_{\times} = \mathcal{B} \times \mathcal{A}$, with enabled actions $\mathcal{A}(z, s)$ in state (z, s);*
- *Initial joint state $s_{\times}^I = (z_I, s_I)$;*
- *For each $(z, s), (z', s') \in \mathcal{Z} \times \mathcal{S}$ and $(b, a) \in \mathcal{A}(z, s)$, the probability interval*

$$\mathcal{P}_{\times}\big((z, s), (b, a)\big)\big((z', s')\big) = \\ [\underline{t}(z, b)(z') \cdot \underline{p_z}(s, a)(s'), \ \overline{t}(z, b)(z') \cdot \overline{p_z}(s, a)(s')]. \tag{12}$$

Here $\underline{p_z}(s, a)(s')$ and $\overline{p_z}(s, a)(s')$ are, respectively, the lower and upper bound state transition probability of β-iMDP $\mathcal{M}_{\mathbb{I}}^z$ for mode $z \in \mathcal{Z}$, and $[\underline{t}(z, b)(z'), \overline{t}(z, b)(z')]$ are the intervals in the transition function $\mathcal{T}$ of the jump process iMDP.

By construction, the product iMDP merges the individual mode abstractions and the mode-switching iMDP in a sound manner, thus avoiding the issues with spurious actions described in Sect. 4.1. The product iMDP depends on NV samples (N sets of V samples, one for each mode), hence the abstraction is a random variable on the space NV. The $\mathbb{P}^{NV}$ appearing in these theorems denotes the product measure $\mathbb{P}_{z_1}^V \otimes \mathbb{P}_{z_2}^V \cdots \mathbb{P}_{z_N}^V$ (note that the noise distribution can differ between modes). We extend Theorem 1 to the product iMDP as follows.

Theorem 2 (iMDP abstraction of controlled MJLS). *The product iMDP defined by Definition 4 is a β'-iMDP abstraction with confidence $\beta' = \beta \cdot |\mathcal{Z}|$ for the MJLS in (1), which captures the mode switching iMDP $\mathcal{M}_{\mathbb{I}}$. In particular,*

$$\mathbb{P}^{NV}\big\{(\mathcal{M}_{\mathbb{I}}^{\times} \models_{\pi} \Phi) \implies (\mathfrak{J} \models_F \Phi)\big\} \geq 1 - \beta'. \tag{13}$$

We provide an outline of the proof here, whilst for a detailed proof we refer to [47, Appendix 1]. The key observation is that the product iMDP is defined as the product between $|Z|$ β-iMDPs (having intervals that are "correct" with probability at least $1 - \frac{\beta}{|\mathcal{A}| \cdot |\mathcal{S}|}$, cf. (10)) and the mode switching iMDP (which is "correct" with probability one). These $|\mathcal{Z}|$ individual-mode iMDPs have $|\mathcal{A}| \cdot |\mathcal{S}| \cdot |\mathcal{Z}|$ unique intervals in total. Thus, the probability for all intervals to be correct (and thus for the product iMDP to be sound) is at least $1 - \frac{\beta \cdot |\mathcal{A}| \cdot |\mathcal{S}| \cdot |\mathcal{Z}|}{|\mathcal{A}| \cdot |\mathcal{S}|} = 1 - \beta'$. Finally, analogously to Theorem 1, the iMDP is a probabilistic simulation relation [28], such that the satisfaction of general PCTL formulae in the discrete abstraction guarantees the satisfaction of the same formulae in the concrete MJLS system.

4.3 Abstraction Under Unknown Markov Jumps (Assumption B)

Under Assumption B, the mode transition probabilities are now completely unknown. Thus, in contrast with Sect. 4.2, we generate an abstraction that is robust to any mode we may be in.

Robustifying Enabled Actions. First, we modify the computation of the backward reachable set in (7) to introduce a backward reachable set across all possible modes (i.e., the set that can reach d regardless of which mode we are in – note the universal quantification $\forall z \in \mathcal{Z}$ in the following equation):

$$\mathcal{G}^{-1}(d) = \{x \in \mathbb{R}^n \mid d = A_z x + B_z u + q_z, u \in \mathcal{U}, \forall z \in \mathcal{Z}\}$$
$$= \bigcap_{z \in \mathcal{Z}} \{x \in \mathbb{R}^n \mid d = A_z x + B_z u + q_z, u \in \mathcal{U}\} = \bigcap_{z \in \mathcal{Z}} \mathcal{R}_z^{-1}(d), \qquad (14)$$

where $\mathcal{R}_z^{-1}(d)$ is the backward reachable set for mode $z \in \mathcal{Z}$, as defined in (7).

Similar to Sect. 3, we use backward reachable set computation to define the set of enabled actions, now denoted by $\mathcal{A}_{\forall z}$, in the iMDP. Indeed, for a given partition of the state space $X = \{X_1, \ldots, X_p\}$, an action a is enabled at a state s if the backward reachable set $\mathcal{G}^{-1}(d)$ defined in (14) contains the corresponding element X_i of the partition X, i.e., a is enabled if $X_i \subseteq \mathcal{G}^{-1}(d)$. Thus, by definition, the action a is realisable on the concrete dynamical model, regardless of the current mode of operation.

Robustifying Probability Intervals. We render the transition probability intervals robust against any mode in two steps. First, we compute the transition probability intervals $\mathcal{P}_z$ for the iMDP $\mathcal{M}_{\mathbb{I}}^z$ for each individual mode $z \in \mathcal{Z}$. For each transition (s, a, s'), the robust interval $\mathcal{P}_{\forall z}(s, a)(s')$ is then obtained as the *smallest probability interval* that contains the intervals in $\mathcal{P}_z$ for all modes $z \in \mathcal{Z}$:

$$\mathcal{P}_{\forall z}(s, a)(s') = \left[\min_{z \in \mathcal{Z}} \underline{p_z}(s, a)(s'), \; \max_{z \in \mathcal{Z}} \overline{p_z}(s, a)(s') \right], \qquad (15)$$

where $\underline{p_z}(s, a)(s')$ and $\overline{p_z}(s, a)(s')$ are again the lower/upper bound probabilities of iMDP $\mathcal{M}_{\mathbb{I}}^z$. Using (15) we obtain probability intervals that are, by construction, a sound overapproximation of the probability intervals *under any mode* $z \in \mathcal{Z}$. We use this key observation to state the correctness of the resulting iMDP.

Theorem 3 (Robust iMDP with unknown mode jumps). *The robust iMDP $\mathcal{M}_{\mathbb{I}}^{\forall z} = (\mathcal{S}, \mathcal{A}_{\forall z}, s_I, \mathcal{P}_{\forall z})$ with actions defined through (14) and intervals defined by (15) is a β'-iMDP abstraction for the MJLS in (1), which models state transitions robustly against any mode transition. In particular,*

$$\mathbb{P}^{NV}\left\{ (\mathcal{M}_{\mathbb{I}}^{\forall z} \models_\pi \varPhi) \implies (\mathfrak{J} \models_F \varPhi) \right\} \geq 1 - \beta'. \qquad (16)$$

We again provide the full proof in [47, Appendix 1], while only providing an outline here. The robust iMDP is composed of intervals that contain the true transition probabilities, with a probability of at least $1 - \frac{\beta}{|\mathcal{A}|\cdot|\mathcal{S}|}$. Thus, every probability interval of the robust iMDP contains the intervals for all modes $z \in \mathcal{Z}$ with probability at least $1 - \frac{\beta}{|\mathcal{A}|\cdot|\mathcal{S}|} \cdot |\mathcal{Z}|$. Since the robust iMDP has $|\mathcal{S}| \cdot |\mathcal{A}|$ unique intervals, it follows that all intervals of the iMDP are correct with probability at least $1 - \frac{\beta\cdot|\mathcal{A}|\cdot|\mathcal{S}|\cdot|\mathcal{Z}|}{|\mathcal{A}|\cdot|\mathcal{S}|} = 1 - \beta'$. Analogous to the proof of Theorem 2, it is then straightforward to prove that this abstraction is also a β'-iMDP.

5 Synthesis for General PCTL Formulae

To synthesise optimal policies in our discrete abstraction, we use the probabilistic model checker PRISM [33]. We handle complex and nested PCTL formulae by defining a *parse tree* [7], whose leaves are atomic propositions, and whose branches are logical, temporal, or probabilistic operators. The complete formula can be verified using a bottom-up approach. As an example, consider the formula

$$\Phi = \mathsf{P}_{\geq 0.6}[\mathsf{X}\mathsf{P}_{\leq 0.5}(\neg T_C \mathsf{U}^{\leq K-1}(T_L \vee T_C))] \wedge \mathsf{P}_{\geq 0.9}[\neg T_C \mathsf{U}^{\leq K} T_G], \qquad (17)$$

with atomic propositions T_C, T_L, T_G. The parse tree for this formula is shown in Fig. 2. We will explain and use this formula in the temperature control experiment in Sect. 6. When considering multiple PCTL fragments, we find a policy associated with each fragment (our example above will find two policies, one satisfying $\mathsf{P}_{\geq 0.6}[\mathsf{X}\mathsf{P}_{\leq 0.5}(\neg T_C \mathsf{U}^{\leq K-1}(T_L \vee T_C))]$ the other $\mathsf{P}_{\geq 0.9}[\neg T_C \mathsf{U}^{\leq K} T_G]$). At runtime, we choose which PCTL fragment to satisfy and apply its associated policy as $\pi^\star$.

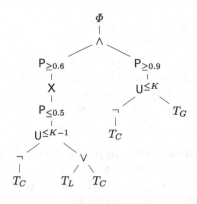

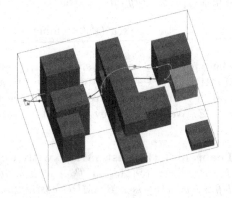

Fig. 2. Parse tree for PCTL formula (17).

Fig. 3. Simulated paths under weak (blue) and strong (orange) wind for the drone.

5.1   Unsatisfied Formulae

If the PCTL formula is not satisfied by the iMDP, we refine our abstraction by increasing the number of samples used to compute the probability intervals (shown by the dashed line in Fig. 1). As also discussed in more detail and shown experimentally by [4], this refinement tightens the probability intervals, which in turn improves the probability of satisfying the property. We iteratively refine our abstraction until the formula is satisfied or until a maximum number of iterations is exceeded (which we fix a priori), in which case nothing is returned. In this way, our method is sound, but not complete: if the formula is not satisfied after the maximum number of iterations, this in general does not imply that the formula cannot be satisfied at all. However, for any policy that is returned by our algorithm, the correctness result of Theorem 3 holds.

5.2   Controller Synthesis via Policy Refinement

We refine the optimal policy $\pi^\star$ to obtain a hybrid-state feedback controller F for the MJLS, as follows. Given the current continuous state $x \in \mathcal{X}$, mode $z \in \mathcal{Z}$ and time step $k \in \mathbb{N}$, we first find the element X_i of partition X containing state x, such that $x \in X_i$. Depending on whether we consider abiding by modelling Assumption A or Assumption B, we then proceed as follows:

- For Assumption A, we find the product state $s_\times = (z, s)$ associated with the current mode $z \in \mathcal{Z}$ and state s. We then look up the optimal product action $a_\times = \pi^\star(s_\times, k) = (b, a)$ from policy $\pi^\star$, with corresponding switching action b and continuous action a.
- For Assumption B, it suffices to know state s associated with X_i only, and we directly obtain action $a = \pi^\star(s, k)$, with no switching action.

Finally, we compute the continuous control input u associated with action a by calculating the control input that drives us to the associated target point d, using $u = B_z^+(d - A_z x - q_z)$, with B_z^+ representing the pseudoinverse of B_z.

6   Numerical Experiments

We have implemented our techniques in Python, using the probabilistic model checker PRISM [33] to verify the satisfaction of PCTL formulae on iMDPs. The codebase is available at https://github.com/lukearcus/ScenarioAbstraction. Experiments were run on a computer with 6 3.7 GHz cores and 32 GB of RAM. We demonstrate our techniques on two models: (1) a UAV motion control problem with two possible levels of noise, and (2) a building temperature regulation problem, in line with our running example from Sect. 2. Details on the UAV model and additional experimental results can be found in [47, Appendix 2].

6.1   UAV Motion Planning

We consider a more refined, *hybrid* version of the unmanned aerial vehicle (UAV) motion planning problem from [4]. We consider two discrete modes, which reflect different levels of noise, namely low and high wind speeds. We use our framework considering Assumption A. The PCTL specification $\Phi = P_{\geq 0.5}[\neg O U^{\leq K} G]$ requires reaching a goal set G (highlighted in green in Fig. 3), whilst avoiding obstacles O (highlighted in red). We choose a finite time horizon $K = 64$. While our theoretical contributions hold for any probability distribution for the additive noise, in this particular experiment we sample from a Gaussian.

Scalability. The number of iMDP states equals the number of partitions, multiplied by the number of discrete modes, here resulting in 51,030 states. The number of transitions depends on the number of samples: with 100 samples, we generate an iMDP with 92.7 million transitions; with 200 samples, 154 million transitions. Computing the iMDP actions enabled in the abstraction is independent of sampling and takes 8.5 min; computing the transition probability intervals of the iMDP takes 70 min; formal synthesis of the optimal policy takes 40 s, and control refinement occurs online.

Variable Noise Affects Decisions. With our techniques, we synthesise a controller that accounts for different noise levels at runtime and reasons about the probability of the noise level changing. Thus, our framework makes use of the information available regarding the jump process, while at the same time reasoning explicitly over the stochastic noise affecting the continuous dynamics in each mode.

6.2   Temperature Regulation in a Building

We consider again the 2-room building temperature control problem [3] introduced in Example 1. Recall that the state $x = [T_1, T_2]^\top \in \mathbb{R}^2$ models the temperature in both rooms, and the control input (modelling the power supply to the heaters) is constrained to $u \in \mathbb{R}^2$. The values of the constants in (2) are $a_{12} = 0.022$, $b_1 = b_2 = 0.0167$, $k_f = 0.8$, $k_r = 0.4$, $x_a = 6$. The noise is distributed according to a zero-mean Gaussian with a standard deviation of 0.2.

Uncertain Mode Transitions

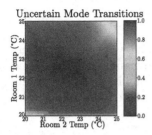
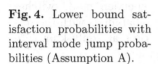

Unknown Mode Transitions

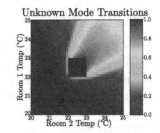

More Complex Formula

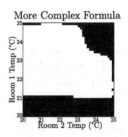

Fig. 4. Lower bound satisfaction probabilities with interval mode jump probabilities (Assumption A).

Fig. 5. Lower bound satisfaction probabilities with unknown mode jump probabilities (Assumption B).

Fig. 6. States that satisfy (in white) the general PCTL formula in (17) (Assumption A).

We wish to optimise the probability of satisfying the path formula $\psi = (\neg T_C)\mathsf{U}^{\leq K}(T_G)$, with goal temperature T_G between 22 and 23 °C, and critical temperature T_C less than 20 °C or greater than 25 °C. We partition the state space into 1600 regions, using a time horizon $K = 32$. We look into two setups, one fulfilling Assumption A (Fig. 4) and the other Assumption B (Fig. 5). We show the results for all initial continuous states, and in Fig. 4 we consider starting in mode 1 (whereas the bounds in Fig. 5 hold for any initial mode).

Assumptions Affect Conservativism and Scalability. When we wish to be robust to all possible modes (cf. Assumption B), our generated iMDP is much smaller (with about 18 times fewer transitions), since we have a single robust iMDP, compared to a product iMDP. However, as expected and seen in Fig. 5, the obtained probability lower bounds are much more conservative. Thus, compared to Assumption B, Assumption A reduces the level of conservatism (because we exploit the probability intervals of the Markov jump process) at the cost of increasing the size of the abstraction.

6.3   Controller Synthesis for General PCTL Formulae

We now consider the general PCTL formula in (17) to show the applicability of our techniques beyond reach-avoid specifications. This formula requires (1) heating both rooms to a goal temperature while avoiding critical temperatures; and (2) reaching a state at the next time step, which is able to avoid entering an unwanted or critical temperature. The new atomic proposition T_L specifies that temperatures should be kept below 21 °C in room 1. In Fig. 6, we show the set of iMDP states that satisfy the PCTL formula (shown in white), if the fan heater is initially in room 1 (see Example 1). Thus, we can compute a feedback controller for the MJLS satisfying the PCTL formula, unless the initial room temperature is (approximately) below 21 °C, or if both initial temperatures are too high.

7   Conclusions and Future Work

We have presented a new method for synthesizing certifiably correct controllers for MJLSs with hybrid, stochastic and partly unknown dynamics. We considered both the case where an estimate of the switching probabilities across discrete operation modes is known, and the alternative instance where these probabilities are not known at all. Our experiments have demonstrated the efficacy of our methods on a number of realistic problems.

Future research directions include considering state-dependent mode switches (e.g. for models in [2,39]), estimating mode-switching probabilities with the scenario approach, and dealing with a setting where matrices are only known to belong to a convex polytope, as in [5] for non-hybrid systems.

References

1. Abate, A., D'Innocenzo, A., Di Benedetto, M.D., Sastry, S.S.: Markov set-chains as abstractions of stochastic hybrid systems. In: Egerstedt, M., Mishra, B. (eds.) HSCC 2008. LNCS, vol. 4981, pp. 1–15. Springer, Heidelberg (2008). https://doi.org/10.1007/978-3-540-78929-1_1
2. Abate, A., Katoen, J., Lygeros, J., Prandini, M.: Approximate model checking of stochastic hybrid systems. Eur. J. Control **16**(6), 624–641 (2010)
3. Abate, A., Prandini, M., Lygeros, J., Sastry, S.: Probabilistic reachability and safety for controlled discrete time stochastic hybrid systems. Autom. **44**(11), 2724–2734 (2008)
4. Badings, T.S., Abate, A., Jansen, N., Parker, D., Poonawala, H.A., Stoelinga, M.: Sampling-based robust control of autonomous systems with non-gaussian noise. In: AAAI, pp. 9669–9678. AAAI Press (2022)
5. Badings, T.S., Romao, L., Abate, A., Jansen, N.: Probabilities are not enough: formal controller synthesis for stochastic dynamical models with epistemic uncertainty. In: AAAI, pp. 14701–14710. AAAI Press (2023)
6. Badings, T.S., et al.: Robust control for dynamical systems with non-gaussian noise via formal abstractions. J. Artif. Intell. Res. **76**, 341–391 (2023)
7. Baier, C., Katoen, J.: Principles of Model Checking. MIT Press, Cambridge (2008)
8. Belta, C., Yordanov, B., Aydin Gol, E.: Formal Methods for Discrete-Time Dynamical Systems. SSDC, vol. 89. Springer, Cham (2017). https://doi.org/10.1007/978-3-319-50763-7
9. Benbrahim, M., Kabbaj, M., Benjelloun, K.: Robust control under constraints of linear systems with markovian jumps. Int. J. Control Autom. Syst. **14**(6), 1447–1454 (2016)
10. Blackmore, L., Ono, M., Bektassov, A., Williams, B.C.: A probabilistic particle-control approximation of chance-constrained stochastic predictive control. IEEE Trans. Robot. **26**(3), 502–517 (2010)
11. Boucheron, S., Lugosi, G., Massart, P.: Concentration Inequalities - A Nonasymptotic Theory of Independence. Oxford University Press, Oxford (2013)
12. Boukas, E.K., Benzaouia, A.: Stability of discrete-time linear systems with markovian jumping parameters and constrained control. IEEE Trans. Autom. Control. **47**(3), 516–521 (2002)

13. Cai, H., Li, P., Su, C., Cao, J.: Robust model predictive control for a class of discrete-time markovian jump linear systems with operation mode disordering. IEEE Access **7**, 10415–10427 (2019)
14. Campi, M.C., Garatti, S.: The exact feasibility of randomized solutions of uncertain convex programs. SIAM J. Optim. **19**(3), 1211–1230 (2008)
15. Campi, M.C., Garatti, S.: A sampling-and-discarding approach to chance-constrained optimization: feasibility and optimality. J. Optim. Theory Appl. **148**(2), 257–280 (2011)
16. Campi, M.C., Garatti, S., Prandini, M.: The scenario approach for systems and control design. Annu. Rev. Control. **33**(2), 149–157 (2009)
17. Cauchi, N., Laurenti, L., Lahijanian, M., Abate, A., Kwiatkowska, M., Cardelli, L.: Efficiency through uncertainty: scalable formal synthesis for stochastic hybrid systems. In: HSCC, pp. 240–251. ACM (2019)
18. Clarke, E., Fehnker, A., Han, Z., Krogh, B., Stursberg, O., Theobald, M.: Verification of hybrid systems based on counterexample-guided abstraction refinement. In: Garavel, H., Hatcliff, J. (eds.) TACAS 2003. LNCS, vol. 2619, pp. 192–207. Springer, Heidelberg (2003). https://doi.org/10.1007/3-540-36577-X_14
19. Cunha, R.F., Gabriel, G.W., Geromel, J.C.: Robust partial sampled-data state feedback control of markov jump linear systems. Int. J. Syst. Sci. **50**(11), 2142–2152 (2019)
20. Do Costa, O.L.V., Marques, R.P., Fragoso, M.D.: Discrete-Time Markov Jump Linear Systems. Springer, London (2005). https://doi.org/10.1007/b138575
21. de Farias, D.P., Geromel, J.C., do Val, J.B.R., Costa, O.L.V.: Output feedback control of markov jump linear systems in continuous-time. IEEE Trans. Autom. Control **45**(5), 944–949 (2000)
22. Gabriel, G.W., Geromel, J.C.: Performance evaluation of sampled-data control of markov jump linear systems. Autom. **86**, 212–215 (2017)
23. Gécseg, F.: Products of Automata, EATCS Monographs on Theoretical Computer Science, vol. 7. Springer (1986)
24. Givan, R., Leach, S.M., Dean, T.L.: Bounded-parameter markov decision processes. Artif. Intell. **122**(1–2), 71–109 (2000)
25. González-Trejo, J.I., Hernández-Lerma, O., Reyes, L.F.H.: Minimax control of discrete-time stochastic systems. SIAM J. Control. Optim. **41**(5), 1626–1659 (2002)
26. Hahn, E.M., Han, T., Zhang, L.: Synthesis for PCTL in parametric markov decision processes. In: Bobaru, M., Havelund, K., Holzmann, G.J., Joshi, R. (eds.) NFM 2011. LNCS, vol. 6617, pp. 146–161. Springer, Heidelberg (2011). https://doi.org/10.1007/978-3-642-20398-5_12
27. Hansson, H., Jonsson, B.: A logic for reasoning about time and reliability. Formal Aspects Comput. **6**(5), 512–535 (1994)
28. Hermanns, H., Parma, A., Segala, R., Wachter, B., Zhang, L.: Probabilistic logical characterization. Inf. Comput. **209**(2), 154–172 (2011)
29. Hespanha, J.P., Naghshtabrizi, P., Xu, Y.: A survey of recent results in networked control systems. Proc. IEEE **95**(1), 138–162 (2007)
30. Hu, L., Shi, P., Frank, P.M.: Robust sampled-data control for markovian jump linear systems. Autom. **42**(11), 2025–2030 (2006)
31. Jiang, B., Wu, Z., Karimi, H.R.: A traverse algorithm approach to stochastic stability analysis of markovian jump systems with unknown and uncertain transition rates. Appl. Math. Comput. **422**, 126968 (2022)
32. Knight, J.C.: Safety critical systems: challenges and directions. In: ICSE, pp. 547–550. ACM (2002)

33. Kwiatkowska, M., Norman, G., Parker, D.: PRISM 4.0: verification of probabilistic real-time systems. In: Gopalakrishnan, G., Qadeer, S. (eds.) CAV 2011. LNCS, vol. 6806, pp. 585–591. Springer, Heidelberg (2011). https://doi.org/10.1007/978-3-642-22110-1_47

34. Lahijanian, M., Andersson, S.B., Belta, C.: Formal verification and synthesis for discrete-time stochastic systems. IEEE Trans. Autom. Control. **60**(8), 2031–2045 (2015)

35. Lavaei, A., Soudjani, S., Abate, A., Zamani, M.: Automated verification and synthesis of stochastic hybrid systems: a survey. Autom. **146**, 110617 (2022)

36. Li, W., Xu, Y., Li, H.: Robust l_2-l_∞ filtering for discrete-time markovian jump linear systems with multiple sensor faults, uncertain transition probabilities and time-varying delays. IET Signal Process. **7**(8), 710–719 (2013)

37. Lindemann, L., et al.: Learning hybrid control barrier functions from data. In: CoRL. Proceedings of Machine Learning Research, vol. 155, pp. 1351–1370. PMLR (2020)

38. Lun, Y.Z., Wheatley, J., D'Innocenzo, A., Abate, A.: Approximate abstractions of markov chains with interval decision processes. In: ADHS. IFAC-PapersOnLine, vol. 51, pp. 91–96. Elsevier (2018)

39. Lunze, J., Lamnabhi-Lagarrigue, F. (eds.): Handbook of Hybrid Systems Control: Theory, Tools, Applications. Cambridge University Press, Cambridge (2009)

40. Mazo, M., Davitian, A., Tabuada, P.: PESSOA: a tool for embedded controller synthesis. In: Touili, T., Cook, B., Jackson, P. (eds.) CAV 2010. LNCS, vol. 6174, pp. 566–569. Springer, Heidelberg (2010). https://doi.org/10.1007/978-3-642-14295-6_49

41. Moggi, E., Farjudian, A., Duracz, A., Taha, W.: Safe & robust reachability analysis of hybrid systems. Theor. Comput. Sci. **747**, 75–99 (2018)

42. Morais, C.F., Palma, J.M., Peres, P.L.D., Oliveira, R.C.L.F.: An LMI approach for H2 and H∞ reduced-order filtering of uncertain discrete-time markov and bernoulli jump linear systems. Automatica **95**, 463–471 (2018)

43. Nejati, A., Soudjani, S., Zamani, M.: Compositional construction of control barrier functions for continuous-time stochastic hybrid systems. Automatica **145**, 110513 (2022)

44. Platzer, A.: Logics of dynamical systems. In: LICS, pp. 13–24. IEEE Computer Society (2012)

45. Puggelli, A., Li, W., Sangiovanni-Vincentelli, A.L., Seshia, S.A.: Polynomial-time verification of PCTL properties of MDPs with convex uncertainties. In: Sharygina, N., Veith, H. (eds.) CAV 2013. LNCS, vol. 8044, pp. 527–542. Springer, Heidelberg (2013). https://doi.org/10.1007/978-3-642-39799-8_35

46. Ramponi, F., Chatterjee, D., Summers, S., Lygeros, J.: On the connections between PCTL and dynamic programming. In: HSCC, pp. 253–262. ACM (2010)

47. Rickard, L., Badings, T.S., Romao, L., Abate, A.: Formal controller synthesis for markov jump linear systems with uncertain dynamics. Technical Report (2023). https://www.lukerickard.co.uk/RBRA23.pdf

48. Robey, A., Lindemann, L., Tu, S., Matni, N.: Learning robust hybrid control barrier functions for uncertain systems. In: ADHS. IFAC-PapersOnLine, vol. 54, pp. 1–6. Elsevier (2021)

49. Romao, L., Papachristodoulou, A., Margellos, K.: On the exact feasibility of convex scenario programs with discarded constraints. IEEE Trans. Autom. Control. **68**(4), 1986–2001 (2023)

50. Soudjani, S.E.Z., Abate, A.: Adaptive and sequential gridding procedures for the abstraction and verification of stochastic processes. SIAM J. Appl. Dyn. Syst. **12**(2), 921–956 (2013)
51. Tabuada, P.: Verification and Control of Hybrid Systems - A Symbolic Approach. Springer (2009)
52. Tian, E., Yue, D., Wei, G.: Robust control for markovian jump systems with partially known transition probabilities and nonlinearities. J. Frankl. Inst. **350**(8), 2069–2083 (2013)
53. Tkachev, I., Abate, A.: Characterization and computation of infinite horizon specifications over markov processes. Theor. Comput. Sci. **515**, 1–18 (2014)
54. do Valle Costa, O.L., Fragoso, M.D.: Discrete-time LG-optimal control problems for infinite markov jump parameter systems. IEEE Trans. Autom. Control. **40**(12), 2076–2088 (1995)
55. do Valle Costa, O.L., Fragoso, M.D., Todorov, M.G.: A detector-based approach for the h_2 control of markov jump linear systems with partial information. IEEE Trans. Autom. Control. **60**(5), 1219–1234 (2015)
56. Zhang, L., Boukas, E.K.: Stability and stabilization of markovian jump linear systems with partly unknown transition probabilities. Automatica **45**(2), 463–468 (2009)

Jajapy: A Learning Library for Stochastic Models

Raphaël Reynouard[1]($^\boxtimes$) , Anna Ingólfsdóttir[1], and Giovanni Bacci[2]

[1] Reykjavík University, Reykjavik, Iceland
raphal20@ru.is
[2] Aalborg University, Aalborg, Denmark

Abstract. We present JAJAPY, a Python library that implements a number of methods to aid the modelling process of Markov models from a set of partially-observable executions of the system. Currently, JAJAPY supports different types of Markov models such as discrete and continuous-time Markov chains, as well as Markov decision processes.

JAJAPY can be used both to learn the model from scratch or to estimate parameter values of a given model so that it fits the observed data the best. To this end, the tool offers different learning techniques, either based on expectation-maximization or state-merging methods, each adapted to different types of Markov models. One key feature of JAJAPY consists in its compatibility with the model checkers STORM and PRISM.

The paper briefly presents JAJAPY's functionalities and reports an empirical evaluation of their performance and accuracy. We conclude with an experimental comparison of JAJAPY against AALPY, which is the current state-of-the-art Python library for learning automata. JAJAPY and AALPY complement each other, and the choice of the library should be determined by the specific context in which it will be used.

Keywords: Machine Learning · Expectation-Maximization · Model Checking · Markov models · Python

1 Introduction

Markov models are a very popular formalism. Discrete-time Markov chains (MCs) and continuous-time Markov chains (CTMCs) have wide applications in performance and dependability analysis, whereas Markov decision processes (MDPs) are key models for stochastic decision-making and planning which find numerous applications in the design and analysis of cyber-physical systems.

PRISM [1] and STORM [2] are two widely-used model checking tools that provide an efficient and reliable way to verify the correctness of probabilistic systems. They both accept models written in the PRISM language, an expressive

Anna Ingólfsdóttir and Raphaël Reynouard are supported by the *Learning and Applying Probabilistic Systems* project (Project nr. 206574-051) funded by the Icelandic Research Fund.

state-based language based on [3]. PRISM is a powerful tool for modeling and analysing MCs, MDPs, and probabilistic timed automata. It has a user-friendly interface and supports a variety of analysis techniques, including model checking, parameter synthesis, and probabilistic model checking. STORM, on the other hand, is a highly scalable and efficient tool for analysing probabilistic systems with continuous-time and hybrid dynamics [4]. It supports both explicit and symbolic model representation, and provides state-of-the-art algorithms for model checking and synthesis tasks. Both tools have been extensively used in academia and industry to analyse a wide range of systems, including communication protocols, cyber-physical systems, and biological systems.

The standard assumption of model checking tools is that the model is known precisely. For many application domains, this assumption is too strong. Often the model is not available, or at best is partially known. In such cases, the model is typically estimated empirically from a set of partially-observable executions (a.k.a. traces). Depending on the system under consideration, traces may be collected offline in the form of time series or (possibly continuous) streams of system logs, or the modeller can actively query the system and stir the exploration of its dynamics. In the latter situation, interaction with the system may be limited due to safety critical concerns, or simply to comply with the budget allocated for the task.

To effectively exploit the characteristics of different learning scenarios it is convenient to have a single library that provides a variety of learning algorithms, which can handle different learning scenarios and model types seamlessly, while integrating well with the model-and-verification workflow of PRISM and STORM.

In this paper, we present JAJAPY [5], a free open-source Python library that offers a number of techniques to learn Markov models from traces and is interoperable with PRISM and STORM. JAJAPY implements the following machine-learning techniques:

(i) ALERGIA [6,7] and IOALERGIA [8,9], passive learning procedures that learn respectively MCs and (deterministic) MDPs from a set of traces by successively merging compatible states;

(ii) a number of adaptations of the Baum-Welch algorithm [10] to learn MCs, MDPs [11,12], and CTMCs [13] by estimating their transition probabilities given a set of traces and the size of the resulting model;

(iii) active learning strategies to enhance the quality of the MDPs learned using the Baum-Welch algorithm [11,12] when the user has the possibility to interact with the system;

(iv) MM algorithms [13] for estimating parameter value in parametric CTMCs (pCTMCs) from a set of (possibly non-timed) traces.

JAJAPY implements also metrics to independently evaluate the output model against a test set. This is particularly useful to measure the degree of generalisation that the output model offers on top of the training set and assess whether the output model overfits the training data or not. Interoperability with PRISM and STORM is achieved by supporting import and export functions for PRISM models as well as STORMPY sparse models.

JAJAPY's source code follows a modular architecture design and can therefore be extended to other modelling formalisms and learning algorithms. JAJAPY's documentation can be found on Read the Docs [14] that is complemented with a short video-introduction available on Zenodo [15].

Related Work. AALPY [16] is a recent Python library that can learn both non-stochastic and stochastic models. In particular, AALPY can learn MDPs using L^*_{MDP} [17], an extension of Angluin's L^* algorithm [18], and MCs using Alergia [6,7]. In Sect. 5.3, we compare JAJAPY and AALPY performance.

Other automata learning frameworks have been developed as well. For example, Learnlib [19] and libalf [20], which learn non-stochastic models. In contrast with these tools, as of now, JAJAPY primary focus is on learning Markov models.

In this paper, we present the different learning methods implemented in Jajapy and AALpy. However, other methods also exist, such as the MDI algorithm [21,22], two state-merging based approaches, or the Bayesian method using Gibbs sampling [23] proposed by Neal in [24].

MDPs are extensively used in *reinforcement learning* as in [25–27], and in *robust reinforcement learning* [28] as in [29–31]. In this context, the objective is to learn an optimal policy that maximises long-term rewards in a given environment.

A related line of research is model synthesis. Counterexample-guided inductive synthesis (CEGIS) [32] study the problem of completing a given program sketch (i.e., a probabilistic program with holes) so that it satisfies a given set of quantitative specifications. Another approach is parameter synthesis [33,34], where the objective is to find some (or all) instances of a given parametric Markov model satisfying a logic formula. In [35], the authors combine parameter synthesis and parametric inference techniques to synthesize feasible parameter valuations and quantify the confidence that the corresponding model satisfies a given property of interest.

Paper Outline. We start with a quick introduction to JAJAPY's functionalities in Sect. 2, then we explain JAJAPY's features from a theoretical perspective in Sect. 3. In Sect. 4, we present some technical aspects of JAJAPY, and in Sect. 5 we evaluate our tool and compare it to AALPY.

2   Jajapy in a Nutshell

JAJAPY offers learning methods to construct an accurate model of a system under learning (SUL) from a set of traces and export it to a format that can be directly used in STORM and PRISM for analysis (Fig. 1).

In the following, we call *training set* (resp. *test set*) the collection of traces used to learn the SUL model (resp. to evaluate the learning output model). Depending on the nature of the training set, JAJAPY learns different types of models: (i) MCs are learned from sequences of labels (i.e., sequences of atomic propositions), (ii) CTMCs and pCTMCs are learned from times series of labels, and (iii) MDPs are learned from alternating sequences of actions and labels.

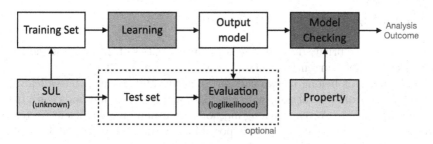

Fig. 1. A complete modeling and verification workflow using JAJAPY . The phases where JAJAPY is employed are highlighted in green, whereas the phase in blue is assumed to be performed with STORM or PRISM (Color figure online).

A trace denotes, depending on the context, a sequence of labels, a time series of labels, and an alternating sequence of actions and labels. The length of a trace is always the number of labels it contains.

The first and main learning algorithm offered by JAJAPY is the Baum-Welch (BW) algorithm. It takes as input a training set and an initial hypothesis, i.e. a Markov model. During the BW execution, the transition probabilities of this initial hypothesis will be updated but no state will be added/removed from it. Therefore, the number of states in the resulting model will be equal to the number of states in the initial hypothesis. By default JAJAPY generates a random initial hypothesis (given as input the number of states) but the user can also provide one explicitly. This enables the user to exploit his knowledge of the SUL to enhance the learning process. Such an initial hypothesis can be a STORMPY sparse model, a model saved in a PRISM file or a JAJAPY model. Given a training set and an initial hypothesis, the BW algorithm constructs an approximate representation of the SUL, called *output model* (Fig. 2).

```
from jajapy import BW
type(training_set) # list
output_model = BW().fit(training_set, nb_states=10)
type(output_model) # stormpy.SparseDtmc
```

Fig. 2. Simple execution of JAJAPY BW to learn an MC with 10 states.

As an alternative to the BW algorithm, JAJAPY offers implementations of Alergia and IOAlergia to learn respectively MCs and MDPs. These algorithms take as input the training set and a *confidence* parameter.

Once JAJAPY has produced the output model, the user can use STORMPY to verify the model against some properties of interest supported by STORM. The output model can also be exported to a PRISM model and analysed with the PRISM model checker.

3   Learning Probabilistic Models

In this section, we briefly describe the key characteristics of the learning methods for Markov models currently available in JAJAPY and AALPY.

These methods belong to two categories, active and passive. *Active* learning methods learn from interactions with the SUL, while *passive* methods learn from the training set only. Active learning methods are usually more efficient (in terms of data), but can be used only if it is possible to interact with the SUL.

Some learning methods allow the user to decide the size (i.e. the number of states) of the output model, preventing the algorithm from generating models too large to be efficiently analysed. The downside of such a feature consists in the fact that, if the number of states requested is too large (resp. small), the output model may overfit (resp. underfit) the training set.

A Markov model is *deterministic* if, for any state s and label ℓ, there exists at most one transition leaving s to a state labelled with ℓ. Some of the learning methods described below assume the SUL to be deterministic. When such methods are exercised with a SUL that is non-deterministic, they are not guaranteed to converge to the true model, instead, they will return a deterministic model that approximates the SUL. Typically, the approximated model is larger than the SUL.

Expectation Maximisation Approach. The Baum-Welch (BW) algorithm is an iterative maximum likelihood estimation method to estimate the parameters of Markov models [36]. This technique is an application of the Expectation Maximisation algorithm. Originally designed for Hidden Markov Models [10], it has been adapted to MCs, CTMCs and MDPs [12,13].

Given a set of traces $\mathcal{O}$ (the training set) and an initial hypothesis $\mathcal{H}_0$, the BW algorithm iteratively updates $\mathcal{H}_0$ such that the likelihood that the hypothesis generates $\mathcal{O}$ has increased with respect to the previous step. The algorithm stops when the likelihood difference between two successive hypotheses is lower than a fixed threshold ϵ. In JAJAPY, the user can also set an upper bound on the number of BW iterations. BW converges to a local optimum [37].

The BW algorithm is a passive learning approach, it allows the user to decide the size of the output model, and can learn non-deterministic models.

Active Learning with Sampling Strategy. JAJAPY implements an active learning extension of the BW algorithm for MDPs [11,12]. This method uses a sampling strategy to generate new training samples that are most informative for the current model hypothesis. With this method, the user decides the size of the output model. This algorithm is able to learn non-deterministic models.

Currently, JAJAPY only supports the sampling strategy described in [11,12].

State-merging Approach. Both JAJAPY and AALPY provide an implementation of the Alergia algorithm [6,7] to learn MCs and its extension IOAlergia [9] to learn MDPs. These algorithms use a state-merging approach. Starting from a maximal tree-shaped probabilistic automaton representing the training set, they iteratively merge states that are "similar enough" according to an Hoeffding test [38]. The accuracy of the Hoeffding test is provided as input. These algorithms are passive, they do not allow the user to choose the number of states in the output model, and they assume the SUL to be deterministic.

Active Learning with Membership and Equivalence Queries. AALPY provides an implementation of L^*_{MDP} [17], an extension of Angluin's L^* algorithm [18] to learn MDPs. As for Alergia, this method assumes the SUL to be deterministic, and the size of the output model cannot be chosen in advance.

Table 1 summarises the key characteristics of the learning methods discussed above. The 5th column indicates whether or not the user can choose the number of states in the output model, and the 6th column indicates whether the algorithm is able to generate non-deterministic models or not.

Table 1. Key characteristics of the selected learning algorithms for Markov models.

Algorithm	Model	Reference	Active	# states	Non-det.	JAJAPY	AALPY
BW-MC	MC	[11,12]	✗	✓	✓	✓	✗
BW-CTMC	CTMC	[13]	✗	✓	✓	✓	✗
MM-pCTMC	pCTMC	[13]	✗	✓	✓	✓	✗
BW-MDP	MDP	[11,12]	✗	✓	✓	✓	✗
Active-BW	MDP	[11,12]	✓	✓	✓	✓	✗
Alergia	MC	[6,7]	✗	✗	✗	✓	✓
IOAlergia	MDP	[8,9]	✗	✗	✗	✓	✓
L^*_{MDP}	MDP	[17]	✓	✗	✗	✗	✓

4 Architecture and Technical Aspects

In this section, we describe some internal aspects of JAJAPY.

Jajapy models. In JAJAPY, each kind of model is represented by a class, which inherits from an abstract class `Model`. This makes JAJAPY modular and easy to extend to other model formalisms.

The abstract class `Model` implements the methods to run the model, generate traces and compute the loglikelihood[1] of a set of traces under the model. Currently, all models in JAJAPY use an explicit state-space representation.

Every JAJAPY model has an attribute `matrix` which contains the transition probabilities. This `matrix` is a *Numpy ndarray* [39] of floats. Each JAJAPY model has also an attribute `labelling` containing the label associated to each state. This attribute is a Python list whose length equals to the number of states of the model. Finally, JAJAPY uses *Sympy* [40] to represent symbolic expressions used for transition rate expressions in pCTMCs.

---

[1] The logarithm of the likelihood function.

Learning with BW. BW executions are handled by the `BW` class.

The `BW.fit` method starts by determining which model formalism should be used according to the given initial model (if provided) and the training set.

Then, it selects the appropriate update procedure and runs the BW algorithm.

An execution of the BW algorithm resolves into a sequence of matrix operations that are handled by Numpy. In addition, if JAJAPY is executed on a Linux machine, it supports multithreading to speed up the BW algorithm: at each BW iteration, JAJAPY executes one thread for each unique trace in the training set.

Output Models. The output format of any JAJAPY learning methods can be chosen among the following: STORMPY sparse model or JAJAPY model. The output model can also be exported to a PRISM file by setting the `output_file_prism` parameter of the `BW.fit` method.

Representing Training Sets and Test Sets. JAJAPY uses its own `Set` class to represent training and test sets. This class has two attributes: (i) `sequences`, the set of all unique traces in the training set, and (ii) `times`, which contains, for each trace in `sequences`, the number of times this trace has been observed. This reduces significantly the number of computations during the learning process when traces appear several times in the training set. Nevertheless, the training set can be given as a Python list (or *Numpy ndarray*) to the `BW.fit` method.

In Jajapy, training sets and test sets are not represented through a prefix tree (as is normal in other libraries) since this is only advantageous when Jajapy is used in single thread mode. The training sets/test sets are sorted by Jajapy only in this case.

5   Experimental Evaluation and Comparison

In this section, we first test JAJAPY validity. Secondly, we empirically evaluate how the different learning methods scale with the size of the output model and the training set. Finally, we compare it with AALPY.

All the experiments were run on a Linux machine with an AMD Ryzen 9 3900X 12-Core processor and 32 GB of memory.

In the experiments, we use the *loglikelihood distance* (abbreviated as *ll. distance*) as a metric to compare two models: given two models $\mathcal{M}$ and $\mathcal{M}'$ and n traces $\mathcal{O}$, the loglikelihood distance w.r.t. $\mathcal{O}$ is $\frac{1}{n}|\ln L(\mathcal{M}, \mathcal{O}) - \ln L(\mathcal{M}', \mathcal{O})|$, where $L(\mathcal{M}, \mathcal{O})$ denotes the likelihood of $\mathcal{O}$ under $\mathcal{M}$.

5.1   JAJAPY validation testing

We test JAJAPY validity as follows: (i) we translate a STORMPY model $\mathcal{M}$ representing the Yao-Knuth's die [41] to a JAJAPY one; (ii) we use it to generate a training set of 10,000 traces of length 10: 10 being big enough to reach the final state with a decent probability and 10,000 being small enough to learn the model in few seconds, but sufficiently big to learn a correct approximation of the SUL; (iii) we learn, using JAJAPY BW and Alergia implementations, two new STORMPY models $\mathcal{M}'$ and $\mathcal{M}''$, and finally (iv) $\mathcal{M}'$, $\mathcal{M}''$ are compared both w.r.t. their outcomes on some relevant model checking queries and their loglikelihood distance on a test set relative to the true model $\mathcal{M}$.

The first three queries correspond to the probability that the die roll gives us 1, 2 or 3. The next three queries indicate the probability that the die gives us 4, 5 or 6 without ever going through the same state (except the final one) more than once. Finally the last query corresponds to the probability that 10 throws of the coin are enough to simulate the roll of the die.

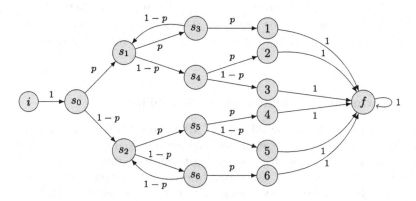

Fig. 3. The Yao-Knuth's die from [41]

We run these experiments on a Markov chain modelling the Yao-Knuth's die represented in Fig. 3 once with $p = 0.5$ (i.e. with a unbiased coin) and once with $p = 0.9$. Table 2 and 3 show that JAJAPY learned a valid representation of the source model regardless of the algorithm used. When the coin is unbiased, Alergia learns a bigger model than BW (23 states against 14) which is better in terms of loglikelihood distance but worst for the model checking queries. This is explained by the fact that, in this case, Alergia is not able to merge a large number of states and, therefore, generates a model close to a PTA, which is efficient in terms of ll distance (especially when the sequences in the test set are the same length as those in the training set). When the coin is biased, some possible traces do not appear or appear very little in the training set. Therefore, the training set being composed of much more similar traces, the initial PTA is much smaller as well as the model generated by Alergia. On the other hand, the likelihood of the sequences present in the test set and not in the training set can be fairly different between the model generated by Alergia and the SUL. In other words, for the same training set, a model generated by Alergia will often be less general than one generated by BW, because Alergia is more sensitive to overfitting.

Table 2. Results for an unbiased Yao-Knuth's die ($p = 0.5$).

	true	BW	Alergia
# states	14	**14**	23
$\mathbb{P}(F(1))$	0.167	0.168	0.168
$\mathbb{P}(F(2))$	0.167	0.170	0.169
$\mathbb{P}(F(3))$	0.167	0.163	0.163
$\mathbb{P}(F^{\leq 4}(4))$	0.125	**0.130**	0.143
$\mathbb{P}(F^{\leq 4}(5))$	0.125	**0.124**	0.136
$\mathbb{P}(F^{\leq 4}(6))$	0.125	0.107	**0.129**
$\mathbb{P}(F^{\leq 10}(f))$	0.996	0.979	**0.973**
ll. distance	0.0	1.700	**1.616**
learning time (s)	–	1.039	**0.003**

Table 3. Results for a biased Yao-Knuth's die ($p = 0.9$).

	true	BW	Alergia
# states	14	**14**	12
$\mathbb{P}(F(1))$	0.801	0.797	0.797
$\mathbb{P}(F(2))$	0.089	0.092	0.092
$\mathbb{P}(F(3))$	0.010	0.008	0.008
$\mathbb{P}(F^{\leq 4}(4))$	0.081	0.088	**0.076**
$\mathbb{P}(F^{\leq 4}(5))$	0.009	0.010	**0.009**
$\mathbb{P}(F^{\leq 4}(6))$	0.001	0.002	0.002
$\mathbb{P}(F^{\leq 10}(f))$	1.0	**0.999**	0.992
ll. distance	0.0	**0.511**	1.569
learning time (s)	–	1.060	**0.001**

5.2   Experimental Evaluation of the Scalability

Scalability Evaluation for MCs, CTMCs and MDPs. To evaluate the scalability of our software, we report the running time and the memory footprint required to learn models with an increasing number of states.

We use JAJAPY to learn randomly generated transition-labeled MCs and CTMCs ranging from 10 to 200 states, corresponding to models with up to 100 to 40,000 parameters (the number of parameters is at most s^2, where s is the number of states). We perform the same experiment for MDPs with 5 to 100 states and 4 actions, thus having at most 100 to 40,000 parameters (here the number of parameters is at most $s^2 \cdot a$, where s and a are respectively the number of states and actions). We employ training sets containing 1,000 traces of length 10. These two values offer a good compromise between accuracy and running time. We set the size of the initial hypothesis equal to that of the SUL. The results are shown in Fig. 4.

The running time for all type of SULs increases exponentially, but at a larger rate for CTMCs: while one BW iteration for an MDP with 200 states and an MC with 400 states took around two minutes in this setting, one BW iteration for a CTMC with 200 states took 97 min. This is due to the computational difficulty of calculating rates of exponential distributions when learning CTMCs parameters. Memory usage also increases exponentially for all types of Markov models. These exponential growths were expected, since the number of parameters to estimate increases exponentially with the number of states.

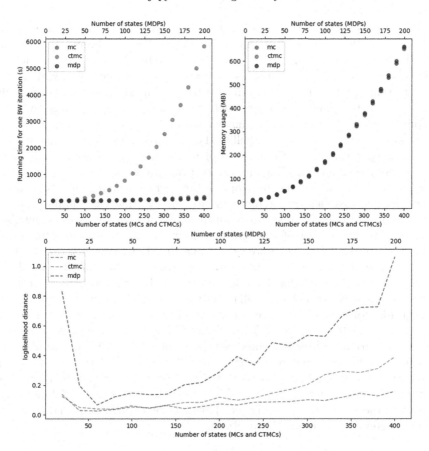

Fig. 4. JAJAPY running time, memory usage and loglikelihood distance w.r.t. the number of parameters of the hypothesis.

Finally, as the complexity of the model increases, the loglikelihood distance grows. This issue is usually mitigated by increasing the length and number of traces in the training set.

Scalability Evaluation for pCTMCs. To evaluate the scalability of our software on pCTMCs, we use the tandem queueing network model from [42] (*cf.* Fig. 5) as a benchmark for our evaluation.

The experiments have been designed according to the following setup. We assume that the state of `serverC` is fully observable —i.e., its state variables `sc` and `ph` are— as well as the size `c` of the queue and the value of `lambda`. In contrast, we assume that the state of `serverM` is not observable.

Each experiment consists in estimating the value of the parameters `mu1a`, `mu1b`, `mu2`, and `kappa` from a training set consisting of 100 traces of length 30, generated by simulating the PRISM model depicted in Fig. 5. When the value of c is large, it is necessary to have lengthy traces to cover the state space of the SUL. As a result, traces with a length of 30 are utilised. However, in order to restrict the amount of time taken for execution, only 100 traces are employed. We perform this experiment both using timed and non-timed observations, by increasing the size c of the queue until the running time of the estimation exceeds a time-out set to 1 hour. We repeat each experiment 10 times by randomly re-sampling the initial values of each unknown parameter x_i in the interval $[0.1, 5.0]$. We annotate the running time as well as the relative error δ_i for each parameter x_i, calculated according to the formula $\delta_i = |e_i - r_i|/|r_i|$, where e_i and r_i are respectively the estimated value and the real value of x_i.

Figure 6 (bottom) depicts the graph of the average running time relative to the model size together with error bars. We observe that the running time is quadratic in the number of states (equivalently, linear in the size the number of states plus the number of non-zero transitions of the model) both for timed and non-timed observations. However, for non-timed observations, the variance of the measured running times tends to grow with the size of the model. In this respect, we observed that large models required more iterations than small models to converge. Nevertheless, all experiments required at most 20 iterations.

Figure 6 (top) details the average L_1-norm (resp. L_∞-norm) of the vector $\delta = (\delta_i)$, calculated as $\|\delta\|_1 = \sum_i |\delta_i|$ (resp. $\|\delta\|_\infty = \max_i |\delta_i|$). As one may expect, the variance of the measured relative errors is larger in the experiments performed with non-timed observations, and the quality of the estimation is better when employing timed observations. Notably, for timed observations, the quality of the estimation remained stable despite the size of the model increased relative to the size of the training set. This may be explained by the fact that the parameters occur in many transitions.

```
ctmc
  // Tandem Queuing Network [Hermanns, Meyer-Kayser & Siegle]
  const int c;  // queue capacity
  const double lambda = 4 * c;
  // model parameters
  const double mu1a = 0.2; const double mu1b = 1.8; const double mu2 = 2; const double kappa = 4;

module serverC
    sc :  [0..c] init 0;
    ph :  [1..2] init 1;

    [] (sc<c)  →  lambda : (sc'=sc + 1);
    [route] (sc>0) & (ph=1)  →  mu1b : (sc'=sc - 1);
    [] (sc>0) & (ph=1)  →  mu1a : (ph'=2);
    [route] (sc>0) & (ph=2)  →  mu2 : (ph'=1) & (sc'=sc - 1);
endmodule

module serverM
    sm :  [0..c] init 0;

    [route]    (sm<c)  →  1 : (sm'=sm + 1);
    [] (sm>0)  →  kappa : (sm'=sm - 1);
endmodule
```

Fig. 5. Prism model for the tandem queueing network from [42].

5.3   Comparison with AALPY

AALPY is an active automata-learning Python library that implements several learning algorithms to learn various families of automata. Since JAJAPY learns stochastic models only, our comparison will focus on these models. However, AALPY is also able to Deterministic Finite Automata and Mealy Machines.

AALPY implements L^*_{MDP}, an extension of Angluin's L^* algorithm [17,18] to learn MDPs, and the Alergia algorithm to learn MCs.

Table 1 summarises the learning algorithms available in JAJAPY and AALPY for stochastic models. On the one hand, JAJAPY can learn non-deterministic models and CTMCs, which AALPY cannot; on the other hand, AALPY can learn non-stochastic models, which JAJAPY cannot. In contrast to AALPY, JAJAPY's output models are immediately usable in Stormpy.

We compare AALPY L^* and JAJAPY BW algorithms on learning two variants of the grid-worlds presented in [12] and illustrated in Fig. 7. A robot is moving in the grid, starting from the top-left corner. Its objective is to reach the

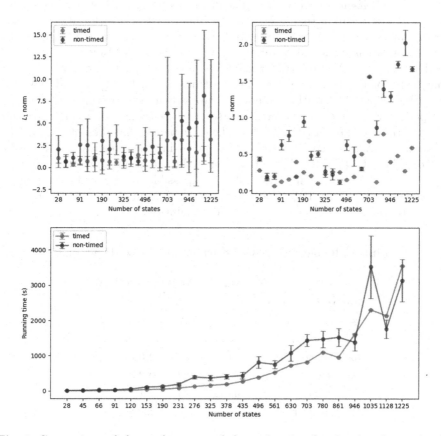

Fig. 6. Comparison of the performance of the estimation for timed and non-timed observations on the tandem queueing network with different size of the queue.

bottom-right corner. The actions are the four directions —north, east, south, and west— and the observed labels are the different terrains. Depending on the target terrain, the robot may make errors in movement, e.g. move southwest instead of south. By construction, the 3×3 world is a deterministic MDP, and the 4×4 world is a non-deterministic one.

We run, for both models, AALPY for 200 L^* learning iterations and JAJAPY for 200 BW iterations. We emphasize the fact that the two tools are using two different learning algorithms that are, in the author's opinion, complementary.

Table 4 and 5 show the results respectively for the 3×3 grid and the 4×4 grid. In both cases, the loglikelihood distance is computed for a test set containing 10,000 traces of length 20. First, we observe that, when the SUL is deterministic, the two output models are similar. Actually, JAJAPY output is slightly closer to the SUL, but AALPY ran faster. However, when the SUL is non-deterministic, the difference between the two output models is more important. AALPY ran faster but produced a model with almost 8 times more states. Indeed L^*_{MDP}, by property, learned a deterministic approximation of the SUL, that is much bigger than the SUL itself. In terms of loglikelihood distance, AALPY output model is slightly less accurate than JAJAPY one. Finally, we notice that JAJAPY uses far less information than AALPY, and does not require any interaction with the SUL (using a passive learning approach), in contrast to AALPY.

The fact that AALPY runs faster than JAJAPY can be explained by the complexity of the two algorithms involved here, namely L^*_{MDP} and the BW algorithm. The BW algorithm is known to be costly in terms of time and memory complexity. Khreich et al. [43] point out several cases where, due to its cost, the BW algorithm could not be applied. In the same paper, they present a variant of it, requiring fewer memory resources while achieving the same results. However, Bartolucci et al. [44] show that this variant suffers from numerical problems.

In general, when learning MDPs, if it is impossible to interact with the SUL, we recommend JAJAPY BW. Otherwise, we recommend using AALPY L^*_{MDP}, especially when the SUL is known to be deterministic.

Table 4. Results for the 3×3 deterministic grid-world model.

	true	AALPY	JAJAPY BW
overall # of labels	–	$74,285$	$74,285$
# of traces	-	$15,218$	$3,714$
$\|S\|$ (# of states)	17	18	**17**
loglikelihood distance	0.0	0.7305	**0.3352**
$\mathbb{P}_{\max}[F^{\leq 4}(\text{goal})]$	0.336	0.322	**0.347**
$\mathbb{P}_{\max}[\neg G\ U^{\leq 4}(\text{goal})]$	0.072	0.074	0.074
Running time	–	**1.15 s**	290.8 s

Fig. 7. Grid worlds models. (Left) a 3×3 deterministic model; (Right) a 4×4 non-deterministic model.

Table 5. Results for the 4×4 non-deterministic grid-world model.

	true	AALPY	JAJAPY BW
overall # of labels	–	16,232,244	**200,000**
# of traces	–	$2,174,167$	**10,000**
$|S|$ (# of states)	28	207	**28**
loglikelihood distance	0.0	0.4963	**0.4680**
$\mathbb{P}_{\max}[F^{\leq 7}(\text{goal})]$	0.687	0.680	**0.692**
$\mathbb{P}_{\max}[F^{\leq 12}(\text{goal})]$	0.996	0.995	**0.996**
$\mathbb{P}_{\max}[\neg(\text{C} \mid \text{W}) \ U^{\leq 7}(\text{goal})]$	0.520	**0.514**	0.504
Running time	–	**290.65 s**	15,303.83 s

6   Conclusions and Future Work

We presented JAJAPY, a Python learning library for Markov models, and discussed its key features, implementation, usage, and performance evaluation. JAJAPY is designed to be interoperable with PRISM and STORM, and offers a variety of learning methods, both active and passive. We compared JAJAPY and AALPY and argued that the two libraries complement each other, thus the choice of which library to use depends on the learning scenario.

As a future work, we consider implementing GPU-accelerated methods to speed-up the forward-backward computations required at each iteration of the BW algorithms borrowing ideas from [45,46].

Data Availability. An artifact allowing one to reproduce the experiments from this paper has been submitted to the QEST 2023 artifact evaluation.

References

1. Kwiatkowska, M., Norman, G., Parker, D.: PRISM 4.0: verification of probabilistic real-time systems. In: Gopalakrishnan, G., Qadeer, S. (eds.) CAV 2011. LNCS, vol. 6806, pp. 585–591. Springer, Heidelberg (2011). https://doi.org/10.1007/978-3-642-22110-1_47

2. Hensel, C., Junges, S., Katoen, J.-P., Quatmann, T., Volk, M.: The probabilistic model checker storm. Int. J. Softw. Tools Technol. Transfer **24**, 1–22 (2022)

3. Alur, R., Henzinger, T.A.: Reactive modules. Formal Methods Syst. Des. **15**(1), 7–48 (1999). https://doi.org/10.1023/A:1008739929481

4. Budde, C.E., et al.: On correctness, precision, and performance in quantitative verification. In: Margaria, T., Steffen, B. (eds.) ISoLA 2020. LNCS, vol. 12479, pp. 216–241. Springer, Cham (2021). https://doi.org/10.1007/978-3-030-83723-5_15

5. Reynouard, R.: Jajapy github repository (2022). https://github.com/Rapfff/jajapy

6. Carrasco, R.C., Oncina, J.: Learning stochastic regular grammars by means of a state merging method. In: Carrasco, R.C., Oncina, J. (eds.) ICGI 1994. LNCS, vol. 862, pp. 139–152. Springer, Heidelberg (1994). https://doi.org/10.1007/3-540-58473-0_144

7. Carrasco, R.C., Oncina, J.: Learning deterministic regular grammars from stochastic samples in polynomial time. RAIRO - Theor. Inf. Appl. (RAIRO: ITA) **33**(1), 1–20 (1999)

8. Mao, H., Chen, Y., Jaeger, M., Nielsen, T., Larsen, K., Nielsen, B.: Learning probabilistic automata for model checking, pp. 111–120, October 2011

9. Mao, H., Chen, Y., Jaeger, M., Nielsen, T.D., Larsen, K.G., Nielsen, B.: Learning deterministic probabilistic automata from a model checking perspective. Mach. Learn. **105**(2), 255–299 (2016)

10. Rabiner, L.R.: A tutorial on hidden markov models and selected applications in speech recognition. Proc. IEEE **77**(2), 257–286 (1989)

11. Bacci, G., Ingólfsdóttir, A., Larsen, K.G., Reynouard, R.: Active learning of markov decision processes using baum-welch algorithm (extended), CoRR, vol. abs/2110.03014 (2021). https://arxiv.org/abs/2110.03014

12. Bacci, G., Ingólfsdóttir, A., Larsen, K.G., Reynouard, R.: Active learning of markov decision processes using baum-welch algorithm. In: 2021 20th IEEE International Conference on Machine Learning and Applications (ICMLA), pp. 1203–1208 (2021)

13. Bacci, G.: Mm algorithms to estimate parameters in continuous-time markov chains (2023). https://arxiv.org/abs/2302.08588

14. Reynouard, R.: Jajapy's documentation (2022). https://jajapy.readthedocs.io/en/latest/

15. Reynouard, R.: A short introduction to jajapy, March 2023. https://doi.org/10.5281/zenodo.7695105

16. Muškardin, E., Aichernig, B., Pill, I., Pferscher, A., Tappler, M.: Aalpy: an active automata learning library. Innov. Syst. Softw. Eng. **18**, 1–10 (2022)

17. Tappler, M., Muškardin, E., Aichernig, B.K., Pill, I.: Active model learning of stochastic reactive systems. In: Calinescu, R., Păsăreanu, C.S. (eds.) SEFM 2021. LNCS, vol. 13085, pp. 481–500. Springer, Cham (2021). https://doi.org/10.1007/978-3-030-92124-8_27

18. Angluin, D.: Learning regular sets from queries and counterexamples. Inf. Comput. **75**(2), 87–106 (1987)

19. Biere, A., Bloem, R. (eds.): CAV 2014. LNCS, vol. 8559. Springer, Cham (2014). https://doi.org/10.1007/978-3-319-08867-9

20. Bollig, B., Katoen, J.-P., Kern, C., Leucker, M., Neider, D., Piegdon, D.R.: The automata learning framework. In: Touili, T., Cook, B., Jackson, P. (eds.) CAV 2010. LNCS, vol. 6174, pp. 360–364. Springer, Heidelberg (2010). https://doi.org/10.1007/978-3-642-14295-6_32

21. Thollard, F., Dupont, P., de la Higuera, C.: Probabilistic DFA inference using kullback-leibler divergence and minimality. In: International Conference on Machine Learning (2000)

22. Stolcke, A.: Bayesian learning of probabilistic language models (1994)

23. Gelfand, A., Smith, A.: Sampling-based approaches to calculate marginal densities. J. Am. Stat. Assoc. **85**, 398–409 (1990)

24. Neal, R.: Markov chain sampling methods for dirichlet process mixture models. J. Comput. Graph. Stat. **9**, 249–265 (2000)

25. Rashidinejad, P., Zhu, B., Ma, C., Jiao, J., Russell, S.J.: Bridging offline reinforcement learning and imitation learning: a tale of pessimism. IEEE Trans. Inf. Theor. **68**, 8156–8196 (2021)

26. Jin, Y., Yang, Z., Wang, Z.: Is pessimism provably efficient for offline rl? In: International Conference on Machine Learning (2020)

27. Buckman, J., Gelada, C., Bellemare, M.G.: The importance of pessimism in fixed-dataset policy optimization, ArXiv, vol. abs/2009.06799 (2020)

28. Morimoto, J., Doya, K.: Robust reinforcement learning. Neural Comput.**17**(2), 335–359 (2005). https://doi.org/10.1162/0899766053011528

29. Lim, S.H., Xu, H., Mannor, S.: Reinforcement learning in robust markov decision processes. In: Burges, C., Bottou, L., Welling, M., Ghahramani, Z., Weinberger, K. (eds.), Advances in Neural Information Processing Systems, vol. 26. Curran Associates Inc, (2013). https://proceedings.neurips.cc/paper_files/paper/2013/file/0deb1c54814305ca9ad266f53bc82511-Paper.pdf

30. Suilen, M., Simão, T.D., Parker, D., Jansen, N.: Robust anytime learning of markov decision processes (2023)

31. Derman, E., Mankowitz, D.J., Mann, T.A., Mannor, S.: A bayesian approach to robust reinforcement learning. In: Conference on Uncertainty in Artificial Intelligence (2019)

32. Češka, M., Hensel, C., Junges, S., Katoen, J.-P.: Counterexample-guided inductive synthesis for probabilistic systems. Form. Asp. Comput. **33**(4–5), 637–667 (2021). https://doi.org/10.1007/s00165-021-00547-2

33. Quatmann, T., Dehnert, C., Jansen, N., Junges, S., Katoen, J.: Parameter synthesis for markov models: faster than ever, CoRR, vol. abs/1602.05113 (2016). http://arxiv.org/abs/1602.05113

34. Jansen, N., Junges, S., Katoen, J.: Parameter synthesis in markov models: a gentle survey. In: Raskin, J., Chatterjee, K., Doyen, L., Majumdar, R. (eds.), Principles of Systems Design - Essays Dedicated to Thomas A. Henzinger on the Occasion of His 60th Birthday, ser. LNCS, vol. 13660, pp. 407–437. Springer, Cham (2022). https://doi.org/10.1007/978-3-031-22337-2_20

35. Polgreen, E., Wijesuriya, V.B., Haesaert, S., Abate, A.: Data-efficient bayesian verification of parametric markov chains. In: Agha, G., Van Houdt, B. (eds.) QEST 2016. LNCS, vol. 9826, pp. 35–51. Springer, Cham (2016). https://doi.org/10.1007/978-3-319-43425-4_3

36. Baum, L., Petrie, T., Soules, G., Weiss, N.: A maximization technique occurring in the statistical analysis of probabilistic functions of markov chains (1970)

37. Yang, F., Balakrishnan, S., Wainwright, M.J.: Statistical and computational guarantees for the baum-welch algorithm. J. Mach. Learn. Res. **18**(125), 1–53 (2017). http://jmlr.org/papers/v18/16-093.html

38. Hoeffding, W.: Probability inequalities for sum of bounded random variables (1961)
39. Harris, C.R., et al.: Array programming with NumPy. Nature **585**(7825), 357–362 (2020). https://doi.org/10.1038/s41586-020-2649-2
40. Meurer, A., et al.: Sympy: symbolic computing in python. PeerJ Comput. Sci. **3**, e103 (2017). https://doi.org/10.7717/peerj-cs.103
41. Knuth, D., Yao, A.: Algorithms and Complexity: New Directions and Recent Results. Academic Press, Cambridge (1976), ch. The complexity of nonuniform random number generation
42. Hermanns, H., Meyer-Kayser, J., Siegle, M.: Multi terminal binary decision diagrams to represent and analyse continuous time Markov chains. In: Plateau, B., Stewart, W., Silva, M. (eds.) Proceedings of 3rd International Workshop on Numerical Solution of Markov Chains (NSMC 1999), Prensas Universitarias de Zaragoza, pp. 188–207 (1999)
43. Khreich, W., Granger, E., Miri, A., Sabourin, R.: On the memory complexity of the forward-backward algorithm. Pattern Recogn. Lett. **31**(2), 91–99 (2010). https://www.sciencedirect.com/science/article/pii/S0167865509002578
44. Bartolucci, F., Pandolfi, S.: Comment on the paper on the memory complexity of the forward-backward algorithm. In: khreich, W., Granger, E., Miri, A., Sabourin, R. (eds.), Pattern Recognition Letters, vol. 38, pp. 15–19 (2014). https://www.sciencedirect.com/science/article/pii/S0167865513003863
45. Shao, Y., Wang, Y., Povey, D., Khudanpur, S.: Pychain: a fully parallelized pytorch implementation of LF-MMI for end-to-end ASR. In: Meng, H., Xu, B., Zheng, T.F. (eds.), Interspeech 2020, 21st Annual Conference of the International Speech Communication Association, Virtual Event, Shanghai, China, 25–29 October 2020, ISCA, pp. 561–565 (2020). https://doi.org/10.21437/Interspeech.2020-3053
46. Ondel, L., Lam-Yee-Mui, L., Kocour, M., Corro, C.F., Burget, L.: Gpu-accelerated forward-backward algorithm with application to lattice-free MMI. In: IEEE International Conference on Acoustics, Speech and Signal Processing, ICASSP 2022, Virtual and Singapore, 23–27 May 2022, pp. 8417–8421. IEEE (2022). https://doi.org/10.1109/ICASSP43922.2022.9746824

Introducing Asynchronicity
to Probabilistic Hyperproperties

Lina Gerlach[1]([✉])🆔, Oyendrila Dobe[2]🆔, Erika Ábrahám[1]🆔, Ezio Bartocci[3]🆔,
and Borzoo Bonakdarpour[2]🆔

[1] RWTH Aachen University, Aachen, Germany
{gerlach,abraham}@cs.rwth-aachen.de
[2] Michigan State University, East Lansing, MI, USA
{dobeoyen,borzoo}@msu.edu
[3] Technische Universität Wien, Vienna, Austria
ezio.bartocci@tuwien.ac.at

Abstract. Probabilistic hyperproperties express probabilistic relations between different executions of systems with uncertain behavior. Hyper-PCTL [3] allows to formalize such properties, where quantification over probabilistic schedulers resolves potential non-determinism. In this paper we propose an extension named AHyperPCTL to additionally introduce *asynchronicity* between the observed executions by quantifying over *stutter-schedulers*, which may randomly decide to delay scheduler decisions by idling. To our knowledge, this is the first asynchronous extension of a probabilistic branching-time hyperlogic. We show that AHyperPCTL can express interesting information-flow security policies, and propose a model checking algorithm for a decidable fragment.

1 Introduction

Consider the following simple multi-threaded program [29] consisting of two threads with a secret input h and a public output l:

$$th\colon \textbf{while } h > 0 \textbf{ do } \{h \leftarrow h - 1\}; l \leftarrow 2 \quad \| \quad th'\colon l \leftarrow 1$$

Assuming that this program is executed under a probabilistic scheduler, the probability of observing $l = 1$ decreases for increasing initial values of h. Hence, this program does not satisfy scheduler-specific probabilistic observational determinism (SSPOD) [27], which requires that no information about the private data is leaked through the publicly visible data, for any scheduling of the threads. In fact, the scheduler is creating a probabilistic side channel that leaks the value of the secret. Probabilistic hyperlogics such as HyperPCTL [2,3,20] and PHL [17] are able to express and verify requirements such as SSPOD.

Interestingly, there is a way to mitigate this side channel similar to the padding mechanism that counters timing side channels. In the above example, for any two executions of the program under the same scheduler with different

N. Jansen and M. Tribastone (Eds.): QEST 2023, LNCS 14287, pp. 47–64, 2023.
https://doi.org/10.1007/978-3-031-43835-6_4

initial h, we can find *stuttering variations* of the program such that the probability of reaching any specific final value of l is the same for both executions. For example, for two different values of h, say h_1 and h_2, where $h_1 < h_2$, letting thread th' initially stutter $(h_2 - h_1)$ times (i.e., repeating the current state) in the execution starting with h_1 will equalize the probability of reaching $l = 1$.

While there have been efforts to incorporate stuttering semantics in non-probabilistic logics (e.g., A-HLTL [8]), in the probabilistic setting, neither Hyper-PCTL nor PHL allow reasoning about stuttering behaviors, i.e., their semantics are "synchronous" in the sense that all computation trees are evaluated in lock-step. In this paper, we propose an asynchronous extension of HyperPCTL that allows to reason about stuttering computations and whether we can find stuttering variations of programs such that a probabilistic hyperproperty is satisfied.

Related Work. HyperPCTL [3] was the first logic for specifying probabilistic hyperproperties over DTMCs, by providing state-based quantifiers over the computation trees of the DTMC. This logic was further extended [1,2,18–20] with the possibility to specify quantifiers over schedulers for model checking Markov decision processes (MDPs). The probabilistic hyperlogic PHL [17] can also handle analysis of MDPs. In general, the (exact) model checking problem for both HyperPCTL and PHL is undecidable unless we restrict the class of schedulers or we rely on some approximating methods. HyperPCTL* [30,31] extends PCTL* [5] with quantifiers over execution paths and is employed in statistical model checking. All three logics are synchronous and lock-step. To the best of our knowledge, our work is the first to consider asynchronicity in the probabilistic setting.

In the non-probabilistic setting, asynchronicity has already been studied [7,9, 10,12,13,23]. In [7], the authors study the expressivity of HyperLTL [15] showing the impossibility to express the "two-state local independence" asynchronous hyperproperty, where information flow is allowed only after a change of state (for example in the case of declassification [28]). To cope with this limitation, several asynchronous extensions of HyperLTL have been proposed.

For example, *Asynchronous HyperLTL* [10] extends HyperLTL with quantification over "trajectories" that enable the alignment of execution traces from different runs. *Stuttering HyperLTL* [12] relates only stuttering traces where two consecutive observations are different. *Context HyperLTL* [12] instead allows to combine synchronous and asynchronous hyperproperties. All three logics are in general undecidable, but there are useful decidable fragments that can be model-checked. The expressiveness of these logics has been compared in [13].

Contributions. Our main contribution is a new logic, called AHyperPCTL, which is an asynchronous extension of HyperPCTL and allows to reason about probabilistic relations between stuttering variations of probabilistic and potentially non-deterministic systems. To our knowledge, this is the first asynchronous extension of a probabilistic branching-time hyperlogic. Our goal is to associate several executions with independent stuttering variations of the same program and compare them. We implement this by extending HyperPCTL with quantification over *stutter-schedulers*, which specify when the program should stutter.

We show that AHyperPCTL is useful to express whether information leaks can be avoided via suitable stuttering. In the context of our introductory example, the following AHyperPCTL formula expresses SSPOD under stuttering:

$$\forall \hat{\sigma}. \ \forall \hat{s}(\hat{\sigma}). \ \forall \hat{s}'(\hat{\sigma}). \ \exists \hat{\tau}(\hat{s}). \ \exists \hat{\tau}'(\hat{s}').$$
$$(h_{\hat{\tau}} \neq h_{\hat{\tau}'} \wedge init_{\hat{\tau}} \wedge init_{\hat{\tau}'}) \Rightarrow (\bigwedge_{k \in \{1,2\}} \mathbb{P}(\lozenge(l=k)_{\hat{\tau}}) = \mathbb{P}(\lozenge(l=k)_{\hat{\tau}'})),$$

where $\hat{\sigma}$ represents a probabilistic scheduler that specifies which thread is allowed to execute in which program state, $\hat{s}$ and $\hat{s}'$ represent initial states, and $\hat{\tau}$ and $\hat{\tau}'$ are stutter-scheduler variables for the computation trees rooted at $\hat{s}$ and $\hat{s}'$ under the scheduler $\hat{\sigma}$. This formula specifies that under any probabilistic scheduling $\hat{\sigma}$ of the two threads, if we consider two computation trees starting in states $\hat{s}$ and $\hat{s}'$ with different values for the secret variable h, there should exist stutterings for the two experiments such that the probabilities of observing any specific final value of l are the same for both.

We propose a model checking algorithm for AHyperPCTL under restrictions on the classes of schedulers and stutter-schedulers. Our method generates a logical encoding of the problem in real arithmetic and uses a satisfiability modulo theories (SMT) solver, namely Z3 [25], to determine the truth of the input statement. We experimentally demonstrate that the model checking problem for asynchronous probabilistic hyperproperties is a computationally highly complex synthesis problem at two levels: both for synthesizing scheduler policies and for synthesizing stutter schedulers. This poses serious problems for model checking: our current implementation does not scale beyond a few states. We discuss some insights about this complexity and suggest possible future directions.

Organization. We discuss preliminary concepts in Sect. 2 and introduce AHyper-PCTL in Sect. 3. We dedicate Sect. 4 to applications and Sect. 5 to our algorithm. We discuss results of our prototype implementation in Sect. 6. We conclude in Sect. 7 with a summary and future work.

2   Preliminaries

We denote the real (non-negative real) numbers by $\mathbb{R}$ ($\mathbb{R}_{\geq 0}$), and the natural numbers including (excluding) 0 by $\mathbb{N}$ ($\mathbb{N}_{>0}$). For any $n \in \mathbb{N}$, we define $[n]$ to be the set $\{0, \ldots, n-1\}$. We use () to denote the empty tuple and $\circ$ for concatenation.

Definition 1. *A* discrete-time Markov chain (DTMC) *is a tuple* $\mathcal{D}=(S, \mathsf{AP}, L, \mathbf{P})$ *where (1) S is a non-empty finite set of states, (2) AP is a set of atomic propositions, (3) $L: S \to 2^{AP}$ is a labeling function and (4) $\mathbf{P}: S \times S \to [0,1]$ is a transition probability function such that $\sum_{s' \in S} \mathbf{P}(s, s') = 1$ for all $s \in S$.*

An *(infinite)* path *is a sequence* $s_0 s_1 s_2 \ldots \in S^\omega$ of states with $\mathbf{P}(s_i, s_{i+1}) > 0$ for all $i \geq 0$. Let $Paths_s^{\mathcal{D}}$ denote the set of all paths of $\mathcal{D}$ starting in $s \in S$, and $fPaths_s^{\mathcal{D}}$ denote the set of all non-empty finite prefixes of paths from $Paths_s^{\mathcal{D}}$, which we call *finite paths*. For a finite path $\pi = s_0 \ldots s_k \in \mathit{fPaths}_{s_0}^{\mathcal{D}}$, $k \geq 0$, we define $|\pi| = k$. A state $t \in S$ is *reachable* from $s \in S$ if there exists a finite path in $\mathit{fPaths}_s^{\mathcal{D}}$ that ends in t. The *cylinder set* $\mathit{Cyl}^{\mathcal{D}}(\pi)$ of a finite path π is the set of all infinite paths with π as a prefix. The *probability space* for $\mathcal{D}$ and $s \in S$ is

$$\left(\mathit{Paths}_s^{\mathcal{D}}, \left\{ \bigcup_{\pi \in R} \mathit{Cyl}^{\mathcal{D}}(\pi) \mid R \subseteq \mathit{fPaths}_s^{\mathcal{D}} \right\}, \Pr_s^{\mathcal{D}} \right),$$

where the *probability* of the cylinder set of $\pi \in \mathit{fPaths}_s^{\mathcal{D}}$ is $\Pr_s^{\mathcal{D}}(\mathit{Cyl}^{\mathcal{D}}(\pi)) = \Pi_{i=1}^{|\pi|} \mathbf{P}(\pi_{i-1}, \pi_i)$. These concepts have been discussed in detail in [5].

Markov decision processes extend DTMCs to allow the modeling of environment interaction or user input in the form of non-determinism.

Definition 2. *A* Markov decision process (MDP) *is defined as a tuple* $\mathcal{M} = (S, \mathsf{AP}, L, \mathit{Act}, \mathbf{P})$, *where (1)* S *is a non-empty finite set of* states, *(2)* AP *is a set of* atomic propositions, *(3)* $L \colon S \to 2^{\mathsf{AP}}$ *is a* labeling function, *(4)* Act *is a non-empty finite set of* actions, *(5)* $\mathbf{P} \colon S \times \mathit{Act} \times S \to [0,1]$ *is a* transition probability function *such that for all* $s \in S$ *the set of its* enabled actions

$$\mathit{Act}(s) = \left\{ \alpha \in \mathit{Act} \mid \textstyle\sum_{s' \in S} \mathbf{P}(s, \alpha, s') = 1 \right\}$$

is non-empty and $\sum_{s' \in S} \mathbf{P}(s, \alpha, s') = 0$ *for all* $\alpha \in \mathit{Act} \setminus \mathit{Act}(s)$.

Let $\mathbb{M}$ be the set of all MDPs. For every execution step of an MDP, a *scheduler* resolves the non-determinism by selecting an enabled action to be executed.

Definition 3. *For an MDP* $\mathcal{M} = (S, \mathsf{AP}, L, \mathit{Act}, \mathbf{P})$, *a* scheduler *is a tuple* $\sigma = (Q, \mathit{mode}, \mathit{init}, \mathit{act})$, *where (1)* Q *is a non-empty countable set of* modes, *(2)* $\mathit{mode} \colon Q \times S \to Q$ *is a* mode transition function, *(3)* $\mathit{init} \colon S \to Q$ *selects the starting mode* $\mathit{init}(s)$ *for each state of* $s \in S$, *and (4)* $\mathit{act} \colon Q \times S \times \mathit{Act} \to [0,1]$ *is a function with* $\sum_{\alpha \in \mathit{Act}(s)} \mathit{act}(q, s, \alpha) = 1$ *and* $\sum_{\alpha \in \mathit{Act} \setminus \mathit{Act}(s)} \mathit{act}(q, s, \alpha) = 0$ *for all* $s \in S$ *and* $q \in Q$.

We use $\Sigma^{\mathcal{M}}$ to denote the set of all schedulers for an MDP $\mathcal{M}$. A scheduler is called *finite-memory* if Q is finite, *memoryless* if Q is a singleton, and *deterministic* if $\mathit{act}(q, s, \alpha) \in \{0, 1\}$ for all $(q, s, \alpha) \in Q \times S \times \mathit{Act}$. If a scheduler is memoryless, we sometimes omit its only mode.

3 Asynchronous HyperPCTL

Probabilistic hyperproperties specify probabilistic relations between different executions of one or several probabilistic models. In previous work, we introduced HyperPCTL [3] to reason over non-determinism [20] and rewards [19] for *synchronous* executions, i.e., where all executions make their steps simultaneously. In this work, we propose an extension to reason about *asynchronous* executions, where some of the executions may also *stutter* (i.e., stay in the same state without observable changes) while others execute.

This is useful if, for example, the duration of some computations depend on some secret input and we thus might wish to make the respective duration unobservable. A typical application area are multi-threaded programs, like the one presented in Sect. 1, where we want to relate the executions of the different threads. If there is a single processor available, each execution step allows one of the threads to execute, while the others idle. The executing thread, however, might also decide to stutter in order to hide its execution duration. To be able to formalize such behavior, the decision whether an execution stutters or not must depend not only on the history but also on the chosen action (in our example, corresponding to which of the threads may execute).

In this section, we first introduce a novel scheduler concept that supports stuttering, followed by the extension of HyperPCTL that we henceforth refer to as AHyperPCTL. To improve readability, we assume that all executions run in the same MDP; an extension to different MDPs is a bit technical but straightforward.

3.1   Stutter Schedulers

We define a *stutter-scheduler* as an additional type of scheduler that only distinguishes between stuttering, represented by ε, or proceeding, represented by $\bar{\varepsilon}$, for every state $s \in S$ and action $\alpha \in Act$.

Definition 4. *A stutter-scheduler for an MDP $\mathcal{M} = (S, \mathsf{AP}, L, Act, \mathbf{P})$ is a tuple $\tau = (Q^\varepsilon, mode^\varepsilon, init^\varepsilon, act^\varepsilon)$ where (1) Q^ε is a non-empty countable set of* modes, *(2) $mode^\varepsilon : Q^\varepsilon \times S \times Act \to Q^\varepsilon$ is a mode transition function, (3) $init^\varepsilon : S \to Q^\varepsilon$ is a function selecting a starting mode $init(s)$ for each state $s \in S$ and (4) $act^\varepsilon : Q^\varepsilon \times S \times Act \times \{\varepsilon, \bar{\varepsilon}\} \to [0,1]$ is a function with*

$$act^\varepsilon(q^\varepsilon, s, \alpha, \varepsilon) + act^\varepsilon(q^\varepsilon, s, \alpha, \bar{\varepsilon}) = 1$$

for all $(q^\varepsilon, s, \alpha) \in Q^\varepsilon \times S \times Act$.

We use $\mathcal{T}^{\mathcal{M}}$ to denote the set of all stutter-schedulers for an MDP $\mathcal{M}$. When reasoning about asynchronicity, we consider an MDP $\mathcal{M}$ in the context of a scheduler and a stutter-scheduler for $\mathcal{M}$. At each state, first the scheduler chooses an action α, followed by a decision of the stutter-scheduler whether to execute α or to stutter (i.e., stay in the current state). Thus, a stutter-scheduler makes its decisions based on not only its mode and the MDP state, but also depending on the action chosen by the scheduler.

Definition 5. *For an MDP $\mathcal{M} = (S, \mathsf{AP}, L, Act, \mathbf{P})$, a scheduler σ for $\mathcal{M}$ and a stutter-scheduler τ for $\mathcal{M}$, the DTMC induced by σ and τ is defined as $\mathcal{M}^{\sigma,\tau} = (S^{\sigma,\tau}, \mathsf{AP}, L^{\sigma,\tau}, \mathbf{P}^{\sigma,\tau})$, where $S^{\sigma,\tau} = Q \times Q^\varepsilon \times S$, $L^{\sigma,\tau}(q, q^\varepsilon, s) = L(s)$ and*

$$\mathbf{P}^{\sigma,\tau}((q, q^\varepsilon, s), (q', q^{\varepsilon'}, s')) = \begin{cases} stut & \text{if } q' = q \neq mode(q, s) \land s' = s \\ cont & \text{if } q' = mode(q, s) \land (q' \neq q \lor s' \neq s) \\ stut + cont & \text{if } q' = q = mode(q, s) \land s' = s \\ 0 & \text{otherwise} \end{cases}$$

with $stut = \sum_{\alpha \in Act, mode^\varepsilon(q^\varepsilon, s, \alpha) = q^{\varepsilon'}} act(q, s, \alpha) \cdot act^\varepsilon(q^\varepsilon, s, \alpha, \varepsilon)$

and $cont = \sum_{\alpha \in Act, mode^\varepsilon(q^\varepsilon, s, \alpha) = q^{\varepsilon'}} act(q, s, \alpha) \cdot act^\varepsilon(q^\varepsilon, s, \alpha, \bar{\varepsilon}) \cdot \mathbf{P}(s, \alpha, s')$.

The properties *finite-memory, memoryless,* and *deterministic* for stutter sched-ulers are defined analogously as for schedulers. A stutter-scheduler τ for an MDP $\mathcal{M}$ is *fair* for a scheduler $\sigma \in \Sigma^\mathcal{M}$ if the probability of taking infinitely many con-secutive stuttering steps is 0. The different executions, whose relations we want to analyze, will be evaluated in the *composition* of the induced models. The composition we use is the standard product of DTMCs with the only difference that we annotate atomic propositions with an index, indicating the execution in which they appear.

Definition 6. *For $n \in \mathbb{N}$ and DTMCs $\mathcal{D}_1, \dots, \mathcal{D}_n$ with $\mathcal{D}_i = (S_i, \mathsf{AP}_i, L_i, \mathbf{P}_i)$ for $i = 1, \dots, n$, we define the composition $\mathcal{D}_1 \times \dots \times \mathcal{D}_n$ to be the DTMC $\mathcal{D} = (S, \mathsf{AP}, L, \mathbf{P})$ with $S = S_1 \times \dots \times S_n$, $\mathsf{AP} = \cup_{i=1}^n \{a_i \mid a \in \mathsf{AP}_i\}$, $L(s_1, \dots, s_n) = \cup_{i=1}^n \{a_i \mid a \in L_i(s_i)\}$ and $P((s_1, \dots, s_n), (s'_1, \dots, s'_n)) = \Pi_{i=1}^n P_i(s_i, s'_i)$.*

Definition 7. *For an MDP $\mathcal{M}$, $n \in \mathbb{N}_{>0}$, a tuple $\boldsymbol{\sigma} = (\sigma_1, \dots, \sigma_n) \in (\Sigma^\mathcal{M})^n$ of schedulers, and a tuple $\boldsymbol{\tau} = (\tau_1, \dots, \tau_n) \in (\mathcal{T}^\mathcal{M})^n$ of stutter-schedulers, we define the induced DTMC $\mathcal{M}^{\sigma, \tau} = \mathcal{M}^{\sigma_1, \tau_1} \times \dots \times \mathcal{M}^{\sigma_n, \tau_n}$.*

Later we will make use of *counting* stutter-schedulers. These are deterministic bounded-memory stutter-schedulers which specify for each state $s \in S$ and action $\alpha \in Act(s)$ a stuttering duration $j_{s,\alpha}$. Intuitively, $j_{s,\alpha}$ determines how many successive stutter-steps need to be made in state s before α can be executed.

Definition 8. *An m-bounded counting stutter-scheduler for an MDP $\mathcal{M} = (S, \mathsf{AP}, L, Act, \mathbf{P})$ and $m \in \mathbb{N}_{>0}$ is a stutter-scheduler $\tau = ([m], mode^\varepsilon, init^\varepsilon, act^\varepsilon)$ such that for all $s \in S$ and $\alpha \in Act(s)$ there exists $j_{s,\alpha} \in [m]$ with (1) $init^\varepsilon(s) = 0$ and (2) for each $j \in [m]$, if $j < j_{s,\alpha}$ then $mode^\varepsilon(j, s, \alpha) = j + 1$ and $act^\varepsilon(j, s, \alpha, \varepsilon) = 1$, and otherwise (if $j \geq j_{s,\alpha}$) $mode^\varepsilon(j, s, \alpha) = 0$ and $act^\varepsilon(j, s, \alpha, \bar{\varepsilon}) = 1$.*

Example 1. Consider the MDP $\mathcal{M}$ from Fig. 1 as well as the DTMC $\mathcal{M}^{\sigma, \tau}$ induced by a probabilistic memoryless scheduler σ with $\sigma(s_0, \alpha) = p$, $\sigma(s_0, \beta) = 1 - p$ for some $p \in [0, 1]$ and a 3-counting stutter-scheduler τ on $\mathcal{M}$ with $j_{s_0, \alpha} = 2$, $j_{s_0, \beta} = 0$. The modes $q \in [3]$ of τ store how many times we have stuttered since actually executing the last action. For each state (s, j) of $\mathcal{M}^{\sigma, \tau}$, first σ chooses an action $\alpha \in Act(s)$ probabilistically, and then j is compared with the stutter-ing duration $j_{s,\alpha}$ stipulated by τ. If we choose β in state $(s_0, 0)$, then we move to a β-successor of s_0. However, if we choose α, then we move to the state $(s_0, 1)$ and then choose an action again. If we choose β at $(s_0, 1)$, then we move to a β-successor of s_0, but if we choose α then we move to $(s_0, 2)$. In particular, choosing α at $(s_0, 0)$ does not mean that we stutter twice in s_0. We stutter twice only if we also choose α in $(s_0, 1)$.

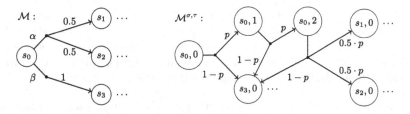

Fig. 1. The interplay of a probabilistic memoryless scheduler σ and a counting stutter-scheduler τ from Ex. 1. (The mode of σ is omitted.)

3.2 Syntax

To formalize relations of different executions, we begin an AHyperPCTL formula as in HyperPCTL [20] by first quantifying over the possible schedulers and then over the states of the MDP in which the respective execution under the chosen scheduler starts. In contrast to [20], we now additionally quantify over stutter-schedulers in dependence on the chosen schedulers and initial states. Hence, the *non-quantified* part of an AHyperPCTL formula is evaluated on the DTMC(s) induced by not only the schedulers but also the stutter-schedulers, in accordance with Definition 5. Formally, we inductively define AHyperPCTL scheduler-quantified formulas as follows:

$$
\begin{aligned}
scheduler - quantified: \quad &\varphi^{sch} ::= \forall\hat{\sigma}.\varphi^{sch} \mid \exists\hat{\sigma}.\varphi^{sch} \mid \varphi^{s} \\
state - quantified: \quad &\varphi^{s} ::= \forall\hat{s}(\hat{\sigma}).\varphi^{s} \mid \exists\hat{s}(\hat{\sigma}).\varphi^{s} \mid \varphi^{st} \\
stutter - quantified: \quad &\varphi^{st} ::= \forall\hat{\tau}(\hat{s}).\varphi^{st} \mid \exists\hat{\tau}(\hat{s}).\varphi^{st} \mid \varphi^{nq} \\
non - quantified: \quad &\varphi^{nq} ::= \textbf{true} \mid a_{\hat{\tau}} \mid \varphi^{nq} \wedge \varphi^{nq} \mid \neg\varphi^{nq} \mid \varphi^{pr} \sim \varphi^{pr} \\
probability \ expression: \quad &\varphi^{pr} ::= \mathbb{P}(\varphi^{path}) \mid f(\varphi_1^{pr}, \ldots, \varphi_k^{pr}) \\
path \ formula: \quad &\varphi^{path} ::= \bigcirc\varphi^{nq} \mid \varphi^{nq} \, \mathcal{U} \, \varphi^{nq}
\end{aligned}
$$

where $\hat{\sigma}$ is a *scheduler variable* from an infinite set $\hat{\Sigma}$, $\hat{s}$ is a *state variable* from an infinite set $\hat{S}$, $\hat{\tau}$ is a *stutter-scheduler variable* from an infinite set $\hat{T}$, $a \in \mathsf{AP}$ is an atomic proposition, $\sim\; \in \{\leq, <, =, \neq, >, \geq\}$, and $f\colon [0,1]^k \to \mathbb{R}$ is a k-ary arithmetic operator over probabilities, where a constant c is viewed as a 0-ary function. $\mathbb{P}$ refers to the probability operator and '$\bigcirc$', '$\mathcal{U}$' refer to the temporal operators 'next' and 'until', respectively.

An AHyperPCTL scheduler-quantified formula φ^{sch} is *well-formed* if each occurrence of any $a_{\hat{\tau}}$ for $\hat{\tau} \in \hat{T}$ is in the scope of a stutter quantifier for $\hat{\tau}(\hat{s})$ for some $\hat{s} \in \hat{S}$, any quantifier for $\hat{\tau}(\hat{s})$ is in the scope of a *state quantifier* for $\hat{s}(\hat{\sigma})$ for some $\hat{\sigma} \in \hat{\Sigma}$, and any quantifier for $\hat{s}(\hat{\sigma})$ is in the scope of a *scheduler quantifier* for $\hat{\sigma}$. AHyperPCTL *formulas* are well-formed AHyperPCTL scheduler-quantified formulas, where we additionally allow standard syntactic sugar: $\textbf{false} = \neg\textbf{true}$, $\varphi_1 \vee \varphi_2 = \neg(\neg\varphi_1 \wedge \neg\varphi_2)$, $\varphi_1 \Rightarrow \varphi_2 = \neg(\varphi_1 \wedge \neg\varphi_2)$, $\Diamond\varphi = \textbf{true} \, \mathcal{U} \, \varphi$, and $\mathbb{P}(\Box\varphi) = 1 - \mathbb{P}(\Diamond\neg\varphi)$.

Example 2. The well-formed AHyperPCTL formula

$$
\exists\hat{\sigma}.\, \forall\hat{s}(\hat{\sigma}).\, \forall\hat{s}'(\hat{\sigma}).\, \exists\hat{\tau}(\hat{s}).\, \exists\hat{\tau}'(\hat{s}').\, \big(init_{\hat{\tau}} \wedge init_{\hat{\tau}'}\big) \Rightarrow \big(\mathbb{P}(\Diamond a_{\hat{\tau}}) = \mathbb{P}(\Diamond a_{\hat{\tau}'})\big)
$$

states that there exists an assignment for $\hat{\sigma}$, such that, if we start two independent *experiments* from any state assignment to $\hat{s}$ and $\hat{s}'$, there exist independent
possible stutter-schedulers for the two experiments, under which the probability
of reaching a state labeled a is equal, provided the initial states of the experiments are labeled *init*. Further examples will be provided in Sect. 4.

We restrict ourselves to quantifying first over schedulers, then over states, and
finally over stutter-schedulers. This choice is a balance between the expressivity
required in our applications and understandable syntax and semantics. Note that
different state variables can share the same scheduler, but they cannot share the
same stutter-scheduler. Further, several different quantified stutter-schedulers in
a formula are not allowed to depend on each other.

3.3 Semantics

The semantic judgement rules for AHyperPCTL closely mirror the rules for Hyper-
PCTL [2]. AHyperPCTL state formulas are evaluated in the context of an MDP
$\mathcal{M}$, a sequence $\boldsymbol{\sigma} \in (\Sigma^{\mathcal{M}})^n$ of schedulers, a sequence $\boldsymbol{\tau} \in (\mathcal{T}^{\mathcal{M}})^n$ of stutter-
schedulers and a sequence $\boldsymbol{s} \in S^{\sigma,\tau}$ of $\mathcal{M}^{\sigma,\tau}$-states. The length n of these tuples
corresponds to the number of stutter-schedulers in the given formula, which
determines how many experiments run in parallel. The elements of these tuples
are instantiations of the corresponding variables in the formula. We assume the
stutter-schedulers to be fair and the variables used to refer to each of these
quantifiers in the formula to be unique to avoid ambiguity. In the following,
we use $\mathbb{Q}$ to refer to quantifiers $\{\forall, \exists\}$. We recursively evaluate the formula
by instantiating the quantifier variables with concrete schedulers, states, and
stutter-schedulers, and store them in sequences $\boldsymbol{\sigma}, \boldsymbol{s}, \boldsymbol{\tau}$. We begin by initializing
each of these sequences as empty. An MDP $\mathcal{M} = (S, \mathsf{AP}, L, Act, \mathbf{P})$ satisfies an
AHyperPCTL formula φ, denoted by $\mathcal{M} \models \varphi$, iff $\mathcal{M}, (), (), () \models \varphi$.

When instantiating a scheduler quantifier $\mathbb{Q}\hat{\sigma}.\varphi$ by a scheduler σ, we syn-
tactically replace all occurrences of $\hat{\sigma}$ in φ by σ and denote the result by
$\varphi[\hat{\sigma} \rightsquigarrow \sigma]$[1]. The instantiation of a state quantifier $\mathbb{Q}\hat{s}(\hat{\sigma}).\varphi$ works similarly but
it also remembers the respective scheduler: $\varphi[\hat{s} \rightsquigarrow s_\sigma]$ denotes the result of syn-
tactically replacing all occurrences of $\hat{s}$ in φ by s_σ. Finally, for instantiating the
n^{th} stutter-scheduler quantifier $\mathbb{Q}\hat{\tau}(s_\sigma).\varphi$, we replace all occurrences of $a_{\hat{\tau}}$ by a_n
and denote the result by $\varphi[\hat{\tau} \overset{n}{\rightsquigarrow} \tau]$. The semantics judgement rules for quantified
and non-quantified state formulas, as well as probability expressions are defined

---

[1] Note that we substitute a syntactic element with a semantic object in order to reduce
notation; alternatively one could store respective mappings in the context.

as follows:

$$\mathcal{M}, \boldsymbol{\sigma}, \boldsymbol{s}, \boldsymbol{\tau} \models \forall \hat{\sigma}. \; \varphi \qquad \textit{iff} \; \forall \sigma \in \Sigma^{\mathcal{M}}. \; \mathcal{M}, \boldsymbol{\sigma}, \boldsymbol{s}, \boldsymbol{\tau} \models \varphi[\hat{\sigma} \rightsquigarrow \sigma]$$

$$\mathcal{M}, \boldsymbol{\sigma}, \boldsymbol{s}, \boldsymbol{\tau} \models \exists \hat{\sigma}. \; \varphi \qquad \textit{iff} \; \exists \sigma \in \Sigma^{\mathcal{M}}. \; \mathcal{M}, \boldsymbol{\sigma}, \boldsymbol{s}, \boldsymbol{\tau} \models \varphi[\hat{\sigma} \rightsquigarrow \sigma]$$

$$\mathcal{M}, \boldsymbol{\sigma}, \boldsymbol{s}, \boldsymbol{\tau} \models \forall \hat{s}(\sigma). \; \varphi \qquad \textit{iff} \; \forall s \in S. \; \mathcal{M}, \boldsymbol{\sigma}, \boldsymbol{s}, \boldsymbol{\tau} \models \varphi[\hat{s} \rightsquigarrow s_{\sigma}]$$

$$\mathcal{M}, \boldsymbol{\sigma}, \boldsymbol{s}, \boldsymbol{\tau} \models \exists \hat{s}(\sigma). \; \varphi \qquad \textit{iff} \; \exists s \in S. \; \mathcal{M}, \boldsymbol{\sigma}, \boldsymbol{s}, \boldsymbol{\tau} \models \varphi[\hat{s} \rightsquigarrow s_{\sigma}]$$

$$\mathcal{M}, \boldsymbol{\sigma}, \boldsymbol{s}, \boldsymbol{\tau} \models \forall \hat{\tau}(s_{\sigma}). \; \varphi \qquad \textit{iff} \; \forall \tau \in T^{\mathcal{M}}. \; \mathcal{M}, \boldsymbol{\sigma} \circ \sigma, \boldsymbol{s} \circ (init(s), init^{\varepsilon}(s), s),$$
$$\boldsymbol{\tau} \circ \tau \models \varphi[\hat{\tau} \overset{n}{\rightsquigarrow} \tau]$$

$$\mathcal{M}, \boldsymbol{\sigma}, \boldsymbol{s}, \boldsymbol{\tau} \models \exists \hat{\tau}(s_{\sigma}). \; \varphi \qquad \textit{iff} \; \exists \tau \in T^{\mathcal{M}}. \; \mathcal{M}, \boldsymbol{\sigma} \circ \sigma, \boldsymbol{s} \circ (init(s), init^{\varepsilon}(s), s),$$
$$\boldsymbol{\tau} \circ \tau \models \varphi[\hat{\tau} \overset{n}{\rightsquigarrow} \tau]$$

$$\mathcal{M}, \boldsymbol{\sigma}, \boldsymbol{s}, \boldsymbol{\tau} \models \mathbf{true}$$

$$\mathcal{M}, \boldsymbol{\sigma}, \boldsymbol{s}, \boldsymbol{\tau} \models a_i \qquad\qquad \textit{iff} \; a_i \in L^{\sigma, \tau}(\boldsymbol{s})$$

$$\mathcal{M}, \boldsymbol{\sigma}, \boldsymbol{s}, \boldsymbol{\tau} \models \varphi_1 \wedge \varphi_2 \qquad \textit{iff} \; \mathcal{M}, \boldsymbol{\sigma}, \boldsymbol{s}, \boldsymbol{\tau} \models \varphi_1 \text{ and } \mathcal{M}, \boldsymbol{\sigma}, \boldsymbol{s}, \boldsymbol{\tau} \models \varphi_2$$

$$\mathcal{M}, \boldsymbol{\sigma}, \boldsymbol{s}, \boldsymbol{\tau} \models \neg \varphi \qquad\qquad \textit{iff} \; \mathcal{M}, \boldsymbol{\sigma}, \boldsymbol{s}, \boldsymbol{\tau} \not\models \varphi$$

$$\mathcal{M}, \boldsymbol{\sigma}, \boldsymbol{s}, \boldsymbol{\tau} \models \varphi_1^{pr} < \varphi_2^{pr} \qquad \textit{iff} \; [\![\varphi_1^{pr}]\!]_{\mathcal{M}, \boldsymbol{\sigma}, \boldsymbol{s}, \boldsymbol{\tau}} < [\![\varphi_2^{pr}]\!]_{\mathcal{M}, \boldsymbol{\sigma}, \boldsymbol{s}, \boldsymbol{\tau}}$$

$$[\![\mathbb{P}(\varphi^{path})]\!]_{\mathcal{M}, \boldsymbol{\sigma}, \boldsymbol{s}, \boldsymbol{\tau}} \;\; = \; Pr_{\boldsymbol{s}}^{\mathcal{M}^{\sigma, \tau}}(\{\pi \in Paths_{\boldsymbol{s}}^{\mathcal{M}^{\sigma, \tau}} \mid \mathcal{M}, \boldsymbol{\sigma}, \pi, \boldsymbol{\tau} \models \varphi^{path}\})$$

$$[\![f(\varphi_1^{pr}, \dots, \varphi_k^{pr})]\!]_{\mathcal{M}, \boldsymbol{\sigma}, \boldsymbol{s}, \boldsymbol{\tau}} \;\; = \; f([\![\varphi_1^{pr}]\!]_{\mathcal{M}, \boldsymbol{\sigma}, \boldsymbol{s}, \boldsymbol{\tau}}, \dots, [\![\varphi_k^{pr}]\!]_{\mathcal{M}, \boldsymbol{\sigma}, \boldsymbol{s}, \boldsymbol{\tau}})$$

where the tuples $\boldsymbol{\sigma}$, $\boldsymbol{s}$ and $\boldsymbol{\tau}$ are of length $n-1$. The semantics of path formulas is defined as follows for a path $\pi = \boldsymbol{s}_0 \boldsymbol{s}_1 \dots$ of $\mathcal{M}^{\sigma, \tau}$ with $\boldsymbol{s}_i \in S^{\sigma_1, \tau_1} \times \dots \times S^{\sigma_n, \tau_n}$:

$$\mathcal{M}, \boldsymbol{\sigma}, \pi, \boldsymbol{\tau} \models \bigcirc \varphi \qquad \textit{iff} \; \mathcal{M}, \boldsymbol{\sigma}, \boldsymbol{s}_1, \boldsymbol{\tau} \models \varphi$$

$$\mathcal{M}, \boldsymbol{\sigma}, \pi, \boldsymbol{\tau} \models \varphi_1 \, \mathcal{U} \, \varphi_2 \quad \textit{iff} \; \exists j \geq 0. \; (\mathcal{M}, \boldsymbol{\sigma}, \boldsymbol{s}_j, \boldsymbol{\tau} \models \varphi_2 \wedge$$
$$\forall i \in [0, j). \; \mathcal{M}, \boldsymbol{\sigma}, \boldsymbol{s}_i, \boldsymbol{\tau} \models \varphi_1)$$

Lemma 1. *AHyperPCTL is strictly more expressive than HyperPCTL.*

Proof (Sketch). For every MDP $\mathcal{M}$ and HyperPCTL formula φ, we can construct an MDP $\mathcal{M}'$ and an AHyperPCTL formula φ' such that $\mathcal{M} \models_{\mathsf{HyperPCTL}} \varphi$ iff $\mathcal{M}' \models_{\mathsf{AHyperPCTL}} \varphi'$. The MDP $\mathcal{M}'$ is constructed from $\mathcal{M}$ by transforming each self-loop to a two-state-loop and then adding a unique label a_s to each state s. For this MDP, we define a formula $trivial_{\hat{\tau}_1, \dots, \hat{\tau}_m}$ that checks whether the given stutter-schedulers are trivial by requiring that the probability of seeing the same state label a_s in the current and in the next step must always be 0. We construct φ' by adding a universal stutter-quantifier for each state quantifier and requiring that if these stutter-schedulers are all trivial, then the original non-quantified formula must hold.

AHyperPCTL is thus at least as expressive as HyperPCTL, and since HyperPCTL cannot express stutter quantification, AHyperPCTL is strictly more expressive.

Hence, since the model checking problem for HyperPCTL is already undecidable [3], it follows that AHyperPCTL model checking is undecidable as well.

Theorem 1. *The AHyperPCTL model checking problem is undecidable.*

4   Applications of AHyperPCTL

ACDB   Consider the code snippet [22] in Fig. 2, where two threads synchronize across a critical region realized by the `await semaphore` command. Different interleavings of the threads can yield different sequences of observable outputs (i.e., permutations of abcd). Assume this program
is executed according to a probabilistic sched-
uler. Since the behavior of thread T2 depends
on a secret input h, under synchronous semantics, the program leaks information about the
secret input: the probability of observing output sequence acdb is 0 if h = 0 and non-
zero for h = 1. However, stuttering after line 8
until b is printed would prevent an information leak. The following AHyperPCTL formula expresses a variation of SSPOD, requiring that the probability of observing any specific output should be the same regardless of
the value of h:

```
1    Thread T1(){
2      await semaphore{
3        print('a');
4        v = v+1;
5        print('b');}
6    }
7    Thread T2(){
8      print('c');
9      if h=1{
10       await semaphore{
11         v = v+2;}}
12     print('d');
13   }
```

Fig. 2. Information leak.

$$\forall \hat{\sigma}. \; \forall \hat{s}(\hat{\sigma}). \; \forall \hat{s}'(\hat{\sigma}). \; \exists \hat{\tau}(\hat{s}). \; \exists \hat{\tau}'(\hat{s}').$$
$$(h_{\hat{\tau}} \neq h_{\hat{\tau}'} \wedge init_{\hat{\tau}} \wedge init_{\hat{\tau}'}) \Rightarrow \big(\mathbb{P}(\Box \bigwedge_{a \in obs} \mathbb{P}(\bigcirc a_{\hat{\tau}}) = \mathbb{P}(\bigcirc a_{\hat{\tau}'})) = 1\big)$$

Side-Channel Timing Leaks are a kind of information leak where an attacker can infer the approximate value of the secret input based on the difference in execution time for different inputs to the algorithm. Stuttering could hide these differences. Consider the code snippet
in Fig. 3 representing the modular expo-
nentiation algorithm, which is part of the
RSA public-key encryption protocol. It
computes the value of a^b mod n where a
(integer) is the plaintext and b (integer)
is the encryption key. In [2], we verified
that we can notice the timing difference
using a synchronous logic. We formalize in
AHyperPCTL that for *any* possible scheduling of the two threads there exists possible
stuttering that prevents the timing leak:

```
1    void mexp(){
2      c = 0; d = 1; i = k;
3      while (i >= 0){
4        i = i-1; c = c*2;
5        d = (d*d) % n;
6        if (b(i) = 1){
7          c = c+1;
8          d = (d*a) % n;}}
9    }
10   ...
11   t = new Thread(mexp());
12   j = 0; m = 2 * k;
13   while (j < m & !t.stop){
14     j++;}
```

Fig. 3. Modular exponentiation.

$$\forall \hat{\sigma}. \; \forall \hat{s}(\hat{\sigma}). \; \forall \hat{s}'(\hat{\sigma}). \; \exists \hat{\tau}(\hat{s}). \exists \hat{\tau}'(\hat{s}').$$
$$(h_{\hat{\tau}} \neq h_{\hat{\tau}'} \wedge init_{\hat{\tau}} \wedge init_{\hat{\tau}'}) \Rightarrow (\bigwedge_{l=0}^{m} \mathbb{P}(\Diamond(j = l)_{\hat{\tau}}) = \mathbb{P}(\Diamond(j = l)_{\hat{\tau}'}))$$

Algorithm 1: Main SMT encoding algorithm

Input: $\mathcal{M} = (S, Act, \mathbf{P}, AP, L)$: MDP; m: Memory size for the stutter-schedulers;
$\exists \hat{\sigma} Q_1 \hat{s}_1(\hat{\sigma}) \dots Q_l \hat{s}_l(\hat{\sigma}) \exists \hat{\tau}_1(\hat{s}_{k_1}) \dots \exists \hat{\tau}_n(\hat{s}_{k_n}).\varphi^{nq}$: AHyperPCTL formula.
Output: Whether $\mathcal{M}$ satisfies the input formula.

1 **Function** Main($\mathcal{M}, m, \exists \hat{\sigma} Q_1 \hat{s}_1(\hat{\sigma}) \dots Q_l \hat{s}_l(\hat{\sigma}) \exists \hat{\tau}_1(\hat{s}_{k_1}) \dots \exists \hat{\tau}_n(\hat{s}_{k_n}).\varphi^{nq}$)

2 $\quad \varphi_{sch} := \bigwedge_{\emptyset \neq A \subseteq Act} \left(\bigwedge_{\alpha \in A} 0 \leq \sigma_{A,\alpha} \leq 1 \right) \wedge \sum_{\alpha \in A} \sigma_{A,\alpha} = 1$ // scheduler choice

3 $\quad \wedge \bigwedge_{i=1}^{n} \bigwedge_{s \in S} \bigwedge_{\alpha \in Act(s)} (\bigvee_{j=0}^{m-1} \tau_{i,s,\alpha} = j)$ // stuttering choice

4 $\quad \wedge \bigwedge_{i=1}^{n} \bigwedge_{(s,j) \in S \times [m]} \bigwedge_{\alpha \in Act(s)} \bigwedge_{(s',j') \in succ(s,\alpha)} \varphi_{go_{i,(s,j),\alpha,(s',j')}}$

5 $\quad \wedge \bigwedge_{i=1}^{n} \bigwedge_{(s,j) \in S \times [m]} \bigwedge_{\alpha \in Act(s)} \bigwedge_{(s',j') \in succ(s,\alpha)} \varphi_{tr_{i,(s,j),\alpha,(s',j')}}$

6 $\quad \varphi_{sem} := \mathsf{Sem}(\mathcal{M}, n, \varphi^{nq})$ // semantics of φ^{nq}

7 $\quad$ **foreach** $i = 1, \dots, l$ **do**   // encode state quantifiers

8 $\quad\quad$ **if** $Q_i = \forall$ **then** $A_i := \text{``}\bigwedge_{s_i \in S}\text{''}$ **else** $A_i := \text{``}\bigvee_{s_i \in S}\text{''}$

9 $\quad \varphi_{tru} := A_1 \dots A_l \left(h_{((s_{k_1},0),\dots,(s_{k_n},0)),\varphi^{nq}} \right)$ // truth of input formula

10 $\quad$ **if** $check(\varphi_{sch} \wedge \varphi_{sem} \wedge \varphi_{tru}) = SAT$ **then return** $TRUE$ **else return** $FALSE$

---

5 Model Checking AHyperPCTL

Due to general undecidability, we propose a model checking algorithm for a practically useful semantical fragment of AHyperPCTL: (1) we restrict scheduler quantification to probabilistic memoryless schedulers such that the same probabilistic decisions are made in states with identical enabled action sets, i.e., if $Act(s) = Act(s')$, then $act(s, \alpha) = act(s', \alpha)$ for all $\alpha \in Act(s)$, and (2) stutter quantification ranges over m-bounded counting stutter-schedulers.

These restrictions were chosen to achieve decidability but still be expressive enough for our applications.

For simplicity, here we describe the case for a single scheduler quantifier, but the algorithm can be extended to an arbitrary number of scheduler quantifiers. Additionally, we only describe the algorithm for existential scheduler and stutter quantification. The extension to purely universal quantification is straightforward; we will discuss the handling of quantifier alternation in Sect. 6.

Our AHyperPCTL model checking method adapts the HyperPCTL algorithm [2] with two major extensions: (1) we consider probabilistic memoryless schedulers instead of deterministic memoryless ones and (2) we support stuttering. Assume in the following an MDP $\mathcal{M} = (S, Act, \mathbf{P}, AP, L)$, a memory bound m for stutter-schedulers, and an input AHyperPCTL formula φ. Our method generates a quantifier-free real-arithmetic formula $\varphi_{sch} \wedge \varphi_{tru} \wedge \varphi_{sem}$ that is satisfiable if and only if $\mathcal{M} \models \varphi$ (under the above restrictions on the domains of schedulers and stutter-schedulers). The main method (Algorithm 1) generates this encoding. 1) In φ_{sch} (Lines 2–5) we encode the scheduler probabilities and counting stutter-scheduler choices. We use real-valued variables $\sigma_{A,\alpha}$ to encode the probability of choosing α in state s with $Act(s) = A$, and variables $\tau_{i,s,\alpha}$ with domain $[m]$ (m being the stutter-scheduler memory bound) to represent the stuttering duration for state s and action α under the ith stutter-scheduler quantifier.

Algorithm 2: SMT encoding for the meaning of the non-quantified formula

---

Input: $\mathcal{M} = (S, Act, \mathbf{P}, AP, L)$: MDP; n: number of experiments;
$\quad\quad\varphi$: quantifier-free AHyperPCTL formula or expression.

Output: SMT encoding of the meaning of φ for $\mathcal{M}$.

1 **Function** Sem$(\mathcal{M}, n, \varphi)$

2 $\quad$ **if** φ is $\mathbb{P}(\varphi_1 \, \mathcal{U} \, \varphi_2)$ **then**

3 $\quad\quad$ $E := \text{Sem}(\mathcal{M}, \varphi_1, n) \wedge \text{Sem}(\mathcal{M}, \varphi_2, n)$

4 $\quad\quad$ **foreach** $\mathbf{s} = ((s_1, j_1), \ldots, (s_n, j_n)) \in (S \times [m])^n$ **do**

5 $\quad\quad\quad$ $E := E \wedge (h_{\mathbf{s},\varphi_2} \Rightarrow pr_{\mathbf{s},\varphi}=1) \wedge ((\neg h_{\mathbf{s},\varphi_1} \wedge \neg h_{\mathbf{s},\varphi_2}) \Rightarrow pr_{\mathbf{s},\varphi}=0)$

6 $\quad\quad\quad$ $E := E \wedge \left[\left[h_{\mathbf{s},\varphi_1} \wedge \neg h_{\mathbf{s},\varphi_2} \right] \Rightarrow \left[pr_{\mathbf{s},\varphi} = \right.\right.$

$$\sum_{\alpha \in Act(\mathbf{s})} \sum_{\mathbf{s}' \in succ(\mathbf{s},\alpha)} \left(\prod_{i=1}^{n} \sigma_{Act(s_i),\alpha_i} \cdot go_{i,s_i,\alpha_i,s_i'} \cdot tr_{i,s_i,\alpha_i,s_i'} \right) \cdot pr_{\mathbf{s}',\varphi} \wedge$$

$$\left[pr_{\mathbf{s},\varphi} > 0 \Rightarrow \bigvee_{\alpha \in Act(\mathbf{s})} \bigvee_{\mathbf{s}' \in succ(\mathbf{s},\alpha)} \left(\prod_{i=1}^{n} \sigma_{Act(s_i),\alpha_i} \cdot go_{i,s_i,\alpha_i,s_i'} > 0 \wedge (h_{\mathbf{s}',\varphi_2} \vee \right.\right.$$

$$\left.\left.\left.\left. d_{\mathbf{s},\varphi_2} > d_{\mathbf{s}',\varphi_2} \right) \right) \right] \right]$$

7 $\quad$ **else if** $\ldots$

8 $\quad$ **return** E

For $\mathbf{s} = ((s_1, j_1), \ldots, (s_n, j_n)) \in (S \times [m])^n$ we define $Act(\mathbf{s}) = Act(s_1) \times \ldots \times Act(s_n)$. The calculation of successor states for the encoding of the temporal operators depends on the chosen stutterings. To describe possible successors, we use two mechanisms: (i) For each $\mathbf{s} \in (S \times [m])^n$ and $\alpha \in Act(\mathbf{s})$ we define $succ(\mathbf{s}, \alpha)$ to be the set of all $\mathbf{s}' = ((s_1', j_1'), \ldots, (s_n', j_n')) \in (S \times [m])^n$ which under *some* stutter-scheduler could be successors of $\mathbf{s}$ under α, i.e., such that

$$\forall 1 \leq i \leq n. \, ((j_i < m - 1 \wedge s_i = s_i' \wedge j_i' = j_i + 1) \vee (\mathbf{P}(s_i, \alpha_i, s_i') > 0 \wedge j_i' = 0)).$$

(ii) For each $1 \leq i \leq n$, $(s, j) \in (S \times [m])$, $\alpha \in Act(s)$, and $(s', j') \in succ(s, \alpha)$ we define a pseudo-Boolean variable $go_{i,(s,j),\alpha,(s',j')}$ as well as a real variable $tr_{i,(s,j),\alpha,(s',j')}$ and define the formulas

$$\varphi go_{i,(s,j),\alpha,(s',j')} = (go_{i,(s,j),\alpha,(s',j')}=0 \vee go_{i,(s,j),\alpha,(s',j')}=1) \wedge (go_{i,(s,j),\alpha,(s',j')}=1$$
$$\leftrightarrow ((j < \tau_{i,s,\alpha} \wedge j' = j + 1) \vee (j \geq \tau_{i,s,\alpha} \wedge j' = 0)))$$
$$\varphi tr_{i,(s,j),\alpha,(s',j')} = (j' = j + 1 \wedge tr_{i,(s,j),\alpha,(s',j')} = 1) \vee$$
$$(j' = 0 \wedge tr_{i,(s,j),\alpha,(s',j')} = \mathbf{P}(s, \alpha, s')) .$$

2) In φ_{sem} (Line 6) we encode the semantics of the quantifier-free part φ^{nq} of the input formula by calling Algorithm 2. The truth of each Boolean subformula φ' of φ^{nq} at state sequence $\mathbf{s}$ is encoded in a Boolean variable $h_{\mathbf{s},\varphi'}$. We also define variables $hInt_{\mathbf{s},\varphi'}$ for the integer encoding (i.e., 0 or 1) of $h_{\mathbf{s},\varphi'}$, and real-valued variables $pr_{\mathbf{s},\varphi''}$ for values of probability expressions φ''.

3) In φ_{tru} (Lines 7–9) we state the truth of the input formula by first encoding the state quantifiers (Lines 7–8) and then stating the truth of the quantifier-free part φ^{nq} under all necessary state quantifier instantiations $(s_1,\ldots,s_l)$, i.e., where experiment $i \in \{1,\ldots,n\}$ starts in state s_{k_i} and stutter-scheduler mode 0 (Line 9).

In Algorithm 2, we recursively encode the meaning of atomic propositions and Boolean, temporal, arithmetic and probabilistic operators. Due to space constraints, here we present only the encoding for the temporal operator "until", and refer to the extended version [21] for the full algorithm.

For the encoding of the probability $\mathbb{P}(\varphi_1 \, \mathcal{U} \, \varphi_2)$ that $\varphi_1 \, \mathcal{U} \, \varphi_2$ is satisfied along the executions starting in state $s \in (S \times [m])^n$, the interesting case, where φ_1 holds in s but φ_2 does not, is in Line 6. The probability is a sum over all possible actions and potential successor states. Each summand is a product over (i) the probability of choosing the given action tuple, (ii) pseudo-Boolean values which encode whether the potential successor state is indeed a successor state under the encoded stutter-schedulers, (iii) real variables encoding the probability of moving to the given successors under the encoded stutter-schedulers, and (iv) the probability to satisfy the until formula from the successor state. We use real variables $d_{s,\varphi,\tau}$ to assure that finally a φ_2-state will be reached on all paths whose probabilities we accumulate.

Our SMT encoding is a Boolean combination of Boolean variables, non-linear real constraints, and linear integer constraints. Our linear integer constraints can be implemented as linear real constraints. The non-linear real constraints stem from the encoding of the probabilistic schedulers, not from the encoding of the stutter-schedulers. We check whether there exists an assignment of the variables such that the encoding is satisfied. SMT solving for non-linear real arithmetic without quantifier alternation has been proven to be solvable in exponential time in the number of variables [11, 24]. However, all available tools run in doubly exponential time in the number of variables [4, 16, 26]. The number of variables of our encoding is exponential in the number of stutter quantifiers, and polynomial in the size of the formula, the number of states and actions of the model, and the memory size for the stutter-schedulers. Hence, in practice, our implementation is triple exponential in the size of the input. This yields an upper bound on the complexity of model checking the considered fragment.

The size of the created encoding is exponential in the number of stutter-schedulers and polynomial in the size of the AHyperPCTL formula, the number of states and actions of the model and the memory-size for the stutter-schedulers.

6   Implementation and Evaluation

We implemented the model checking algorithm described in Sect. 5 based on the existing implementation for HyperPCTL, using the SMT-solver Z3 [25]. We performed experiments on a PC with a 3.60 GHz i7 processor and 32 GB RAM. Our implementation and case studies are available at https://github.com/carolinager/A-HyperProb. It is important to note that checking the constructed SMT formula is more complicated than in the case for HyperPCTL, since the SMT formula contains non-linear real constraints due to the probabilistic schedulers, whereas the SMT formula for HyperPCTL contains only linear real arithmetic.

We optimized our implementation by reducing the number of variables as described in [20] based on the quantifiers relevant for the encoding of the considered subformula. However, for interesting properties like the properties presented in Sect. 4, where we want to compare probabilities in different executions, we nevertheless have to create a variable encoding the validity of a subformula at a state for $(|S| \cdot m)^n$ combinations of states for each subformula.

All example applications presented in Sect. 1 and 4 consist of universal scheduler and state quantification, and existential stutter quantification. However, our implementation is restricted to existential scheduler and stutter quantification. Allowing arbitrary quantifiers for scheduler and stutter quantification is also possible in theory, but we found it to be infeasible in practice. For universal quantifiers we would need to check whether the SMT encoding holds for all possible assignments of the variables encoding the schedulers and stutter-schedulers, while for all other variables we only check for existence. This would yield an SMT instance with quantifier alternation, which is considerably more difficult than the current encoding with purely existential quantification [14]. Alternatively, we can encode the semantics for each possible combination of stutter-schedulers separately, meaning that we have to use variables $h_{s,\varphi,\tau}$ encoding the truth of φ at s under τ. This makes the number of variables exponential in the number of stutter quantifiers and the number of states and actions of the model, and polynomial in the size of the formula and the memory-size for stuttering. As a result, the implementation scales very badly, even after decreasing the number of variables via an optimization based on the relevant quantifiers. For all our applications, creating the semantic encoding exceeded memory after 30 min at the latest in this case.

Since neither option is a viable solution, for our case studies we instead consider the presented formulas with existential instead of universal scheduler quantification. In future work, it would be worth to explore how one could employ quantifier elimination to generate a set of possible schedulers from one scheduler instance satisfying the existential quantification.

The results of our case studies are presented in Table 1. Our first case study, **CE**, is the classic example presented in Sect. 1. We compare executions with different initial values h_1 and h_2, denoted by $h = (h_1, h_2)$. For $h = (0,1)$, the property can already be satisfied for the smallest non-trivial memory size $m = 2$. For higher values of h_2, however, the SMT solver does not finish solving after 1 h. The second case study, **TL**, is the side-channel timing leak described in Sect. 4. We found that already for encryption key length $k = 1$, and memory size $m = 2$, the SMT solver did not finish after 1 h even for a smaller formula, where we restrict the conjunction to the case $l = 0$. Our third case study, **ACDB**, is the output information leak presented in Sect. 4. Here, the SMT solving exceeds memory after 18 min, even if we check only part of the conjunction.

Table 1. Experimental results. **CE**: Classic example, **TL**: Timing leakage, **ACDB**: Output information leak. DNF: did not finish, OOM: out of memory.

	Case	Running time (s)			SMT Solver			Model	
	study	Enc.	Solving	Total	result	#variables	#subform	#states	#transitions
CE	$m = 2, h = (0,1)$	0.25	15.82	16.07	sat	829	341	7	9
	$m = 2, h = (0,2)$	0.38	DNF	–	–	1287	447	9	12
TL	$m = 2, k = 1$	0.99	DNF	–	–	3265	821	15	23
ACDB	$m = 2$	9.17	OOM	–	–	14716	1284	24	36

We see several possibilities to improve the scalability of our implementation. Firstly, we could experiment with different SMT solvers, like cvc5 [6]. Secondly, we could parallelize the construction of encodings for different stutter-schedulers, and possibly re-use sub-encodings that are the same for multiple stutter-schedulers. Another possibility would be to turn towards less accurate methods, and employ Monte Carlo or statistical model checking approaches.

7 Conclusions

We proposed a new logic, called AHyperPCTL, which is, to our knowledge, the first asynchronous probabilistic hyperlogic. AHyperPCTL extends HyperPCTL by quantification over stutter-schedulers, which allow to specify when the execution of a program should stutter. This allows to verify whether there exist stuttering variants of programs that would prevent information leaks, by comparing executions of different stuttering variations of a program. AHyperPCTL subsumes HyperPCTL. Therefore, the AHyperPCTL model checking problem on MDPs is, in general, undecidable. However, we showed that the model checking is decidable if we restrict the quantification to probabilistic memoryless schedulers and deterministic stutter-schedulers with bounded memory. Since our prototype implementation does not scale well, future work could investigate the use of other SMT solvers, statistical model checking, or Monte Carlo methods, as well as a feasible extension to quantifier alternation for scheduler and stutter quantifiers.

Acknowledgements. Lina Gerlach is supported by the DFG RTG 2236/2 *UnRAVeL*. Ezio Bartocci is supported by the Vienna Science and Technology Fund (WWTF) [10.47379/ICT19018]. This work is also partially sponsored by the United States NSF SaTC awards 2245114 and 2100989.

References

1. Ábrahám, E., Bartocci, E., Bonakdarpour, B., Dobe, O.: Parameter synthesis for probabilistic hyperproperties. In: Proceedings of LPAR-23. EPiC Series in Computing, vol. 73, pp. 12–31. EasyChair (2020). https://doi.org/10.29007/37lf

2. Ábrahám, E., Bartocci, E., Bonakdarpour, B., Dobe, O.: Probabilistic hyperproperties with nondeterminism. In: Hung, D.V., Sokolsky, O. (eds.) ATVA 2020. LNCS, vol. 12302, pp. 518–534. Springer, Cham (2020). https://doi.org/10.1007/978-3-030-59152-6_29

3. Ábrahám, E., Bonakdarpour, B.: HyperPCTL: a temporal logic for probabilistic hyperproperties. In: McIver, A., Horvath, A. (eds.) QEST 2018. LNCS, vol. 11024, pp. 20–35. Springer, Cham (2018). https://doi.org/10.1007/978-3-319-99154-2_2

4. Ábrahám, E., Davenport, J.H., England, M., Kremer, G.: Deciding the consistency of non-linear real arithmetic constraints with a conflict driven search using cylindrical algebraic coverings. J. Log. Algebraic Methods Program. **119**, 100633 (2021). https://doi.org/10.1016/j.jlamp.2020.100633

5. Baier, C., Katoen, J.: Principles of Model Checking. MIT Press, Cambridge (2008)

6. Barbosa, H., et al.: cvc5: a versatile and industrial-strength SMT solver. In: TACAS 2022. LNCS, vol. 13243, pp. 415–442. Springer, Cham (2022). https://doi.org/10.1007/978-3-030-99524-9_24

7. Bartocci, E., Ferrère, T., Henzinger, T.A., Nickovic, D., da Costa, A.O.: Flavors of sequential information flow. In: Finkbeiner, B., Wies, T. (eds.) VMCAI 2022. LNCS, vol. 13182, pp. 1–19. Springer, Cham (2022). https://doi.org/10.1007/978-3-030-94583-1_1

8. Baumeister, J., Coenen, N., Bonakdarpour, B., Sánchez, B.F.C.: A temporal logic for asynchronous hyperproperties. In: Proceedings of the 33rd International Conference on Computer-Aided Verification (CAV), pp. 694–717 (2021)

9. Baumeister, J., Coenen, N., Bonakdarpour, B., Finkbeiner, B., Sánchez, C.: A temporal logic for asynchronous hyperproperties. In: Silva, A., Leino, K.R.M. (eds.) CAV 2021. LNCS, vol. 12759, pp. 694–717. Springer, Cham (2021). https://doi.org/10.1007/978-3-030-81685-8_33

10. Beutner, R., Finkbeiner, B.: A logic for hyperproperties in multi-agent systems. CoRR abs/2203.07283 (2022). https://doi.org/10.48550/arXiv.2203.07283

11. Biere, A.: Bounded model checking. In: Biere, A., Heule, M., van Maaren, H., Walsh, T. (eds.) Handbook of Satisfiability - Second Edition, Frontiers in Artificial Intelligence and Applications, vol. 336, pp. 739–764. IOS Press (2021). https://doi.org/10.3233/FAIA201002

12. Bozzelli, L., Peron, A., Sánchez, C.: Asynchronous extensions of HyperLTL. In: Proceedings of LICS 2021: the 36th Annual ACM/IEEE Symposium on Logic in Computer Science, pp. 1–13. IEEE (2021). https://doi.org/10.1109/LICS52264.2021.9470583

13. Bozzelli, L., Peron, A., Sánchez, C.: Expressiveness and decidability of temporal logics for asynchronous hyperproperties. In: Klin, B., Lasota, S., Muscholl, A. (eds.) Proceedings of CONCUR 2022: the 33rd International Conference on Concurrency Theory. LIPIcs, vol. 243, pp. 27:1–27:16. Schloss Dagstuhl - Leibniz-Zentrum für Informatik (2022). https://doi.org/10.4230/LIPIcs.CONCUR.2022.27

14. Brown, C.W., Davenport, J.H.: The complexity of quantifier elimination and cylindrical algebraic decomposition. In: Wang, D. (ed.) Symbolic and Algebraic Computation, International Symposium, ISSAC 2007, Waterloo, Ontario, Canada, July 28 - August 1, 2007, Proceedings. pp. 54–60. ACM (2007). https://doi.org/10.1145/1277548.1277557

15. Clarkson, M.R., Finkbeiner, B., Koleini, M., Micinski, K.K., Rabe, M.N., Sánchez, C.: Temporal logics for hyperproperties. In: Proceedings of the 3rd Conference on Principles of Security and Trust POST, pp. 265–284 (2014)

16. Collins, G.E.: Quantifier elimination for real closed fields by cylindrical algebraic decompostion. In: Brakhage, H. (ed.) GI-Fachtagung 1975. LNCS, vol. 33, pp. 134–183. Springer, Heidelberg (1975). https://doi.org/10.1007/3-540-07407-4_17

17. Dimitrova, R., Finkbeiner, B., Torfah, H.: Probabilistic hyperproperties of markov decision processes. In: Hung, D.V., Sokolsky, O. (eds.) ATVA 2020. LNCS, vol. 12302, pp. 484–500. Springer, Cham (2020). https://doi.org/10.1007/978-3-030-59152-6_27

18. Dobe, O., Ábrahám, E., Bartocci, E., Bonakdarpour, B.: HYPERPROB: a model checker for probabilistic hyperproperties. In: Huisman, M., Păsăreanu, C., Zhan, N. (eds.) FM 2021. LNCS, vol. 13047, pp. 657–666. Springer, Cham (2021). https://doi.org/10.1007/978-3-030-90870-6_35

19. Dobe, O., Wilke, L., Ábrahám, E., Bartocci, E., Bonakdarpour, B.: Probabilistic hyperproperties with rewards. In: Deshmukh, J.V., Havelund, K., Perez, I. (eds.) Proceedings of NFM 2022: the 14th International Symposium on NASA Formal Methods. LNCS, vol. 13260, pp. 656–673. Springer, Cham (2022). https://doi.org/10.1007/978-3-031-06773-0_35

20. Dobe, O., Ábrahám, E., Bartocci, E., Bonakdarpour, B.: Model checking hyperproperties for markov decision processes. Inf. Comput. **289**, 104978 (2022). https://doi.org/10.1016/j.ic.2022.104978, special Issue on 11th International Symposium on Games, Automata, Logics and Formal Verification

21. Gerlach, L., Dobe, O., Ábrahám, E., Bartocci, E., Bonakdarpour, B.: Introducing asynchronicity to probabilistic hyperproperties. CoRR abs/2307.05282 (2023)

22. Guernic, G.L.: Automaton-based confidentiality monitoring of concurrent programs. In: Proceedings of CSF 2007: the 20th IEEE Computer Security Foundations Symposium, pp. 218–232. IEEE Computer Society (2007). https://doi.org/10.1109/CSF.2007.10

23. Hsu, T., Bonakdarpour, B., Finkbeiner, B., Sánchez, C.: Bounded model checking for asynchronous hyperproperties. CoRR abs/2301.07208 (2023). https://doi.org/10.48550/arXiv.2301.07208

24. Kroening, D., Strichman, O.: Decision Procedures - An Algorithmic Point of View, Second Edition. Texts in Theoretical Computer Science. An EATCS Series, Springer (2016). https://doi.org/10.1007/978-3-662-50497-0

25. de Moura, L., Bjørner, N.: Z3: an efficient SMT solver. In: Ramakrishnan, C.R., Rehof, J. (eds.) TACAS 2008. LNCS, vol. 4963, pp. 337–340. Springer, Heidelberg (2008). https://doi.org/10.1007/978-3-540-78800-3_24

26. de Moura, L., Jovanović, D.: A model-constructing satisfiability calculus. In: Giacobazzi, R., Berdine, J., Mastroeni, I. (eds.) VMCAI 2013. LNCS, vol. 7737, pp. 1–12. Springer, Heidelberg (2013). https://doi.org/10.1007/978-3-642-35873-9_1

27. Minh Ngo, T., Stoelinga, M., Huisman, M.: Confidentiality for probabilistic multithreaded programs and its verification. In: Jürjens, J., Livshits, B., Scandariato, R. (eds.) ESSoS 2013. LNCS, vol. 7781, pp. 107–122. Springer, Heidelberg (2013). https://doi.org/10.1007/978-3-642-36563-8_8

28. Sabelfeld, A., Sands, D.: Declassification: dimensions and principles. J. Comput. Secur. **17**(5), 517–548 (2009). https://doi.org/10.3233/JCS-2009-0352

29. Smith, G.: Probabilistic noninterference through weak probabilistic bisimulation. In: Proceedings of (CSFW-16 2003: the 16th IEEE Computer Security Foundations Workshop, pp. 3–13. IEEE Computer Society (2003). https://doi.org/10.1109/CSFW.2003.1212701

30. Wang, Y., Nalluri, S., Bonakdarpour, B., Pajic, M.: Statistical model checking for hyperproperties. In: Proceedings of CSF 2021: the 34th IEEE Computer Secu-

rity Foundations Symposium, pp. 1–16. IEEE (2021). https://doi.org/10.1109/CSF51468.2021.00009

31. Wang, Y., Zarei, M., Bonakdarpour, B., Pajic, M.: Statistical verification of hyperproperties for cyber-physical systems. ACM Trans. Embed. Comput. Syst. **18**(5s), 92:1–92:23 (2019). https://doi.org/10.1145/3358232

A Bounded Model Checking Technique
for Discrete-Time Nonlinear Systems

YoungMin Kwon[1(✉)], Eunhee Kim[2], and Gul Agha[3]

[1] Department of Computer Science, The State University of New York Korea,
Incheon, South Korea
youngmin.kwon@sunykorea.ac.kr
[2] Department of Library and Information Science, Yonsei University, Seoul, South Korea
eunhee.kim@yonsei.ac.kr
[3] Department of Computer Science, University of Illinois at Urbana, Champaign, USA
agha@illinois.edu

Abstract. Validating properties of nonlinear systems is difficult given the complexity of solving nonlinear dynamics equations. In this paper, we develop an LTL-based model checking technique to specify and validate properties of a nonlinear system. The technique can hide the difficulties in handling nonlinear difference equations, simplify the specification of properties of physical systems, and provide a high level abstraction to analyze and control nonlinear systems. We apply the proposed technique to controller synthesis problems: determining a design for a simple nonlinear control system, and finding an appropriate dosage to prescribe for drug administration.

1 Introduction

Many systems around us have nonlinearities, but analyzing nonlinear behaviors is still a challenging task. Traditional analysis techniques have been centered around the overall properties such as the Lyapunov-based stability of a system [32,34]. They are often approximated by a more tractable system such as a linear system or a hybrid system with linear system modes [11]. However, these simplified models typically have an operation range where the approximation errors are tolerable. Moreover, the approximation itself is not a simple task. Therefore, it is often desirable to examine behaviors of nonlinear systems without linearizing them and without solving their complex dynamics equations manually. In this paper, we developed a tool that can describe properties of a nonlinear system and can check whether the system actually satisfies the properties. We define a temporal logic called *Linear Temporal Logic for Nonlinear systems* (LTLN) which enables us to express properties of a nonlinear dynamic system at a high level of abstraction rather than in terms of its complex nonlinear dynamics equations. A key difference between LTLN and standard LTL is that in LTLN, it is possible that neither a given formula nor its negation hold due to the bound. We discuss this further in Sect. 3.1.

Y. Kwon—This work was supported by NRF of MSIT, Korea (2021R1F1A104859812) and by the MSIT, Korea, under the ICTCCP (IITP-2020-2011-1-00783) supervised by the IITP.

N. Jansen and M. Tribastone (Eds.): QEST 2023, LNCS 14287, pp. 65–81, 2023.
https://doi.org/10.1007/978-3-031-43835-6_5

Model checking techniques have been developed to validate properties of finite state models such as a Kripke structure [9, 18]. They have been successfully utilized in validating hardware and software designs such as an electric circuit design, proving the correctness of a concurrent system design, and so on [3, 9]. These techniques have been extended to validate properties of continuous state models like Markov chains, timed automata and hybrid automata [2, 15]. HyTech, UPPAAL, BACH, HySAT, iLTL and LTLC [6, 12, 16, 22, 23, 25, 27] are some of their model checking and bounded reachability checking tools. For continuous system trajectories, STL is a linear temporal logic that can monitor LTL properties of a single continuous trajectory of a hybrid system [29]. Tools such as RTL [30], Flow* [7], dReach [20] use overapproximation techniques to handle uncountable trajectories. Model checking techniques have been utilized in automatic control systems as well [14, 19, 21, 28, 33], where state space is partitioned into finite polytopes. Against a transition system built on the polytopes, a control strategy is computed such that a goal can be satisfied.

We develop a model checking tool for LTLN which differs from the above techniques in the following significant ways:

- Reachability checkers solve constraints from the guard and the invariant conditions of a system. The LTLN-Checker solves constraints expressed in LTLN formulas and the constraints implied by the dynamics equations of a system.
- Several tools (e.g., STL, RTL, Flow*) check a single trajectory or its nearby region using overapproximation. The LTLN-checker chooses inputs so that all possible trajectories of a system are explored without overapproximation.
- Unlike the controller synthesis techniques that pre-partition the state space, LTLN partitions the state space as necessary. A feedback control loop can be constructed using the receding horizon scheme [8].
- Many systems have parameters that are not known *a priori* but are estimated by periodic sampling. For such systems, LTLN-checker can be used directly without the need to discretize the system.

The key contributions of the paper are: 1) we developed a bounded LTL model checking algorithm for a nonlinear system using the DMS technique; 2) the algorithm checks not only a single trajectory and its nearby regions but all trajectories of a system generated by all initial states and all inputs; 3) for a solvable set of constraints, one can soundly and completely check bounded properties of a system (in general, the constraints may not be solveable; in this case we have soundness but not completeness); 4) a model checker for the algorithm is implemented and the technique is applied to practical problems of controlling an inverted pendulum and computing a prescription.

The outline of the paper is as follows. Section 2 defines a discrete-time nonlinear system model. Section 3 provides the syntax and semantics of LTLN, illustrating it with an example (controlling an inverted pendulum). Section 4 explains two LTLN model checking algorithms: a DNF-based intuitive algorithm, and a more efficient Büchi automaton-based one. Finally, in Sect. 5, we apply the proposed model checking techniques to a control system and to Pharmaceutics.

2 Discrete-Time Nonlinear System Model

We model the dynamics of nonlinear systems by nonlinear difference equations. We express these equations in a *state space* form because we specify properties of a system using the states. The system dynamics equations in a state space form are:

$$\mathbf{x}(t+1) = \mathbf{f}(\mathbf{x}(t), \mathbf{u}(t)), \quad \mathbf{y}(t) = \mathbf{h}(\mathbf{x}(t), \mathbf{u}(t)), \tag{1}$$

where $\mathbf{u} : \mathbb{N} \to \mathbb{R}^{nu}$, $\mathbf{y} : \mathbb{N} \to \mathbb{R}^{ny}$, and $\mathbf{x} : \mathbb{N} \to \mathbb{R}^{nx}$ are discrete-time input, output, and state trajectories of a system respectively, $\mathbf{f} : \mathbb{R}^{nx} \times \mathbb{R}^{nu} \to \mathbb{R}^{nx}$ is a function that relates the current state and the current input to the next state, and $\mathbf{h} : \mathbb{R}^{nx} \times \mathbb{R}^{nu} \to \mathbb{R}^{ny}$ is a function that relates the state and the input to the output at the same time index. (As usual, $\mathbb{N}$ represents the set of Natural numbers, and $\mathbb{R}$ represents the set of real numbers). In the difference equation, the initial state $\mathbf{x}(0)$ and the input trajectory $\mathbf{u}(t)$ for $t \geq 0$ are independent variables that determine the state trajectory $\mathbf{x}(t)$ for $t > 0$ and the output trajectory $\mathbf{y}(t)$ for $t \geq 0$.

Combining the variables and the dynamics equations between them, we define a *discrete-time nonlinear system model* as a five tuple:

$$M = \langle U, Y, X, \mathbf{f}, \mathbf{h} \rangle, \tag{2}$$

where $U = \{u_1, \ldots, u_{nu}\}$, $Y = \{y_1, \ldots, y_{ny}\}$, and $X = \{x_1, \ldots, x_{nx}\}$ are the set of input, output, and state variables respectively; $\mathbf{f} : \mathbb{R}^{nx} \times \mathbb{R}^{nu} \to \mathbb{R}^{nx}$ and $\mathbf{h} : \mathbb{R}^{nx} \times \mathbb{R}^{nu} \to \mathbb{R}^{ny}$ are the two functions of Eq. (1) that mathematically model the discrete-time nonlinear dynamics of the system.

A *computation path* $\pi : \mathbb{N} \to (\mathbb{R}^{nu} \times \mathbb{R}^{ny} \times \mathbb{R}^{nx})$ of a discrete-time nonlinear system model M is a tuple composed of the input trajectory $\mathbf{u}$, the output trajectory $\mathbf{y}$ and the state trajectory $\mathbf{x}$ such that $\pi(t) = (\mathbf{u}(t), \mathbf{y}(t), \mathbf{x}(t))$. We call $\mathbf{x}(t)$ a *state* and $\pi(t)$ a *system state*, but when their meanings are clear from the context, we simply call them state.

2.1 Example: An Inverted Pendulum Model

As an example of a nonlinear system, consider an inverted pendulum model, an unstable nonlinear system. The goal is to keep the pendulum standing vertically by moving the cart. We use this model to illustrate the complexity of analyzing nonlinear systems rather than to explain the dynamics equations *per se*. We will show how we avoid dealing directly with the complexity by using our proposed model checking method later.

Figure 1 shows a diagram of an inverted pendulum system: a pendulum is attached to a cart upside down and the goal is to keep the pendulum at the upright position by moving the cart. The dynamics of the inverted pendulum system can be modeled by the following nonlinear differential Eqs. [11].

$$(M + m)\ddot{x} + m\ell\ddot{\theta}\cos(\theta) - m\ell\dot{\theta}^2 \sin(\theta) = F, \quad \ell\ddot{\theta} + \ddot{x}\cos(\theta) - g\sin(\theta) = -r\dot{\theta}. \tag{3}$$

We rewrite Eq. (3) into a state space form with four state variables: $x_1 = x$, $x_2 = \dot{x}$, $x_3 = \theta$, and $x_4 = \dot{\theta}$. Rearranging the equation using the state variables, the dynamics

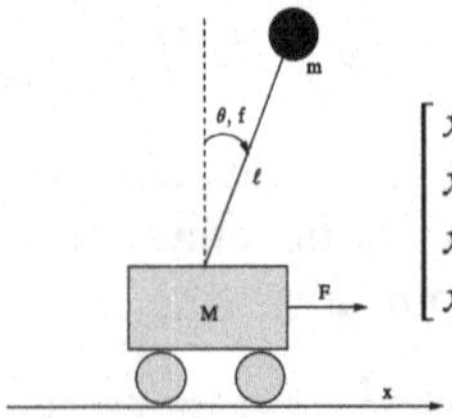

$$\begin{bmatrix}\dot{x}_1\\ \dot{x}_2\\ \dot{x}_3\\ \dot{x}_4\end{bmatrix} = \begin{bmatrix} x_2 \\ f_1(x_1,x_2,x_3,x_4,F) \triangleq \frac{-mg\sin(x_3)\cos(x_3)+m\ell x_4^2\sin(x_3)+rmx_4\cos(x_3)+F}{M+m(1-\cos^2(x_3))} \\ x_4 \\ f_2(x_1,x_2,x_3,x_4,F) \triangleq \frac{(M+m)(g\sin(x_3)-rx_4)-(\ell m x_4\sin(x_3)+F)\cos(x_3)}{\ell M+\ell m(1-\cos^2(x_3))} \end{bmatrix}$$

Fig. 1. A diagram of an inverted pendulum system. In the equation, $M = 1Kg$ is the weight of the cart, $m = 0.1Kg$ is the weight of the pendulum, $r = 0.1Kg/sec$ is the friction of the pendulum, $\ell = 1m$ is the length of the pendulum, $g = 9.8m/sec^2$ is the gravity constant, and $\Delta t = 0.1sec$ is the sampling time.

equations can be rewritten in the state space form as in Fig. 1 and using Euler's method, the continuous-time equations can be discretized as in Fig. 2. Embedding the discrete-time equations, the system model *IP* is

$$\begin{aligned} IP &= \langle U, \emptyset, X, \mathbf{f}, \emptyset\rangle, \\ U &= \{F\}, \\ X &= \{x_1, x_2, x_3, x_4\}, \end{aligned} \qquad \mathbf{f}(\mathbf{x},\mathbf{u}) = \begin{bmatrix} \mathrm{x}_1 + \Delta\mathrm{t}\cdot \mathrm{x}_2 \\ \mathrm{x}_2 + \Delta\mathrm{t}\cdot f_1(\mathrm{x}_1,\mathrm{x}_2,\mathrm{x}_3,\mathrm{x}_4,\mathrm{u}_1) \\ \mathrm{x}_3 + \Delta\mathrm{t}\cdot \mathrm{x}_4 \\ \mathrm{x}_4 + \Delta\mathrm{t}\cdot f_2(\mathrm{x}_1,\mathrm{x}_2,\mathrm{x}_3,\mathrm{x}_4,\mathrm{u}_1) \end{bmatrix}, \tag{4}$$

where $\mathbf{x} \in \mathbb{R}^4$ is $[x_1, x_2, x_3, x_4]^T$, and $\mathbf{u} \in \mathbb{R}^1$ is $[F]$.

3 Specification on Nonlinear System Models

We define a temporal logic called *Linear Temporal Logic for Nonlinear systems* (LTLN) to specify the behaviors of a nonlinear system model. LTLN specifies how the properties of a system should change along each of the computation paths of a system. Using LTLN, we can specify and examine properties of a nonlinear system without having to solve its complex nonlinear dynamics equations.

3.1 LTLN: The Specification Logic

LTLN is a quantitative variant of *Linear Temporal Logic* (LTL) in which the atomic propositions are (in)equalities about system states.

Definition 1. *The syntax of an LTLN formula* ϕ *is*

$$\begin{aligned} &\phi ::= \mathtt{T} \mid \mathtt{F} \mid AP \mid \neg\phi \mid \phi\wedge\varphi \mid \phi\vee\varphi \mid \phi\rightarrow\varphi \mid \phi\leftrightarrow\varphi \mid \mathbf{X}\,\phi \mid \Diamond\,\phi \mid \Box\,\phi \mid \phi\,\mathbf{U}\,\varphi \mid \phi\,\mathbf{R}\,\varphi, \\ &AP ::= c(v_1, v_2, \ldots, v_n) \bowtie 0, \end{aligned}$$

where $c : \mathbb{R}^n \rightarrow \mathbb{R}$ *is a nonlinear function,* $v_i \in U \cup Y \cup X$ *for* $i = 1, \ldots, n$ *is one of the input, output, and state variables of a model, and* $\bowtie \in \{\leq, <, =, \neq, >, \geq\}$ *is the usual comparison operator.* □

The meaning of an LTLN formula ϕ is interpreted over a computation path. Let the variable assignment function $\theta_{\pi(t)} : U \cup Y \cup X \to \mathbb{R}$ at time t on a computation path π be $\theta_{\pi(t)}(v) = \mathbf{u}(t)_i$ if $v = u_i$ for a $u_i \in U$; $\theta_{\pi(t)}(v) = \mathbf{y}(t)_i$ if $v = y_i$ for a $y_i \in Y$; and $\theta_{\pi(t)}(v) = \mathbf{x}(t)_i$ if $v = x_i$ for an $x_i \in X$. Note that since we are dealing with real numbers, equality will be meaningful only in cases where an exact value can be computed and represented in the implementation.

The meaning of an LTLN formula can be defined by the bounded *ternary satisfaction relation* $\models_b \subset \Pi \times \mathbb{N} \times \Phi$ and the bounded *binary satisfaction relation* $\models_b \subset \mathcal{M} \times \Phi$, where Π is the set of all computation paths, Φ is the set of all LTLN formulas, and $\mathcal{M}$ is the set of all nonlinear system models. For simplicity, we write $\pi, t \models_b \phi$ for $(\pi, t, \phi) \in \models_b$ and $M \models_b \phi$ for $(M, \phi) \in \models_b$.

Definition 2. *Given a bound $b \in \mathbb{N}$, the* bounded ternary satisfaction relation $\models_b \subset \Pi \times \mathbb{N} \times \Phi$ is

$$\pi, t \models_b \mathrm{T},$$
$$\pi, t \not\models_b \mathrm{F},$$
$$\pi, t \models_b c(v_1, v_2, \ldots, v_n) \bowtie 0 \Leftrightarrow c\left(\theta_{\pi(t)}(v_1), \theta_{\pi(t)}(v_2), \ldots, \theta_{\pi(t)}(v_n)\right) \bowtie 0,$$
$$\pi, t \models_b \phi \wedge \varphi \Leftrightarrow \pi, t \models_b \phi \text{ and } \pi, t \models_b \varphi,$$
$$\pi, t \models_b \mathrm{X}\, \phi \quad \Leftrightarrow t < b \text{ and } \pi, t+1 \models_b \phi,$$
$$\pi, t \models_b \phi \,\mathrm{U}\, \varphi \Leftrightarrow \pi, i \models_b \varphi \text{ for some } i \in [t, b] \text{ and } \pi, j \models_b \phi \text{ for all } j \in [t, i),$$
$$\pi, t \models_b \phi \,\mathrm{R}\, \varphi \Leftrightarrow \pi, i \models_b \phi \text{ for some } i \in [t, b] \text{ and } \pi, j \models_b \varphi \text{ for all } j \in [t, i].$$

Using the ternary relation $\models_b$, the bounded binary satisfaction relation $\models_b \subset \mathcal{M} \times \Phi$ is

$$M \models_b \phi \Leftrightarrow \pi, 0 \models_b \phi \text{ for all computation paths } \pi \text{ of } M.$$

$\square$

The ternary satisfaction relation $\models_b$ is true *if and only if* (iff) a given computation path π of a nonlinear system model M satisfies the specification ϕ from time step t onward. Observe that $\pi, 0 \models_b \phi$ means π satisfies ϕ using its prefix up to time b regardless of the suffix thereafter. The binary satisfaction relation $\models_b$ is true iff all computation paths of a model M satisfy ϕ from time step 0 onward. That is, if all behaviors of a model M satisfy the specification ϕ, we call M a model of ϕ.

Following [4], the $\neg$ operators are removed from an LTLN formula by converting the formula to its *Negation Normal Form* (NNF) [9] and changing the comparators of atomic propositions. In NNF, all $\neg$ operators of an LTLN formula are moved to the front of atomic propositions using the equivalences of the standard LTL: $\neg \mathrm{T} \equiv \mathrm{F}$, $\neg \mathrm{F} \equiv \mathrm{T}$, $\neg\neg\phi \equiv \phi$, $\neg(\phi \wedge \varphi) \equiv \neg\phi \vee \neg\varphi$, $\neg(\phi \vee \varphi) \equiv \neg\phi \wedge \neg\varphi$, $\neg \mathrm{X}\, \phi \equiv \mathrm{X}\,\neg\phi$, $\neg(\phi \mathrm{U}\varphi) \equiv \neg\phi \mathrm{R}\neg\varphi$, and $\neg(\phi \mathrm{R}\varphi) \equiv \neg\phi \mathrm{U}\neg\varphi$. For example, $\mathrm{X}\neg\mathrm{X}(p > 0 \vee p < 1)$ in NNF is $\mathrm{XX}(\neg(p > 0) \wedge \neg(p < 1))$ and the $\neg$ operators can be removed as $\mathrm{XX}(p \leq 0 \wedge p \geq 1)$.

After the removal of $\neg$ operators, $\models_b$ is a subset of $\models$, the usual unbounded ternary satisfaction relation of LTL. That is, if $\pi, t \models_b \phi$ then $\pi, t \models \phi$. However, because of the bound, $M \not\models_b \phi$ does not mean $M \models_b \neg\phi$. Instead both $M \not\models_b \phi$ and $M \not\models_b \neg\phi$ are possible at the same time. The following equivalences hold for the other missing operators in the ternary relation: $\phi \vee \varphi \equiv \neg(\neg\phi \wedge \neg\varphi)$, $\phi \rightarrow \varphi \equiv \neg\phi \vee \varphi$, $\phi \leftrightarrow \varphi \equiv (\phi \rightarrow \varphi) \wedge (\varphi \rightarrow \phi)$, and $\Diamond \phi \equiv T \cup \phi$.

An always formula $\Box \phi$ is equivalent to $F R \phi$. However, $M \not\models_b \Box \phi$ because $\models_b$ is a bounded relation. There are other equivalence relations of the unbounded LTL that do not hold in LTLN because of the bound. For example, $\Box\phi \not\equiv \neg\Diamond\neg\phi$, $\phi R\varphi \not\equiv \neg(\neg\phi U\neg\varphi)$ and so on. In LTL_f [10], the equivalence $\Box \phi \equiv \neg\Diamond \neg\phi$ holds because $\Box \phi$ is true iff ϕ is true up to the bound. The $\Box \phi$ formula of LTL_f can be similarly defined in LTLN as $last R \phi$, where $last$ is an atomic proposition $\tau = b$ and $\tau(t)$ is a clock system with a dynamics equation of $\tau(t + 1) = \tau(t) + 1$ and $\tau(0) = 0$.

3.2   Example: Specification on the Inverted Pendulum Model

To demonstrate how we can specify properties of a system using LTLN formulas without involving the details of the system, we build several formulas about controlling the inverted pendulum model of Sect. 2.1. The formulas use the input, output, and state variables of a system, but they do not rely on the dynamics equations of the system. Using the constructed formulas, we will find a sequence of input $\mathbf{u}(t)$ that can achieve the goal while satisfying other conditions.

Let us express the following conditions in LTLN: (1) initially the pendulum was tilted by $10\,^\circ$ with the angular velocity of $0\,^\circ/sec$, i.e. $\theta = 10$, $\dot{\theta} = 0$. In addition, the cart was stopped at the origin, i.e. $x = 0$, $\dot{x} = 0$; (2) the goal is to bring the pendulum to the upright pose and keep the cart stationary, i.e. $\theta = 0$, $\dot{\theta} = 0$, $\dot{x} = 0$; and (3) from the initial state until the goal is achieved, the cart position should remain within ± 1 (m) range from the origin and the external force should be within ± 20 (N) range.

Using the eventually operator ($\Diamond$), the $goal$ can be expressed as $\phi'_g = \Diamond \phi_g$, where $\phi_g = (x_2 = 0 \wedge x_3 = 0 \wedge x_4 = 0)$. The $last condition$ requires that the cart position and the external force should remain within their respective limits. In standard LTL, the condition can be expressed as $\phi'_\ell = \Box \phi_\ell$, where $\phi_\ell = (-1 < x_1 \wedge x_1 < 1 \wedge -20 < F \wedge F < 20)$. However, $M \not\models_b \Box \phi$ in the bounded semantics. To fix the problem, we combined the goal condition and the last condition together as $\phi''_g = \phi_g R \phi_\ell$. The formula ϕ''_g means that until the moment (inclusively) when the goal is achieved, the conditions about the cart position and the external force are maintained. Because we are interested in achieving the goal by a finite step b, we do not need to enforce the limit conditions forever. The $initial condition$ can be specified without any temporal operators as $\phi_i = (x_1 = 0 \wedge x_2 = 0 \wedge x_3 = 10 \cdot \pi/180 \wedge x_4 = 0)$. Combining the subformulas together, we can build our overall specification. Our $objective$ is to find a sequence of control inputs that can achieve the goal from the initial condition. As such, we negate the conditions we described so far and let the model checking algorithm find a counterexample that can satisfy the objective.

$$\phi = \phi_i \rightarrow \neg(\phi_g R \phi_\ell). \tag{5}$$

Model checking $IP \models_b \phi$ is a process of finding a computation path π such that $\pi, 0 \models_b \phi_i \wedge (\phi_g R \phi_\ell)$. If a counterexample π is found, it contains a sequence of input that can bring the pendulum system from the initial state to the goal state at some time $t \leq b$ while satisfying the bound conditions.

4 LTLN Model Checking Algorithm

LTLN model checking is a process of validating whether all computation paths of a nonlinear system model satisfy a specification written in an LTLN formula. In the model checking process, we look for a *counterexample*, a computation path that violates a specification. Like many model checking algorithms, we search for a computation path that satisfies the negation of a specification. However, with an infinite state space of a model, we cannot build an intersection automaton between a model and a specification. In this section, we explain the LTLN model checking algorithm and its implementation. We begin with an intuitive algorithm that converts an LTLN model checking problem into a feasibility checking problem in a *Disjunctive Normal Form* (DNF). Explanations about a more efficient algorithm using a Büchi automaton will follow.

4.1 DNF-based Model Checking Algorithm

We describe a DNF-based model checking algorithm in this section. Because we are performing a bounded model checking, a computation path π that does not satisfy a specification within the bound b is regarded as violating the specification as stated in Definition 2. Hence, using the equivalences $\phi \, \mathrm{U} \, \varphi \equiv \varphi \vee (\phi \wedge \mathrm{X} \, (\phi \, \mathrm{U} \, \varphi))$ and $\phi \, \mathrm{R} \, \varphi \equiv (\phi \wedge \varphi) \vee (\varphi \wedge \mathrm{X} \, (\phi \, \mathrm{R} \, \varphi))$ an LTLN formula can be converted to a formula without temporal operators other than X immediately preceding atomic propositions. Treating atomic propositions with X as a single unit, the formula can be further converted to a DNF, which is satisfiable if any of its conjunctive terms are satisfiable.

The next example shows how temporal operators other than X are removed.

Example 1. Let a nonlinear system model M be $M = \langle \{u\}, \emptyset, \{x\}, \mathbf{f}, \emptyset \rangle$ with $x(t + 1) = \sin(x(t) + u(t))$ and $\phi = p \, \mathrm{U} \, q$ with $p = x > 0$ and $q = x > 0.9$. Suppose that we want to check whether $M \models_2 \neg \phi$.

Using the equivalence relation $p \, \mathrm{U} \, q \equiv q \vee (p \wedge \mathrm{X} \, (p \, \mathrm{U} \, q))$, the model checking problem is to find a counterexample π that can satisfy the DNF with $\mathrm{X} : \pi, 0 \models_2$ $(q) \vee (p \wedge \mathrm{X} q) \vee (p \wedge \mathrm{X} p \wedge \mathrm{X} \mathrm{X} q)$. In other words, $M \not\models_2 \neg \phi$ if there is a π that can satisfy any of the conjunctive terms of $\pi, 0 \models_2 q$; $\pi, 0 \models_2 p \wedge \mathrm{X} q$; or $\pi, 0 \models_2 p \wedge \mathrm{X} p \wedge \mathrm{X} \mathrm{X} q$. □

Now, we remove the X operators. Between input, output, and state variables at different time steps, the system dynamics of Eq. (1) should be respected. We call this condition a *trajectory constraint*. One way to enforce the constraint is to express all dependent variables in terms of the independent variables. Because $\mathbf{x}(0)$ and $\mathbf{u}(t)$ for $t \in [0, b]$ are the only independent variables in π, if the others can be rewritten using these variables, the trajectory constraint can be satisfied. However, this conversion is not always possible: it is infeasible to solve some nonlinear equations. As an alternative, we use the *Direct Multiple Shooting* (DMS) technique [5]. In DMS, fresh variables are added to each time step such that the (in)equalities at the steps are rewritten in terms of the fresh variables. In addition, Eq. (1) is enforced between the fresh variables as a trajectory constraint. This step can be summarized as follows: (1) prepare fresh variables for each of input, output, and state variables at each time step up to the bound b; (2) substitute the input, output, and state variables in the (in)equalities with the fresh variables corresponding to the time step, i.e., the number of next operators preceding the (in)equalities; (3) for each conjunctive term of a DNF, add the set of equality constraints of Eq. (1) on fresh variables for different time steps.

As an illustration of removing X operators, let us continue with Example 1.

Example 2. The condition $\pi, 0 \models_2 q$ is true iff q can be satisfied at time 0 by some computation path π. Introducing a fresh variable x_0, $\pi, 0 \models_2 q$ is satisfiable iff
$$\{x_0 \in \mathbb{R} \mid x_0 > 0.9\} \neq \emptyset.$$
Using fresh variables x_0, x_1, and u_0, the condition $\pi, 0 \models_2 p \wedge X q$ is satisfiable iff
$$\{x_0, x_1, u_0 \in \mathbb{R} \mid x_0 > 0 \wedge x_1 > 0.9 \wedge x_1 = \sin(x_0 + u_0)\} \neq \emptyset.$$
Observe that applying the DMS technique, the inequality constraint at $t = 1$ is expressed in terms of x_1. Moreover, the equality constraint $x_1 = \sin(x_0 + u_0)$ ensures that x_0, x_1, and u_0 are on a system state trajectory, i.e. satisfy the system dynamics. Similarly, using fresh variables x_0, x_1, x_2, u_0 and u_1, the condition $\pi, 0 \models_2 p \wedge X p \wedge X X q$ is satisfiable iff
$$\{x_0, x_1, x_2, u_0, u_1 \in \mathbb{R} \mid x_0 > 0 \wedge x_1 > 0 \wedge x_2 > 0.9 \wedge x_1 = \sin(x_0 + u_0) \wedge$$
$$x_2 = \sin(x_1 + u_1)\} \neq \emptyset.$$
Once again, the atomic propositions at $t = 1$ and $t = 2$ are expressed in terms of the fresh variables x_1 and x_2. The equality constraints $x_1 = \sin(x_0 + u_0)$ and $x_2 = \sin(x_1 + u_1)$ ensure the trajectory constraint.   □

4.2   Büchi Automaton-based Model Checking Algorithm

Converting an LTLN formula to a DNF may result in exponentially many conjunctive terms. For example the DNF of $(p \vee q) \, U \, r$ have sequences of p or q ending with r up to length b. To make the model checking process efficient, we utilize Büchi automata.

Before explaining the details of a Büchi automaton-based model checking algorithm, let us define some related terms. A *Büchi automaton* for an LTLN formula ϕ is a quintuple $\mathcal{B}_\phi = \langle 2^{AP}, Q, \Delta, \{\imath\}, F \rangle$, where AP is the set of atomic propositions; Q is the set of nodes; $\Delta \subset Q \times Q \times 2^{AP}$ is a transition relation; $\imath \in Q$ is the initial node; and $F \subseteq Q$ is a set of accepting nodes. We write $p \xrightarrow{L} q$ for $(p, q, L) \in \Delta$. Given a computation path π, a run $\rho : \mathbb{N} \cup \{-1\} \to Q$ is a path on $\mathcal{B}_\phi$ such that if $\rho(-1) = \imath$ and if $\rho(t-1) \xrightarrow{L} \rho(t)$ then $\forall_{ap \in L} . \pi, t \models_b ap$ for $t \geq 0$. An ϵ-*transition* is a transition with $L = \emptyset$. An ϵ-*neighbor* of a node q is the set of nodes that can reach q by ϵ-transitions. Let $S' \subset F$ be the set of accepting nodes of $\mathcal{B}_\phi$ that can reach themselves by ϵ-transitions. The *sink-set* $S \subset Q$ is the union of the ϵ-neighbors of all S'.

Example 3. We use a Büchi automaton $\mathcal{B}_\phi$ as a generator of a DNF of ϕ. The figure below is a Büchi automaton $\mathcal{B}_\phi$ for $\phi = (p \vee q) \, \mathsf{U} \, r$.

$$\mathcal{B}_\phi = \langle 2^{\{p,q,r\}}, \{\imath, P, Q, R, F\}, \Delta, \{\imath\}, \{R, F\} \rangle, \text{ where}$$

$$\Delta = \left\{ \begin{array}{l} \imath \xrightarrow{p} P, \ \imath \xrightarrow{q} Q, \ \imath \xrightarrow{r} R, \ R \xrightarrow{\emptyset} F \\ P \xrightarrow{p} P, \ P \xrightarrow{q} Q, \ P \xrightarrow{r} R, \ F \xrightarrow{\emptyset} F \\ Q \xrightarrow{p} P, \ Q \xrightarrow{q} Q, \ Q \xrightarrow{r} R, \end{array} \right\}.$$

In the figure, any run that can reach the sink-set $S = \{R, F\}$ within the bound b is a candidate for an accepting run. For example $\imath R$, $\imath P R$, $\imath P P R$, or $\imath P Q R$ can be an accepting run if their respective conjunctive terms r, $p \wedge \mathsf{X} r$, $p \wedge \mathsf{X} p \wedge \mathsf{X} \mathsf{X} r$, or $p \wedge \mathsf{X} q \wedge \mathsf{X} \mathsf{X} r$ have a feasible solution. Observe that if $p \wedge \mathsf{X} p$ is infeasible, all paths with this prefix are infeasible and can be skipped. For instance, $p \wedge \mathsf{X} p \wedge \mathsf{X} \mathsf{X} r$, $p \wedge \mathsf{X} p \wedge \mathsf{X} \mathsf{X} p \wedge \mathsf{X} \mathsf{X} r$, $p \wedge \mathsf{X} p \wedge \mathsf{X} \mathsf{X} q \wedge \mathsf{X} \mathsf{X} r$, *etc.* can be skipped.   □

Lemma 1. *If $\mathcal{B}_\phi$ accepts a bounded computation path, there is a sink-set S in $\mathcal{B}_\phi$.*

Proof. In the bounded model checking with a bound b, if a computation path is accepted by a Büchi automaton, then it is accepted by the prefix of length b regardless of its suffix. Hence, the run corresponding to the suffix of the computation path must cycle a loop containing all the accepting states without requiring any conditions, i.e. by the ϵ-transitions. That means, $\mathcal{B}_\phi$ has a sink-set S.

Theorem 1. *If $\pi, 0 \models_b \phi$ for a computation path π, then there is a run ρ for π on $\mathcal{B}_\phi$ that reaches the sink-set S in b steps.*

Proof. Because $\pi, 0 \models_b \phi$, there is an accepting run ρ for π on $\mathcal{B}_\phi$. Moreover, because $\pi, 0 \models_b \phi$ means π satisfies ϕ for any $\pi(t)$ after b steps, there is a run ρ on $\mathcal{B}_\phi$ such that $\rho(t)$ visits the accepting nodes of $\mathcal{B}_\phi$ infinitely often for all $\pi(t)$ for $t > b$. Suppose that $\rho(b) \notin S$, then for some $i > 0$, conditions in L of $\rho(b + i - 1) \xrightarrow{L} \rho(b + i)$ can be violated depending on $\pi(b + i)$. Therefore, $\rho(b) \in S$.

   Overall, our LTLN model checking process is as follows. First, a Büchi automaton $\mathcal{B}_{\neg\phi}$ for an LTLN formula ϕ is built using the *expand algorithm* [17]. Then, a *Depth First Search* (DFS)[1] on $\mathcal{B}_{\neg\phi}$ is performed to find the sink-set. On each step of the DFS, the atomic propositions are collected along each search path and their feasibility is tested before the search continues to a next node. If the path is infeasible, the entire subtree sharing the path as a prefix is pruned. Employing the pruning technique, the issue of exponentially many conjunctive terms in a DNF can be relieved.

   Algorithm 1 shows a main function of the Büchi automaton-based model checking algorithm. The function check is a recursive function performing the DFS on $\mathcal{B}_\phi$ with the maximum search depth of b. In the algorithm, the parameters C, n, and t of check are a set of (in)equality constraints collected along a path during the DFS, a node of $\mathcal{B}_\phi$,

---

[1] During the DFS, the same node may be visited multiple times because $\pi(t)$ may be different.

Algorithm 1: Büchi automaton-based LTLN model checking algorithm.

1 **function** check(C *{set of constraints}, n {node of $\mathcal{B}_\phi$}, t {time}*):
2   **if** $t > b$ **then**
3 | **return** $\emptyset$ *{DFS reached the bound}*
4   **foreach** *outbound neighbor n' of n {n' is a next node from n during DFS}* **do**
5   *{DMS: add the dynamics of Eq. (1) as equality constraints to C}*
6        $C' = C \cup \{ \mathbf{r}_{t+1} = \mathbf{f(x, u)} \mid_{\mathbf{x}=\mathbf{r}_t, \mathbf{u}=\mathbf{p}_t} \} \cup \{ \mathbf{q}_{t+1} =$
         $\mathbf{h(x, u)} \mid_{\mathbf{x}=\mathbf{r}_{t+1}, \mathbf{u}=\mathbf{p}_{t+1}} \}$
7   *{add all atomic propositions to satisfy before visiting n' to C'}*
8   **foreach** $c(\mathbf{u}, \mathbf{y}, \mathbf{x}) \bowtie 0$ *in between n and n'* **do**
9           $\lfloor \ C' = C' \cup \{ c(\mathbf{u}, \mathbf{y}, \mathbf{x}) \mid_{\mathbf{u}=\mathbf{p}_{t+1}, \mathbf{y}=\mathbf{q}_{t+1}, \mathbf{x}=\mathbf{r}_{t+1}} \bowtie 0 \}$
10   **if** $\bigwedge_{ap \in C'} ap$ *is feasible {if not feasible, prune the subtree rooted at n'}* **then**
11   **if** $n' \in S$ *{if DFS found a counterexample}* **then**
12 | **return** *a feasible solution of* $\bigwedge_{ap \in C'} ap$
13          $Z = $ check$(C', n', t + 1)$ *{continue DFS to n' with C'}*
14   **if** $Z \neq \emptyset$ **then**
15 | **return** Z *{DFS found a counterexample}*
16   **return** $\emptyset$

---

and the time step respectively. check returns a counterexample if found; otherwise it returns the empty set. In the algorithm $|$ is a substitution operator: $\mathbf{f(x, u)} \mid_{\mathbf{x}=\mathbf{r}_t, \mathbf{u}=\mathbf{p}_t}$ means substituting each occurrence of variables $\mathbf{x}$ and $\mathbf{u}$ in $\mathbf{f}$ with fresh variables $\mathbf{r}_t$ and $\mathbf{p}_t$ respectively. $\mathbf{q}_{t+1} = \mathbf{h(x, u)}|_{\mathbf{x}=\mathbf{r}_{t+1}, \mathbf{u}=\mathbf{p}_{t+1}}$ and $c(\mathbf{u}, \mathbf{y}, \mathbf{x})|_{\mathbf{u}=\mathbf{p}_{t+1}, \mathbf{y}=\mathbf{q}_{t+1}, \mathbf{x}=\mathbf{r}_{t+1}}$ are similarly defined. In Algorithm 1, S is the sink-set of $\mathcal{B}_\phi$. The set C that check builds can be regarded as a conjunctive term of a DNF. The algorithm checks the feasibility of C on every step and prunes the subtree if C becomes infeasible.

With the continuous state space of a model, LTLN model checking algorithm cannot rely on building an intersection automaton as is done in many other model checking algorithms. Instead, on every step of a DFS, the trajectory constraint is enforced as equality constraints using the DMS (see Example 2). Specifically, let C be an initially empty set of (in)equality constraints. On each step of DFS, (1) the system dynamics of Eq. (1) are added to C as equality constraints on fresh variables; (2) the label–the atomic propositions to satisfy to move to the next node [9], rewritten in terms of the fresh variables–of the edge to the next node of $\mathcal{B}_\phi$ are added to C; and (3) before DFS continues to the next node, the feasibility of C is checked.

Büchi automata can be regarded as a generator of a DNF of Sect. 4.1. Let T be a DFS tree on $\mathcal{B}_\phi$ of height b and further reduced by pruning the paths that do not end in the sink-set S. T is composed only of candidates for accepting runs. For each path ρ from the root node $\imath$ to a leaf node of T, let us consider a conjunctive formula $\bigwedge_{ap \in C(\rho)} ap$, where $C(\rho)$ is the (in)equality constraints collected along ρ. This conjunctive formula corresponds to a conjunctive term in a DNF. Moreover, the disjunctions of all such conjunctive formulas $\bigvee_\rho$ of $T \left(\bigwedge_{ap \in C(\rho)} ap \right)$ are a DNF itself.

Theorem 2. *If the sink-set S can be reached by* check *in b steps, then* $\pi, 0 \models_b \phi$, *where* π *is the computation path constructed from the feasible solution returned from* check.

Proof. Let $C(\rho)$ be the constraints collected along a DFS path ρ when reaching S and z be its feasible solution. Because $C(\rho)$ contains all atomic propositions along the run and the system dynamics as equality constraints in each step, the fresh variables $\mathbf{u}_t$, $\mathbf{y}_t$, $\mathbf{x}_t$ from z are on a valid computation path $\pi(t)$ and they satisfy all $ap \in L$ of $\rho(t-1) \xrightarrow{L} \rho(t)$ for $t \in [0, b]$ at the same time. Because cycling in S does not require any constraints, the run can visit accepting states infinitely often by repeatedly making ϵ-transitions.

Corollary 1. *Model checking* $M \models_b \phi$ *is sound and assuming a solution oracle for a system of constraints, it is complete.*

Proof. Theorem 2 shows the validity of a counterexample found by check function and Theorem 1 shows the existence of a reachable sink-set that the DFS of Algorithm 1 will find when a counterexample exists. Hence, assuming that an oracle solves the systems of constraints, one can *completely* and *correctly* decide $M \models_b \phi$ by checking both $M \models_b \phi$ and $M \models_b \neg\phi$. Specifically, the three cases of $M \models_b \phi$, $M \models_b \neg\phi$, and both $M \not\models_b \phi$ and $M \not\models_b \neg\phi$ can be decided by checking both formulas.

We assumed a solution oracle for a set of constraints to identify the element that restricts the completeness condition of LTLN model checking. However, a system of nonlinear equations is solvable only under certain conditions [13]. Because a system of nonlinear equations is not always solvable, the LTLN model checking is *not complete* in general.

Apart from solving the constraints using the assumed solution oracle, the worst case time complexity of check is $O(d^b)$, where d is the maximum out-degree of the nodes of $\mathcal{B}_\phi$. However, LTLN model checking process can be accelerated by *pruning* infeasible search trees. A path from the node ι to a non-leaf node n of T is a prefix shared by all paths in the subtree starting from n. Hence, C collected along the prefix can be regarded as a common formula in the conjunctive terms corresponding to the paths in the subtree. If C for a node is infeasible, searching its subtree can be skipped as the runs in the subtree are all infeasible. To expedite the model checking process, the feasibility of the constraints C is checked on every step of DFS. This is equivalent to checking a common sub-formula of conjunctive terms of a DNF and skipping the conjunctive terms if the sub-formula is infeasible.

4.3 Model Checker Implementation

To check the feasibility of the (in)equality constraints, we use a nonlinear solver provided by MATLAB® through Matlab Engine API. As a result LTLN can use built-in Matlab functions as well as custom functions in the model and specification.

Most constraint solvers do not distinguish strict and non-strict inequalities. To distinguish them, we convert inequality constraints to equality constraints using slack variables rather than using Matlab functions such as *fmincon*. The conversion is:

$$\mathbf{g}(\mathbf{v}) \geq 0 \Rightarrow \mathbf{g}(\mathbf{v}) - s^2 = 0, \quad \mathbf{g}(\mathbf{v}) > 0 \Rightarrow \mathbf{g}(\mathbf{v}) - \tfrac{1}{s^2} = 0, \tag{6}$$
$$\mathbf{g}(\mathbf{v}) \leq 0 \Rightarrow \mathbf{g}(\mathbf{v}) + s^2 = 0, \quad \mathbf{g}(\mathbf{v}) < 0 \Rightarrow \mathbf{g}(\mathbf{v}) + \tfrac{1}{s^2} = 0,$$

where $s \in \mathbb{R}$ is a slack variable. Observe that s^2 is non-negative and s^{-2}, used for strict inequalities, is positive when a finite solution exists.

After the conversion, *fsolve* suffices our needs. *fsolve* function solves the problem $\mathbf{g}(\mathbf{v}) = \mathbf{0}$, where $\mathbf{g}$ is a nonlinear function that takes a vector argument and returns a vector value. In our implementation, each equality constraint takes an entry in the vector of $\mathbf{g}$'s return value. The input parameter $\mathbf{v}$ comprises an initial state vector $\mathbf{x}(0)$, an input vector sequence $\mathbf{u}$ up to the time step t of check function, and slack variables for inequality constraints.

Continuing from Example 1 and Example 2, let us consider an example about checking the feasibility of constraints.

Example 4. We showed that any feasible $x_0, x_1, x_2, u_0, u_1 \in \mathbb{R}$ that can satisfy any of the three constraints of Example 2 are a counterexample witnessing $M \models_2 \neg(p \mathsf{U} q)$. The three conditions are converted to checking the feasibility of $\mathbf{g}_0(\mathbf{v}_0) = \mathbf{0}$, $\mathbf{g}_1(\mathbf{v}_1) = \mathbf{0}$, and $\mathbf{g}_2(\mathbf{v}_2) = \mathbf{0}$, where

$$\mathbf{g}_0(\mathbf{v}_0) = x_0 - 0.9 - \frac{1}{s_0^2}, \qquad \mathbf{g}_1(\mathbf{v}_1) = \begin{bmatrix} x_0 - \frac{1}{s_0^2} \\ x_1 - 0.9 - \frac{1}{s_1^2} \end{bmatrix}, \qquad \mathbf{g}_2(\mathbf{v}_2) = \begin{bmatrix} x_0 - \frac{1}{s_0^2} \\ x_1 - \frac{1}{s_1^2} \\ x_2 - 0.9 - \frac{1}{s_2^2} \end{bmatrix},$$

$$x_1 = \sin(x_0 + u_0) \qquad\qquad \begin{aligned} x_1 &= \sin(x_0 + u_0) \\ x_2 &= \sin(x_1 + u_1) \end{aligned}$$

where $\mathbf{v}_0 = [x_0, s_0]^T$, $\mathbf{v}_1 = [x_0, u_0, s_0, s_1]^T$, and $\mathbf{v}_2 = [x_0, u_0, u_1, s_0, s_1, s_2]^T$. If *fsolve* can find a vector solution $\mathbf{v}_0$, $\mathbf{v}_1$, or $\mathbf{v}_2$ that can make $\mathbf{g}_0$, $\mathbf{g}_1$, or $\mathbf{g}_2$ the zero vector respectively, constructing a counterexample π witnessing $\pi, 0 \models_2 p \mathsf{U} q$ from the vector solution is a straightforward process.   □

5   LTLN Model Checking Examples

In this section we demonstrate our model checking tool LTLN-Checker [1] using two examples. The first example is about finding a control input to control the inverted pendulum model of Sect. 2.1 and the second one is about computing a prescription for a pharmacokinetic model.

5.1   Inverted Pendulum

Figure 2 shows an LTLN-Checker description of the inverted pendulum model of Sect. 2.1 and the two control objectives of Sect. 3.2. In the description, the text from # to the end of the line is comments. The description has two main blocks: model block that describes a nonlinear system model and specification block that describes a property, such as a control objective, about the model in an LTLN formula. model block begins with the name of the model, IP, followed by an optional list of constants. After a var tag, input, output, and state variables are defined with their types. System's dynamics equations such as the four equations of Fig. 2 are defined in dyneq block. specification block begins with optional condition definitions, where a list of LTLN formulas is defined. Finally, check statement concludes an LTLN description by

```
model IP :
    const M = 1, m = 0.1, f = 0.1, l = 1, g = 9.8, dt = 0.1, pi=3.141592;
    var x[4]: state; # [x, dot x, theta, dot theta]
    var F: input;    # external force
    dyneq
        x[1] = x[1] + dt * x[2],
        x[2] = x[2] + dt * (-m*g*sin(x[3])*cos(x[3]) + m*l*x[4]^2*sin(x[3])
                        + f*m*x[4]*cos(x[3]) + F) / (M + m*(1 - cos(x[3])^2)),
        x[3] = x[3] + dt * x[4],
        x[4] = x[4] + dt * ((M + m)*(g*sin(x[3]) - f*x[4]) - (l*m*x[4]^2*sin(x[3])
                        + F)*cos(x[3])) / (l*M + l*m*(1-cos(x[3])^2));
specification:
    condition
        init: x[1] = 0 /\ x[2] = 0 /\ x[3] = pi*10/180 /\ x[4] = 0,
        goal: x[2] = 0 /\ x[3] = 0 /\ x[4] = 0,
        cond:  -1 < x[1] /\ x[1] < 1 /\ -20 < F /\ F < 20;
        #cond2: -1 < x[1] /\ x[1] < 1 /\ -10 < F /\ F < 10;
    check IP |= init -> ~(goal R cond) in 10;              # goal
    #check IP |= init -> ~((goal R cond2) R cond) in 10; # goal2
```

Fig. 2. LTLN-Checker description of the inverted pendulum model and specifications.

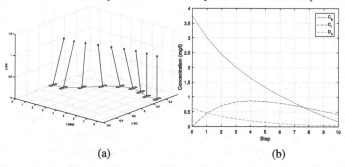

(a)   (b)

Fig. 3. (a) responses of the inverted pendulum system after applying the input found. (b) drug concentration level changes of the two-compartment model.

defining the model checking problem. In Fig. 2, the statement check IP |= ~(goal R cond) in 10 means $IP \models_{10} \neg(goal\ R\ cond)$.

For the goal, $IP \models_{10} init \rightarrow \neg(goal\ R\ cond)$, the LTLN-Checker reported false with a counter example of $\mathbf{x}(0) = [0, 0, 0.174, 0]^T$ and $\mathbf{u} = [$ 18.041, 0.194, -9.521, -11.620, -6.547, 9.439, 0.411 $]$. The force remains within ±20 (N) range and by recursively applying Eq. (1) to the input sequence and the initial state, the goal is achieved. The commented goal2 requires using strong forces until the goal can be achieved by using weak forces. Figure 3 (a) shows how the inverted pendulum changes when the counterexample found for goal2 is applied.

5.2 Nonlinear Pharmacokinetics

When a drug is administered, it goes through the *Absorption, Distribution, Metabolism,* and *Elimination* (ADME) processes [31]. A *Compartment model* is a common mathematical model to describe a drug ADME process. When administering a drug, it is important to ensure that the drug is effective and at the same time it does not turn toxic. Particularly, the drug concentration level should remain above a *Minimum Effective Concentration* (MEC) for a certain duration to be effective and the level should never exceed a *Minimum Toxic Concentration* (MTC) to be non-toxic. In this section we will find a drug prescription using an LTLN model checking like [24].

The diagram below shows a two-compartment model, where P and T represent plasma and tissue compartments respectively, K_{pt} and K_{tp} are rate constants from P to T and T to P respectively, K is a rate constant for the drug elimination, and K_a is a rate constant for the drug absorption. In the hepatic drug elimination process, the drug molecules are removed from the liver by some enzymes. When there are more drug molecules than the enzymes can handle, the elimination process shows a saturation behavior. This saturable enzyme process is modeled by a nonlinear differential equation called Michaelis-Menten kinetics [31] as below.

$$\dot{C}_p = -\frac{V_m C_p}{K_m + C_p} - K_{pt}C_p + K_{tp}C_t + K_a D_o,$$

$$\dot{C}_t = K_{pt}C_p - K_{tp}C_t, \tag{7}$$

$$\dot{D}_o = -K_a D_o,$$

where C_p and C_t are drug concentration levels at the plasma and at the tissue compartments respectively, V_m is the maximum elimination rate, K_m is the Michaelis constant and D_o is the amount of orally administered drug. The first term of the first equation corresponds to a drug elimination. When $C_p \gg K_m$, the elimination rate becomes a constant V_m. That is, the elimination process is saturated and it becomes irrelevant to the drug concentration level C_p. The continuous time model can be discretized using Euler's method as in Fig. 4.

Now, let us specify the requirements for the drug administration in LTLN. We want to find a prescription that can satisfy the conditions that (1) the drug should be active (the concentration level is larger than MEC) for an hour; (2) the drug should be cleared in the end; and (3) it should never be toxic (the concentration level is lower than MTC). As a first step, the formal model of the drug kinetics is $M = \langle \emptyset, \emptyset, X, \mathbf{f}, \emptyset \rangle$, where $X = \{C_p, C_t, D_o\}$ and $\mathbf{f}$ is in Fig. 4. In this example, we do not need input and output variables, because the prescription is about the doses for IV bolus and oral administration at time 0 and we are only interested in the trajectory of C_p.

As a prescription we would like to find the initial drug concentrations at C_p for the IV bolus and at D_o for the oral administration. Hence, we can write the initial condition as $\phi_{init} = C_p \geq 0 \ \wedge \ C_t = 0 \ \wedge \ D_o \geq 0$. Let the sampling time Δt be 0.5 h and mec be the drug's MEC, the first requirement can be written as $\Diamond \phi_{mec}$, where $\phi_{mec} = C_p \geq mec \ \wedge \ \mathsf{X} C_p \geq mec$. The second condition is about the drug clearance, which can be written as $\Diamond \Box \phi_{clear}$, where $\phi_{clear} = C_p \leq mcl$ and mcl is the maximum cleared level. The last condition is about the drug toxicity, which can be described as $\Box \phi_{mtc}$, where $\phi_{mtc} = C_p \leq mct$ when mtc is the drug's MTC.

```
model Prescription :
    const dt = 0.5, Km  = 0.1, Vm  = 0.7, Kpt = 0.3, Ktp = 0.5, Ko = 0.5;
    const mec = 3, mtc = 5, mcl = 1;
    var Cp, Ct, Do: state;
    dyneq
        Cp = Cp + dt * (-Vm * Cp / (Km + Cp) - Kpt * Cp + Ktp * Ct + Ko * Do),
        Ct = Ct + dt * (Kpt * Cp - Ktp * Ct),
        Do = Do + dt * (-Ko * Do);
specification:
    condition
        init:      Cp >= 0 /\ Ct = 0 /\ Do >= 0,
        active:    (Cp >= mec) /\ X (Cp >= mec),
        cleared:   Cp <= mcl,
        nontoxic:  Cp <= mtc;
    check Prescription |= init -> ˜(active R (cleared R nontoxic)) in 10;
```

Fig. 4. LTLN-Checker description for the drug prescription example.

Now we need to adjust the conditions for the bounded model checking: ϕ_{mec} should be satisfied before ϕ_{clear}, whereas ϕ_{mtc} should be satisfied until the drug is cleared (ϕ_{clear}). Hence, combining the conditions, the model checking problem to find a prescription is $M \models_b \neg\phi$, where

$$\phi = \phi_{init} \rightarrow \neg (\phi_{mec} \; R \; (\phi_{clear} \; R \; \phi_{mtc})). \tag{8}$$

Any counterexample satisfies ϕ_{init} first, then it satisfies ϕ_{mec}, and then ϕ_{clear} is satisfied. In addition, ϕ_{mtc} is satisfied until the moment ϕ_{clear} is satisfied. Figure 4 shows an LTLN description for the drug kinetic model and the prescription. When $K_{pt}=0.3$, $K_{tp}=0.5$, $K_a=0.5$, $V_m=0.7$, $K_m=0.1$, $mec=3$, $mtc=5$, and $mcl=1$, the model checker found a counterexample of $C_p=3.75$, $C_t=0$, and $D_o=0.614$. Figure 3 (b) shows the trajectories of C_p, C_t, and D_o when the doses found in the counterexample are administered: the conditions about active duration, clearance, and non-toxicity are all satisfied.

Model checking the first and the second goals of Fig. 2 took about 54 (*sec*) and 59 (*sec*) respectively on a Xeon machine (hyper-threaded 6 core running at 2.30 GHz). On the other hand, checking the pharmaceutical example of Fig. 4 took about 3,110 (*sec*), about 55 times larger than the inverted pendulum example. Frequent inclusive termination of *fsolve* was the cause of the large latency of the latter: when *fsolve* ended inconclusively about solving the constraints for a node, DFS continued on the next neighboring nodes, which made the pruning strategy ineffective.

6   Conclusion

We developed a temporal logic, called LTLN, to describe properties of nonlinear systems and its bounded model checking algorithm. The model checking algorithm is sound and complete for a set of solvable constraints, but not complete for a set of general nonlinear constraints. One of the main utilities of the LTLN model checking technique is to compute inputs for a control objective written in LTLN.

A major performance bottleneck in the LTLN model checking process is solving nonlinear equations. Because LTLN model checking algorithm serializes the constraints to solve, if any of the feasibility checking step is blocked, the whole process is blocked. We are redesigning the model checking process to utilize GPU devices. As demonstrated in [26], with thousands of parallel threads solving feasibility problems, a small fraction of threads with blocking issues would not block the whole process.

One of the merits of LTLN model checking is that the input and the initial state of the system are free variables. This enables our method to be applicable to both model checking and control systems. LTLN-checker can facilitate the design of systems as well as find vulnerabilities in a system. Given the increasingly safety critical applications of cyberphysical systems, ensuring their safe and reliable operation is an important concern. We believe LTLN-checker provides a versatile method that can help ensure the safety of cyberphysical systems.

References

1. LTLN-Checker: http://www3.cs.stonybrook.edu/youngkwon/sw/LTLN.v0.8.190125.zip
2. Alur, R., Dill, D.L.: A theory of timed automata. Theor. Comput. Sci. **126**, 183–235 (1994)
3. Baier, C., Katoen, J.P.: Principles of Model Checking. 1st edn., MIT Press, Cambridge (2008)
4. Biere, Armin, Cimatti, Alessandro, Clarke, Edmund, Zhu, Yunshan: Symbolic model checking without BDDs. In: Cleaveland, W. Rance. (ed.) TACAS 1999. LNCS, vol. 1579, pp. 193–207. Springer, Heidelberg (1999). https://doi.org/10.1007/3-540-49059-0_14
5. Bock, H.G., Plitt, K.J.: A multiple shooting algorithm for direct solution of optimal control problems. In: IFAC World Congress: A Bridge Between Control Science and Technology, vol. 17, pp. 1603–1608. IFAC (1984)
6. Bu, L., Li, Y., Wang, L., Li, X.: BACH: Bounded reachability checker for linear hybrid automata. In: Formal Methods in Computer Aided Design, pp. 65–68. IEEE Computer Society (2008)
7. Chen, X., Abraham, E., Sankaranarayanan, S.: Taylor model flowpipe construction for nonlinear hybrid systems. In: Real-Time Systems Symposium, pp. 183–192. IEEE (2012)
8. Clarke, D., Mohtai, C., Tuffs, P.: Generalized predictive control. Automatica. **23**, 137–160 (1987)
9. Clarke, E., Grumberg, O., Peled, D.: Model Checking. 1st edn. MIT Press, Cambridge (2000)
10. De Giacomo, G., Vardi, M.Y.: Linear temporal logic and linear dynamic logic on finite traces, pp. 854–860. IJCAI 2013, AAAI Press (2013)
11. Franklin, G.F., Powell, J.D., Emami-Naeini, A.: Feedback Control of Dynamic Systems. 3rd edn. Addison Wesley, Boston (1994)
12. Fränzle, M., Herde, C.: HySAT: an efficient proof engine for bounded model checking of hybrid systems. Formal Methods Syst. Des. **30**, 179–198 (2007)
13. Fucik, S.: Solvability of Nonlinear Equations and Boundary Value Problems. Springer, Dordrecht (1980)
14. Girard, A., Pola, G., Tabuada, P.: Approximately bisimilar symbolic models for incrementally stable switched systems. Trans. Autom. Control. **55**, 116–126 (2010)
15. Henzinger, T.A.: The theory of hybrid automata. In: Annual Symposium on Logic in Computer Science, pp. 278–292. IEEE Computer Society (1996)
16. Henzinger, T.A., Ho, P.-H., Wong-Toi, H.: HYTECH: a model checker for hybrid systems. In: Grumberg, O. (ed.) CAV 1997. LNCS, vol. 1254, pp. 460–463. Springer, Heidelberg (1997). https://doi.org/10.1007/3-540-63166-6_48

17. Holzmann, G.J.: The model checker spin. IEEE Trans. Softw. Eng. **23**, 279–295 (1997)
18. Hughes, G., Creswell, M.: Introduction to Modal Logic. Methuen (1997)
19. Kloetzer, M., Belta, C.: A fully automated framework for control of linear systems from temporal logic specifications. Trans. Autom. Control **53**, 287–297 (2008)
20. Kong, S., Gao, S., Chen, W., Clarke, E.: dReach: δ-reachability analysis for hybrid systems. In: Baier, C., Tinelli, C. (eds.) TACAS 2015. LNCS, vol. 9035, pp. 200–205. Springer, Heidelberg (2015). https://doi.org/10.1007/978-3-662-46681-0_15
21. Kress-Gazit, H., Fainekos, G.E., Pappas, G.J.: Temporal-logic-based reactive mission and motion planning. Trans. Robot. **25**, 1370–1381 (2009)
22. Kwon, Y., Agha, G.: iLTLChecker: a probabilistic model checker for multiple DTMCs. In: International Conference on the Quantitative Evaluation of Systems (QEST), IEEE Computer Society (2005)
23. Kwon, Y., Agha, G.: Verifying the evolution of probability distributions governed by a DTMC. In: IEEE Transactions on Software Engineering, vol. 37, pp. 126–141. IEEE Computer Society (2011)
24. Kwon, Y., Kim, E.: Specification and verification of pharmacokinetic models. In: Arabnia, H. (eds) Advances in Computational Biology. Advances in Experimental Medicine and Biology, vol. 680, pp. 465–472. Springer, New York, NY (2010). https://doi.org/10.1007/978-1-4419-5913-3_52
25. Kwon, Y., Kim, E.: Bounded model checking of hybrid systems for control. IEEE Trans. Autom. Control **60**, 2961–2976 (2015)
26. Kwon, Y.M., Kim, E.: A design of GPU-based quantitative model checking. In: Henglein, F., Shoham, S., Vizel, Y. (eds.) VMCAI 2021. LNCS, vol. 12597, pp. 441–463. Springer, Cham (2021). https://doi.org/10.1007/978-3-030-67067-2_20
27. Larsen, K.G., Pettersson, P., Yi, W.: UPPAAL in a nutshell. Int. J. Softw. Tools Technol. Transfer **1**, 134–152 (1997)
28. Liu, J., Ozay, N., Topcu, U., Murray, R.M.: Synthesis of reactive switching protocols from temporal logic specifications. Trans. Autom. Control **58**(7), 1771–1785 (2013)
29. Maler, O., Nickovic, D., Pnueli, A.: Checking temporal properties of discrete, timed and continuous behaviors. In: Avron, A., Dershowitz, N., Rabinovich, A. (eds.) Pillars of Computer Science. LNCS, vol. 4800, pp. 475–505. Springer, Heidelberg (2008). https://doi.org/10.1007/978-3-540-78127-1_26
30. Roehm, H., Oehlerking, J., Heinzer, T., Althoff, M.: Stl model checking of continuous and hybrid systems. In: ATVA, pp. 412–427 (2016)
31. Shargel, L., Yu, A.B.: Applied Biopharmaceutics and Pharmacokinetics. 2nd edn. Appleton-Century-Crofts, Norwalk (1985)
32. Slotine, J.J.E., Li, W.: Applied Nonlinear Control. 1st edn. Prentice Hall, Hoboken (1991)
33. Tabuada, P., Pappas, G.J.: Linear time logic control of discrete-time linear systems. Trans. Autom. Control **51**, 1862–1877 (2006)
34. Vidyasagar, M.: Nonlinear Systems Analysis. 2nd edn. Prentice Hall, London (1993)

An MM Algorithm to Estimate Parameters in Continuous-Time Markov Chains

Giovanni Bacci[1]([✉])(iD), Anna Ingólfsdóttir[2], Kim G. Larsen[1],
and Raphaël Reynouard[2](iD)

[1] Department of Computer Science, Aalborg University, Aalborg, Denmark
{giovbacci,kgl}@cs.aau.dk
[2] Department of Computer Science, Reykjavík University, Reykjavík, Iceland
{annai,raphal20}@ru.is

Abstract. PRISM and STORM are popular model checking tools that provide a number of powerful analysis techniques for Continuous-time Markov chains (CTMCs). The outcome of the analysis is strongly dependent on the parameter values used in the model which govern the timing and probability of events of the resulting CTMC. However, for some applications, parameter values have to be empirically estimated from partially-observable executions.

In this work, we address the problem of estimating parameter values of CTMCs expressed as PRISM models from a number of partially-observable executions which might possibly miss some dwell time measurements. The semantics of the model is expressed as a parametric CTMC (pCTMC), i.e., CTMC where transition rates are polynomial functions over a set of parameters. Then, building on a theory of algorithms known by the initials MM, for minorization–maximization, we present an iterative maximum likelihood estimation algorithm for pCTMCs. We present an experimental evaluation of the proposed technique on a number of CTMCs from the quantitative verification benchmark set. We conclude by illustrating the use of our technique in a case study: the analysis of the spread of COVID-19 in presence of lockdown countermeasures.

Keywords: MM Algorithm · Continuous-time Markov chains · Maximum likelihood estimation

1 Introduction

A continuous-time Markov chain (CTMC) is a model of a dynamical system that, upon entering some state, remains in that state for a random real-valued amount of time—called the dwell time or sojourn time—and then transitions

K.G. Larsen and G. Bacci were supported by the S40S Villum Investigator Grant (37819) from Villum Fonden; R. Reynouard and A. Ingólfsdóttir were supported by the project *Learning and Applying Probabilistic Systems* (206574-051) of the Icelandic Research Fund.

N. Jansen and M. Tribastone (Eds.): QEST 2023, LNCS 14287, pp. 82–100, 2023.
https://doi.org/10.1007/978-3-031-43835-6_6

```
ctmc
  // SIR model paramaters
  const double beta; const double gamma;
  const double plock;
  const int SIZE = 100000;  // population size

  module SIR
  s : [0..SIZE] init 99936;
  i : [0..SIZE] init 48;
  r : [0..SIZE] init 16;

  [] i>0 & i<SIZE & s>0 →
     beta * s * i * plock/SIZE : (s'=s − 1)&(i'=i + 1);
  [] i>0 & r<SIZE →
     gamma * i * plock : (i'=i − 1)&(r'=r + 1);
  endmodule
```

```
ctmc
  // SIR model paramaters
  const double beta; const double gamma;
  const double plock;
  const int SIZE = 100000;  // population size

  module Susceptible
  s : [0..SIZE] init 99936;
  [infection] s>0 → s : (s'=s − 1);
  endmodule

  module Infected
  i : [0..SIZE] init 48;
  [infection] i>0 & i<SIZE → i : (i'=i + 1);
  [recovery] i>0 → i : (i'=i − 1);
  endmodule

  module Recovered
  r : [0..SIZE] init 16;
  [recovery] r<SIZE → 1 : (r'=r + 1);
  endmodule

  module Rates
  [infection] true → beta * plock/SIZE : true;
  [recovery] true → gamma * plock : true;
  endmodule
```

Fig. 1. (Left) SIR model with lockdown from [36], (Right) Semantically equivalent formulation of the model to the left where different individuals are modeled as distinct modules interacting with each other via synchronization.

probabilistically to another state. CTMCs are popular models in performance and dependability analysis. They have wide application and constitute the underlying semantics for real-time probabilistic systems such as queuing networks [33], stochastic process algebras [24], and calculi for systems biology [13, 29].

Model checking tools such as PRISM [30] and STORM [14] provide a number of powerful analysis techniques for CTMCs. Both tools accept models written in the PRISM language, a state-based language based on [1] that supports compositional design via a uniform treatment of synchronous and asynchronous components.

For example, consider the PRISM model depicted in Fig. 1 (left) implementing a variant of the Susceptible-Infected-Recovered (SIR) model proposed in [36] to describe the spread of disease in presence of lockdown restrictions. The model distinguishes between three types of individuals: susceptible, infected, and recovered respectively associated with the state variables s, i, and r. Susceptible individuals become infected through contact with another infected person and can recover without outside interference. The SIR model is parametric in beta, gamma, and plock. beta is the *infection coefficient*, describing the probability of infection after the contact of a susceptible individual with an infected one; gamma is the *recovery coefficient*, describing the rate of recovery of an infected individual (in other words, 1/gamma is the time one individual requires to recover); and plock ∈ [0, 1] is used to scale down the infection coefficient modeling restrictions to reduce the spread of disease.

Clearly, the outcome of the analysis of the above SIR model is strongly dependent on the parameter values used, as they govern the timing and probability of events of the CTMC describing its semantics. However, in some application domains, parameter values have to be empirically evaluated from a number of partially-observable executions of the model. A paradigmatic example is the modeling pipeline described in [36], where the parameters of the SIR model in Fig. 1 (left) are estimated based on a definition of the model as ODEs, and later used in an approximation of the original SIR model designed to reduce the state space of the SIR model in Fig. 1 (left). Such modeling pipelines require high technical skills, are error-prone, and are time-consuming, thus limiting the applicability and the user base of model checking tools.

In this work, we address the problem of estimating parameter values of CTMCs expressed as PRISM models from a number of partially-observable executions. The expressive power of the PRISM language brings two technical challenges: (i) the classic state-space explosion problem due to modular specification, and (ii) the fact that the transition rates of the CTMCs result from the algebraic composition of the rates of different (parallel) modules which are themselves defined as arithmetic expressions over the parameters (*cf.* Fig. 1). We address the second aspect of the problem by considering a class of *parametric* CTMCs (pCTMC) [10, 21], which are CTMCs where transition rates are polynomial functions over a fixed set of parameters. In this respect, pCTMCs have the advantage to cover a rich subclass of PRISM models and to be closed under the operation of parallel composition implemented by the PRISM language.

Following the standard approach, we pursue the maximum likelihood estimate (MLE), i.e., we look for the parameter values that achieve the maximum joint likelihood of the observed execution sequences. However, given the non-convex nature of the likelihood surface, computing the global maximum that defines the MLE is computationally intractable [42].

To deal with this issue we employ a theoretical iterative optimization principle known as MM algorithm [31, 32]. The well-known EM algorithm [15] is an instance of MM optimization framework and is a versatile tool for constructing optimization algorithms. MM algorithms are typically easy to design, numerically stable, and in some cases amenable to accelerations [25, 44]. The versatility of the MM principle consists in the fact that is built upon a simple theory of inequalities, allowing one to derive optimization procedures. The MM principle is useful to derive iterative procedures for maximum likelihood estimation which increase the likelihood at each iteration and converge to some local optimum.

The main technical contribution of the paper consists in devising a novel iterative maximum likelihood estimation algorithm for pCTMCs. Crucially, our technique is robust to missing data. In contrast with [18, 41], where state labels and dwell times are assumed to be observable at each step of the observations while only state variables are hidden, our estimation procedure accepts observations to have information to be missing at some steps.

Notably, when state labels and dwell times are observable and only state variables are hidden, our learning procedure results in a generalization of the Baum-Welch algorithm [38]—an EM algorithm that estimates transition probabilities in hidden Markov models—to pCTMCs.

We demonstrate the effectiveness of our estimation procedure on a case study taken from [36] and show that our technique can be used to simplify modeling pipelines that involve a number of modifications of the model—possibly introducing approximations—and the re-estimation of its parameters.

2   Preliminaries and Notation

We denote by $\mathbb{R}$, $\mathbb{Q}$, and $\mathbb{N}$ respectively the sets of real, rational, and natural numbers, and by Σ^n, Σ^* and, Σ^ω respectively the set of words of length $n \in \mathbb{N}$, finite length, and infinite length, built over the finite alphabet Σ.

We use $\mathcal{D}(\Omega)$ to denote the set of discrete probability distributions on Ω, i.e., functions $\mu \colon \Omega \to [0,1]$, such that $\mu(X) = 1$, where $\mu(E) = \sum_{x \in E} \mu(x)$ for $E \subseteq X$. For a proposition p, we write $\llbracket p \rrbracket$ for the Iverson bracket of p, i.e., $\llbracket p \rrbracket = 1$ if p is true, otherwise 0.

A labelled continuous-time Markov chain (CTMC) is defined as follows.

Definition 1. *A labelled CTMC is a tuple* $\mathcal{M} = (S, R, s_0, \ell)$ *where S is a finite set of states,* $R \colon S \times S \to \mathbb{R}_{\geq 0}$ *is the transition rate function, $s_0 \in S$ the initial states, and* $\ell \colon S \to 2^{AP}$ *is a labelling function which assigns to each state a subset of atomic propositions that the state s satisfies.*

The transition rate function assigns rates $r = R(s, s')$ to each pair of states $s, s' \in S$ which are to be seen as transitions of the form $s \xrightarrow{r} s'$. A transition $s \xrightarrow{r} s'$ can only occur if $r > 0$. In this case, the probability of this transition to be triggered within $\tau \in \mathbb{R}_{>0}$ time-units is $1 - e^{-r\,\tau}$. When, from a state s, there are more than one outgoing transition with positive rate, we are in presence of a *race condition*. In this case, the first transition to be triggered determines which label is observed as well as the next state of the CTMC. According to these dynamics, the time spent in state s before any transition occurs, called *dwell time*, is exponentially distributed with parameter $E(s) = \sum_{s' \in S} R(s, s')$, called *exit-rate* of s. A state s is called *absorbing* if $E(s) = 0$, that is, s has no outgoing transition. Accordingly, when the CTMC ends in an absorbing state it will remain in the same state indefinitely. The probability that the transition $s \xrightarrow{r} s'$ is triggered from s is $r/E(s)$ and is independent from the time at which it occurs. Accordingly, from the CTMC $\mathcal{M}$, we construct a (labelled) discrete-time Markov chain $emb(\mathcal{M}) = (S, P, s_0, \ell)$ with transition probability function $P \colon S \times S \to [0,1]$ defined as

$$P(s, s') = \begin{cases} R(s, s')/E(s) & \text{if } E(s) \neq 0 \\ 1 & \text{if } E(s) = 0 \text{ and } s = s' \\ 0 & \text{otherwise} \end{cases}$$

Remark 1. A CTMC can be equivalently described as a tuple $(S, \to, s_0, \ell)$ where $\to \; \subseteq S \times \mathbb{R}_{\geq 0} \times S$ is a transition *relation*. The transition rate function R induced by $\to$ is obtained as, $R(s, s') = \sum \{r \mid s \xrightarrow{r} s'\}$ for arbitrary $s, s' \in S$.

An *infinite path* of a CTMC $\mathcal{M}$ is a sequence $s_0 \tau_0 s_1 \tau_1 s_2 \tau_2 \cdots \in (S \times \mathbb{R}_{>0})^\omega$ where $R(s_i, s_{i+1}) > 0$ for all $i \in \mathbb{N}$. A *finite path* is a sequence $s_0 \tau_0 \cdots s_{k-1} \tau_{k-1} s_k$

where $R(s_i, s_{i+1}) > 0$ and $\tau_i \in \mathbb{R}_{>0}$ for all $i \in \{1, \ldots, k-1\}$ and s_k is absorbing. The meaning of a path is that the system started in state s_0, where it stayed for time τ_0, then transitioned to state s_1 where it stayed for time τ_1, and so on. For a finite path the system eventually reaches an absorbing state s_k, where it remains. We denote by $\mathbf{Path}_{\mathcal{M}}$ the set of all (infinite and finite) paths of $\mathcal{M}$. The formal definition of the probability space over $\mathbf{Path}_{\mathcal{M}}$ induced by $\mathcal{M}$ can be given by following the classical cylinder set construction (see e.g., [6,28]).

Finally, we define the random variables S_i, L_i, and T_i ($i \in \mathbb{N}$) that respectively indicate the i-th state, its label, and i-th dwell time of a path.

The MM Algorithm. The MM algorithm is an iterative optimization method. The acronym MM has a double interpretation: in minimization problems, the first M stands for majorize and the second for minimize; dually, in maximization problems, the first M stands for minorize and the second for maximize. In this paper we only focus on maximizing an objective function $f(\mathbf{x})$, hence we tailor the presentation of the general principles of the MM framework to maximization problems. The MM algorithm is based on the concept of *surrogate function*. A surrogate function $g(\mathbf{x} \mid \mathbf{x}_m)$ is said to *minorize* a function $f(\mathbf{x})$ at $\mathbf{x}_m$ if

$$f(\mathbf{x}_m) = g(\mathbf{x}_m \mid \mathbf{x}_m), \tag{1}$$

$$f(\mathbf{x}) \geq g(\mathbf{x} \mid \mathbf{x}_m) \quad \text{for all } \mathbf{x} \neq \mathbf{x}_m. \tag{2}$$

In the MM optimization framework, we maximize the surrogate minorizing function $g(\mathbf{x} \mid \mathbf{x}_m)$ rather than the actual function $f(\mathbf{x})$. If $\mathbf{x}_{m+1}$ denotes the maximum of the surrogate $g(\mathbf{x} \mid \mathbf{x}_m)$, then the next iterate $\mathbf{x}_{m+1}$ forces $f(\mathbf{x})$ uphill, Indeed, the inequalities

$$f(\mathbf{x}_m) = g(\mathbf{x}_m \mid \mathbf{x}_m) \leq g(\mathbf{x}_{m+1} \mid \mathbf{x}_m) \leq f(\mathbf{x}_{m+1})$$

follow directly from the definition of $\mathbf{x}_{m+1}$ and the axioms (1) and (2).

Because piecemeal composition of minorization works well, the derivations of surrogate functions are typically achieved by applying basic minorizations to strategic parts of the objective function, leaving other parts untouched. Finally, another aspect that can simplify the derivation of MM algorithms comes from the fact that the iterative maximization procedure hinges on finding $\mathbf{x}_{m+1} = \arg\max_{\mathbf{x}} g(\mathbf{x} \mid \mathbf{x}_m)$. Therefore, $g(\mathbf{x} \mid \mathbf{x}_m)$ can be replaced by any other surrogate function $g'(\mathbf{x} \mid \mathbf{x}_m)$ satisfying $\arg\max_{\mathbf{x}} g(\mathbf{x} \mid \mathbf{x}_m) = \arg\max_{\mathbf{x}} g'(\mathbf{x} \mid \mathbf{x}_m)$ for all $\mathbf{x}_m$. This is for instance the case when $g(\mathbf{x} \mid \mathbf{x}_m)$ and $g'(\mathbf{x} \mid \mathbf{x}_m)$ are equal up to some (irrelevant) constant c, that is $g(\mathbf{x} \mid \mathbf{x}_m) = g'(\mathbf{x} \mid \mathbf{x}_m) + c$.

3   Parametric Continuous-Time Markov Chains

As mentioned in the introduction, the PRISM language offers constructs for the modular design of CTMCs within a uniform framework that represents synchronous and asynchronous module interaction. For example, consider the PRISM models depicted in Fig. 1. The behavior of each module is described by a set of commands which take the form [action] guard → rate: update representing a set of transitions of the module. The guard is a predicate over the state variables in the model. The update and the rate describe a transition that the module can make if the guard is true. The command optionally includes an action used to force two or more modules to make transitions simultaneously (i.e., to synchronize). For example, in the model in Fig. 1 (right), in state $(50, 20, 5)$ (i.e., $s = 50$, $i = 20$, and $r = 5$), the model can move to state $(49, 21, 5)$ by synchronizing over the action infection. The rate of this transition is equal to the product of the individual rates of each module participating in an infection transition, which in this case amounts to $0.01 \cdot$ beta $\cdot$ plock. Commands that do not have an action represent asynchronous transitions that can be taken independently (i.e., asynchronously) from other modules.

By default, all modules are combined following standard parallel composition in the sense of the parallel operator from Communicating Sequential Processes algebra (CPS), that is, modules synchronize over all their common actions. The PRISM language offers also other CPS-based operators to specify the way in which modules are composed in parallel.

Therefore, a parametric representation of a CTMC described by a PRISM model shall consider *transition rate expressions* which are closed under finite sums and finite products: sums deal with commands with overlapping guards and updates, while products take into account synchronization. In line with [10, 21] we employ *parametric* CTMCs (pCTMCs).

Let $\mathbf{x} = (x_1, \ldots, x_n)$ be a vector of parameters. We write $\mathcal{E}$ for the set of polynomial maps $f \colon \mathbb{R}^n_{\geq 0} \to \mathbb{R}_{\geq 0}$ of the form $f(\mathbf{x}) = \sum_{i=1}^m b_i \prod_{j=1}^n x_j^{a_{ij}}$, where $b_i \in \mathbb{R}_{\geq 0}$ and $a_{ij} \in \mathbb{N}$ for $i \in \{1, \ldots, m\}$ and $j \in \{1, \ldots, n\}$. $\mathcal{E}$ is a commutative semiring satisfying the above-mentioned requests for transition rate expressions.

Definition 2. *A pCTMC is a tuple* $\mathcal{P} = (S, R, s_0, \ell)$ *where* S, s_0, *and* ℓ *are defined as for CTMCs, and* $R \colon S \times S \to \mathcal{E}$ *is a parametric transition rate function.*

Intuitively, a pCTMC $\mathcal{P} = (S, R, s_0, \ell)$ defines a family of CTMCs arising by plugging in concrete values for the parameters $\mathbf{x}$. Given a parameter evaluation $\mathbf{v} \in \mathbb{R}^n_{\geq 0}$, we denote by $\mathcal{P}(\mathbf{v})$ the CTMC associated with $\mathbf{v}$, and $R(\mathbf{v})$ for its rate transition function. Note that by construction $R(\mathbf{v})(s, s') \geq 0$ for all $s, s' \in S$, therefore $\mathcal{P}(\mathbf{v})$ is a proper CTMC.

As for CTMCs, parametric transitions rate functions can be equivalently described by means of a transition relation $\to \subseteq S \times \mathcal{E} \times S$, where the parametric transition rate from s to s' is $R(s, s')(\mathbf{x}) = \sum \{f(\mathbf{x}) \mid s \xrightarrow{f} s'\}$.

Example 1. Consider the model in Fig. 1 parametric in beta, gamma, and plock. The semantics of this model is a pCTMC with states $S = \{(s, i, r) \mid s, i, r \in \{0, \ldots, 10^5\}\}$ and initial state $(99936, 48, 16)$. For example, the initial state has two outgoing transitions: one that goes to $(99935, 49, 16)$ with rate $47.96928 \cdot$ beta$\cdot$plock, and the other that goes to $(99935, 48, 17)$ with rate $49 \cdot$gamma$\cdot$plock.

One relevant aspect of the class of pCTMCs is the fact that it is closed under parallel composition in the sense described above. This justifies the study

of parameter estimation of PRISM models from observed data via maximum
likelihood estimation for pCTMCs.

4   Estimating Parameters from Partial Observations

In this section, we present an algorithm to estimate the parameters of a pCTMC
$\mathcal{P}$ from a collection of i.i.d. observation sequences $\mathcal{O} = \mathbf{o}_1, \ldots, \mathbf{o}_J$. Notably, the
algorithm is devised to be robust to missing dwell time values. In this line,
we consider *partial observations* of the form $p_{0:k}, \tau_{0:k-1}$ representing a finite
sequence $p_0 \tau_0 \cdots \tau_{k-1} p_k$ of consecutive dwell time values and atomic propositions
observed during a random execution of $\mathcal{M}$. Here, for uniformity of treatment,
the dwell times τ_t that are missing are denoted as $\tau_t = \emptyset$.

We follow a maximum likelihood approach: the parameters $\mathbf{x}$ are estimated
to maximize the joint likelihood $\mathcal{L}(\mathcal{P}(\mathbf{x})|\mathcal{O}) = \prod_{j=1}^{J} l(\mathbf{o}_j|\mathcal{P}(\mathbf{x}))$ of the observed
data. When $\mathcal{P}$ and $\mathcal{O}$ are clear from the context, we write $\mathcal{L}(\mathbf{x})$ for the joint
likelihood and $l(\mathbf{o}|\mathbf{x})$ for the likelihood of the observation $\mathbf{o}$.

According to the assumption that some dwell time values may be missing,
the likelihood of a partial observation $\mathbf{o} = p_{0:k}, \tau_{0:k-1}$ for a generic CTMC $\mathcal{M}$ is

$$l(\mathbf{o}|\mathcal{M}) = \sum_{s_{0:k}} P[S_{0:k} = s_{0:k}, L_{0:k} \ni p_{0:k}|\mathcal{M}] \cdot l(S_{0:k} = s_{0:k}, T_{0:k-1} = \tau_{0:k-1}|\mathcal{M})$$

$$= \sum_{s_{0:k}} \left([\![\ell(s_{0:k}) \ni p_{0:k}]\!] \prod_{t=0}^{k-1} R(s_t, s_{t+1})/E(s_t) \right) \left(\prod_{t \in \mathcal{T}(\mathbf{o})} E(s_t) e^{-E(s_t)\tau_t} \right), \quad (3)$$

where $\mathcal{T}(\mathbf{o}) = \{t \mid 1 \leq t < k, \tau_t \neq \emptyset\}$ denotes the subset of indices of the
observation $\mathbf{o}$ that correspond to actual dwell time measurement.

Our solution to the maximum likelihood estimation problem builds on the
MM optimization framework [31,32]. In this line, our algorithm starts with an
initial hypothesis $\mathbf{x}_0$ and iteratively improves the current hypothesis $\mathbf{x}_m$, in the
sense that the likelihood associated with the next hypothesis $\mathbf{x}_{m+1}$ enjoys the
inequality $\mathcal{L}(\mathbf{x}_m) \leq \mathcal{L}(\mathbf{x}_{m+1})$. The procedure terminates when the improvement
does not exceed a fixed threshold ϵ, namely when $\mathcal{L}(\mathbf{x}_m) - \mathcal{L}(\mathbf{x}_{m-1}) \leq \epsilon$.

Before proceeding with the formulation of the surrogate function, we find it
convenient to introduce some notation. Let $\mathcal{P} = (S, \rightarrow, s_0, \ell)$, we write f_ρ for the
rate function of a transition $\rho \in \rightarrow$, and write $s \rightarrow \cdot$ for the set of transitions
departing from $s \in S$.

Without loss of generality, we assume that the rate function f_ρ of a transition
is either a constant map, i.e., $f_\rho(\mathbf{x}) = c_\rho$ for some $c_\rho \geq 0$ or a map of the form
$f_\rho(\mathbf{x}) = c_\rho \prod_{i=1}^{n} x_i^{a_{\rho i}}$ for some $c_\rho > 0$ and $a_{\rho i} > 0$ for some $i \in \{1, \ldots, n\}$; we
write a_ρ for $\sum_{i=1}^{n} a_{\rho i}$. We denote by $\xrightarrow{c}$ the subset of transitions with constant
rate function and $\xrightarrow{\mathbf{x}}$ for the remaining transitions.

To maximize $\mathcal{L}(\mathbf{x})$ we propose to employ an MM algorithm based on the following surrogate function $g(\mathbf{x}|\mathbf{x}_m) = \sum_{i=1}^{n} g(x_i|\mathbf{x}_m)$ where

$$g(x_i|\mathbf{x}_m) = \sum_{\rho \in \xrightarrow{\mathbf{x}}} \xi_\rho a_{\rho i} \ln x_i - \sum_{s} \sum_{\rho \in s \xrightarrow{\mathbf{x}}} \frac{f_\rho(\mathbf{x}_m) a_{\rho i} \gamma_s}{a_\rho (x_{mi})^{a_\rho}} x_i^{a_\rho}. \tag{4}$$

Here the coefficients γ_s and ξ_ρ are respectively defined as

$$\gamma_s = \sum_{j=1}^{J} \sum_{t=0}^{k_j-1} \gamma_s^j(t) \left([\![\tau_t^j \neq \emptyset]\!] \tau_t^j + [\![\tau_t^j = \emptyset]\!] E_m(s)^{-1} \right) \tag{5}$$

$$\xi_\rho = \sum_{j=1}^{J} \sum_{t=0}^{k_j-1} \xi_\rho^j(t) \tag{6}$$

where $\gamma_s^j(t)$ denotes the likelihood that having observed $\mathbf{o}_j$ on a random execution of $\mathcal{P}(\mathbf{x}_m)$ the state $S_t = s$; and $\xi_\rho^j(t)$ is the likelihood that for such random execution the transition performed from state S_t is ρ.

The following theorem states that the surrogate function $g(\mathbf{x}|\mathbf{x}_m)$ is a minorizer of the log-likelihood relative to the observed dataset $\mathcal{O}$.

Theorem 1. *The surrogate function $g(\mathbf{x}|\mathbf{x}_m)$ minorizes $\ln \mathcal{L}(\mathbf{x})$ at $\mathbf{x}_m$ up to an irrelevant constant.*

By Theorem 1 and the fact that the logarithm is an increasing function, we obtain that the parameter valuation that achieves the maximum of $g(\mathbf{x}|\mathbf{x}_m)$ improves the current hypothesis $\mathbf{x}_m$ relative to likelihood function $\mathcal{L}(\mathbf{x})$.

Corollary 1. *Let $\mathbf{x}_{m+1} = \arg\max_{\mathbf{x}} g(\mathbf{x}|\mathbf{x}_m)$, then $\mathcal{L}(\mathbf{x}_m) \leq \mathcal{L}(\mathbf{x}_{m+1})$.*

The surrogate function $g(\mathbf{x}|\mathbf{x}_m)$ is easier to maximize than $\mathcal{L}(\mathbf{x})$ because its parameters are separated. Indeed, maximization of $g(\mathbf{x}|\mathbf{x}_m)$ is done by point-wise maximization of each univariate function $g(x_i|\mathbf{x}_m)$. This has two main advantages: first, it is easier to handle high-dimensional problems [31,32]; second, one can choose to fix the value of some parameters and perform the maximization of $g(\mathbf{x}|\mathbf{x}_m)$ only on the corresponding subexpressions $g(x_i|\mathbf{x}_m)$.

The maxima of $g(x_i|\mathbf{x}_m)$ are found among the *non-negative* roots[1] of the polynomial function $P_i \colon \mathbb{R} \to \mathbb{R}$

$$P_i(y) = \sum_{s} \sum_{\rho \in s \xrightarrow{\mathbf{x}}} \frac{f_\rho(\mathbf{x}_m) a_{\rho i} \gamma_s}{(x_{mi})^{a_\rho}} y^{a_\rho} - \sum_{\rho \in \xrightarrow{\mathbf{x}}} \xi_\rho a_{\rho i} \tag{7}$$

Remark 2. There are some cases when (7) admits a closed-form solution. For instance, when the parameter index i satisfies the property $\forall \rho \in \xrightarrow{\mathbf{x}}. a_{\rho i} > 0 \implies a_\rho = C$ for some constant $C \in \mathbb{N}$, then maximization of $g(x_i|\mathbf{x}_m)$ leads to the following update

$$x_{(m+1)i} = \left[\frac{(x_{mi})^C \sum_{\rho \in \xrightarrow{\mathbf{x}}} \xi_\rho a_{\rho i}}{\sum_{s} \sum_{\rho \in s \xrightarrow{\mathbf{x}}} f_\rho(\mathbf{x}_m) a_{\rho i} \gamma_s} \right]^{1/C}$$

---

[1] Note that P_i always admits non-negative roots. Indeed, $P_i(0) \leq 0$ and $P_i(M) > 0$ for $M > 0$ sufficiently large. Therefore, by the intermediate value theorem, there exists $y_0 \in [0, M]$ such that $P_i(y_0) = 0$.

A classic situation when the above condition is fulfilled occurs when all transitions ρ where x_i appear (i.e., $a_{\rho i} > 0$), the transition rate is $f_\rho(\mathbf{x}) = c_\rho x_i$ (i.e., $a_{\rho i} = a_\rho = 1$). In that case, the above equation simplifies to

$$x_{(m+1)i} = \frac{\sum_{\rho \xrightarrow{x} } \xi_\rho}{\sum_s \sum_{\rho \in s \xrightarrow{x} } c_\rho \gamma_s}$$

For example, the pCTMC associated with the SIR models in Fig. 1 satisfies the former property for all parameters, because all transition rates are expressions either of the form $c \cdot \mathtt{plock} \cdot \mathtt{beta}$ or the form $c \cdot \mathtt{plock} \cdot \mathtt{gamma}$ for some constant $c > 0$. Furthermore, if we fix the value of the parameter $\mathtt{plock}$ the remaining parameters satisfy the latter property. In Sect. 6, we will take advantage of this fact for our calculations. $\qquad\square$

Finally, we show how to compute $\gamma_s^j(t)$ and $\xi_\rho^j(t)$ w.r.t. the observation $\mathbf{o}_j = p_0^j \tau_0^j \cdots \tau_{k_j-1}^j p_{k_j}^j$ by using standard forward and backward procedures. We define the forward function $\alpha_s^j(t)$ and the backward function $\beta_s^j(t)$ respectively as

$$\alpha_s^j(t) = l(L_{0:t} \ni p_{0:t}^j, T_{0:t} = \tau_{0:t}^j, S_t = s | \mathcal{P}(\mathbf{x}_m)), \text{ and}$$
$$\beta_s^j(t) = l(L_{t+1:k_j} \ni p_{t+1:k_j}^j, T_{t+1:k_j-1} = \tau_{t+1:k_j-1}^j | S_t = s, \mathcal{P}(\mathbf{x}_m)).$$

These can be computed using dynamic programming according to the following recurrences. Let $\mathcal{P}(\mathbf{x}_m) = (S, R, s_0, \ell)$, then

$$\alpha_s^j(t) = \begin{cases} [\![s = s_0]\!] \, \omega_s^j(t) & \text{if } t = 0 \\ \omega_s^j(t) \sum_{s' \in S} \frac{R(s',s)}{E(s')} \, \alpha_{s'}^j(t-1) & \text{if } 0 < t \le k_j \end{cases} \qquad (8)$$

$$\beta_s^j(t) = \begin{cases} 1 & \text{if } t = k_j \\ \sum_{s' \in S} \frac{R(s,s')}{E(s)} \, \beta_{s'}^j(t+1) \, \omega_{s'}^j(t+1) & \text{if } 0 \le t < k_j \end{cases} \qquad (9)$$

where

$$\omega_s^j(t) = \begin{cases} [\![\ell(s) \in p_t^j]\!] E(s) e^{-E(s)\tau_t^j} & \text{if } 0 \le t < k_j \text{ and } \tau_t^j \ne \emptyset \\ [\![\ell(s) \in p_t^j]\!] & \text{if } t = k_j \text{ or } \tau_t^j = \emptyset. \end{cases} \qquad (10)$$

Finally, for $s \in S$ and $\rho = (s \xrightarrow{f_\rho} s')$, $\gamma_s^j(t)$ and $\xi_\rho^j(t)$ are related to the forward and backward functions as follows

$$\gamma_s^j(t) = \frac{\alpha_s^j(t) \, \beta_s^j(t)}{\sum_{s' \in S} \alpha_{s'}^j(t) \, \beta_{s'}^j(t)}, \quad \xi_\rho^j(t) = \frac{\alpha_s^j(t) f_\rho(\mathbf{x}_m) \, \omega_{s'}^j(t+1) \, \beta_{s'}^j(t+1)}{E(s) \sum_{s'' \in S} \alpha_{s''}^j(t) \, \beta_{s''}^j(t)}. \qquad (11)$$

The Case of Non-timed Observations. Consider the limit situation when dwell time variables are not observable (i.e., $\tau_t^j = \emptyset$ for all $j = 1 \ldots J$ and $t = 1 \ldots k_j - 1$). Under this assumption, two CTMCs $\mathcal{M}_1$ and $\mathcal{M}_2$ having the same embedded Markov chain satisfy $\mathcal{L}(\mathcal{M}_1 | \mathcal{O}) = \mathcal{L}(\mathcal{M}_2 | \mathcal{O})$. In other words, when dwell time variables are not observable the MLE objective does not fully capture the continuous-time aspects of the model under estimation.

The next section provides experimental evidence that, when the number of parametric transitions is sufficiently small relative to that of constant transitions, our algorithm can hinge on the value of the transition rates that are fixed, leading the procedure to converge to the real parameter values.

5 Experimental Evaluation

We implemented the algorithm from Sect. 4 as an extension of the `Jajapy` Python library [39], which has the advantage of being compatible with PRISM models. In this section, we present an empirical evaluation of the efficiency of our algorithm as well as the quality of their outcome. To this end, we employ a selection of CTMCs from the QComp benchmark set [22]. Experiments on each model have been designed according to the following setup.

For each model, we selected a set of parameters to be estimated as well as the set of observable atomic propositions[2]. We then estimated the parameter values from a training set consisting of 100 observation sequences of length 30, generated by simulating the original benchmark model. After the estimation, we verify all the formulas associated with the given benchmark model and compare the result with the expected one.

We perform experiments both using timed and non-timed observations. Each experiment is repeated 10 times by randomly re-sampling the initial parameter values $\mathbf{x}_0$ in the range $[0.00025, 0.0025]$. We annotate the running time, the relative error δ_i for each parameter x_i, and the relative error Φ_i for each formula[3].

Table 1. Performance comparison on selected QComp benchmarks [22].

Model	$|S|$	$|\rightarrow|$	Timed Observations				Non-timed Observations			
			Time(s)	Iter	avg δ	avg Φ	Time(s)	Iter	avg δ	avg Φ
polling	240	800	136.430	4	0.053	0.421	33.743	12	1.000	7.146
cluster	276	1120	132.278	3	0.089	1.293	279.853	12	0.313	3.827
tandem	780	2583	1047.746	3	0.043	0.544	4302.197	74	0.161	1.354
philosophers (i)	1065	4141	2404.803	3	0.043	0.119	2232.706	6	0.263	0.235
philosophers (ii)	1065	4141	9865.645	12	0.032	0.026	33265.151	200	0.870	2.573

Table 1 reports the aggregated results of the experiments. The columns $|S|$ and $|\rightarrow|$ provide respectively the number of states and transitions of the model;

---

[2] The models are available at https://github.com/Rapfff/MM-PCTMC-benchmark-models. The source files contain a description of the parameters and what is observable.

[3] The relative error is $|e - r|/|r|$, where e (resp. r) is the estimated (resp. real) value.

the columns "Time" and "Iter" respectively report the average running time[4] and number of iterations; and the columns "avg δ" and "avg Φ" respectively report the average relative error of the estimated parameters and model checking outcomes. Unsurprisingly, the quality of the estimation is higher for timed observations. Despite in most cases the initial parameter valuation $\mathbf{x}_0$ being picked far from the real parameter values, our method is capable to get close to the expected parameter values by using relatively few observation sequences. Most of the formulas employed in the experiments compute expected accumulated rewards for a time horizon exceeding that of the used training set, as a consequence, also the error tends to build up. The issue can be tamed by having longer observations in the training set. Notably, for timed observations, each iteration is more expensive than non-timed ones, but the additional overhead is largely compensated by a consistently smaller number of iterations.

To understand how, for non-timed observation, the quality of the estimation varies based on the number of constant transitions we ran our algorithm on two variants of the philosophers model: (i) with the variable gammax as a constant; and (ii) with gammax as a parameter. The algorithm clearly benefits from the presence of constant transitions and it converges way faster to better estimates.

Figure 2 reports the results of the experiments performed on the tandem queueing network model from [23] for different sizes of the queue. Each experiment was repeated 10 times by randomly re-sampling the initial valuation $\mathbf{x}_0$ in the interval $[0.1, 5.0]$. Accordingly, measurements are presented together with their respective error bars. The graph of the running time (cf. Fig. 2 bottom) follows a quadratic curve in the number of states both for timed and non-timed observations. However, for non-timed observations, the variance of the measured running times tends to grow with the size of the model. Figure 2 (top) shows how the L_1-norm (resp. L_∞-norm) of the vector $\delta = (\delta_i)$ may vary for different size of the model. The variance of the measured relative errors is larger in the experiments performed with non-timed observations. Notably, for timed observations, the quality of the estimation remained stable despite the size of the model increased relative to the size of the training set. This may be explained by the fact that, in the tandem model, the parameters occur in many transitions.

6   Case Study: SIR Modeling of Pandemic

In this section, we take as a case study the modeling pipeline proposed by Milazzo [36] for the analysis and simulation in PRISM of the spread of COVID-19 in presence of lockdown countermeasures. The modeling pipeline includes: (i) parameter estimation from real data based on a modified SIR model described by means of a system of ODEs; (ii) encoding of the modified SIR model into as a PRISM model; and (iii) stochastic simulation and model checking with PRISM.

The model devised in step (ii) is depicted in Fig. 1 (left). However, to perform the analysis, Milazzo had to apply "a couple of modeling tricks (variable prun-

---

[4] Experiments were performed on a Linux machine with an AMD-Ryzen 9 3900X 12-Core processor and 32 GB of RAM.

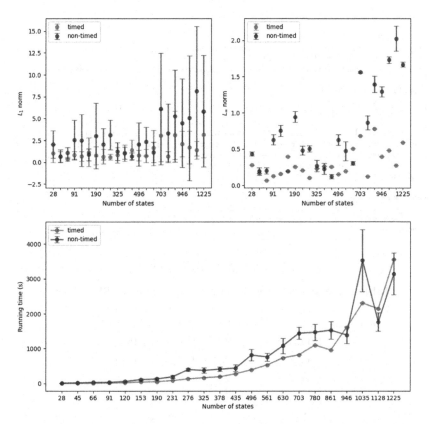

Fig. 2. Comparison of the performance of the estimation for timed and non-timed observations on the tandem queueing network with different size of the queue.

ing and upper bounds) that allowed state space of the model [..] to be reduced by several orders of magnitude. " [36]. These kinds of modeling tricks are not uncommon in formal verification, but they require the modeler to ensure that the parameter values estimated for the original model are still valid in the approximated one. In this section, we showcase the use of our algorithm to simplify this task. Specifically, we generate two training sets by simulating the SIR model in Fig. 1 using PRISM and, based on that, we re-estimate beta, gamma, and plock on an approximated version of the model (*cf.* Fig. 3).

The first training set represents the spread of the disease without lockdown (i.e., plock = 1), while the second one is obtained by fixing the value of plock estimated in [36] (i.e., plock = 0.472081). In line with the data set used in [36], both training sets consist of one (timed) observation reporting the number of infected individuals for a period of 30 days.

The estimation of the parameters beta, gamma and plock is performed on the model depicted in Fig. 3. As in [36], we use an approximated version of the original SIR model (*cf.* Fig. 1) obtained by employing a few modeling tricks:

```
ctmc
  // bounds
  const int ubound_i; const int lbound_i; const int nb_r = 10;
  const double size_r = 500/nb_r; const int SIZE = 100000;
  const double beta; const double gamma; const double plock; // SIR model parameters

  module SIR
      i : [lbound_i..ubound_i] init 48;
      r : [0..nb_r - 1] init 0;

      [infection] i>0 & i <ubound_i  →  i * (SIZE - (i + (r + 0.5) * size_r)) : (i'=i + 1);
      [recovery] i>0 & r<nb_r - 1  →  i * ((size_r) - 1)/(size_r) : (i'=i - 1);
      [recovery] i>0 & r<nb_r - 1  →  i * 1/(size_r) : (r'=r + 1) & (i'=i - 1);
      [recovery] i>0 & r=nb_r - 1  →  i : (i'=i - 1);
  endmodule

  module Rates
      [infection] true  →  beta * plock/SIZE : true;
      [recovery] true  →  gamma * plock : true;
  endmodule
```

Fig. 3. Approximated SIR model.

Table 2. Parameter estimation on the approximated SIR model.

Parameter	Expected Value	Estimated Value	Absolute Error
beta	0.122128	0.135541	0.013413
gamma	0.127283	0.128495	0.001212
plock	0.472081	0.437500	0.034581

variable pruning, set upper bounds on the state variable i, and re-scaling of the variable r in the interval $[0, nb\_r - 1]$. These modeling tricks have the effect to reduce the state space of the underlying CTMC, speeding-up in this way parameter estimation and the following model analysis.

We perform the estimation in two steps. First, we estimate the values of beta and gamma on the first training set with plock set to 1 (i.e., with no restrictions). Then, we estimate the value of plock on the second training set with beta and gamma set to the values estimated in the first step. Each step was repeated 10 times by randomly re-sampling the initial values of each unknown parameter in the interval $[0, 1]$. Table 2 reports the average estimated values and absolute errors relative to each parameter. The running time of each estimation was on average 89.94 seconds[5].

Notably, we were able to achieve accurate estimations of all the parameters from training sets consisting of a single partially-observable execution of the original SIR model. As observed in Sect. 5, this may be due to the fact that each parameter occurs in many transitions.

This case study demonstrates that our estimation procedure can be effectively used to simplify modeling pipelines that involve successive modifications of the model and the re-estimation of its parameter values.

---

[5] Experiments were performed on a Linux machine with an AMD-Ryzen 9 3900X 12-Core processor and 32 GB of RAM.

7   Related Work

Literature on parameter estimation for CTMCs follows two approaches. The first approach is based on Bayesian inference and assumes a probability distribution over parameters which in turn produces an *uncertain* CTMC [5,20]. In this line of work, Georgoulas et al. [19,20] proposed ProPPA, a stochastic process algebra with inference capabilities. Using probabilistic inference, the ProPPA model is combined with the observations to derive updated probability distributions over rates. Uncertain pCTMCs require dedicated model checking techniques [5].

The second approach aims at estimating parameter values producing concrete CTMCs via maximum likelihood estimation. In this line, Geisweiller proposed EMPEPA [18], an expectation-maximization algorithm that estimates the rate values inside a PEPA model. Wei et al. [43] learn the infinitesimal generator of a continuous-time hidden Markov model by first employing the Baum-Welch algorithm [38] to estimate the transition probability matrix of its (embedded) hidden Markov model from a set of periodic observations.

A large body of literature studies parameter estimation for stochastic reaction networks (SRN) (*cf.* [35,40] and references therein). According to Gillespie's theory of stochastic chemical kinetics, SRNs can be represented using CTMCs with states in $\mathbb{N}^d$. An SRN describes the dynamics of a population of d chemical species by means of a number of *chemical reaction rules*. Notably, the SIR model in Fig. 1 was encoded from an SRN. The parameter estimation problem for SRNs focuses on estimating the rate values associated with each reaction rule. Due to the nature of the models, estimation algorithms for SRNs aim at being able to scale on the number of species as well as the size of the population. In this respect, (i) Andreychenko et al. [2] employs numerical approximations of the likelihood function (and its derivatives) w.r.t. reaction rate constants by dynamically truncating the state space in an on-the-fly fashion, considering only those states that significantly contribute to the likelihood in a given time interval, while (ii) Bayer et al. [8] combines the Monte Carlo version of the expectation-maximization algorithm [34] with the forward-reverse technique developed in [9] to efficiently simulate SRN bridges conditional on the observed data. Compared with our method, the estimation algorithms of [2,8] scale better in the number of species and population size. The main limitation of the efficiency of our method is the computation of the forward and backward functions, whose complexity grows quadratically in the number of states. In contrast with our technique both [2] and [8] assume to observe *all* the coordinates of the state (in [2] these are additionally assumed to be subject to Gaussian noise). In our opinion, this limits the applicability of their methods to scenarios where states are partially observed. Such an example is the case study of Sect. 6 where the available data set was only reporting the number of infected individuals (i.e., two components out of three were not observable). Daigle et al. [27] developed an efficient version of the Monte Carlo expectation-maximization technique which employs modified cross-entropy methods to account for rare events. Notably, their technique is executable also on data sets with missing species but, as for our algorithm, such flexibility comes at the expense of efficiency. As pointed out in [8], the techniques used in [27] have some analogies with those of [8], but its run-time performance is comparably slower than that of [8]: the algorithm of [27] took 8.7 days for the parameter estimation on an SRN describing an auto-regulatory gene network with five species, while the method of [8] took 2 days[6].

Sen et al. [41] presented a variant of ALERGIA [11] to learn a (transition-labeled) CTMC from timed observations. In contrast with our work, [41] does not perform parameter estimation over structured models, but learns an unstructured CTMC. Hence, [41] suits better for learning a single component CTMC when no assumption can be made on the structure or the size of the model.

Another related line of research is parameter synthesis of Markov models [26]. In particular, [12,21] consider parametric CTMCs, but are generally restricted to a few parameters. In contrast with our work, parameter synthesis revolves around the problem of finding (some or all) parameter instantiations of the model that satisfy a given logical specification.

8   Conclusion and Future Work

We presented a novel technique to estimate parameter values of CTMCs expressed as PRISM models from partially-observable executions. We demonstrated, with a case study, that our solution is a concrete aid in applications involving modeling and analysis, especially when the model under study requires successive approximations that require re-estimation of the parameters. The major strengths of our algorithm are (i) its interoperability with the model checking tools PRISMand STORM, and (ii) the fact that it accepts partially-observable data sets where both state and dwell times can be missing; However, the generality of our approach comes at the expense of efficiency. The computations of the forward and backward functions which are required to update the coefficients of the surrogate function (4) have a time and space complexity that grows quadratically in the number of states of the pCTMC, thus limiting the number of components that our implementation can currently handle. In future work, we consider investigating how to speed up the computation of the forward and backward functions either by integrating GPU-accelerated techniques from [37] or by replacing their exact computation in favor of numerical approximations obtained through Monte Carlo simulations in line with the idea employed in Monte Carlo EM algorithm [34].

Notably, the algorithm presented in this paper was devised following simple optimization principles borrowed from the MM optimization framework. We suggest that similar techniques can be employed to other modeling languages (e.g., Markov automata [16,17]) and metric-based approximate minimization [3, 7]. An interesting future direction of research consists in extending our techniques to Markov decision processes by integrating the active learning strategies [4].

---

[6] Details on the experiments can be found in the respective papers.

References

1. Alur, R., Henzinger, T.A.: Reactive modules. Formal Methods Syst. Des. **15**(1), 7–48 (1999). https://doi.org/10.1023/A:1008739929481
2. Andreychenko, A., Mikeev, L., Spieler, D., Wolf, V.: Parameter identification for Markov models of biochemical reactions. In: Gopalakrishnan, G., Qadeer, S. (eds.) CAV 2011. LNCS, vol. 6806, pp. 83–98. Springer, Heidelberg (2011). https://doi.org/10.1007/978-3-642-22110-1_8
3. Bacci, G., Bacci, G., Larsen, K.G., Mardare, R.: On the metric-based approximate minimization of Markov chains. In: Chatzigiannakis, I., Indyk, P., Kuhn, F., Muscholl, A. (eds.) 44th International Colloquium on Automata, Languages, and Programming, ICALP 2017, 10–14 July 2017, Warsaw, Poland. LIPIcs, vol. 80, pp. 104:1–104:14. Schloss Dagstuhl - Leibniz-Zentrum für Informatik (2017). https://doi.org/10.4230/LIPIcs.ICALP.2017.104
4. Bacci, G., Ingólfsdóttir, A., Larsen, K.G., Reynouard, R.: Active learning of markov decision processes using Baum-Welch algorithm. In: Wani, M.A., Sethi, I.K., Shi, W., Qu, G., Raicu, D.S., Jin, R. (eds.) 20th IEEE International Conference on Machine Learning and Applications, ICMLA 2021, pp. 1203–1208. IEEE (2021). https://doi.org/10.1109/ICMLA52953.2021.00195
5. Badings, T.S., Jansen, N., Junges, S., Stoelinga, M., Volk, M.: Sampling-based verification of CTMCs with uncertain rates. In: Shoham, S., Vizel, Y. (eds.) CAV 2022, Part II. LNCS, vol. 13372, pp. 26–47. Springer, Cham (2022). https://doi.org/10.1007/978-3-031-13188-2_2
6. Baier, C., Haverkort, B.R., Hermanns, H., Katoen, J.: Model-checking algorithms for continuous-time Markov chains. IEEE Trans. Software Eng. **29**(6), 524–541 (2003). https://doi.org/10.1109/TSE.2003.1205180
7. Balle, B., Lacroce, C., Panangaden, P., Precup, D., Rabusseau, G.: Optimal spectral-norm approximate minimization of weighted finite automata. In: Bansal, N., Merelli, E., Worrell, J. (eds.) 48th International Colloquium on Automata, Languages, and Programming, ICALP 2021, 12–16 July 2021, Glasgow, Scotland (Virtual Conference). LIPIcs, vol. 198, pp. 118:1–118:20. Schloss Dagstuhl - Leibniz-Zentrum für Informatik (2021). https://doi.org/10.4230/LIPIcs.ICALP.2021.118
8. Bayer, C., Moraes, A., Tempone, R., Vilanova, P.: An efficient forward-reverse expectation-maximization algorithm for statistical inference in stochastic reaction networks. Stoch. Anal. Appl. **34**(2), 193–231 (2016). https://doi.org/10.1080/07362994.2015.1116396
9. Bayer, C., Schoenmakers, J.: Simulation of forward-reverse stochastic representations for conditional diffusions. Ann. Appl. Probab. **24**(5), 1994–2032 (2014). https://doi.org/10.1214/13-AAP969
10. Calinescu, R., Ceska, M., Gerasimou, S., Kwiatkowska, M., Paoletti, N.: Efficient synthesis of robust models for stochastic systems. J. Syst. Softw. **143**, 140–158 (2018). https://doi.org/10.1016/j.jss.2018.05.013
11. Carrasco, R.C., Oncina, J.: Learning stochastic regular grammars by means of a state merging method. In: Carrasco, R.C., Oncina, J. (eds.) ICGI 1994. LNCS, vol. 862, pp. 139–152. Springer, Heidelberg (1994). https://doi.org/10.1007/3-540-58473-0_144

12. Češka, M., Dannenberg, F., Paoletti, N., Kwiatkowska, M., Brim, L.: Precise parameter synthesis for stochastic biochemical systems. Acta Informatica **54**(6), 589–623 (2016). https://doi.org/10.1007/s00236-016-0265-2
13. Ciocchetta, F., Hillston, J.: Bio-PEPA: a framework for the modelling and analysis of biological systems. Theor. Comput. Sci. **410**(33–34), 3065–3084 (2009). https://doi.org/10.1016/j.tcs.2009.02.037
14. Dehnert, C., Junges, S., Katoen, J.-P., Volk, M.: A storm is coming: a modern probabilistic model checker. In: Majumdar, R., Kunčak, V. (eds.) CAV 2017. LNCS, vol. 10427, pp. 592–600. Springer, Cham (2017). https://doi.org/10.1007/978-3-319-63390-9_31
15. Dempster, A.P., Laird, N.M., Rubin, D.B.: Maximum likelihood from incomplete data via the EM algorithm. J. Roy. Stat. Soc. **39**(1), 1–38 (1977)
16. Eisentraut, C., Hermanns, H., Zhang, L.: Concurrency and composition in a stochastic world. In: Gastin, P., Laroussinie, F. (eds.) CONCUR 2010. LNCS, vol. 6269, pp. 21–39. Springer, Heidelberg (2010). https://doi.org/10.1007/978-3-642-15375-4_3
17. Eisentraut, C., Hermanns, H., Zhang, L.: On probabilistic automata in continuous time. In: Proceedings of the 25th Annual IEEE Symposium on Logic in Computer Science, LICS 2010, 11–14 July 2010, Edinburgh, United Kingdom, pp. 342–351. IEEE Computer Society (2010). https://doi.org/10.1109/LICS.2010.41
18. Geisweiller, N.: Finding the most likely values inside a PEPA model according to partially observable executions. Ph.D. thesis, LAAS (2006)
19. Georgoulas, A., Hillston, J., Milios, D., Sanguinetti, G.: Probabilistic programming process algebra. In: Norman, G., Sanders, W. (eds.) QEST 2014. LNCS, vol. 8657, pp. 249–264. Springer, Cham (2014). https://doi.org/10.1007/978-3-319-10696-0_21
20. Georgoulas, A., Hillston, J., Sanguinetti, G.: Proppa: probabilistic programming for stochastic dynamical systems. ACM Trans. Model. Comput. Simul. **28**(1), 3:1–3:23 (2018). https://doi.org/10.1145/3154392
21. Han, T., Katoen, J., Mereacre, A.: Approximate parameter synthesis for probabilistic time-bounded reachability. In: Proceedings of the 29th IEEE Real-Time Systems Symposium, RTSS 2008, Barcelona, Spain, 30 November–3 December 2008, pp. 173–182. IEEE Computer Society (2008). https://doi.org/10.1109/RTSS.2008.19
22. Hartmanns, A., Klauck, M., Parker, D., Quatmann, T., Ruijters, E.: The quantitative verification benchmark set. In: Vojnar, T., Zhang, L. (eds.) TACAS 2019. LNCS, vol. 11427, pp. 344–350. Springer, Cham (2019). https://doi.org/10.1007/978-3-030-17462-0_20
23. Hermanns, H., Meyer-Kayser, J., Siegle, M.: Multi terminal binary decision diagrams to represent and analyse continuous time Markov chains. In: Plateau, B., Stewart, W., Silva, M. (eds.) Proceedings of 3rd International Workshop on Numerical Solution of Markov Chains (NSMC 1999), pp. 188–207. Prensas Universitarias de Zaragoza (1999)
24. Hillston, J.: A compositional approach to performance modelling. Ph.D. thesis, University of Edinburgh, UK (1994). http://hdl.handle.net/1842/15027
25. Jamshidian, M., Jennrich, R.I.: Acceleration of the EM algorithm by using quasi-newton methods. J. Roy. Stat. Soc. Ser. B (Methodol.) **59**(3), 569–587 (1997). http://www.jstor.org/stable/2346010

26. Jansen, N., Junges, S., Katoen, J.: Parameter synthesis in Markov models: a gentle survey. In: Raskin, J., Chatterjee, K., Doyen, L., Majumdar, R. (eds.) Principles of Systems Design. LNCS, vol. 13660, pp. 407–437. Springer, Cham (2022). https://doi.org/10.1007/978-3-031-22337-2_20

27. Daigle, B.J., Roh, M.K., Petzold, L.R., Niemi, J.: Accelerated maximum likelihood parameter estimation for stochastic biochemical systems. BMC Bioinform. **13**, 68 (2012). https://doi.org/10.1186/1471-2105-13-68

28. Kwiatkowska, M., Norman, G., Parker, D.: Stochastic model checking. In: Bernardo, M., Hillston, J. (eds.) SFM 2007. LNCS, vol. 4486, pp. 220–270. Springer, Heidelberg (2007). https://doi.org/10.1007/978-3-540-72522-0_6

29. Kwiatkowska, M.Z., Norman, G., Parker, D.: Using probabilistic model checking in systems biology. SIGMETRICS Perform. Evaluation Rev. **35**(4), 14–21 (2008). https://doi.org/10.1145/1364644.1364651

30. Kwiatkowska, M., Norman, G., Parker, D.: PRISM 4.0: verification of probabilistic real-time systems. In: Gopalakrishnan, G., Qadeer, S. (eds.) CAV 2011. LNCS, vol. 6806, pp. 585–591. Springer, Heidelberg (2011). https://doi.org/10.1007/978-3-642-22110-1_47

31. Lange, K.: Optimization, 2nd edn. Springer, New York (2013). https://doi.org/10.1007/978-1-4614-5838-8

32. Lange, K.: MM Optimization Algorithms. SIAM (2016). http://bookstore.siam.org/ot147/

33. Lazowska, E.D., Zahorjan, J., Graham, G.S., Sevcik, K.C.: Quantitative System Performance - Computer System Analysis Using Queueing Network Models. Prentice Hall, Hoboken (1984)

34. Levine, R.A., Casella, G.: Implementations of the Monte Carlo EM algorithm. J. Comput. Graph. Stat. **10**(3), 422–439 (2001). http://www.jstor.org/stable/1391097

35. Loskot, P., Atitey, K., Mihaylova, L.: Comprehensive review of models and methods for inferences in bio-chemical reaction networks. Front. Genet. **10** (2019). https://doi.org/10.3389/fgene.2019.00549

36. Milazzo, P.: Analysis of COVID-19 data with PRISM: parameter estimation and SIR modelling. In: Bowles, J., Broccia, G., Nanni, M. (eds.) DataMod 2020. LNCS, vol. 12611, pp. 123–133. Springer, Cham (2021). https://doi.org/10.1007/978-3-030-70650-0_8

37. Ondel, L., Lam-Yee-Mui, L.M., Kocour, M., Corro, C.F., Burget, L.: GPU-accelerated forward-backward algorithm with application to lattice-free mmi. In: ICASSP 2022-2022 IEEE International Conference on Acoustics, Speech and Signal Processing (ICASSP), pp. 8417–8421 (2022). https://doi.org/10.1109/ICASSP43922.2022.9746824

38. Rabiner, L.R.: A tutorial on hidden Markov models and selected applications in speech recognition. Proc. IEEE **77**(2), 257–286 (1989). https://doi.org/10.1109/5.18626

39. Reynouard, R.: Jajapy (v 0.10) (2022). https://github.com/Rapfff/jajapy

40. Schnoerr, D., Sanguinetti, G., Grima, R.: Approximation and inference methods for stochastic biochemical kinetics-a tutorial review. J. Phys. A Math. Theor. **50**(9), 093001 (2017). https://doi.org/10.1088/1751-8121/aa54d9

41. Sen, K., Viswanathan, M., Agha, G.: Learning continuous time Markov chains from sample executions. In: 1st International Conference on Quantitative Evaluation of Systems (QEST 2004), pp. 146–155. IEEE Computer Society (2004). https://doi.org/10.1109/QEST.2004.1348029

42. Terwijn, S.A.: On the learnability of hidden Markov models. In: Adriaans, P., Fernau, H., van Zaanen, M. (eds.) ICGI 2002. LNCS (LNAI), vol. 2484, pp. 261–268. Springer, Heidelberg (2002). https://doi.org/10.1007/3-540-45790-9_21
43. Wei, W., Wang, B., Towsley, D.F.: Continuous-time hidden Markov models for network performance evaluation. Perform. Evaluation **49**(1/4), 129–146 (2002)
44. Zhou, H., Alexander, D.H., Lange, K.: A quasi-Newton acceleration for high-dimensional optimization algorithms. Stat. Comput. **21**(2), 261–273 (2011). https://doi.org/10.1007/s11222-009-9166-3

STAMINA in C++: Modernizing an Infinite-State Probabilistic Model Checker

Joshua Jeppson[1]([⊠])(ID), Matthias Volk[2](ID), Bryant Israelsen[1](ID), Riley Roberts[1](ID), Andrew Williams[1](ID), Lukas Buecherl[3](ID), Chris J. Myers[3](ID), Hao Zheng[4](ID), Chris Winstead[1](ID), and Zhen Zhang[1]([⊠])(ID)

[1] Utah State University, Logan, UT, USA
{joshua.jeppson,bryant.israelsen,
chris.winstead,zhen.zhang}@usu.edu
[2] University of Twente, Enschede, The Netherlands
m.volk@utwente.nl
[3] University of Colorado Boulder, Boulder, CO, USA
{lukas.buecherl,chris.myers}@colorado.edu
[4] University of South Florida, Tampa, FL, USA
haozheng@usf.edu

Abstract. Improving the scalability of probabilistic model checking (PMC) tools is crucial to the verification of real-world system designs. The STAMINA infinite-state PMC tool achieves scalability by iteratively constructing a partial state space for an unbounded continuous-time Markov chain model, where a majority of the probability mass resides. It then performs time-bounded transient PMC. It can efficiently produce an accurate probability bound to the property under verification. We present a new software architecture design and the C++ implementation of the STAMINA 2.0 algorithm, integrated with the STORM model checker. This open-source STAMINA implementation offers a high degree of modularity and provides significant optimizations to the STAMINA 2.0 algorithm. Performance improvements are demonstrated on multiple challenging benchmark examples, including hazard analysis of infinite-state combinational genetic circuits, over the previous STAMINA implementation. Additionally, its design allows for future customizations and optimizations to the STAMINA algorithm.

Keywords: Probabilistic Model Checking · Infinite-state Systems · Markov Chains

1 Introduction

Continuous-time Markov Chain (CTMC) can represent real-time probabilistic systems (e.g., genetic circuits [15], *Dynamic Fault Trees* (DFTs) [20]). Unfortunately, *probabilistic model checking* (PMC) may not always be feasible due to their infinite or finite but large state spaces. For instance, one might prefer to have unbounded species molecule count in a genetic circuit model due to insufficient information at design time, which results in an infinite state space. Furthermore, the required explicit-state representation challenges scalable CTMC numerical analysis for verifying time-bounded transient properties. Existing PMC tools such as PRISM [13] and STORM [9] can efficiently analyze reasonably-sized *finite-state* CTMCs. INFAMY [8] approximates state-spaces with breadth-first search to some depth k and truncates any state beyond it. STAR [14] and SeQuaiA [5] approximate the most probable behavior for population Markov models of biochemical reaction networks. The STORM-DFT library [20] implements an approximation algorithm for DFTs based on partial state space generation of CTMCs.

STAMINA [16,17,19] performs *on-the-fly* state truncation using estimated state reachability probability to enable efficient PMC of CTMCs with an extremely large or infinite state space. STAMINA differentiates from the aforementioned techniques as follows: It can analyze both bounded and unbounded CTMCs and is not restricted to specific input models such as DFTs or biochemical population models. Specifically, DFTs typically incur an acyclic state space, whereas cycles commonly exist in genetic circuit models. Also, STAMINA does not truncate the state space with a fixed depth.

Contributions. This paper presents a new software architecture of STAMINA and a first re-implementation in C++ that interfaces the STORM model checker. Modularity plays a central role in STAMINA at both the core model builder class and other specialized classes. It allows optimized heuristics for state-space truncation and clean compartmentalization of functionality. Additionally, it includes memory optimizations tailored to the STAMINA 2.0 algorithm [19]. STAMINA demonstrated marked performance improvements on multiple challenging benchmarks, including hazard analysis of infinite-state genetic circuits, over the previous STAMINA implementation, due to both the speed of STORM's internals, and increased opportunities for optimization within STAMINA during its integration with STORM. STAMINA and STORM are both licensed under the GPLv3 license.

2 Overview of STAMINA

STAMINA takes an *unbounded* CTMC model in the PRISM modeling language and generates and truncates its state space where the probability mass resides. It calculates a probability bound $[P_{min}, P_{max}]$ based on a partial state space. STAMINA 1.0 [16,17] truncates the state space by preventing expansion of states whose reachability fall below a threshold κ; and 2.0 [19] supports state re-exploration to obtain more accurate probability estimates. Both are written in Java and interface with the PRISM model checker. STAMINA has now been rewritten in C++ to integrate with STORM. It is available on GitHub and https://staminachecker.org with extensive documentation. It is now available via REST API for web-based usage.

The state truncation algorithm in STAMINA 1.0 [16,17] performs breadth-first state expansion in multiple iterations. In every iteration, it terminates a state-transition path exploration *on-the-fly* when the estimated state reachability probability $\hat{\pi}(s)$ of state s, also called *terminal state*, falls below a user-specified bound κ. It walks through the explored state space at the end of each iteration to find all terminal states to be re-explored in the next iteration. It repeats this process until the change in state space size between iterations becomes sufficiently small. This method has the following main drawbacks: (1) Inefficiency in repeatedly re-exploring the state space to find terminal states, and (2) inaccurate state space truncation due to its inability to update state reachabilities of previously visited non-terminal states in each iteration. STAMINA then interfaces the PRISM CTMC transient analysis engine to compute a probability bound $[P_{min}, P_{max}]$ that encloses the actual probability of the property under verification. STAMINA accepts non-nested time-bounded until CSL formulas. It computes P_{max} by assuming the unexplored state space, abstracted as one artificial absorbing state, *satisfies* the property and P_{min} by assuming it *violates* the property. If $P_{max} - P_{min} \nleq w$, STAMINA reduces κ in order to further expand the state space until $P_{max} - P_{min} \leq w$. The user-specified probability window w allows the user to control the tightness of the result.

The STAMINA 2.0 algorithm [19] balances re-exploring states to distribute their estimated state reachability probabilities $\hat{\pi}$ and the number of states to re-explore. Specifically, it addresses both drawbacks in STAMINA 1.0 by repeatedly pushing the updated $\hat{\pi}(s)$ of state s to its successor states sufficiently frequently so that it prevents $\hat{\pi}$ from being ignored for states in a cycle in the state space. To optimize the performance, it limits the number of states to re-explore by setting a smaller amount to reduce κ in each iteration. It also reduces the expensive CTMC model checking procedure from two to one. Additionally, STAMINA 2.0 provides reasonable defaults for both κ and w, relieving the end-user's burden in guessing their appropriate values. The STAMINA tool presented in this work interfaces the STORM model checker, described next.

The STORM Model Checker. STORM [9] is a probabilistic model checker that supports the analysis of discrete- and continuous-time variants of both Markov chains and Markov decision processes. STORM is written in C++, open-source, and publicly available at stormchecker.org. It is one of the state-of-the-art tools for probabilistic model checking as witnessed in the Quantitative Verification Competition (QComp) [4]. The performance of STORM stems from its modular design where solvers, model representations, and model checking algorithms can easily be exchanged and combined. Lastly, STORM provides a rich C++ API which allows fine-granular access to its model checking algorithms and underlying datastructures. This allows other tools such as STAMINA to tightly integrate with STORM and benefit from its model checking performance.

Models such as CTMCs can be provided to STORM in many modelling languages including the PRISM and JANI modelling languages, DFTs, or generalized stochastic Petri nets. The Markov models can be built both in an explicit matrix-based or a symbolic decision-diagram-based representation. STAMINA uses the explicit representation based on an efficient sparse matrix data structure. State space generation within STORM is performed via the `NextStateGenerator` interface which iteratively generates successors states for a given state. Analysis of CTMC models can be performed with respect to CSL using efficient algorithms [3]. Time-bounded properties are solved via transient analysis of the CTMC using uniformisation and the approach by Fox and Glynn [7] for approximating Poisson probabilities. Time-unbounded properties are solved on the underlying discrete-time model using algorithms such as value iteration or via linear equation system solvers.

3 Software Architecture

STAMINA has been modularized as shown in Fig. 1. It has a CLI entry point, but STAMINA's modularity allows alternate entry points into the program, which in the future will facilitate rapid development of a GUI. This entry point instantiates an instance of the Stamina class, which contains all of the information needed to run STAMINA. The static Options class allows the CLI and GUI to specify input parameters, such as w and κ. From there, the Stamina class instantiates one ModelModify and StaminaModelChecker objects, whose functions are explained below.

Model Building. The core of STAMINA is the StaminaModelBuilder base class. Although only one type is presented in this paper, this polymorphism allows multiple types of model builders, which reuse code for shared functionality. Since STORM requires that the CTMC transition rate matrix entries to be inserted in order, this base class orders transitions, connects terminal states to the artificial absorbing state, and keeps track of all other data needed for model checking with STORM. It employs a ProbabilityState data type to store the estimated state reachability probability and the state status, including whether the state is new, terminal, or deadlock.

To support different types of model builders, StaminaModelBuilder easily allows sub-classes to explore (and re-explore) states without enforcing an exploration order. For example, StaminaReExploringModelBuilder implements the algorithm in STAMINA 2.0 discussed in this paper. While it's exploration order is similar to breadth-first, other sub-classes, such as StaminaIterativeModelBuilder and its cousin StaminaPriorityModelBuilder explore states in a different order by nature of their heuristics. Additionally, generic design of the StaminaModelBuilder can eventually allow multithreading, which is under active development.

Specialized Classes. A number of specialized classes are delegated certain actions in order to compartmentalize functionality and increase modularity. The ModelModify class takes the non-nested time-bounded CSL property on time interval I, $\mathsf{P}_{=?}(\varphi U^I \phi)$ under verification, extracts the path formula ϕ, and automatically generates $\phi \wedge \neg \hat{s}$ and $\phi \vee \hat{s}$, respectively. $\hat{s}$ is the artificial absorbing state created by STAMINA. These are passed to the StaminaModelChecker class responsible for calling STORM to compute $\mathsf{P}_{min} = \mathsf{P}_{=?}(\varphi U^I \phi \wedge \neg \hat{s})$ and $\mathsf{P}_{max} = \mathsf{P}_{=?}(\varphi U^I \phi \vee \hat{s})$, respectively. Finally, there are static StateSpaceInformation and StaminaMessages classes which read information from the generated state-space and write log messages respectively.

Memory Optimization. Throughout its execution, STAMINA creates many small state-probability objects in memory. While in Java this is fairly fast (due to the built-in memory management), in C++ this is delegated to the OS, making it slower. One solution to this is to use a memory pool. Rather than using a memory-pool library data structure, STAMINA is cognisant of object lifetime. Since each state exists for the lifetime of the entire state space, the memory pool does not have a method to deallocate a single state, reducing bookkeeping and CPU usage.

The custom memory pool (StateMemoryPool) is optimized for the STAMINA algorithm. It makes one allocation request to the OS per "page" of size $B_{mp} =$

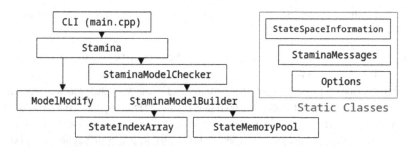

Fig. 1. Architecture of STAMINA: Arrows indicate object instantiations and data ownership. Classes in the red box are static and accessible everywhere. (Color figure online)

$2^{b_{mp}} \times$ sizeof(s)—where b_{mp} is an integer chosen at compile time based on system resources—and allocating the next B_{mp} instances from that block. These requests only occur when a particular "page" is full. B_{mp} is chosen to be reasonably large so that such requests do not happen frequently during state-space construction when state instances are allocated. All states are deallocated the end of StateMemoryPool's lifetime. This, in tandem with the StateIndexArray class, allows STAMINA to access states by index in $O(N/B_{mp})$. Due to the high value of B_{mp}, while linear, this averages close to constant time on most models.

While some optimizations, such as keeping active tally of terminal states for w-estimate, are more general, many are possible due to differences in STORM and PRISM. PRISM provides an easy-to-use Java API which does most of the work under the hood, and STORM exposes more of its internals, allowing STAMINA to optimize more extensively. We acknowledge the importance of both tools during STAMINA's development; PRISM's coherent Java API allowed the first two versions of STAMINA to be developed quickly, and STORM's modularity was the inspiration for STAMINA presented in this paper.

4   Results

All results were obtained on an AMD Ryzen Threadripper machine with a 16-core 3.5 GHz processor with 64 GB of RAM, running Debian 11 (Linux 5.10). Default parameters were used for all user-specifiable variables except where indicated. Both versions of STAMINA limit the probability window ($w = P_{max} - P_{min}$) to be $w \leqslant 10^{-3}$ for all models, as was used in [19]. The STAMINA 2.0 algorithm was tested in both STAMINA/PRISM and STAMINA/STORM for comparison. Note that STAMINA 1.0 was not implemented in STAMINA/STORM, because of the marked advantage of STAMINA 2.0 already demonstrated in [19]. The source code for STAMINA is available at [1].

Hazard Analysis in Genetic Circuits. The emerging *genetic design automation* (GDA) tools enable the design of genetic circuits. However, due to the inherent noisy behavior of biological systems, the predictability of genetic circuits remains largely unaddressed. The state space of models of genetic circuit designs are infinite, and therefore

Table 1. Probability comparisons. Digits in bold font show differences in STAMINA/PRISM and STAMINA/STORM.

Model	STAMINA/PRISM 2.0			STAMINA/STORM 2.0		
	P_{min}	P_{max}	w	P_{min}	P_{max}	w
010 to 100	0.395**048522**	0.395**115209**	6.66874E-05	0.3950**5093**	0.395**100628**	4.96987E-05
010 to 111	0.016**594466**	0.016**78343**	0.000188964	0.016**474768**	0.0**20078199**	0.003603431
100 to 111	0.016**627243**	0.016**822966**	0.000195723	0.016**504701**	0.0**20246672**	0.003741971
111 to 010	0.694**656523**	0.694**808764**	0.000152241	0.694**661415**	0.694**745396**	8.39815E-05
100 to 010	0.455**029708**	0.455**120325**	9.06176E-05	0.455**034656**	0.455**085702**	5.10458E-05
111 to 100	0.735**722567**	0.735**846036**	0.00012347	0.735**725864**	0.735**790289**	6.44251E-05
000 to 011	0.82**6019153**	0.826**159295**	0.000140142	0.825**468467**	0.827**041855**	0.001573389
011 to 101	0.989**516179**	0.989**742961**	0.000226783	0.989**23085**	0.990**289532**	0.001058682
000 to 101	0.990**235688**	0.990**472905**	0.000237217	0.990**019346**	0.990**893279**	0.000873933
101 to 000	0.864**410883**	0.864**464495**	5.36124E-05	0.864**417637**	0.864**456068**	3.84307E-05
011 to 000	0.857436475	0.857504873	6.83981E-05	0.857436475	0.857504873	6.83981E-05
101 to 011	0.989**490726**	0.989**802643**	0.000311917	0.989**025011**	0.990**641893**	0.001616882
TQN2047	0.498962403	0.498968523	6.11917E-06	0.498962404	0.498968523	6.11917E-06
TQN4095	0.499263342	0.499269661	6.3189E-06	0.499263342	0.499269661	6.3189E-06
Polling20	1	1	0	1	1	0
JQN 4/5	0.865**393696**	0.865**546456**	0.00015276	0.865**436981**	0.865**452685**	1.57041E-05
JQN 5/5	0.819**441133**	0.820**159054**	0.000717921	0.819**599345**	0.819**810381**	0.000211036

great case studies to test STAMINA. For this work, circuit 0x8E was selected to verify STAMINA's functionality. The circuit is part of 60 combinational genetic circuits designed and built as part of the development of the genetic design automation tool *cello* [18]. It has three inputs and one output, indicated by yellow fluorescence proteins. The circuit was specifically chosen since it exhibits an unwanted output behavior. While the output is supposed to be high throughout an entire input transition, it turns off briefly before returning to a high output. Analysis by Fontanarrosa et al. showed that the behavior is explained due to a *function hazard* (i.e., a property of the function being implemented) [6]. To validate STAMINA, twelve models representing the 12 input transitions, shown as the first 12 rows in Tables 1 and 2, were analyzed to calculate the likelihood of that unwanted switching behavior.

Other Benchmarks. STAMINA was also evaluated on the same subset of examples from the PRISM benchmark suite [12] and the INFAMY case studies [2] as in [19], shown in the last few rows in Tables 1 and 2. One exception is the grid-world robot examples, since STAMINA does not yet support nested properties. The *Tandem Queueing Network* (TQN) models [10] consist of two interconnected queues with capacity 2047 and 4095, respectively, whose property queries the probability that the first queue is full before 0.23 time units. The *Jackson Queuing Network* (JQN) [11] models both have 5 as the arrival rate and contain 4 and 5 interconnected queuing stations, respectively. The property queries the probability that at least 4 and 6 jobs are in the first and

Table 2. State count and runtimes (in seconds) comparison for STAMINA 2.0.

Model	State Count		Runtime		
	STAMINA/P	STAMINA/S	STAMINA/P	STAMINA/S	Improve.(%)
010 to 100	85160	87902	47.79	28.67	40.00
010 to 111	3527020	2479199	3337.96	2843.89	14.8
100 to 111	3568525	2510758	3280.53	2873.05	12.42
111 to 010	467635	490145	283.53	236.78	16.49
100 to 010	165043	175194	100.78	73.61	26.95
111 to 100	406424	425443	216.10	210.31	2.68
000 to 011	2543695	1839612	2393.88	2109.79	11.87
011 to 101	2813395	2234057	2612.57	2262.65	13.39
000 to 101	2829690	2331091	2417.28	2557.20	-5.79
101 to 000	327687	337394	193.00	136.80	29.12
011 to 000	381372	381372	230.41	156.23	32.19
101 to 011	3006113	2302933	2896.84	2452.61	15.33
TQN2047	21293	21293	15.59	3.44	77.90
TQN4095	42469	42469	52.80	6.80	87.12
JQN4/5	187443	257265	33.38	59.88	-79.39
JQN5/5	1480045	1896415	419.47	577.40	-37.65

the second queue, within 10 time units. Note that the property is satisfied in the initial state in the polling model, so the probability is always 1.0 and the runtime is negligible.

5 Discussion

In all but four tests, state count and probabilities, namely P_{min} and P_{max}, were different between both tools. However, all cases are comparable and the probability windows w always overlap as shown in Table 1. That is, $P_{min}^{P} \leq P_{max}^{S}$ where P_{min}^{P} is the lower bound provided by STAMINA/PRISM and P_{max}^{S} the upper bound from STAMINA/STORM. This is also true for the condition $P_{min}^{S} \leq P_{max}^{P}$, the lower bound from STAMINA/S and the upper bound from STAMINA/P, respectively. In order for the actual probability to exist within these bounds, both conditions must—and did—hold. Although the core algorithm is the same, the different but overlapping bounds may be due to differences such as property-based truncation and CTMC numerical analysis that exist in PRISM and STORM. There were several cases which STAMINA/STORM did not meet the requirement, i.e., $w \leq 10^{-3}$, due to a limit of 10 iterations in the outer while-loop in [19]. Since PMC occurs in each iteration of that loop, this limit exists in [19] and both versions of STAMINA to prevent excessive PMC with diminishing returns. However, after lifting this iteration limit, STAMINA/STORM was able to meet the window requirement. Increasing this limit resulted in the windows of the two reactions which transition to 111 to be 8.4×10^{-4} and 8.7×10^{-4}, in 16 iterations (4669.08 and

4754.01 s), respectively. While this increased runtimes, the worst case being the 100 to 111 circuit transition, modifying other parameters, such as a smaller κ, would reduce the number of iterations. All of these results are available online.

Table 2 shows that, STAMINA/STORM outperformed STAMINA/PRISM in most cases, with substantial reductions in runtime, except for the two JQNs, and one circuit model. However, for the JQN models, the runtime improved by over two times as the model scaled from $4/5$ to $5/5$. State counts are generally comparable. More optimization opportunities presented when interfacing STAMINA with STORM, although more considerations such as the order of state insertion into the transition matrix had to be made.

6   Conclusion

This paper presents the first C++ implementation of the infinite-state CTMC model checker STAMINA with increased modularity in software architecture and memory optimizations. By interacting with the CTMC analysis engine in STORM, STAMINA further improved model checking efficiency over the previous implementation, while maintaining comparable scalability. Future improvements to STAMINA may involve additionally algorithmic efficiency, further optimizations with respect to STORM integration, tuning the algorithm for rare events, and more challenging benchmarks or requirements, such as bounding $w \leqslant 10^{-7}$. We also are working on several user-experience improvements to STAMINA, such as its REST API and a GUI, the former currently available for public use and the latter under active development.

Data Availability. An artifact with both versions of STAMINA, as well as STORM and PRISM, and the testing suite used in this paper is provided in the open virtual appliance (OVA) format. It has been submitted to the QEST 2023 artifact evaluation, and is publicly available on STAMINA's website.

Acknowledgment. We thank Tim Quatmann at RWTH Aachen University for his help with interfacing the STORM model checker. This work was supported by the National Science Foundation under Grant No. 1856733, 1856740, and 1900542. Any opinions, findings, and conclusions or recommendations expressed in this material are those of the authors and do not necessarily reflect the views of the funding agencies.

References

1. https://github.com/fluentverification/stamina-storm
2. https://depend.cs.uni-saarland.de/tools/infamy/casestudies/
3. Baier, C., Haverkort, B.R., Hermanns, H., Katoen, J.P.: Model-checking algorithms for continuous-time Markov chains. IEEE Trans. Software Eng. **29**(6), 524–541 (2003). https://doi.org/10.1109/TSE.2003.1205180
4. Budde, C.E., et al.: On correctness, precision, and performance in quantitative verification. In: Margaria, T., Steffen, B. (eds.) ISoLA 2020. LNCS, vol. 12479, pp. 216–241. Springer, Cham (2021). https://doi.org/10.1007/978-3-030-83723-5_15

5. Češka, M., Křetínský, J.: Semi-quantitative abstraction and analysis of chemical reaction networks. In: Dillig, I., Tasiran, S. (eds.) CAV 2019. LNCS, vol. 11561, pp. 475–496. Springer, Cham (2019). https://doi.org/10.1007/978-3-030-25540-4_28

6. Fontanarrosa, P., Doosthosseini, H., Borujeni, A.E., Dorfan, Y., Voigt, C.A., Myers, C.: Genetic circuit dynamics: hazard and glitch analysis. ACS Synth. Biol. 15 (2020)

7. Fox, B.L., Glynn, P.W.: Computing poisson probabilities. Commun. ACM **31**(4), 440–445 (1988). https://doi.org/10.1145/42404.42409

8. Hahn, E.M., Hermanns, H., Wachter, B., Zhang, L.: INFAMY: an infinite-state Markov model checker. In: Bouajjani, A., Maler, O. (eds.) CAV 2009. LNCS, vol. 5643, pp. 641–647. Springer, Heidelberg (2009). https://doi.org/10.1007/978-3-642-02658-4_49

9. Hensel, C., Junges, S., Katoen, J.P., Quatmann, T., Volk, M.: The probabilistic model checker Storm. Int. J. Softw. Tools Technol. Transfer **24**(4), 589–610 (2022). https://doi.org/10.1007/s10009-021-00633-z

10. Hermanns, H., Meyer-Kayser, J., Siegle, M.: Multi terminal binary decision diagrams to represent and analyse continuous time Markov chains. In: Plateau, B., Stewart, W., Silva, M. (eds.) NSMC, pp. 188–207 (1999)

11. Jackson, J.: Networks of waiting lines. Oper. Res. **5**, 518–521 (1957)

12. Kwiatkowsa, M., Norman, G., Parker, D.: The PRISM benchmark suite. In: International Conference on Quantitative Evaluation of Systems (QEST), pp. 203–204 (2012). https://doi.org/10.1109/QEST.2012.14. https://doi.ieeecomputersociety.org/10.1109/QEST.2012.14

13. Kwiatkowska, M., Norman, G., Parker, D.: PRISM 4.0: verification of probabilistic real-time systems. In: Gopalakrishnan, G., Qadeer, S. (eds.) CAV 2011. LNCS, vol. 6806, pp. 585–591. Springer, Heidelberg (2011). https://doi.org/10.1007/978-3-642-22110-1_47

14. Lapin, M., Mikeev, L., Wolf, V.: Shave: stochastic hybrid analysis of Markov population models. In: Proceedings of the 14th International Conference on Hybrid Systems: Computation and Control, HSCC 2011, pp. 311–312. ACM, New York (2011)

15. Madsen, C., Zhang, Z., Roehner, N., Winstead, C., Myers, C.: Stochastic model checking of genetic circuits. J. Emerg. Technol. Comput. Syst. **11**(3), 23:1–23:21 (2014). https://doi.org/10.1145/2644817. http://doi.acm.org/10.1145/2644817

16. Neupane, T., Myers, C.J., Madsen, C., Zheng, H., Zhang, Z.: STAMINA: STochastic approximate model-checker for INfinite-state analysis. In: Dillig, I., Tasiran, S. (eds.) CAV 2019. LNCS, vol. 11561, pp. 540–549. Springer, Cham (2019). https://doi.org/10.1007/978-3-030-25540-4_31

17. Neupane, T., Zhang, Z., Madsen, C., Zheng, H., Myers, C.J.: Approximation techniques for stochastic analysis of biological systems. In: Liò, P., Zuliani, P. (eds.) Automated Reasoning for Systems Biology and Medicine. CB, vol. 30, pp. 327–348. Springer, Cham (2019). https://doi.org/10.1007/978-3-030-17297-8_12

18. Nielsen, A.A.K., et al.: Genetic circuit design automation. Science **352**(6281), aac7341 (2016). https://doi.org/10.1126/science.aac7341. https://www.science.org/doi/abs/10.1126/science.aac7341

19. Roberts, R., Neupane, T., Buecherl, L., Myers, C.J., Zhang, Z.: STAMINA 2.0: improving scalability of infinite-state stochastic model checking. In: Finkbeiner, B., Wies, T. (eds.) VMCAI 2022. LNCS, vol. 13182, pp. 319–331. Springer, Cham (2022). https://doi.org/10.1007/978-3-030-94583-1_16

20. Volk, M., Junges, S., Katoen, J.P.: Fast dynamic fault tree analysis by model checking techniques. IEEE Trans. Ind. Informatics **14**(1), 370–379 (2018). https://doi.org/10.1109/TII.2017.2710316

Skipping and Fetching: Insights on Non-conventional Product-Form Solutions

Diletta Olliaro[1]([✉]) [iD], Gianfranco Balbo[2] [iD], Andrea Marin[1] [iD], and Matteo Sereno[2] [iD]

[1] Università Ca' Foscari Venezia, Venice, Italy
{diletta.olliaro,marin}@unive.it
[2] Università di Torino, Turin, Italy
balbo@di.unito.it,matteo.sereno@unito.it

Abstract. Complex models of computer systems are often difficult to study with numerical or analytical approaches because of the state space explosion problem. The class of product-form models is one of the most significant tools for overcoming this problem, and in many applications, this tool is the only way to perform a quantitative analysis.

In this paper, we study the duality between two different product-form models. The first consists of a queuing network with finite capacity waiting rooms governed by the skip-over policy. The second is a recently presented product-form model in which a job fetching policy is applied. To investigate the relationships between these two models, we first extend the fetching queuing model to allow for finite capacity warehouses and for a Repetitive Service Blocking with Random Destination (RS-RD) discipline. Subsequently, we represent their distinctive features in terms of Generalized Stochastic Petri Nets which precisely specify their semantics in a modular manner and provide clear and intuitive interpretations of these policies. With these two preliminary results, we prove that it is possible to structurally transform a model of one class into one of the other and vice versa, thus choosing the representation that is computationally more convenient to compute the performance measures of interest.

Keywords: Queuing Theory · Product-form Solutions · Generalized Stochastic Petri Nets · Fetching Policy · Blocking Policy · Skip-over policy

1 Introduction

Since their discovery more than 60 years ago, product-form models have played an important role in the quantitative analysis of computer and telecommunication systems as well as in the understanding of some important natural phenomena (see, e.g., [12,18]).

In Markovian models of this class, the joint stationary probability distribution of the system can be calculated as a normalized product of functions

© The Author(s), under exclusive license to Springer Nature Switzerland AG 2023
N. Jansen and M. Tribastone (Eds.): QEST 2023, LNCS 14287, pp. 110–126, 2023.
https://doi.org/10.1007/978-3-031-43835-6_8

that solely rely on the local state of the individual components. This property enables analytical investigations that would have been otherwise infeasible, or computationally very complex, due to the state space explosion problem.

While most product-form results for queuing networks assume a probabilistic and state independent routing, an outstanding example of a more general result is that of queuing networks with finite capacity buffers and skip-over policies. This queuing network class was introduced by Pittel [21] and can be seen as a network of BCMP-like stations [6] with finite capacities. Upon completing service at a station, the destination is chosen probabilistically, based on a routing matrix. In Pittel's networks, if the chosen destination is *saturated* (i.e., its input buffer is full), a new station is picked according to the routing probabilities associated with the saturated station. As a result, while the event of job completion still depends on the state of one station and its realization changes the states of two components, the destination choice depends on the network global state since sequences of saturated stations may be chosen before the destination station is actually picked. The model is even more intriguing because the skip-over policy can generate cycles of visits to saturated stations. Therefore, despite being extensions of BCMP networks with quasi-reversible stations, Pittel's networks do not satisfy this property.

In this paper, we examine the relationship between two product-form model classes. The first consists of closed queuing networks with skip-over policy, while the second refers to a novel result presented in [20]. The models pertaining to this second class are queuing networks in which, upon completion of a job at a queue, the server selects a job from its backlog (similarly to Jackson or BCMP networks) if available. However, if the backlog is empty, the server requests the upstream stations to send it a job. This job request process is recursively performed until the server receives a job to process.

The routing of jobs is probabilistic.

While a formal description of the policy is provided in Sect. 4, here we illustrate an example. Let us consider a hardware component, such as lithium batteries, whose efficiencies decline over time. These batteries are used to power two types of devices. The first category of devices requires fresh batteries with high efficiencies, while the second category of devices has lower efficiency requirements. Additionally, there is a regeneration station where the batteries are restored to their original conditions (for example, through recycling methods as described in [25]). We model the system with three stations as depicted in Fig. 1. Each station includes a warehouse containing batteries for regeneration (Station 1) or use (Stations 2 and 3). In the system, there are $M + 3$ batteries and the regeneration times are exponentially distributed with rate μ_1, while the using times at stations 2 and 3 are exponentially distributed with rates μ_2 and μ_3, respectively. In this model, each device is equipped with a battery that must be present and the regeneration facility always works. Hence, in the warehouses, there are exactly M batteries: in W_1 those that have to be regenerated, in W_2 the new ones, and in W_3 those with lower efficiencies sent by W_2 to W_3. Suppose that, at station 2, a battery is disposed. If its warehouse contains at least one battery, then this is used to supply the device, otherwise, a request is sent to Station 1 to get a battery that has not been entirely regenerated yet. A similar

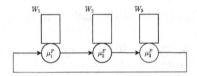

Fig. 1. A closed fetching queuing network with tandem topology.

policy is used by Station 3 which may request Station 2 to get a high efficient battery and even by Station 1 which might ask Station 3 to send a battery to the regeneration station before it was really needed. All requests are instantaneous.

As we may see, similarly to Pittel's networks, the routing depends on the global state, but according to a policy that is very different. The analysis of models with state dependent routing is known to be challenging and the existence of product-form solutions is indeed interesting. Moreover, we notice that the relation between networks with skip-over and job fetching policies is not intuitive, and unveiling this connection is one of the goals of this paper.

The contributions of this paper can be summarized as follows:

- We extend the class of product-form fetching queues with the introduction of finite capacity stations and repetitive service - random destination (RS-RD) blocking discipline.
- We study the semantics of both the skip-over and fetching queuing network models, accounting for the fact that synchronizations are not pairwise, so that some complications may arise. For example, in skip-over networks, is it allowed for a customer who completes the service at a station to return to the same station because of a sequence of skippings? To clarify these and other questions for skip-over and fetching networks, we give modular semantics in terms of Generalized Stochastic Petri Nets (GSPNs) [1,2].
- We investigate the relationships between closed queuing networks with finite capacity and skip-over policy and closed fetching queues with finite capacity and RS-RD policy. Specifically, for each model of one class, we associate a dual model of the other. While the duality relation is somehow unexpected and non-trivial, given the very different network dynamics, the performance indices of one model can be expressed in terms of those of the other in a very simple manner.

The paper is organized as follows. After a brief review of the state of the art in Sect. 2, Sect. 3 provides a concise overview of Pittel's queueing network with finite capacity buffers and skip-over policy. In Sect. 4, we present the generalization of the model introduced in [20]. Section 5 offers a formal and rigorous description of the underlying semantics for both models studied in this work, utilizing the GSPN formalism. We explore the duality between the two models in Sect. 6. Finally, in Sect. 7, we provide concluding remarks and suggest avenues for future research.

2   Related Work

Ever since the first results on product-form models were introduced [16,17], research in this area has been lively, with scholars attempting to precisely identify this class of models. These efforts involve finding sufficient conditions for the product-form solution, which may depend on the properties of individual components rather than on the joint process. However, the state of the art reveals that there is no complete set of necessary and sufficient conditions on the components that fully characterize product-form models.

The most well-known sufficient condition for product-form is the quasi-reversibility [18]. For this condition to hold, components must interact in pairs, meaning that an event can only depend on and change the (local) states of at most two components simultaneously. It is worth noting that, even when limiting models to pairwise interactions, quasi-reversibility is not a necessary condition for product-form, as demonstrated in queuing networks with finite capacity and blocking [4,5]. More specific results are derived in [6] for the so-called BCMP networks where sufficient conditions for the product-form solution are expressed in terms of combinations of queuing disciplines and service time distributions for all stations of the network.

In the literature, there are models that have product-form solution despite not adhering to the pairwise assumption of quasi-reversibility and/or the criteria of the BCMP networks [8,13,15]. For example, networks with state dependent routing are studied in [4,13,20,21,23].

This paper focuses on the relations between the state model studied in [21] and that recently proposed in [20]. They are both queuing network models with state dependent routing but with quite different dynamics. The networks proposed in [21] have been widely investigated. In particular, it is known that the standard formulation of the *Arrival Theorem* [22] does not hold while a Mean Value Analysis has been recently shown to apply [24] to models with specific topologies. The latter result suggests that a formulation of the Arrival Theorem should be possible. For the model studied in [20] only a convolution algorithm is proposed.

The duality result that we propose in this work and the formal GSPN modular semantics of both networks are expected to contribute to the understanding of the state dependent routing processes that characterizes these models and represent the main challenge in their analysis.

3   Background on Skip-Over Queuing Networks

In this section, we summarize the salient characteristics of Pittel's network with skip-over policy [21]. The definition of the fetching networks is postponed to Sect. 4, where we extend the result presented in [20] to finite capacity systems.

We refer to closed queuing networks with skip-over policy to indicate those networks in which every station is composed of a service center and of a corresponding finite capacity buffer where customers are waiting before getting served.

Definition 1 (Closed Skipping Queuing Network with finite capacity Buffers (CSQNB)).   *A CSQNB is defined by a tuple* $\mathcal{Q}^P = (\Omega, \mathbf{P}^P, \boldsymbol{\mu}^P, \mathbf{c}^P, M^P)$ *where:*

- $\Omega = \{1, 2, \ldots, N\}$ *is the set of stations.*
- $\mathbf{P}^P = [p_{ij}^P]$ *is a stochastic matrix that describes the independent probabilistic routing.*
- $M^P > 0$ *is the total number of customers at the stations.*
- $\mathbf{c}^P = [c_i^P]$ *is the vector of positive capacities; specifically c_i^P describes the finite capacity for buffer i.*
- $\boldsymbol{\mu}^P = [\mu_i^P]$ *is the vector of positive service rates.*

Network Dynamics. The state of the network is described by a vector $\mathbf{n} = (n_1, \ldots, n_N) \in \mathcal{S}^P(\Omega, M^P, \mathbf{c}^P)$, where n_i denotes the number of customers in station i and $\mathcal{S}^P(\Omega, M^P, \mathbf{c}^P)$ is the state space of the system defined as follows:

$$\mathcal{S}^P(\Omega, M^P, \mathbf{c}^P) = \{\mathbf{n} = (n_1, \ldots, n_N) \mid \sum_{i \in \Omega} n_i = M^P \wedge n_i \le c_i^P, \ \forall i \in \Omega\} \quad (1)$$

Customers move from one station to another according to a stochastic routing matrix $\mathbf{P}^P$ in which every entry p_{ij}^P denotes the probability that upon service completion at server i a customer tries to enter station j. If the destination j is saturated, the routing process is repeated recursively as if the job had been completed in j. The service time at server i is exponentially distributed with rate μ_i^P.

Theorem 1 (Product-form of CSQNB [21]). *The steady-state distribution of a CSQNB is in product-form:*

$$\pi_{(M^P, \mathbf{c}^P)}^P(\mathbf{n}) = \frac{1}{G^P(\Omega, M^P, \mathbf{c}^P)} \prod_{i=1}^N g_i^P(n_i). \quad (2)$$

where service functions are $g_i^P(n_i) = (\rho_i^P)^{n_i}$, with $\rho_i^P = v_i/\mu_i^P$ where the v_i's are the solutions of the following system of Jackson's traffic equations:

$$v_i = \sum_{j=1}^N v_j p_{ji}^P. \quad (3)$$

$G^P(\Omega, M^P, \mathbf{c}^P)$ *is the normalizing constant, equal to* $\sum_{\mathbf{n} \in \mathcal{S}^P(\Omega, M^P, \mathbf{c}^P)} \prod_{i=1}^N g_i^P(n_i)$.

4   Fetching Queuing Networks with Finite Capacity

In this section, we define and study closed fetching queuing networks with finite capacity. The result strictly generalizes the model introduced in [20], where only warehouses with infinite capacity are considered. According to the literature on finite capacity queuing networks, we define a blocking policy to handle the case of arrivals at saturated buffers. Differently from Pittel's model that adopts the skip-over policy, these models use the Repetitive Service Blocking with Random Destination policy (RS-RD) (see, e.g., [5]) described hereafter.

Definition 2 introduces the model and the formalism used along the paper. Graphically, we represent stations of CFQNWs with the warehouse placed on the top of the service room (the circle) and those of CSQNBs with the buffer on the left-hand side of the service room.

Definition 2 (Closed Fetching Queuing Network with finite capacity Warehouses (CFQNW)). *A closed fetching queuing network with finite capacity warehouses is defined by a tuple* $\mathcal{Q}^F = (\Omega, \mathbf{P}^F, \boldsymbol{\mu}^F, \mathbf{c}^F, M^F)$, *where:*

- $\Omega = \{1, 2, \ldots, N\}$ *is the set of stations.*
- $\mathbf{P}^F = [p_{ij}^F]$ *is a stochastic matrix that describes the independent probabilistic routing. If* $p_{ij} > 0$, *then we say that station* i *is a provider of station* j.
- $\boldsymbol{\mu}^F = [\mu_i^F]$ *is the vector of positive service rates.*
- $\mathbf{c}^F = [c_i^F]$ *is the vector of positive capacities, accordingly* $c_i^F \in \mathbb{N} \cup \{\infty\}$ *describes the finite capacity for warehouse* i.
- $M^F > 0$ *is the total number of customers in the warehouses.*

Network Dynamics. The state of the network is described by a vector $\mathbf{n} = (n_1, \ldots, n_N)$ where $n_i \geq 0$ denotes the number of customers in warehouse i. The state space of the system, $S^F(\Omega, M^F, \mathbf{c}^F)$, is defined as:

$$S^F(\Omega, M^F, \mathbf{c}^F) = \left\{ \mathbf{n} = (n_1, \ldots, n_N) | \sum_{i=1}^{N} n_i = M^F \wedge n_i \leq c_i^F \ \forall i \in \Omega \right\} \quad (4)$$

The system's state does not include the customers being served as every server is always busy. Consequently, the network has $M^F + N$ customers circulating through the stations. State transitions in the system are triggered by service completions. The service times are independent and exponentially distributed, with rate μ_i^F at station $i \in \Omega$. After a customer is served at station i, it moves to station j with probability p_{ij}^F. Additionally, if station i has no jobs in its warehouse, i.e., $n_i = 0$, when a service is completed in i, the server contacts one of its providers to fetch a job from its warehouse. The choice of the provider is probabilistic and defined as:

$$q_{ik} = \frac{\mu_k^F p_{ki}^F}{\sum_{h=1}^{N} \mu_h^F p_{hi}^F}. \quad (5)$$

Intuitively, a provider is chosen with a probability proportional to the intensity of the stream of customers directed to the station that completes a service while its warehouse is empty. If the contacted warehouse is also empty, then its server applies recursively the same policy choosing in this way one of its providers. The recursion stops when a provider with a non-empty warehouse is found. Moreover, if, upon completion of a job at station i, the routing process selects a saturated station j as destination, before starting the fetching process (if needed), the blocking policy RS-RD, is applied, meaning that the job receives a new and independent service at station i, after which it chooses a new destination independently of its previous selection according to $\mathbf{P}^F$.

Thanks to the exponential assumption on the service times, the stochastic process underlying a CFQNW is a Continuous Time Markov Chain (CTMC). It is easy to show that if $\mathbf{P}^F$ is irreducible, then also the CTMC is irreducible and, since the state space is finite, the CTMC is ergodic and admits a unique limiting distribution as stated in Theorem 2. Henceforth, coherently with [6], we will assume the irreducibility of the routing process.

Theorem 2. *The steady-state distribution of a CFQNW is in product-form:*

$$\pi^F_{(M^F, \, \mathbf{c}^F)}(\mathbf{n}) = \frac{1}{G^F(\Omega, M^F, \mathbf{c}^F)} \prod_{i=1}^{N} g_i(n_i). \tag{6}$$

where service functions $g_i^F(n_i) = (\rho_i^F)^{n_i}$, *with* $\rho_i^F = \sum_{k=1}^{N} \mu_k^F p_{ki}^F / \sum_{k=1}^{N} x_{ik}$, *and the* x_{ik}'s *are the solutions of the linear system of rate equations:*

$$x_{ik} = \left(\sum_{j=1}^{N} x_{kj} \right) \frac{\mu_i^F p_{ik}^F}{\sum_{j=1}^{N} \mu_j^F p_{jk}^F}. \tag{7}$$

$G^F(\Omega, M^F, \mathbf{c}^F)$ *is the normalising constant equal to* $\sum_{\mathbf{n} \in \mathcal{S}^F(\Omega, M^F, \mathbf{c}^F)}$ $\prod_{i=1}^{N} g_i^F(n_i)$

Proof. The proof of the product-form is based on the observation that the interactions among the isolated stations satisfy the conditions presented in [15]. Then, we derive the expression of the steady-state distribution. □

Remark 1. Similarly to what happens in the comparison between Gordon-Newell queuing networks and CSQNBs [4,21], the expressions of the *unnormalized* stationary distributions of CFQNWs are the same for infinite and finite capacity systems. Indeed, the differences come from the computation of the normalizing constants since the state spaces of the two networks are different. It is worth remembering that, since the CTMC underlying the system is *not* reversible, a truncation of the joint state space does not trivially bring to a product-form solution, and this is why a new proof of the result must be provided.

Remark 2. We recall that the systems of rate equations (Eq. (7)) and traffic equations (Eq. (3)) are linear and they have full rank minus 1 thanks to the assumption on the irreducibility of the routing matrix. Hence, one has to fix one unknown to a non-zero constant (typically 1) and then compute the values of the remaining unknowns. Notice that the service functions associated with a CFQNW (and CSQNB) are therefore not unique since they depend on the chosen solution of the rate or traffic equations. Nevertheless, the steady-state probability distributions are unique thanks to the computation of the normalizing constant [19].

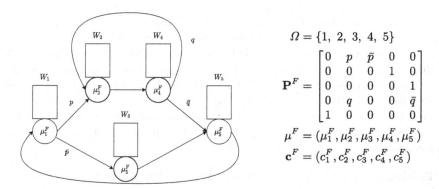

$$\Omega = \{1,\, 2,\, 3,\, 4,\, 5\}$$

$$\mathbf{P}^F = \begin{bmatrix} 0 & p & \bar{p} & 0 & 0 \\ 0 & 0 & 0 & 1 & 0 \\ 0 & 0 & 0 & 0 & 1 \\ 0 & q & 0 & 0 & \bar{q} \\ 1 & 0 & 0 & 0 & 0 \end{bmatrix}$$

$$\mu^F = (\mu_1^F, \mu_2^F, \mu_3^F, \mu_4^F, \mu_5^F)$$

$$\mathbf{c}^F = (c_1^F, c_2^F, c_3^F, c_4^F, c_5^F)$$

Fig. 2. Example of a CFQNW. Warehouses are denoted by labels $W_1, \ldots, W_5$.

Example 1. Let us consider the model depicted in Fig. 2.
Solving the system of rate equations and fixing x_{24} to 1, we obtain:

$$x_{24} = 1, \qquad x_{12} = x_{45} = \frac{\mu_1^F p}{\mu_1^F p + \mu_4^F q}, \qquad x_{13} = x_{35} = \frac{\mu_1^F \mu_3^F p}{\mu_4^F \bar{q}(\mu_1^F p + \mu_4^F q)},$$

$$x_{42} = \frac{\mu_4^F q}{\mu_1^F p + \mu_4^F q}, \qquad x_{51} = \frac{\mu_1^F p(\mu_3^F + \mu_4^F \bar{q})}{\mu_4^F \bar{q}(\mu_1^F p + \mu_4^F q)}.$$

Accordingly, we obtain:

$$g_1^F(n_1) = \left(\frac{(\mu_1^F p + \mu_4^F q)\mu_4^F \mu_5^F \bar{q}}{\mu_1^F p(\mu_3^F + \mu_4^F \bar{q})} \right)^{n_1}, \qquad g_2^F(n_2) = \left(\mu_1^F p + \mu_4^F q \right)^{n_2},$$

$$g_3^F(n_3) = \left(\frac{(\mu_1^F p + \mu_4^F q)\mu_4^F \bar{q} \,\bar{p}}{\mu_3^F p} \right)^{n_3}, \qquad g_4^F(n_4) = (\mu_2^F)^{n_4}, \qquad (8)$$

$$g_5^F(n_5) = \left(\frac{(\mu_1^F p + \mu_4^F q)\mu_4^F \bar{q}}{\mu_1^F p} \right)^{n_5}.$$

5   A GSPN-Based Semantics

In this section, we utilize Generalized Stochastic Petri Nets (GSPN) [1,2,9] to formally describe the semantics of CSQNB and CFQNW, allowing for a clear and modular description of their skipping and fetching policies. Immediate transitions and inhibitor arcs of GSPNs are utilized for this purpose. Through this work, we uncover important characteristics of these models. For instance, we demonstrate that under the skip-over policy, it may happen that a customer that departs from a station and subsequently searches a non-saturated buffer, could return to the station it just left. Similarly, CFQNWs by first placing a customer at the target station and then selecting which customer to serve next, potentially lead to cases in which the fetching policy returns the same customer to the same service station immediately after serving it. Notice that placing the customer at the target station before or after selecting which customer to serve next yields different outcomes so that the precise semantics of this action is needed to obtain the analytical results presented in [20] and the corresponding numerical ones. It is also important to observe that these events may depend on various factors, including network topology and station capacities, but are not immediately evident without the formal specification of the model components. Finally, we will also be able to relate this result with the literature of GSPN in product-form.

Definition 3. *A Generalized Stochastic Petri Net is formally defined by the following tuple:*

$$(P, T, I(.,.), O(.,.), H(.,.), \Pi(.), W(.), \boldsymbol{m}_0)$$

where $P = (p_1, p_2, ..., p_P)$ *is the set of places,* $T = (t_1, t_2, ..., t_T)$ *is the set of transitions (timed and immediate),* $I(t_j, p_i) : T \times P \to \mathbb{N}$ *is the input function,* $O(t_j, p_i) : T \times P \to \mathbb{N}$ *is the output function,* $H(t_j, p_i) : T \times P \to \mathbb{N}$ *is the inhibition function,* $\Pi(t_j) : T \to \mathbb{N}$ *is the priority function* $W : T \to \mathbb{R}$ *is a weight function,* $\boldsymbol{m}_0 = (m_{01}, m_{02}, ..., m_{0P})$ *is the initial marking.*

$\Pi(.)$ *specifies the priority level associated with the transitions of the net. For transitions with priority zero, their firing delays are negative exponentially distributed random variables; such transitions are consequently referred to as timed. For transitions with priority* $n \geq 1$, *delays are deterministically zero; such transitions are referred to as n-immediate.*

$W(.)$ *maps transitions into real positive numbers. The quantity* $W(t_j)$ *(or* w_j*) is called the "rate" of transition* t_j *if* t_j *is timed, and the "weight" of transition* t_j *if* t_j *is immediate.*

In the graphical representation of GSPNs, transitions are drawn as bars or white boxes, depending on whether they are immediate or timed; places are drawn as circles.

The Input and Output functions are represented as directed arcs from places to transitions and vice versa; the inhibition function is represented by circle-headed arcs. When greater than one, the arc multiplicity is written as a number next to the corresponding arc. Finally, a marked place is a place that contains (at least) one token drawn as a black dot. A distribution of tokens over the places of the net identifies one possible marking.

Notice that the stochastic process underlying a GSPN is, in general, a semi-Markov process because of the presence of immediate transitions. This can algo-rithmically be reduced to a CTMC [1] even in presence of cycles of immedi-ate transitions[1]. Figure 3 shows the GSPN representation of a CSQBN station. Places queue, idle, and busy, together with transitions start and service implement the traditional mechanism commonly characterizing Jackson networks and used

---

[1] How to deal with cycles of immediate transitions is described in the original GSPN paper [2], discussed in details in [11], and implemented in SMART [10].

also in the definition of CSQBN [21]. Upon arrival of a job in a queue, the latter is immediately served if the corresponding server is available, meaning it is idle, otherwise it waits in the queue until the job currently in service terminates. New customers arriving at the input place may join the station if space is available in the buffer, thus reducing the number of free slots. If the buffer is full (i.e., no available slots are left) the customer skips the station and proceeds directly to the output place where it decides the station to join next according to the routing mechanism implemented by the different immediate transitions that withdraw tokens from that place.

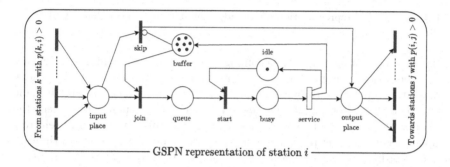

Fig. 3. A GSPN representation of a station in a CSQNB.

Connecting the output place of a block to the input place of another block it is easy to construct a CSQNB network.

The GSPN representation of a station of the CFQNW type is more complex. Figure 4 shows two CFQNW stations connected in series (Station i receiving customers from Station $i - 1$).

When a customer leaves Station $i-1$, it tries to join Station i by first checking if there is available space in Station i's input queue (i.e., the warehouse). If the latter is not saturated, the customer enters Station i and if the server is idle it immediately receives service (i.e., if a token is present in place csr_i - which is the usual acceptance mechanism for a single server station). If instead, the buffer is full, the customer returns to Station i, ready to receive a new and independent service after which it will choose a new destination, unrelated to the previous choice.

Things are instead more complex when the server of Station i completes the service of a customer and its warehouse is empty. This situation is represented in the model of Fig. 4 by a token put in place csr_i when no tokens are in place wh_i. Since the fetching policy assumes that a server cannot be left idle, a request is immediately sent to the server of Station $i-1$ (firing of the immediate transition $fetch_i$) which is interrupted and forced to forward to Station i the customer it was currently serving. This customer is put in the warehouse of Station i and immediately taken by the server of the $i - th$ Station that is now ready to start a new service. Let us call this mechanism a Backward-Request-Mechanism.

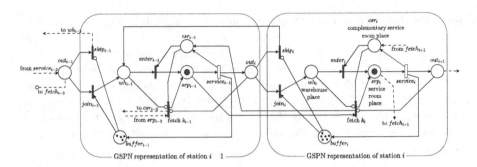

Fig. 4. A GSPN representation of two adjacent stations in a tandem CFQNW.

If the server of Station $i - 1$, becoming idle (token put temporarily in place csr_{i-1}) because of the previous interruption, finds a customer in its warehouse it takes that customer (moving a token from wh_{i-1} to srp_{i-1}) and becomes ready to start the service of this new customer. At this point, the interaction mechanism between the two stations is completed (this complete and relatively complex token game is performed in zero time) and the process continues to evolve according to the time specifications of the timed transitions. When instead wh_{i-1} is empty, it is the server of Station $i-1$ that needs to find a new customer and the procedure described for Station i is repeated with respect to Station $i-1$, possibly affecting Station $i - 2$.

Also in this case, by connecting blocks of this type together it is possible to construct a CFQNW with arbitrary topology. If several stations are connected in the network to feed a specific other station, the Backward-Request-Mechanism must be generalized choosing (with the appropriate probabilities defined by Eq. (5)) the station to be interrupted among those "feeding" stations. This can be modeled by a set of immediate (and conflicting) transitions similar to transition $fetch_i$ of Fig. 4, one for each of the stations sending customers to Station i.

The recursive implementation of the Backward-Request-Mechanism in this more general setting may yield to infinite cycles of customers moved among the different stations in zero time. This is the case of the presence in a GSPN of loops of immediate transitions that must be preliminary solved before constructing the Markov Chains that underlie these models. It is important to point out that the GSPN representations of the two product-form solution models, besides providing a simple alternative graphical representation of the two models, allow to (formally) clarify, using typical results of the GSPNs, issues of the two queueing network models that would otherwise be described using ambiguous and inaccurate natural language.

Remark 3 (Why do we need GSPNs?). It is important to point out that the GSPN representation of the two product-form solution models offers significant benefits. Not only it does provide an alternative graphical description of the two models, but it also allows for the formal clarification of issues that would otherwise be described using ambiguous and inaccurate natural language.

In fact, the model proposed in [20] results in a monolithic definition of the model, as it requires the explicit computation of the absorbing probabilities every time the model parameters change. In contrast, the GSPN representation, with its immediate transition semantic, allows the automatic computation of these absorbing probabilities by solving the cycles of immediate transitions. This approach also enables a modular definition of the model. In addition, the formal definition of the model in [20] also needs an accurate explanation in natural language to highlight the prioritization of certain actions with respect to others within the network dynamics. For instance, before retrieving a customer from an upstream queue, it is important to verify that there are no customers waiting in the queue. This meta-description plays a crucial role in accurately describing the model. However, thanks to GSPNs these aspects are now explicitly represented and incorporated in a rigorous and formal manner, allowing us to bypass the need for such additional descriptions.

A remarkable example of the cross fertilization between the Petri nets world and that of queuing networks concerns the routing process, that is, the Markovian process that describes the customers' flow among the various stations of the network. In the networks we are considering, due to blocking phenomena (in CSQNBs) or lack of customers available for fetching (in CFQNWs), the routing process, becomes dependent on the state (e.g., blocking /non-blocking).

An example of such situations can occur in the network of Fig. 2 when Station 5 starts a fetching event and (probabilistically) selects the upper path (Station 4) or the lower one (Station 3). If we assume that the state of the network is such that in Stations 3 and 1 there are no customers (except those in their service rooms), then an infinite fetching loop could occur. Station 5 picks up the customer from the service room of Station 3, which in turn will have to request it from Station 1, which in turn will request it again from Station 5. This cycle will be terminated with probability 1 when Station 5 chooses Station 4 to request a customer. Cases similar to those just described can be easily formalized in terms of GSPNs using typical concepts of this formalism such as tangible and vanishing states.

6 Duality Results Between CFQNWs and CSQNBs

In this section, we study a duality property between CFQNWs and CSQNBs. Intuitively, we show that the customers in one network correspond to the free spaces in the other and vice versa. As a consequence the routing directions in one network and its dual are inverse. However, it is much less intuitive to understand how the stations' service rates and the routing probabilities are defined.

To simplify the discussion, we consider the case in which every station has finite capacity. This is not a loss of generality of the result since the case of infinite capacity queues can be treated by reducing their capacity to any finite number greater or equal to the total number of customers in the network.

We will show that given a CFQNW, we can define a dual CSQNB. The inverse is also true and easily deducible from the results presented hereafter, but for the sake of brevity, we omit its description. Thus, the quantitative analysis of one of the two models immediately gives the results of the other and vice versa.

Definition 4 (Dual Network from CFQNW to CSQNB). *Given a CFQNW we define its dual CSQNB on the same set of stations Ω, as follows:*

- *the service rate at station i is $\mu_i^P = \sum_{j=1}^{N} p_{ji}^F \mu_j^F$*
- *the capacity of each station is the same for both networks, i.e., $\mathbf{c}^P = \mathbf{c}^F$.*
- *the routing matrix in the CSQNB is such $p_{ij}^F > 0 \Leftrightarrow p_{ji}^P > 0$. More precisely, we have:*

$$p_{ij}^P = \frac{\mu_j^F p_{ji}^F}{\sum_{k=1}^{N} \mu_k^F p_{ki}^F}.$$

- *the state space is $\mathcal{S}^P(\Omega, M^P, \mathbf{c})$ as defined in Eq. (1), where the number of customers in the CSQNB is given by $M^P = \left(\sum_{i \in \Omega} c_i^F\right) - M^F$*

Henceforth, we will simply write $\mathbf{c}$ and the type of network we are referring to will be clear from the context. Notice that the number of customers in the dual CSQNB is equal to the total number of free spaces in the CFQNW.

Lemma 1 (Service functions). *Given a CFQNW and its dual CSQNB, for each pair of stations $i, j \in \Omega$, it holds that: $\rho_i^F / \rho_j^F = \rho_j^P / \rho_i^P$.*

We say that the solution of the rate equations of the CFQNW is compatible with that of the traffic equations of the CSQNB if it holds that $\rho_i^F = 1/\rho_i^P$ for all $i \in \Omega$. Notice that, by Lemma 1 there exists one and only one compatible solution. Therefore, for compatible solutions of the traffic equations, we can write the following relationship among the service functions: $g_i^P(n_i) = 1/g_i^F(n_i)$. In this case, we will say that the service functions are *compatible*.

Remark 4 (Efficient solution of the system of rate equations). Lemma 1 has great practical importance. In fact, we notice that while the system of rate Eq. (7) has a number of unknowns bounded by N^2, the one of traffic equations of the dual CSQNB has only N equations. Therefore, we can compute the service functions of CFQNW with a complexity $\mathcal{O}(N^3)$ instead of $\mathcal{O}(N^6)$ by solving the traffic equations of the dual CSQNB.

Lemma 2 (Bijection between State Spaces). *Let $f : \mathcal{S}^F(\Omega, M^F, \mathbf{c}) \rightarrow \mathcal{S}^P(\Omega, M^P, \mathbf{c})$ be defined as $f(\mathbf{s}) = \mathbf{c} - \mathbf{s}$. f is a bijection between the networks' state spaces.*

Theorem 3 (Normalising Constant). *The normalizing constants of a CFQNW and its dual CSQNB with compatible service functions satisfy the following property:*

$$G^F(\Omega, M^F, \mathbf{c}) = \frac{G^P(\Omega, M^P, \mathbf{c})}{\prod_{i \in \Omega} \left(\rho_i^P\right)^{c_i}}.$$

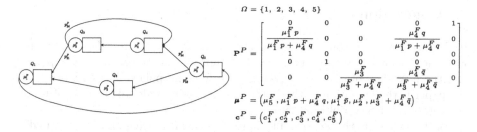

$$\Omega = \{1,\ 2,\ 3,\ 4,\ 5\}$$

$$\mathbf{P}^P = \begin{bmatrix} 0 & 0 & 0 & 0 & 1 \\ \mu_1^F p & 0 & 0 & \mu_4^F q & 0 \\ \overline{\mu_1^F p + \mu_4^F q} & & & \overline{\mu_1^F p + \mu_4^F q} & \\ \mu_1^F p + \mu_4^F q & 0 & 0 & 0 & 0 \\ 1 & 0 & 0 & 0 & 0 \\ 0 & 1 & 0 & 0 & 0 \\ 0 & 0 & \dfrac{\mu_3^F}{\mu_3^F + \mu_4^F \bar{q}} & \dfrac{\mu_4^F q}{\mu_3^F + \mu_4^F \bar{q}} & 0 \end{bmatrix}$$

$$\boldsymbol{\mu}^P = \left(\mu_5^F,\ \mu_1^F p + \mu_4^F q,\ \mu_1^F \bar{p},\ \mu_2^F,\ \mu_3^F + \mu_4^F \bar{q} \right)$$

$$\mathbf{c}^P = \left(c_1^F, c_2^F, c_3^F, c_4^F, c_5^F \right)$$

Fig. 5. Dual CSQNB of the CFQNW depicted in Fig. 2.

The importance of Theorem 3 is not just theoretical. In fact, it is well known that there exists a connection between the definition of the normalizing constant and some important results on the analysis of queuing network models in product-form such as the formulation of the *Arrival Theorem* [14,22]. The formulation of an Arrival Theorem for CFQNW and CSQNB is still an open problem since the standard formulation does not hold as observed in [7] because of the state dependent routing policy of both models. Theorem 3 suggests however that, as soon as an Arrival Theorem is identified and defined for one of the two models, a dual formulation should then be possible for the other too. Theorem 4 states the most important relationship on the performance indices of dual CSQNB and CFQNW.

Theorem 4 (Marginal Probabilities). *Given a CFQNW and its dual CSQNB, we have that $\pi_i^F(m) = \pi_i^P(c_i - m)$, where $\pi_i^F(m)$ is the marginal probability of having m customers at station i in a CFQNW and $\pi_i^P(c_i - m)$ is the marginal probability of having $c_i - m$ customers at station i in the CSQNB.*

Corollary 1 (Mean queue length). *Given a CFQNW and its dual CSQNB, we have that $\overline{N}_i^F = c_i - \overline{N}_i^P$, where $\overline{N}_i^F$ and $\overline{N}_i^P$ are the expected number of jobs at station i in the CFQNW and in the dual CSQNB, respectively.*

Example 2. We start from the CFQNW depicted in Fig. 2 and we derive its dual CSQNB for which we have to redefine routing probabilities and service rates. As far as the routing is concerned, we reverse the direction of the flows with respect to the original model and the new service rates of the different stations are computed as the sum of the flows entering into the same stations in the original network. Thus, we obtain the network depicted in Fig. 5.

Solving the system of traffic equations for the dual CSQNB, we obtain:

$$v_1 = v_5 = \frac{p_{21}^P}{p_{54}^P}, \qquad v_2 = v_4 = 1, \qquad v_3 = \frac{p_{21}^P p_{53}^P}{p_{54}^P}.$$

Accordingly, the $g_i's$ become the reciprocal of the ones in Eq. (8).

7   Conclusion

In this work, we have considered two classes of product-form models, namely, the well known class of queuing networks with finite capacity and skip-over policy and the new class of queuing networks with fetching policy. The first result of the paper is the extension of the latter product-form model to the case of finite capacities and RS-RD blocking policy.

Both models are characterized by state dependent routing policies that make their formal descriptions challenging tasks. In the literature, we can find either informal descriptions of the blocking or fetching policies or a formal description of the joint underlying Markov chains. The latter is not simple since it involves the computation of certain absorption probabilities of a number of Markov chains that, in the worst case, grows exponentially with the number of stations in the model. To cover this gap, we propose two GSPN models that are both modular and formally accurate. GSPNs turn out to be the ideal formalism to describe the state dependent routing in these cases thanks to the use of immediate transitions.

The final contribution of the work is the establishment of a relation between the finite capacity queuing networks with skip-over policies and those with fetching policies. We showed that there exists a non-trivial bijection between the two classes of models which allows interpreting the results in one class in terms of the other and vice versa. As a consequence of this result, the complexity of the unnormalized solution of queuing networks with fetching policy is reduced from $\mathcal{O}(N^6)$ to $\mathcal{O}(N^3)$ since a model with N stations requires the solution of N^2 rate equations, while a skip-over network has a set of only N traffic equations. Finally, we showed a relation between the normalizing constants of the two models and hence of their average performance indices.

Future works will be devoted to the study of the Arrival Theorem for these classes of models. Previously, the problem has been addressed only for skip-over networks and it is well known that the standard formulation of the result does not hold [7]. Nevertheless, the existence of a Mean Value Analysis algorithm for the computation of the performance measures of these models suggests that an interpretation in terms of the Arrival Theorem should be possible [24]. We conjecture that the connection between fetching and skip-over queuing networks presented in this work allows for a formulation of the Arrival Theorem for one of the two models that can be interpreted also in the dual one.

Finally, it is worth mentioning that the GSPN models of the CSQNB and CFQNW networks admit a product-form solution, but do not satisfy the product-form solution criteria that have been proposed in the literature [3]. This suggests that new research efforts must be devoted to the definition of more general characterizations of the product-form solution of models with immediate transitions and inhibitor arcs.

Acknowledgements. This work has been partially supported by project INdAM-GNCS 2023 *RISICO* CUP E53C22001930001.

References

1. Marsan, M.A., Balbo, G., Conte, G., Donatelli, S., Franceschinis, G.: Modelling with Generalized Stochastic Petri Nets. Wiley, Hoboken (1995)
2. Ajmone Marsan, M., Balbo, G., Conte, G.: A class of generalized stochastic petri nets for the performance evaluation of multiprocessor systems. ACM Trans. Comput. Syst. **2**(2), 93–122 (1984)
3. Balbo, G., Bruell, S.C., Sereno, M.: Product form solution for generalized stochastic petri nets. IEEE Trans. Software Eng. **28**(10), 915–932 (2002)
4. Balsamo, S., Harrison, P.G., Marin, A.: A unifying approach to product-forms in networks with finite capacity constraints. In: SIGMETRICS 2010, Proceedings of the 2010 ACM SIGMETRICS International Conference on Measurement and Modeling of Computer Systems, pp. 25–36. ACM (2010)
5. Balsamo, S., de Nitto Persone, V.: A survey of product form queueing networks with blocking and their equivalences. Ann. Oper. Res. **48**(1), 31–61 (1994)
6. Baskett, F., Chandy, K., Muntz, R., Palacios, F.: Open, closed and mixed networks of queues with different classes of customers. J. ACM **22**(2), 248–260 (1975)
7. Boucherie, R.J., van Dijk, N.M.: On the arrival theorem for product form queueing networks with blocking. Perform. Eval. **29**(3), 155–176 (1997)
8. Chao, X., Miyazawa, M., Pinedo, M.: Queueing Networks: Customers, Signals and Product Form Solutions. Wiley, Hoboken (1999)
9. Chiola, G., Ajmone Marsan, M., Balbo, G., Conte, G.: Generalized stochastic petri nets: a definition at the net level and its implications. IEEE Trans. Software Eng. **19**(2), 89–107 (1993)
10. Ciardo, G., Miner, A., Wan, M.: Advanced features in SMART: the stochastic model checking analyzer for reliability and timing. ACM SIGMETRICS Perform. Eval. Rev. **36**(4), 58–63 (2009)
11. Ciardo, G., Muppala, J., Trivedi, K.: On the solution of GSPN reward models. Perform. Eval. **12**(4), 237–253 (1991)
12. Gates, D., Westcott, M.: Stationary states of crystal growth in three dimensions. J. Stat. Phys. **81**, 681–715 (1995)
13. Gelenbe, E.: G-networks: multiple classes of positive customers, signals, and product form results. In: Calzarossa, M.C., Tucci, S. (eds.) Performance 2002. LNCS, vol. 2459, pp. 1–16. Springer, Heidelberg (2002). https://doi.org/10.1007/3-540-45798-4_1
14. Harchol-Balter, M.: Performance Modeling and Design of Computer Systems: Queueing Theory in Action. Cambridge Press, Cambridge (2013)
15. Harrison, P.G., Marin, A.: Product-forms in multi-way synchronizations. Comput. J. **57**(11), 1693–1710 (2014)
16. Jackson, J.R.: Networks of waiting lines. Oper. Res. **5**(4), 518–521 (1957)
17. Jackson, J.R.: Jobshop-like queueing systems. Manage. Sci. **10**(1), 131–142 (1963)
18. Kelly, F.P.: Reversibility and Stochastic Networks. Wiley, Hoboken (1979)
19. Lam, S.S.: Dynamic scaling and growth behavior of queueing network normalization constants. J. ACM **29**(2), 492–513 (1982)
20. Marin, A., Rossi, S., Olliaro, D.: A product-form network for systems with job stealing policies. SIGMETRICS Perform. Eval. Rev. **50**(4), 2–4 (2023)
21. Pittel, B.G.: Closed exponential networks of queues with saturation: the Jackson-type stationary distribution and its asymptotic analysis. Math. Oper. Res. **4**(4), 357–378 (1979)

22. Sevcik, K.C., Mitrani, I.: The distribution of queuing network states at input and output instants. J. ACM **28**(2), 358–371 (1981)
23. Towsley, D.F.: Queuing network models with state-dependent routing. J. ACM **27**(2), 323–337 (1980)
24. Van der Gaast, J., De Koster, R., Adan, I.J.B.F., Resing, J.A.C.: Capacity analysis of sequential zone picking systems. Oper. Res. **68**(1), 161–179 (2020)
25. Zhang, X., Xue, Q., Li, L., Fan, E., Wu, F., Chen, R.: Sustainable recycling and regeneration of cathode scraps from industrial production of lithium-ion batteries. ACS Sustain. Chem. Eng. **4**(12), 7041–7049 (2016)

On the Maximum Queue Length of the Hyper Scalable Load Balancing Push Strategy

Benny Van Houdt[(✉)]

Department of Computer Science, University of Antwerp - imec, Antwerp, Belgium
benny.vanhoudt@uantwerpen.be

Abstract. In this paper we derive explicit and structural results for the steady state probabilities of a structured finite state Markov chain. The study of these steady state probabilities is motivated by the analysis of the hyper scalable load balancing push strategy when using the queue-at-the-cavity approach. More specifically, these probabilities can be used to determine the largest possible arrival rate that can be supported by this strategy without exceeding some predefined maximum queue length. Contrary to prior work, we study the push strategy when the queue length information updates occur according to a phase-type renewal process with non-exponential inter-renewal times.

Keywords: load balancing · hyper scalable push strategy · bounded queue length

1 Introduction

Hyper scalable load balancing strategies for large-scale systems have received considerable attention recently [1–3]. These strategies further reduce the communication overhead of traditional load balancers such as the power-of-d-choices or join-idle-queue strategies [4–7], the overhead of which equals at least one message per job. One of the most fundamental hyper scalable load balancing strategies is the push strategy studied in [1,8]. It operates as follows. There is a single dispatcher that maintains an estimate on the queue length of each server. Jobs arrive at the dispatcher at rate λN, where N is the number of servers. Incoming jobs are assigned in a greedy manner, that is, the dispatcher assigns the job to a server with the smallest estimated queue length among all servers and increases its estimate by one. However, the dispatcher is not informed about job completions. Instead the queue length estimates are updated to their actual values at random points in time (meaning the estimates are upper bounds). Whether these queue length updates are triggered by the dispatcher or the servers does not matter in such case. The mean number of updates that occur per incoming job is a control parameter that can be set well below one.

An effective approach to study the performance of large-scale load balancing strategies is the so-called *queue-at-the-cavity* approach [9], which reflects the

N. Jansen and M. Tribastone (Eds.): QEST 2023, LNCS 14287, pp. 127–142, 2023.
https://doi.org/10.1007/978-3-031-43835-6_9

system behavior as the number of servers tends to infinity assuming asymptotic independence. For the hyper scalable push strategy described above, the corresponding queue-at-the-cavity has a bounded queue length m, the value of which can be determined by studying a structured finite state Markov chain in case of phase-type distributed job sizes [8]. The maximum queue length m grows as a function of the arrival rate λ and explicit results for the largest possible arrival rate $\lambda(m) \in (0, 1)$ that can be supported with a given maximum queue length m were presented in [8] in case of phase-type distributed job sizes with mean one and random server queue length updates. Random updates imply that the time between two updates of the queue length information of a tagged server follows an exponential distribution with some mean $1/\delta$. In this paper we derive explicit results for $\lambda(m)$ when the updates of the queue length of a tagged server follow a renewal process, but the inter-renewal time is not necessarily exponential.

The paper is structured as follows. Section 2 contains the problem statement. Results for the case with exponential job sizes are presented in Sect. 3, while in Sect. 4 we focus on non-exponential job sizes. Conclusions are found in Sect. 5.

2   Problem Statement

We assume that the job size distribution Z follows an order k_S phase type distribution characterized by (α, S) with mean $\alpha(-S)^{-1}e = 1$, that is, $P[Z > t] = \alpha \exp(St)e$, where e is a vector of ones of the appropriate size. The time between two updates X of a tagged server follows an order k_T phase type distribution characterized by (β, T) with mean $\beta(-T)^{-1}e = 1/\delta > 1$, meaning $P[X > t] = \beta \exp(Tt)e$. In other words, whenever the dispatcher updates its queue length information for the tagged server, a phase-type distributed timer is started and the dispatcher receives a new update each time this timer expires. Let $t^* = (-T)e$ and $s^* = (-S)e$. All timers and job sizes are assumed to be independent.

In [8, Section 4] it was shown that using the queue-at-the-cavity approach in case of *random updates*, the maximum queue length for the hyper scalable push strategy is the smallest m value such that the steady-state probability of being in the first state of the $1 + k_S m$ state Markov chain with the following rate matrix is less than $1 - \lambda$:

$$\begin{bmatrix} -\delta & & & & & \delta\alpha \\ s^* & S - \delta I & & & & \delta I \\ & s^*\alpha & S - \delta I & & & \delta I \\ & & \ddots & \ddots & & \vdots \\ & & & s^*\alpha & S - \delta I & \delta I \\ & & & & s^*\alpha & S \end{bmatrix}. \tag{1}$$

This Markov chain captures the evolution of the queue length of a tagged server that serves phase-type distributed jobs with representation (α, S) and that jumps up to length m each time an update occurs, where updates occur according to a Poisson process with rate δ.

It is not hard to see that when phase-type distributed timers are used instead to trigger updates, the maximum queue length of a server equals the smallest m value such that the probability of being in a state part of Ω_0 is less than $1 - \lambda$ in the Markov chain with state space

$$\Omega = \Omega_0 \cup (\cup_{\ell=1}^m \Omega_\ell),$$

where $\Omega_0 = \{(0, j) | j = 1, \ldots, k_T\}$ and $\Omega_\ell = \{(\ell, i, j) | i = 1, \ldots, k_S, j = 1, \ldots, k_T\}$ and rate matrix

$$Q(m) = \begin{bmatrix} T & & & & & \alpha \otimes t^*\beta \\ s^* \otimes I & S \oplus T & & & & I \otimes t^*\beta \\ & s^*\alpha \otimes I & S \oplus T & & & I \otimes t^*\beta \\ & & \ddots & \ddots & & \vdots \\ & & & s^*\alpha \otimes I & S \oplus T & I \otimes t^*\beta \\ & & & & s^*\alpha \otimes I & S \oplus (T + t^*\beta) \end{bmatrix}. \quad (2)$$

If we set $T = -\delta$, meaning $t^* = \delta$, $\beta = 1$ and $k_T = 1$, $Q(m)$ coincides with (1).

We now formally define $\lambda(m) \in (0, 1)$ as the largest possible arrival rate λ such that the maximum queue length of the queue-at-the-cavity of the push strategy is bounded by m given a mean job size equal to one.

Definition 1. *Let* $\lambda(m) = 1 - \pi_0(m)$, *where* $\pi_0(m)$ *is the steady probability to be in a state part of the set* Ω_0 *for the CTMC characterized by* $Q(m)$.

We use the following notations for some special cases:

1. If (β, T) has an Erlang-k_T distribution, we denote this rate as $\lambda_{k_T}(m)$.
2. If the job sizes are exponential, we denote this rate as $\lambda^{(exp)}(m)$.
3. If (β, T) has an Erlang-k_T distribution and job sizes are exponential, hyper exponential, Coxian or Erlang, we denote this rate as $\lambda_{k_T}^{(exp)}(m)$, $\lambda_{k_T}^{(HE)}(m)$, $\lambda_{k_T}^{(Cox)}(m)$ or $\lambda_{k_T}^{(Erl)}(m)$, respectively.

It is possible to develop an algorithm that runs in $O(k_T^3(k_S^3 + \log m))$ time to numerically compute $\pi_0(m)$, and thus $\lambda(m)$, by exploiting the structure of $Q(m)$. The main objective of this paper is however to derive closed from results for $\lambda(m)$ and to establish structural results. A first set of closed form results was presented in [3, Corollary 4.5] for exponential timers, that is,

$$\lambda_1(m) = \frac{\delta(1 - y(1)^m)}{\delta(1 - y(1)^m) + y(1)^{m-1}(1 - y(1))}, \quad (3)$$

where $y(1) = P[Z < X]$ is the probability that the job size Z with $E[Z] = 1$ is less than an exponential timer X with mean $1/\delta > 1$. Note that $\lambda_1(1)$ simplifies to $\delta/(1 + \delta)$ which is independent of the job size distribution Z. The case with $m = 1$ is of particular interest as it corresponds to the case where the queue length is bounded by one in the large-scale limit (assuming asymptotic independence), which corresponds to so-called *vanishing waiting times*.

3 Exponential Job Sizes

In this section we study the arrival rate $\lambda^{(exp)}(m)$ that can be supported such that the maximum queue length is bounded by m in case of exponential job sizes and phase-type distributed timers. We make the following contributions:

1. We prove that $\lambda^{(exp)}(m)$ is maximized over all order k_T phase-type distributions by the Erlang-k_T distribution (due to Theorem 2).
2. We present an explicit formula for $\lambda^{(exp)}_{k_T}(m)$ and prove that $\lambda^{(exp)}_{k_T}(m)$ increases as a function of k_T (see Theorem 3).
3. We derive the limiting expressions as k_T tends to infinity (see Theorem 3).

When the job sizes are exponential the state space reduces to

$$\Omega^{(exp)} = \cup_{\ell=0}^{m} \Omega_\ell^{(exp)} = \cup_{\ell=0}^{m} \{(\ell, j) | j = 1, \ldots, k_T\}.$$

and the rate matrix $Q(m)$ simplifies to

$$Q^{(exp)}(m) = \begin{bmatrix} T & & & & & t^*\beta \\ I & T-I & & & & t^*\beta \\ & I & T-I & & & t^*\beta \\ & & \ddots & \ddots & & \vdots \\ & & & I & T-I & t^*\beta \\ & & & & I & (T+t^*\beta)-I \end{bmatrix}. \tag{4}$$

Theorem 1. *Let $N_1(X)$ denote the number of arrivals of a Poisson process with rate 1 during an (β, T) phase type distributed time X. Denote $\pi_0^{(exp)}(m)$ as the steady state probability that the state of the CTMC with rate matrix $Q^{(exp)}(m)$ is in the set $\Omega_0^{(exp)}$, then*

$$\pi_0^{(exp)}(m) = 1 - \delta E[\min(N_1(X), m)].$$

This implies that $\pi_0^{(exp)}(m)$ is decreasing in m.

Proof. For the CTMC with rate matrix $Q^{(exp)}(m)$ we clearly have a renewal whenever the (β, T) phase-type distributed timer expires. The mean length of a renewal cycle therefore equals $1/\delta$ and its length has an order $(m+1)k_T$ phase-type distribution with initial vector $(\beta, 0, \ldots, 0)$ and subgenerator matrix

$$Q^{(cycle)}(m) = \begin{bmatrix} T-I & I & & & \\ & T-I & I & & \\ & & T-I & I & \\ & & & \ddots & \ddots & \\ & & & & T-I & I \\ & & & & & T \end{bmatrix}, \tag{5}$$

where we reordered the states. In order to express the mean time that the CTMC spends in the set $\Omega_\ell^{(exp)}$ during a cycle, we can focus on the first block row of the matrix $(-Q^{(cycle)}(m))^{-1}$. As $(-Q^{(cycle)}(m))^{-1}$ equals

$$
\begin{bmatrix}
(I-T)^{-1} & \cdots & (I-T)^{-m} & (I-T)^{-m}(-T)^{-1} \\
 & \ddots & \vdots & \vdots \\
 & & (I-T)^{-1} & (I-T)^{-1}(-T)^{-1} \\
 & & & (-T)^{-1}
\end{bmatrix},
$$

the mean time spend in the set $\Omega_\ell^{(exp)}$, with $\ell > 0$, per cycle can be expressed as $\beta(I-T)^{-(m-\ell+1)}e$ and therefore

$$
\pi_0^{(exp)}(m) = \frac{1/\delta - \sum_{\ell=1}^{m} \beta(I-T)^{-(m-\ell+1)}e}{1/\delta} = 1 - \delta \sum_{\ell=1}^{m} \beta(I-T)^{-\ell}e.
$$

Furthermore $\beta(I-T)^{-\ell}e$ is also the probability that $N_1(X)$ equals ℓ or more. Hence,

$$
\pi_0^{(exp)}(m) = 1 - \delta \sum_{\ell=1}^{m} P[N_1(X) > \ell-1] = 1 - \delta E[\min(N_1(X), m)].
$$

$\square$

Theorem 2. *Let $\pi_0^{(exp)}(m)$ be the steady state probability of the CTMC with rate matrix $Q^{(exp)}(m)$ to be in the set $\Omega_0^{(exp)}$, then $\pi_0^{(exp)}(m)$ is minimized over all order k_T phase type distributions (β, T) by the Erlang-k_T distribution. Moreover $\pi_0^{(exp)}(m)$ is decreasing in k_T when (β, T) is an Erlang-k_T distribution.*

Proof. Let X follow any order k_T phase type distribution characterized by (β, T) and let Y have an Erlang-k_T distribution. By Theorem 3 in [10] we have $Y \leq_{cx} X$ where $\leq_{cx}$ is the usual convex ordering between random variables with the same mean (see [11]). By Theorem 3.A.40 in [11] we therefore have $N_1(Y) \leq_{cx} N_1(X)$. Using 3.A.5 in [11], we have

$$
E[\max(N_1(Y), m)] \leq E[\max(N_1(X), m)],
$$

for any m. Clearly,

$$
E[N_1(X)] = E[\max(N_1(X), m)] + E[\min(N_1(X), m)] - m,
$$

and the same holds for Y, while both $E[N_1(X)]$ and $E[N_1(Y)]$ equal $1/\delta$. This allows us to conclude that

$$
E[\min(N_1(Y), m)] \geq E[\min(N_1(X), m)],
$$

and by the previous theorem $\pi_0^{(exp)}(m)$ is minimized by the Erlang-k_T distribution over all order k_T phase-type distributions.

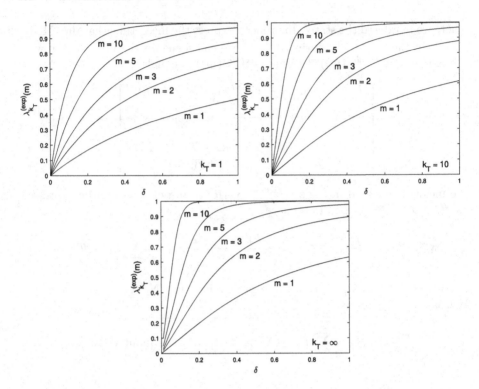

Fig. 1. Illustration of Theorem 3: exponential job sizes and Erlang-k_T timers with $k_T = 1, 10$ and ∞.

The fact that $\pi_0^{(exp)}(m)$ is decreasing in k_T when (β, T) has an Erlang-k_T distribution follows by noting that the Erlang k_T distribution also has an order $k_T + 1$ phase-type representation and therefore $Z_{k_T+1} \leq_{cx} Z_{k_T}$ with Z_k an Erlang-k random variable. Following the same argument as above implies

$$E[\min(N_1(Z_{k_T+1}), m)] \geq E[\min(N_1(Z_{k_T}), m)],$$

which concludes the proof.   □

Theorem 3. *For an Erlang-k_T timer and exponential job sizes we have*

$$\lambda_{k_T}^{(exp)}(m) = \delta m - \delta \left(\frac{\delta k_T}{\delta k_T + 1} \right)^{k_T} \sum_{n=0}^{m-1} \frac{m-n}{(\delta k_T + 1)^n} \binom{k_T + n - 1}{n} \tag{6}$$

$$= 1 - \frac{1}{(\delta k_T + 1)^m} \sum_{j=0}^{k_T-1} \left(1 - \frac{j}{k_T} \right) \binom{m+j-1}{j} \left(\frac{\delta k_T}{\delta k_T + 1} \right)^j \tag{7}$$

Further,

$$\lim_{k_T \to \infty} \lambda_{k_T}^{(exp)}(m) = \delta m - \delta e^{-1/\delta} \sum_{n=0}^{m-1} \frac{m-n}{\delta^n n!} \qquad (8)$$

and $\lim_{m \to \infty} \lambda_{k_T}^{(exp)}(m) = 1$.

Proof. We first establish (6). Using Theorem 1 we know that $\lambda(m)$ can be written as $\delta E[\min(N_1(X), m)]$, which is equivalent to stating that

$$\lambda(m)/\delta = E[\min(N_1(X), m)] = \sum_{j=0}^{m-1} P[N_1(X) > j]$$

$$= m - \sum_{n=0}^{m-1} (m-n) P[N_1(X) = n]$$

$$= m - \left(\frac{\delta k_T}{\delta k_T + 1}\right)^{k_T} \sum_{n=0}^{m-1} \frac{m-n}{(\delta k_T + 1)^n} \binom{k_T + n - 1}{n},$$

as $P[N_1(X) = n]$ if there are n arrivals (these occur at rate 1) and $k_T - 1$ phase changes of the timer (these occur at rate δk_T) in the next $n + k_T - 1$ events, followed by the expiration of the timer.

To obtain (7) we look at the mean time that the order $(m+1)k_T$ phase-type distribution characterized by $(\beta, 0, \dots, 0)$ and $Q^{(cycle)}$ given by (5) spends in the set of states $\Omega_0^{(exp)}$. The probability that this set is reached during a cycle via state $(0, j+1) \in \Omega_0^{(exp)}$ is given by

$$\binom{m+j-1}{j} \left(\frac{1}{\delta k_T + 1}\right)^m \left(\frac{\delta k_T}{\delta k_T + 1}\right)^j,$$

and the mean time that we spend in the set $\Omega_0^{(exp)}$ given that we are in state $(0, j)$ equals $(k_T - j)/(\delta k_T)$. As $\pi_0^{(exp)}(m)$ equals δ times this mean time, we have

$$1 - \lambda(m) = \frac{1}{(\delta k_T + 1)^m} \sum_{j=0}^{k_T - 1} \left(1 - \frac{j}{k_T}\right) \binom{m+j-1}{j} \left(\frac{\delta k_T}{\delta k_T + 1}\right)^j.$$

The expression for the limit in (8) is immediate from (6) as $(\delta k_T/(\delta k_T + 1))^{k_T}$ converges to $e^{-1/\delta}$ and $\binom{k_T + n - 1}{n}/(\delta k_T + 1)^n$ converges to $1/(\delta^n n!)$ as k_T tends to infinity. The second limit is immediate from (7) as $\lim_{m \to \infty} \binom{m-j+1}{j}/(\delta k_T + 1)^m = \lim_{m \to \infty} m^{j-1}/j!(\delta k_T + 1)^m = 0$. $\qquad \square$

Remarks:

1. By (6) we have vanishing waits with Erlang-k_T timers and exponential job sizes if and only if

$$\lambda < \lambda_{k_T}^{(exp)}(1) = \delta - \delta \left(\frac{\delta k_T}{\delta k_T + 1} \right)^{k_T}, \tag{9}$$

   which increases to $\delta(1 - e^{-\delta})$ as k_T tends to infinity.

2. Using (7) we can compute $\lambda(m)$ in $O(k_T + \log m)$ time. Indeed, $(\delta k_T + 1)^m$ can be computed in $\log m$ time and the sum in $O(k_T)$ time as $\binom{m+j-1}{j} = \binom{m+j-2}{j-1}(m+j-1)/j$. Similarly (6) allows us to compute $\lambda(m)$ in $O(m + \log k_T)$ time. Theorem 3 is illustrated in Fig. 1.

4   Vanishing Waiting Times: $m = 1$

When both the job sizes and timers follow a general phase-type distribution, it appears hard to find elegant closed form results. As such we limit ourselves to the case where $m = 1$. Recall that this case is of particular interest as it corresponds to vanishing waiting times in the large-scale limit for the hyper scalable push strategy (assuming asymptotic independence). For exponential timers we know that $\lambda_1(1) = \delta/(\delta + 1)$, which does not depend on the job size distribution. This insensitivity is however lost when timers are not exponential. Given the results in the previous section, we focus on Erlang k_T distributed timers, that is, we focus on $\lambda_{k_T}(1)$ in this section for various job size distributions.

This section contains the following contributions for hyper exponential (HE), Coxian (Cox) and Erlang (Erl) job sizes:

1. We present an explicit formula for $\lambda_{k_T}^{(HE)}(1)$ (see Theorem 4).
2. Tight lower and upper bounds for $\lambda_{k_T}^{(HE)}(1)$ are presented that hold for any HE job size distribution (see Theorem 5).
3. An explicit formula for $\lambda_{k_T}^{(Cox)}(1)$ is derived (see Theorem 6).
4. An explicit formula for $\lambda_{k_T}^{(Erl)}(1)$ is presented (see Theorem 7).

We first present a lemma that is based on the following observation. A renewal occurs for the Markov chain characterized by $Q(m)$ each time the timer expires and the state is in Ω_0. During such a cycle the timer may expire a number of times C_N before the set Ω_0 is reached. Once this set is reached, the cycle ends when the timer expires one more time. Let Y_n denote the service phase of the job when the timer expires for the n-th time during a cycle in the event that $C_N \geq n$.

Lemma 1. *The arrival rate $\lambda(1)$ can be expressed as*

$$\lambda(1) = \delta \left/ \left(1 + \sum_{s=1}^{k_S} \sum_{n \geq 1} P[Y_n = s, C_N \geq n] \right) \right.$$

Proof. A renewal occurs each time that the set Ω_0 is left. For $m = 1$ the time spend in the set Ω_1 clearly equals 1 as the state remains in Ω_1 until the job completes. The value of $\lambda(1)$ can therefore be expressed as 1 divided by the mean length of a cycle, which we denote as $E[C]$.

A cycle ends when the timer expires and the state is in the set Ω_0. As $1/\delta$ is the mean time for the timer to expire, the mean cycle length can be expressed as

$$E[C] = \frac{1}{\delta}(E[C_N] + 1),$$

where C_N reflects the number of times that the timer expires before the job completes. This number can be expressed as

$$E[C_N] = \sum_{n \geq 1} P[C_N \geq n] = \sum_{s=1}^{k_S} \sum_{n \geq 1} P[Y_n = s, C_N \geq n],$$

where Y_n is the phase of the job when the timer expires for the n-th time in a cycle. Note that Y_n is well defined when $C_N \geq n$. □

4.1   Hyper Exponential Job Sizes

The next theorem allows us to compute $\lambda_{k_T}^{(HE)}(1)$ in $O(k_S \log k_T)$ time.

Theorem 4. *For Erlang-k_T timers and HE job sizes we have*

$$\lambda_{k_T}^{(HE)}(1) = \delta \left/ \sum_{i=1}^{k_S} \frac{p_i}{1 - \left(\frac{\delta k_T}{\delta k_T + \mu_i}\right)^{k_T}} \right. , \tag{10}$$

where p_i is the probability that a job has an exponential size with mean $1/\mu_i$.

Proof. We make use of Lemma 1. With probability p_i the service phase equals i and remains the same as long as the job is in service. Hence,

$$P[Y_n = i, C_N \geq n] = p_i \left(\frac{\delta k_T}{\delta k_T + \mu_i}\right)^{n k_T}.$$

Summing over n and i and writing $1 = \sum_{i=1}^{k_S} p_i$ yields the result. □

Lemma 2. *The function $\xi(x)$ given by*

$$\xi(x) = \frac{1}{1 - (\delta k_T)^{k_T}/(\delta k_T + 1/x)^{k_T}}, \tag{11}$$

is convex on $(0, \infty)$.

Proof. We have

$$\xi''(x) = \frac{k_T(\delta k_T + 1/x)^{k_T}(\delta k_T)^{k_T}}{x^2}\frac{\eta(x)}{\delta k_T + x}\left(\frac{1}{(\delta k_T+1/x)^{k_T}-(\delta k_T)^{k_T}}\right)^3,$$

with

$$\eta(x) = (k_T+1)(\delta k_T)^{k_T} + (k_T-1)(\delta k_T + 1/x)^{k_T}$$
$$- 2(\delta k_T)x\left((\delta k_T+1/x)^{k_T}-(\delta k_T)^{k_T}\right).$$

It therefore suffices to show that $\eta(x) \geq 0$ on $(0,\infty)$. Expanding the k_T-th powers implies that $\eta(x)$ equals

$$(k_T+1)(\delta k_T)^{k_T} + (k_T-1)\sum_{j=0}^{k_T}\binom{k_T}{j}\frac{(\delta k_T)^{k_T-j}}{x^j} - 2\sum_{j=1}^{k_T}\binom{k_T}{j}\frac{(\delta k_T)^{k_T-j+1}}{x^{j-1}}$$

$$= \frac{k_T-1}{x^{k_T}} + \sum_{j=1}^{k_T-1}\left((k_T-1)\binom{k_T}{j}-2\binom{k_T}{j+1}\right)\frac{(\delta k_T)^{k_T-j}}{x^j},$$

which is non-negative on $(0,\infty)$ as

$$(k_T-1)\binom{k_T}{j}-2\binom{k_T}{j+1}\geq 0,$$

if and only if $(j-1)(k_T+1) \geq 0$.   □

Theorem 5. *For Erlang-k_T timers and HE job sizes we have for $\lambda_{k_T}^{(HE)}(1)$ that*

$$\frac{\delta}{1+\delta} = \lambda_1^{(exp)}(1) = \lambda_1^{(HE)}(1) \leq \lambda_{k_T}^{(HE)}(1) \leq \lambda_{k_T}^{(exp)}(1) = \delta\left(1-\frac{(\delta k_T)^{k_T}}{(\delta k_T+1)^{k_T}}\right)$$

Proof. The result follows by noting that

$$\lambda_{k_T}^{(HE)}(1) = \delta\Big/\sum_{i=1}^{k_S}p_i\xi(1/\mu_1).$$

Therefore by the convexity of $\xi(x)$ on $(0,\infty)$, we have

$$\lambda_{k_T}^{(HE)}(1) = \delta\Big/\sum_{i=1}^{k_S}p_i\xi(1/\mu_i) \leq \delta/\xi(\sum_{i=1}^{k_S}p_i/\mu_i) = \delta/\xi(1) = \lambda_{k_T}^{(exp)}(1).$$

□

Remarks:

1. The upper bound is clearly tight, while the lower bound for a fixed k_T is also tight by using the 2-phase HE distribution with $p_1 = 1-\epsilon$, $p_2 = \epsilon$, $\mu_1 = (1-\epsilon)/\epsilon$ and $\mu_2 = \epsilon/(1-\epsilon)$ as in such case $\lim_{\epsilon\to 0}\lambda_{k_T}^{(HE)}(1) = \delta/(\delta+1)$.

4.2 Coxian Job Sizes

In this section we consider Coxian job sizes, meaning $\alpha = (1, 0, \ldots, 0)$ and

$$
S = \begin{bmatrix}
-\mu_1 & \mu_1 p_1 & & & \\
& -\mu_2 & \mu_2 p_2 & & \\
& & \ddots & \ddots & \\
& & & \mu_{k_S - 1} & \mu_{k_S - 1} p_{k_S - 1} \\
& & & & \mu_{k_S}
\end{bmatrix},
$$

with μ_i for $i = 1, \ldots, k_S$ and $0 < p_i < 1$ for $i = 1, \ldots, k_S - 1$. We note that any acyclic phase-type distribution, that is, any phase-type distribution where S is upper triangular, can be represented as a Coxian distribution [12].

Lemma 3. *Let $x_1 \neq \ldots \neq x_s \in \mathbb{R}$, then for $n \geq 1$*

$$
\sum_{\substack{j_1, \ldots, j_s \geq 0 \\ j_1 + \ldots + j_s = n-1}} \prod_{i=1}^{s} x_i^{j_i} = \sum_{i=1}^{s} x_i^{n+s-2} \prod_{\substack{\ell=1 \\ \ell \neq i}}^{s} \frac{1}{x_i - x_\ell} \tag{12}
$$

Proof. The equality in (12) is a known identity for complete homogeneous symmetric functions [13, Ex 7.4]. □

Remarks:

1. The result also holds for $n = 0$, that is,

$$
\sum_{i=1}^{s} x_i^{s-2} \prod_{\substack{\ell=1 \\ \ell \neq i}}^{s} \frac{1}{x_i - x_\ell} = 0. \tag{13}
$$

This is easily checked for $s = 2$. For $s > 2$ this follows from Lagrange's interpolation formula as the polynomial $p(x) = x^{s-2}$ interpolates the points (x_i, x_i^{s-2}) for $i = 1, \ldots, s - 1$ and therefore

$$
p(x_s) = x_s^{s-2} = \sum_{i=1}^{s-1} x_i^{s-2} \prod_{\substack{\ell=1 \\ \ell \neq i}}^{s-1} \frac{x_s - x_\ell}{x_i - x_\ell} = -\sum_{i=1}^{s-1} x_i^{s-2} \frac{\prod_{\ell=1}^{s-1}(x_s - x_\ell)}{\prod_{\substack{\ell=1 \\ \ell \neq i}}^{s}(x_i - x_\ell)}.
$$

Theorem 6. *For Erlang-k_T timers and Coxian job sizes with probabilities $p_1, \ldots, p_{k_S - 1}$ and rates $\mu_1 \neq \ldots \neq \mu_{k_S}$ we have*

$$
\lambda_{k_T}^{(Cox)}(1) = \delta \Bigg/ \sum_{i=1}^{k_S} \frac{\hat{p}_i}{1 - \left(\frac{\delta k_T}{\delta k_T + \mu_i}\right)^{k_T}}, \tag{14}
$$

where $\hat{p}_i$ for $i = 1, \ldots, k_S$ is given by

$$
\hat{p}_i = \sum_{s=i}^{k_S} \left(\prod_{j=1}^{s-1} \mu_j p_j \right) \prod_{\substack{\ell=1 \\ \ell \neq i}}^{s} \frac{1}{\mu_\ell - \mu_i}. \tag{15}
$$

138   B. Van Houdt

Proof. We rely on Lemma 1. Clearly, as the initial service phase is one when a cycle starts, we have $P[Y_n = 1, C_N \geq n] = \left(\frac{\delta k_T}{\delta k_T + \mu_1}\right)^{nk_T}$ which yields

$$1 + \sum_{n \geq 1} P[Y_n = 1, C_N \geq n] = 1 \left/ 1 - \left(\frac{\delta k_T}{\delta k_T + \mu_1}\right)^{k_T}\right. .$$

For $s > 1$, we get the more involved expression

$$P[Y_N = s, C_N \geq n]$$

$$= \sum_{\substack{j_1,\ldots,j_s \geq 0 \\ j_1 + \ldots + j_s = nk_T - 1}} \left(\prod_{i=1}^{s-1} \left(\frac{\delta k_T}{\delta k_T + \mu_i}\right)^{j_i} \frac{\mu_i p_i}{\delta k_T + \mu_i}\right) \left(\frac{\delta k_T}{\delta k_T + \mu_s}\right)^{j_s + 1}$$

$$= \frac{\delta k_T}{\delta k_T + \mu_s} \left(\prod_{j=1}^{s-1} \frac{\mu_j p_j}{\delta k_T + \mu_j}\right) \sum_{\substack{j_1,\ldots,j_s \geq 0 \\ j_1 + \ldots + j_s = nk_T - 1}} \prod_{i=1}^{s} x_i^{j_i},$$

$$= \frac{\delta k_T}{\delta k_T + \mu_s} \left(\prod_{j=1}^{s-1} \frac{\mu_j p_j}{\delta k_T + \mu_j}\right) \sum_{i=1}^{s} x_i^{nk_T + s - 2} \prod_{\substack{\ell=1 \\ \ell \neq i}}^{s} \frac{1}{x_i - x_\ell} \quad (16)$$

with $x_i = \delta k_T / (\delta k_T + \mu_i)$ due to Lemma 3. For $s > 1$ we therefore have

$$\sum_{n \geq 1} P[Y_N = s, C_N \geq n]$$

$$= \frac{\delta k_T}{\delta k_T + \mu_s} \left(\prod_{j=1}^{s-1} \frac{\mu_j p_j}{\delta k_T + \mu_j}\right) \sum_{n \geq 0} \sum_{i=1}^{s} x_i^{nk_T + s - 2} \prod_{\substack{\ell=1 \\ \ell \neq i}}^{s} \frac{1}{x_i - x_\ell}, \quad (17)$$

where the sum may start in $n = 0$ due to (13).

By definition of x_i, we have for $i \leq s$

$$x_i^{s-1} \prod_{\substack{\ell=1 \\ \ell \neq i}}^{s} \frac{1}{x_i - x_\ell} = \prod_{\substack{\ell=1 \\ \ell \neq i}}^{s} \frac{\delta k_T + \mu_\ell}{\mu_\ell - \mu_i},$$

which combined with (17) implies for $s > 1$

$$\sum_{n \geq 1} P[Y_N = s, C_N \geq n]$$

$$= \frac{\delta k_T}{\delta k_T + \mu_s} \left(\prod_{j=1}^{s-1} \frac{\mu_j p_j}{\delta k_T + \mu_j}\right) \sum_{n \geq 0} \sum_{i=1}^{s} x_i^{nk_T - 1} \prod_{\substack{\ell=1 \\ \ell \neq i}}^{s} \frac{\delta k_T + \mu_\ell}{\mu_\ell - \mu_i}$$

$$= \left(\prod_{j=1}^{s-1} \mu_j p_j\right) \sum_{i=1}^{s} \left(\sum_{n \geq 0} x_i^{nk_T}\right) \prod_{\substack{\ell=1 \\ \ell \neq i}}^{s} \frac{1}{\mu_\ell - \mu_i}.$$

Therefore,

$$\sum_{n\geq 1} P[Y_N = s, C_N \geq n] = \sum_{i=1}^{s}\left(\prod_{j=1}^{s-1}\mu_j p_j\right)\prod_{\substack{\ell=1\\\ell\neq i}}^{s}\frac{1}{\mu_\ell - \mu_i}\Bigg/\left(1-\left(\frac{\delta k_T}{\delta k_T + \mu_i}\right)^{k_T}\right),$$

and

$$1 + E[C_N] = 1 + \sum_{s=1}^{k_S}\sum_{n\geq 1} P[Y_N = s, C_N \geq n] = \frac{1}{1-\left(\frac{\delta k_T}{\delta k_T + \mu_i}\right)^{k_T}}$$

$$+ \sum_{s=2}^{k_S}\sum_{i=1}^{s}\left(\prod_{j=1}^{s-1}\mu_j p_j\right)\prod_{\substack{\ell=1\\\ell\neq i}}^{s}\frac{1}{\mu_\ell - \mu_i}\Bigg/\left(1-\left(\frac{\delta k_T}{\delta k_T + \mu_i}\right)^{k_T}\right)$$

$$= \sum_{i=1}^{k_S}\hat{p}_i\Bigg/\left(1-\left(\frac{\delta k_T}{\delta k_T + \mu_i}\right)^{k_T}\right)$$

□

Remarks:

1. The sum of the $\tilde{p}_i$ equals one as

$$\sum_{i=1}^{k_S}\tilde{p}_i = \sum_{s=1}^{k_S}\left(\prod_{j=1}^{s-1}\mu_j p_j\right)\sum_{i=1}^{s}\prod_{\substack{\ell=1\\\ell\neq i}}^{s}\frac{1}{\mu_\ell - \mu_i},$$

   as the latter sum equals zero for $s > 1$. However $\tilde{p}_i$ is not necessarily between 0 and 1. For instance, when $\mu_1 = 3/2, p_1 = 1$ and $\mu_2 = 3$, we get $\tilde{p}_1 = 2$ and $\tilde{p}_2 = -1$. This also indicates that Theorem 6 does not have a simple probabilistic interpretation.
2. When $\tilde{p}_i \in [0, 1]$ for $i = 1, \ldots, k_S$ the Coxian job size distribution corresponds to an HE distribution where the job is exponential with parameter μ_i with probability $\tilde{p}_i$. In fact, any HE distribution can be represented as a Coxian distribution [12,14] and as such Theorem 6 can be regarded as a generalization of Theorem 4.
3. Using (14) we can compute $\lambda_{k_T}^{(Cox)}(1)$ in $O(k_S^2 + k_S \log k_T)$ time, where the computation of the $\hat{p}_i$ values require $O(k_S^2)$ time.
4. If some of the μ_i are identical we can still derive an explicit expression by taking limits. For instance, for an order 2 Coxian distribution with service rate μ in both phases this leads to the following formula:

$$\lambda_{k_T}^{(Cox)}(1) = \delta\Bigg/\frac{1}{1-\left(\frac{\delta k_T}{\delta k_T + \mu}\right)^{k_T}} + \frac{p_1\mu k_T}{\delta k_T + \mu}\frac{\left(\frac{\delta k_T}{\delta k_T + \mu}\right)^{k_T}}{\left(1-\left(\frac{\delta k_T}{\delta k_T + \mu}\right)^{k_T}\right)^2}.$$

4.3   Erlang k_S Job Sizes

While results for Erlang distributed job sizes can in principle be derived from Theorem 6 by taking limits, this leads to very involved expressions for larger values of k_S. We therefore present an alternate approach in this section.

Lemma 4. *For $s \geq 1$ and $|x| < 1$, we have*

$$\sum_{n \geq 1} \binom{kn + s - 2}{s - 1} x^{kn} = \frac{1}{k} \sum_{j=1}^{k} \frac{w_k^j x}{(1 - w_k^j x)^s}, \tag{18}$$

with $w_k^j = \cos 2\pi j/k + i \sin 2\pi j/k \in \mathbb{C}$ the k-th roots of unity.

Proof. The orthogonal relation for the k-th roots of unity states that

$$\frac{1}{k} \sum_{j=1}^{k} w_k^{jn} = \begin{cases} 1 \text{ if } n \text{ is a multiple of } k \\ 0 \text{ otherwise.} \end{cases}$$

This implies

$$\sum_{n \geq 1} \binom{kn + s - 2}{s - 1} x^{kn} = \sum_{n \geq 1} \binom{n + s - 2}{s - 1} x^n \left(\frac{1}{k} \sum_{j=1}^{k} w_k^{jn} \right)$$

$$= \frac{1}{k} \sum_{j=1}^{k} \sum_{n \geq 1} \binom{n + s - 2}{s - 1} (x w_k^j)^n = \frac{1}{k} \sum_{j=1}^{k} \frac{w_k^j x}{(1 - w_k^j x)^s},$$

as $\sum_{n \geq 0} \binom{n+s-2}{s-1} x^n = x/(1 - x)^s$.   □

Theorem 7. *For Erlang-k_S job sizes and Erlang-k_T timers, we have*

$$\lambda_{k_T}^{(Erl)}(1) = \delta \left/ \left(1 + \frac{1}{k_T} \sum_{j=1}^{k_T} w_{k_T}^j \sum_{s=1}^{k_S} \frac{x(1 - x)^{s-1}}{(1 - w_{k_T}^j x)^s} \right) \right., \tag{19}$$

with $x = \delta k_T/(\delta k_T + k_S)$ and $w_k^j = \cos 2\pi j/k + i \sin 2\pi j/k \in \mathbb{C}$.

Proof. The proof makes use of Lemma 1. For Erlang k_S job sizes we note that the expression for $P[Y_n = s, C_N \geq n]$ for $s \geq 1$ becomes

$$P[Y_n = s, C_N \geq n] = \binom{nk_T + s - 2}{s - 1} \left(\frac{k_S}{\delta k_T + k_S} \right)^{s-1} \left(\frac{\delta k_T}{\delta k_T + k_S} \right)^{nk_T}.$$

Lemma 4 now suffices to complete the proof.   □

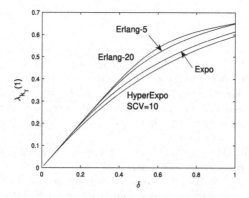

Fig. 2. Illustration of Theorem 4 and 7: Erlang, Exponential and Hyper Exponential job sizes (with balanced means) and Erlang-10 timers.

Remarks:

1. By means of (19) we can compute $\lambda_{k_T}^{(Erl)}(1)$ in $O(k_T k_S)$ time.
2. The result can easily be generalized to mixtures of Erlang distributions (mErl). Suppose that with probability p_i the job size is an order $k_{S,i}$ Erlang with rate μ_i, for $i = 1, \ldots, v$, then

$$\lambda_{k_T}^{(mErl)}(1) = \delta \left/ 1 + \frac{1}{k_T} \sum_{j=1}^{k_T} w_{k_T}^j \sum_{i=1}^{v} p_i \sum_{s=1}^{k_{S,i}} \frac{x_i(1 - x_i)^{s-1}}{(1 - w_{k_T}^j x_i)^s} \right. ,$$

with $x_i = \delta k_T / (\delta k_T + \mu_i)$.
3. Theorem 4 and 7 are illustrated in Fig. 2. Less variable job sizes imply that higher rates can be supported while still having vanishing waiting times.

5   Conclusions

In this paper we studied the steady state probabilities to be in the set Ω_0 of the structured finite state Markov chain with rate matrix $Q(m)$. The study of this Markov chain was motivated by the largest possible arrival rate $\lambda(m)$ that can be supported by the hyper scalable load balancing push strategy such that the queue length is bounded by some predefined maximum m.

More specifically, the following contributions were made. For exponential job sizes we showed that $\lambda(m)$ is maximized among all order k_T phase type distributions by the Erlang k_T distribution and presented explicit formulas for this maximum $\lambda_{k_T}(m)$. For non-exponential job sizes we focussed on the setting with vanishing waiting times, i.e., $m = 1$ and derived closed form expressions for $\lambda_{k_T}(1)$ for various job size distributions such as hyper exponential, Coxian and Erlang distributions.

References

1. van der Boor, M., Borst, S., van Leeuwaarden, J.: Hyper-scalable JSQ with sparse feedback. Proc. ACM Measur. Anal. Comput. Syst. **3**(1), 1–37 (2019)
2. van der Boor, M., Borst, S., van Leeuwaarden, J.: Optimal hyper-scalable load balancing with a strict queue limit. Perform. Eval. **149**, 102217 (2021)
3. Hellemans, T., Kielanski, G., Van Houdt, B.: Performance of load balancers with bounded maximum queue length in case of non-exponential job sizes. IEEE/ACM Trans. Netw. (2022, to appear)
4. Mitzenmacher, M.: The power of two choices in randomized load balancing. IEEE Trans. Parallel Distrib. Syst. **12**, 1094–1104 (2001)
5. Vvedenskaya, N., Tsybakov, B.: Random multiple access of packets to a channel with errors. Problemy Peredachi Informatsii **19**(2), 69–84 (1983)
6. Lu, Y., Xie, Q., Kliot, G., Geller, A., Larus, J.R., Greenberg, A.: Join-idle-queue: a novel load balancing algorithm for dynamically scalable web services. Perform. Eval. **68**, 1056–1071 (2011)
7. Stolyar, A.: Pull-based load distribution in large-scale heterogeneous service systems. Queueing Syst. **80**(4), 341–361 (2015). https://doi.org/10.1007/s11134-015-9448-8
8. Hellemans, T., Kielanski, G., Van Houdt, B.: Performance of load balancers with bounded maximum queue length in case of non-exponential job sizes, arXiv preprint arXiv.org/abs/2201.03905 (2022)
9. Bramson, M., Lu, Y., Prabhakar, B.: Randomized load balancing with general service time distributions. In: ACM SIGMETRICS 2010, pp. 275–286 (2010). http://doi.acm.org/10.1145/1811039.1811071
10. O'Cinneide, C.: Phase-type distributions and majorizations. Ann. Appl. Probab. **1**(2), 219–227 (1991)
11. Shaked, M., Shanthikumar, J.G.: Stochastic Orders and their Applications. Associated Press, New York (1994)
12. Cumani, A.: On the canonical representation of homogeneous markov processes modelling failure - time distributions. Microelectron. Reliab. **22**(3), 583–602 (1982). http://www.sciencedirect.com/science/article/pii/0026271482900336
13. Stanley, R.P., Fomin, S.: Enumerative Combinatorics. Cambridge Studies in Advanced Mathematics, vol. 2. Cambridge University Press, Cambridge (1999)
14. Van Houdt, B.: Global attraction of ODE-based mean field models with hyperexponential job sizes. Proc. ACM Meas. Anal. Comput. Syst. **3**(2), Article 23 (2019). https://doi.org/10.1145/3326137

Data-Driven Inference of Chemical Reaction Networks via Graph-Based Variational Autoencoders

Luca Bortolussi[1], Francesca Cairoli[1], Julia Klein[2,3]([⊠]), and Tatjana Petrov[1,2,3]

[1] Department of Mathematics and Geosciences, University of Trieste, Trieste, Italy
[2] Department of Computer and Information Sciences, University of Konstanz, Konstanz, Germany
julia.klein@uni-konstanz.de
[3] Centre for the Advanced Study of Collective Behaviour, University of Konstanz, Konstanz, Germany

Abstract. We propose a data-driven machine learning framework that automatically infers an explicit representation of a Chemical Reaction Network (CRN) together with its dynamics. The contribution is twofold: on one hand, our technique can be used to alleviate the computational burden of simulating a complex, multi-scale stochastic system; on the other hand, it can be used to extract an interpretable model from data. Our methodology is inspired by Neural Relational Inference and implements a graph-based Variational Autoencoder with the following structure: the encoder maps the observed trajectories into a representation of the CRN structure as a bipartite graph, and the decoder infers the respective reaction rates. Finally, the first two moments of the stochastic dynamics are computed with the standard linear noise approximation algorithm. Our current implementation demonstrates the applicability of the framework to single-reaction systems. Extending the framework towards inferring more complex CRN in a fully automated and data-driven manner involves implementation challenges related to neural network architecture and hyperparameter search, and is a work in progress.

1 Introduction

Complex phenomena in systems and synthetic biology are commonly modelled by Chemical Reaction Networks (CRNs), which are able to capture the underlying nonlinear and stochastic dynamics. Extracting the CRN structure from observed data is a non-trivial task that requires a high level of ingenuity and domain knowledge. Recent works [1–3,5] leverage machine learning approaches to abstract the dynamics of complex, stochastic CRNs from data. However, these approaches provide black-box abstractions unable to inform on the mechanisms underlying the observable dynamics.

This work proposes a data-driven machine learning framework that automatically infers an explicit representation of a CRN, i.e., it simultaneously infers the

N. Jansen and M. Tribastone (Eds.): QEST 2023, LNCS 14287, pp. 143–147, 2023.
https://doi.org/10.1007/978-3-031-43835-6_10

reactions and the accompanying kinetic rates. We build on Neural Relational Inference (NRI) [4], where a graph-based Variational Autoencoder (VAE) [6] simultaneously infers an interaction graph and the dynamical model of a physical system from trajectories. Our inference problem differs in that we infer a static representation of the CRN that best explains the observed trajectories. The encoder infers the CRN reactions and the decoder infers the reaction rates and thereby the dynamics. To reduce the computational effort of simulation and thus the training time, we use the linear noise approximation algorithm to compute the first two moments of the stochastic dynamics.

Finally, we show preliminary results, where the framework is able to automatically infer single-reaction systems from data. The ability of the proposed approach to scale w.r.t. the complexity of the underlying CRN is still under investigation. We hope that improving the neural network architecture and tuning its hyperparameters will allow us to infer more complex CRNs in a fully automated and data-driven manner.

2   Methods

2.1   Background

Chemical Reaction Network (CRN). Consider a CRN with N chemical species $S_1, \ldots, S_N$ that interact according to M chemical reactions $R_1, \ldots, R_M$, and let $\mathbf{x}(t) = (\mathbf{x}_1(t), \ldots, \mathbf{x}_N(t)) \in \mathcal{X} \subseteq \mathbb{N}^N$ denote the state of the population at time t. The propensity function $\rho_j(\mathbf{x})$ determines the rate of reaction R_j when the system is in state $\mathbf{x}$, and includes rate parameter α_j. The update vector ν_j describes the change of the system's state when reaction R_j fires. We denote the total volume of the system by Ω. The dynamics of the CRN is described by the Chemical Master Equation (CME), which consists of a system of infinitely many stochastic differential equations. Since the CME cannot be computed exactly, we use linear noise approximation (LNA) to approximate the CRN's dynamics with a Gaussian distribution [7], where the first two moments can be numerically integrated.

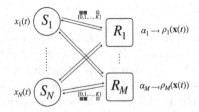

Fig. 1. CRN as coloured bipartite graph.

Representation as Bipartite Graph. The CRN is represented as a coloured bipartite interaction graph (see Fig. 1). Species and reactions represent the first and second type of nodes, respectively. The edge colour $\{0, 1, 2, \ldots, K\}$, where $K \in \mathbb{N}$ is the largest allowed coefficient, denotes the stoichiometric coefficient associated with that edge, i.e., the number of individuals involved in that reaction.

Data. We validate the proposed method over synthetic model-based trajectories. We observe the CRN evolution at equidistant time instants $[t_0, \ldots, t_T]$. Given a fixed time step Δt, $t_{j+1} = t_j + \Delta t$ for every $j \in \{0, \ldots, T-1\}$. The input to the VAE is a pool of n trajectories in dataset $\mathcal{D} = \{\mathbf{x}_{0:T}^1, \ldots, \mathbf{x}_{0:T}^n\}$, where $\mathbf{x}_{0:T}^i = (x^i(t_0), \ldots, x^i(t_T))$, and $x^i(t) = (x_1^i(t), \ldots, x_{N_{obs}}^i(t))$ denotes the state of observable species at time t, i.e., the number of individuals for each species. In order to treat different population levels, we consider concentrations of species, $\hat{\mathbf{x}}(t) = \frac{\mathbf{x}(t)}{\Omega}$.

2.2   Inference

The goal is to infer an explicit representation of a CRN, i.e., the reactions $R_1, \ldots, R_M$, and the rate parameters $\alpha_1, \ldots, \alpha_M$, from observed trajectories using a graph-based VAE (see Fig. 2).

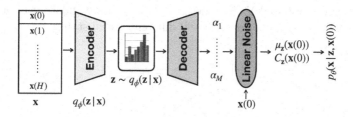

Fig. 2. Diagram of the graph-VAE.

Encoder. The encoder $q_\phi(\mathbf{z}|\mathbf{x})$ infers the interaction graph from the trajectories in $\mathcal{D}$ assuming an initial fully-connected bipartite graph. A graph convolutional neural network (GCNN) maps the trajectories $\mathbf{x}$ located over a fully connected graph to pairwise categorical distributions z_{ij}. The resulting latent categorical distribution $q_\phi(z_{ij}|\mathbf{x}) = softmax(f_{enc,\phi}(\mathbf{x})_{ij,1:T})$ with GCNN $f_{enc,\phi}(\mathbf{x})$ is an explicit representation of the CRN structure given the observations.

Decoder. The decoder $p_\theta(\mathbf{z})$ is composed of a multi-layer perceptron (MLP) that maps the latent distribution $q_\phi(\mathbf{z}|\mathbf{x})$ over CRN formulae into reaction rates that fully determine the dynamics of the system. For each trajectory, we sample a CRN model from the latent distribution and compute the LNA of the dynamics.

Loss. The VAE maximizes the evidence lower bound which is defined as $\mathcal{L}(\theta, \phi) = \mathbb{E}_{q_\phi(\mathbf{z}|\mathbf{x})}[\log(p_\theta(\mathbf{x}|\mathbf{z}))] - KL[q_\phi(\mathbf{z}|\mathbf{x})||p_\theta(\mathbf{z})]$. The first term, the reconstruction error, computes the negative log-likelihood of observing the trajectories given the inferred CRN. It is estimated by $-\sum_{t=1}^{T} \frac{(\hat{\mathbf{x}}(t) - \mu_{\mathbf{z}}(t))^T C_{\mathbf{z}}(t)^{-1} (\hat{\mathbf{x}}(t) - \mu_{\mathbf{z}}(t))}{2}$, where $\hat{\mathbf{x}}(t)$ are the observed species concentrations and $\mu_{\mathbf{z}}(t)$ and $C_{\mathbf{z}}(t)$ are the mean and covariance of the Gaussian distribution estimated by the LNA. The second term is the Kullback-Leibler (KL) divergence and can be computed as the sum of entropies for a categorical prior and softmax distribution, $\sum_{i \neq j} \sum_{k=1}^{K} q_\phi(z_{ij}|y)(k) \cdot [\log(q_\phi(z_{ij}|y)(k))) - \log(p(z_{ij}(k)))]$.

Training and Evaluation. Training is performed by optimising the weights of the VAE through backpropagation. At test time, the most likely CRN graph is used together with the resulting reaction rates to make predictions.

3   Experiments

We demonstrate the applicability of our framework to single-reaction CRNs. As baseline, consider the reaction $R_1 : 1A \xrightarrow{0.2} 1B$ with two species A and B and reaction rate $\alpha_1 = 0.2$. The training dataset consists of 1.500 simulated trajectories with 100 different initial states. We collect data every $\Delta t = 0.5$ time steps with a total of 21 data points. The volume of the system is set to $\Omega = 200$. We train our graph-based VAE as described above and validate it on a dataset consisting of 1.000 trajectories with 100 different initial states.

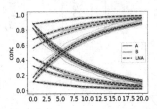

Fig. 3. Simulated trajectories against the inferred dynamics for given initial states.

After 9 epochs with a batch size of 128 trajectories, the VAE converged to the solution: $R_1 : 1A \xrightarrow{0.19} 1B$. Figure 3 shows a pool of simulated trajectories against the estimated dynamics for a few fixed initial states.

4   Conclusion

We propose a data-driven machine learning framework that automatically infers a CRN using a graph-based VAE. Preliminary results show the applicability of our methods on single-reaction networks, where the inferred CRN is very close to the true system and the dynamics are captured well. We aim at extending the framework towards automatically inferring more complex CRNs. We are actively investigating where the scalability and identifiability bottlenecks of the proposed method lie to gain a better understanding of which CRNs can be inferred in practice. Nonetheless, the flexibility of our framework in terms of graph representation allows us to incorporate prior knowledge by fixing parts of the graph before learning, which reduces the number of dimensions for more complex CRNs. To improve performance, we intend to incorporate attention mechanisms to focus on specific parts of the input. In the future, we want to demonstrate the power and effectiveness of our framework on case studies in the fields of epidemics, gene regulation, and signal transduction.

References

1. Bortolussi, L., Palmieri, L.: Deep abstractions of chemical reaction networks. In: Češka, M., Šafránek, D. (eds.) CMSB 2018. LNCS, vol. 11095, pp. 21–38. Springer, Cham (2018). https://doi.org/10.1007/978-3-319-99429-1_2
2. Cairoli, F., Carbone, G., Bortolussi, L.: Abstraction of Markov population dynamics via generative adversarial nets. In: Cinquemani, E., Paulevé, L. (eds.) CMSB 2021. LNCS, vol. 12881, pp. 19–35. Springer, Cham (2021). https://doi.org/10.1007/978-3-030-85633-5_2
3. Gupta, A., Schwab, C., Khammash, M.: DeepCME: a deep learning framework for computing solution statistics of the chemical master equation. PLOS Comput. Biol. **17**(12), e1009623 (2021). https://doi.org/10.1371/journal.pcbi.1009623. https://dx.plos.org/10.1371/journal.pcbi.1009623
4. Kipf, T., Fetaya, E., Wang, K.C., Welling, M., Zemel, R.: Neural Relational Inference for Interacting Systems (2018)
5. Repin, D., Petrov, T.: Automated Deep Abstractions for Stochastic Chemical Reaction Networks (2020)
6. Scarselli, F., Gori, M., Tsoi, A.C., Hagenbuchner, M., Monfardini, G.: The graph neural network model. IEEE Trans. Neural Netw. **20**(1), 61–80 (2009). https://doi.org/10.1109/TNN.2008.2005605. https://ieeexplore.ieee.org/document/4700287/
7. Singh, A., Grima, R.: The linear-noise approximation and moment-closure approximations for stochastic chemical kinetics (2017). https://doi.org/10.48550/ARXIV.1711.07383. https://arxiv.org/abs/1711.07383

Modeling Uncertain Biomass Composition in Genome-Scale Metabolic Models with Flexible Nets

Teresa Joven$^{(\boxtimes)}$, Jorge Lázaro , and Jorge Júlvez

Department of Computer Science and Systems Engineering,
University of Zaragoza, Zaragoza, Spain
{joven,jorgelazaro,julvez}@unizar.es

Abstract. Genome-scale Metabolic Models (GEMs) are mathematical representations of an organism's metabolism that describe mass-balanced relationships between metabolites using gene-protein-reaction associations. These models are used to analyze the metabolic fluxes through reactions involved in the metabolism, with the most commonly used method being Flux Balance Analysis (FBA). The usefulness of GEMs is limited by the presence of uncertain parameters, which can lead to poor predictions. In order to model uncertain biomass composition, a particular class of Flexible Nets (FNs), called ENDI, is proposed. The impact of uncertain biomass composition on the growth rate of the organism can be assessed straightforwardly by a linear programming problem.

1 Introduction

Genome-scale metabolic models (GEMs) are mathematical representations of the metabolism of an organism that describe a whole set of mass-balanced relationships between the metabolites of the organism using gene-protein-reaction associations resulting from the annotation process and experimental information. A key element of a GEM is the biomass reaction which specifies the metabolites consumed and produced by the cellular growth together with their corresponding stoichiometric weights. For simplicity, these weights are *exact* real numbers in most modelling formalisms. Nevertheless, their actual values are usually uncertain. This paper proposes the modelling formalism of Flexible Nets (FNs) to model GEMs with uncertain biomass composition.

2 Flexible Nets for GEMs

A GEM can be represented as a constraint-based model (CBM) with a tuple $C_B = \{R, M, S, lb, ub\}$ where R is the set of reactions, M is the set of metabolites, $S \in \mathbb{R}^{|M| \times |R|}$ is the stoichiometric matrix, and lb, ub: $R \rightarrow \mathbb{R}$ are lower and upper flux bounds of the reactions. Each reaction is associated with a set of reactant metabolites and a set of product metabolites (one of these sets can be

N. Jansen and M. Tribastone (Eds.): QEST 2023, LNCS 14287, pp. 148–151, 2023.
https://doi.org/10.1007/978-3-031-43835-6_11

empty). The stoichiometric matrix S accounts for all the stoichiometric weights of the reactions, i.e. $S[m, r]$ is the stoichiometric weight of metabolite $m \in M$ for reaction $r \in R$. Thus, if $S[m, r] < 0$ then m is consumed when r occurs; if $S[m, r] > 0$ then m is produced when r occurs; and if $S[m, r] = 0$ then m is neither consumed nor produced when r occurs.

Definition 1 (Event net with default intensities (ENDI). *An event net with default intensities is a tuple $N_V = \{P, T, V, E_V, A, B, J, K\}$ where (P, T, V, E_V) is a tripartite graph determining the net structure, (A, B) are matrices determining the stoichiometry, and (J, K) are matrices constraining potential intensities of the net.*

An ENDI has three types of vertices: places (set P), transitions (set T), and event handlers (set V). Each place, $p \in P$, is depicted as a circle and models a metabolite. Each transition, $t \in T$, is depicted as a rectangle and models a reaction. Each event handler, $v \in V$, is depicted as a dot and models the stoichiometry of a reaction. The vertices of the net are connected by the edges in E_V. Each pair of vertices can be connected by at most one edge. The set E_V is partitioned into two sets E_V^P (arc from a place to a handler or vice versa) and E_V^T (edge connecting a transition and a handler). Direct connections among places and transitions are not allowed (see [1]).

The flux of reactions is given by a vector $\lambda \in \mathbb{R}_{\geq 0}^{|R|}$, i.e. $\lambda[t]$ is the flux of reaction t, and the speed at which the amounts of metabolites change is given by a vector $\Delta m_\tau \in \mathbb{R}_{\geq 0}^{|E_V|}$, i.e., $\Delta m_\tau[(v, p)]$ is the speed at which v produces metabolites in p, and $\Delta m_\tau[(p, v)]$ be the speed at which v consumes metabolites from p.

In general, the relationship between reaction fluxes, λ, and speeds at which metabolite change is given by (see Definition 1):

$$A \Delta m_\tau \leq B\lambda \qquad (1)$$

In order to account for uncertain fluxes, all the potential values of λ are assumed to be constrained by:

$$J\lambda \leq K \qquad (2)$$

where J and K are real matrices of appropriate size. Inequality 2 allows the modeller not only to establish individual lower and upper flux bounds as in usual CBMs, but also to state linear constraints among fluxes of different reactions.

3   Modelling Uncertain Biomass Composition

A general expression for the biomass reaction, r_g, is given by:

$$r_g : w_{r1} M_{r1} + ... + w_{rk} M_{rk} \rightarrow w_{p1} M_{p1} + ... + w_{pk'} M_{pk'} \qquad (3)$$

where $\{M_{r1}, ..., M_{rk}\}$ represent the set of reactants, $\{w_{r1}, ..., w_{rk}\}$ are their corresponding coefficients, $\{M_{p1}, ..., M_{pk'}\}$ represent the set of products and their

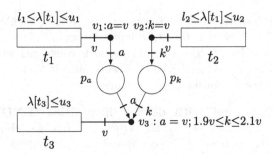

Fig. 1. Event net with default intensities (ENDI) and uncertain stoichiometry.

coefficients are $\{w_{p1}, ..., w_{pk'}\}$. In CBMs, these stoichiometric coefficients are *sharp* real numbers, this makes it difficult to model partially known reactions. An appealing feature of ENDIs is that they can accommodate uncertain stoichiometric coefficients.

The ENDI in Fig. 1 models three reactions: $R_1 : \emptyset \rightarrow p_a$; $R_2 : \emptyset \rightarrow p_k$; $R_3 : p_a + kp_k \rightarrow \emptyset$, where the stoichiometric weight k is uncertain but known to be in the interval $[1.9, 2.1]$. This uncertainty is modelled by the inequalities $1.9v \leq k \leq 2.1v$ associated with v_3. The fluxes of the reactions R_1, R_2, and R_3 are uncertain and constrained to the intervals $[l_1, u_1]$, $[l_2, u_2]$, and $[0, u_3]$ respectively. This is modelled by the inequalities above the reactions.

4   Steady State Bounds

Similarly to FBA, it will be assumed that in the steady state both, the fluxes of reactions and the amounts of metabolites keep constant. This assumption leads to the following constraint that must be satisfied in the steady state:

$$Z\Delta m_\tau = 0 \tag{4}$$

where Z is a matrix with rows indexed by P, columns indexed by E_V^P, and such that $Z[p_i, (p_i, v_k)] = -1 \; \forall \; (p_i, v_k) \in E_V^P$, $Z[p_i, (v_k, p_i)] = 1 \; \forall \; (v_k, p_i) \in E_V^P$ and the rest of the elements in Z are 0.

Proposition 1. *Let N_V be an ENDI, all the potential steady states $(\lambda, \Delta m_\tau)$ belong to SS_{N_V} where:*

$$SS_{N_V} = \{(\lambda, \Delta m_\tau) | A\Delta m_\tau \leq B\lambda; Z\Delta m_\tau = 0; J\lambda \leq K\} \tag{5}$$

Lower and upper bounds for a function of interest can be computed by adding an appropriate objective function to the constraints in (5). For instance, an upper bound for the flux of the biomass reaction, r_g, i.e. an upper bound for the growth rate, can be obtained by solving the following linear programming problem (LPP), where t_g is the reaction modelling biomass composition:

$$max \; \lambda[t_g] \text{ subject to } \{A\Delta m_\tau \leq B\lambda; Z\Delta m_\tau = 0; J\lambda \leq K\} \tag{6}$$

5   Experimental Results

This section presents the results obtained for the constraint-based models of two organisms, namely the GEM of Chinese Hamster Ovary (CHO) cells, *C. griseus* and the unicellular fungus *S. cerevisiae*. Each CMB has been transformed into an ENDI in which a degree of uncertainty x has been added to the original stoichiometric weights of the reactants of the biomass reaction. More precisely, if the original weight is w_{ri}, the ENDI allows any weight, q_{ri}, in the interval $[(1-x)*w_{ri}, (1+x)*w_{ri}]$, with $x \in [0,1]$. Similarly to the inequalities associated with v_3 in Fig. 1, this interval is expressed by two inequalities: $(1-x)*w_{ri}*v \leq q_{ri} \leq (1+x)*w_{ri}*v$ where v is the flux of biomass reaction.

Figure 2 shows the growth rate as a function of the percentage of uncertainty. It can be seen that the growth rate increases monotonically with respect to the percentage of uncertainty. This is an expected result, since higher growth rates can be achieved if less biomass is required.

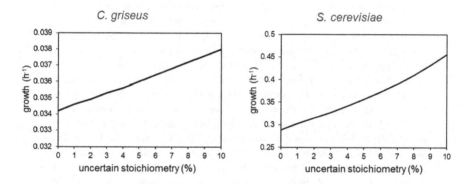

Fig. 2. Steady state fluxes of the biomass reaction.

6   Conclusions

In this work-in-progress paper a particular class of FNs, called ENDI, has been introduced to model uncertain biomass composition of two GEMs. An LPP can be derived from the ENDI to estimate the impact of uncertainty on growth.

References

1. Júlvez, J., Oliver, S.G.: Steady state analysis of flexible nets. IEEE Trans. Autom. Control **65**(6), 2510–2525 (2019). https://doi.org/10.1109/TAC.2019.2931836

On the Trade-Off Between Efficiency and Precision of Neural Abstraction

Alec Edwards[1(✉)], Mirco Giacobbe[2], and Alessandro Abate[1]

[1] University of Oxford, Oxford, UK
alec.edwards@cs.ox.ac.uk
[2] University of Birmingham, Birmingham, UK

Abstract. Neural abstractions have been recently introduced as formal approximations of complex, nonlinear dynamical models. They comprise a neural ODE and a certified upper bound on the error between the abstract neural network and the concrete dynamical model. So far neural abstractions have exclusively been obtained as neural networks consisting entirely of ReLU activation functions, resulting in neural ODE models that have piecewise affine dynamics, and which can be equivalently interpreted as linear hybrid automata. In this work, we observe that the utility of an abstraction depends on its use: some scenarios might require coarse abstractions that are easier to analyse, whereas others might require more complex, refined abstractions. We therefore consider neural abstractions of alternative shapes, namely either piecewise constant or nonlinear non-polynomial (specifically, obtained via sigmoidal activations). We employ formal inductive synthesis procedures to generate neural abstractions that result in dynamical models with these semantics. Empirically, we demonstrate the trade-off that these different neural abstraction templates have vis-a-vis their precision and synthesis time, as well as the time required for their safety verification (done via reachability computation). We improve existing synthesis techniques to enable abstraction of higher-dimensional models, and additionally discuss the abstraction of complex neural ODEs to improve the efficiency of reachability analysis for these models.

Keywords: nonlinear dynamical systems · formal abstractions · safety verification · SAT modulo theory · neural networks

1 Introduction

Abstraction is a fundamental process to advance understanding and knowledge. Constructing abstractions involves separating important details from those that are unimportant, in order to allow for more manageable analysis or computation of that which was abstracted. This enables the drawing of conclusions that would otherwise be unobtainable (whether analytically or computationally). While the features and the properties of abstractions vary between fields and applications, in the context of the analysis and verification of mathematical models of dynamical systems abstractions take on a formal role: namely, abstractions must retain all behaviours of the concrete model that they abstract, while at the same time ought to be easier to analyse or do computations on.

© The Author(s), under exclusive license to Springer Nature Switzerland AG 2023
N. Jansen and M. Tribastone (Eds.): QEST 2023, LNCS 14287, pp. 152–171, 2023.
https://doi.org/10.1007/978-3-031-43835-6_12

These requirements allow for certain formal specifications, such as safety, to be formally provable over the abstractions and, additionally, for these very specifications to hold true for the concrete (original) dynamical model (which was the object of abstraction).

Properties which make an abstraction 'good' cannot be universally declared, and instead are deemed to be 'useful' in relation to given properties that are the object of analysis. Abstractions that are 'simpler' in structure may be easier to analyse, but might be coarser and thus lead to less refined proofs. In contrast, a more refined abstraction, i.e., one that contains fewer additional behaviours relative to the concrete model, might be complex and challenging to analyse. Indeed, for any given problem (a dynamical model and a property), it is unclear at the outset where the optimal middle ground between simplicity and precision lies, leading to abstraction techniques that begin with coarser ones and refine them iteratively [13, 20, 51].

The trade-off between this relationship is the topic of this paper: we study how neural abstractions of different nature and precision can be useful for the formal verification of safety properties of complex (e.g., nonlinear) dynamical models. Recent literature has introduced the use of neural networks as templates to abstract dynamical models [1]. These *neural abstractions* consist of a neural network interpreted as an ODE [15], alongside a formal (bounded) error between abstract and concrete models. This enables safety verification of models with challenging dynamical behaviours, such as non-linear dynamics including non locally-Lipschitz vector fields. While this approach has in principle no constraints on the neural template (e.g., on its shape or on the activation functions employed), in practice they have experimentally been limited to feed-forward neural networks consisting entirely of ReLU-based activation functions [1]. The resulting abstractions are thus piecewise affine models, and can be interpreted as linear hybrid automata [32]. Local linearity is desirable, as the analysis of (piecewise) linear models is a mature area, yet one needs only look at the annual ARCH competition to notice the recent advances in the formal verification of alternative models, such as nonlinear ones, which offer advantages related to their generality. It thus makes sense in this work to explore neural abstractions in a broader context, across different layers of simplicity and precision, as per our discussion above.

Piecewise constant and nonlinear (Lipschitz continuous) models are well studied model semantics with mature tools for their analyses, each with different advantages. In this work, we construct neural-based dynamical models which follow these semantics and cast them as formal abstractions, and show their potential use cases. While piecewise constant and piecewise affine are clearly simpler templates to abstract nonlinear ODEs, it is perhaps initially unclear how the latter can be considered useful. However, not all nonlinear models are alike in complexity: models with complex functional composition or with the presence of functions that are not Lipschitz continuous will be abstracted to models which are still nonlinear, but free of these constraints. The potential of nonlinear neural abstractions is further bolstered by recent advances in developing literature on reachability analysis for neural ODEs [30, 31, 44], which are leveraged in this work.

We summarise our main contributions in the following points:

- we build upon a recent approach that leverages neural networks as abstractions of dynamical models [1], by introducing new neural piecewise constant and nonlinear templates (Sect. 2),
- we extend and improve existing synthesis procedures to generate abstractions and cast them as equivalent, analysable models (Sect. 3, Sect. 4),
- we implement the overall procedure under different abstract models, and empirically demonstrate tradeoffs of different templates with regards to precision and speed (Sect. 5),
- we also demonstrate the ability of our approach to abstract higher-dimensional models, as well as complex neural ODEs, which is significant in light of the growing field of research around neural ODEs (Sect. 5).

Related Work

We observe that the verification of dynamical models with neural network structure is an active area of research. In particular, there is much interest in verifying models with continuous-time dynamics and neural network controllers that take actions at discrete time steps, often referred to as neural network control systems [24,35,36,56,60]. Whilst in this work we focus on continuous-time models, we emphasise that verification of discrete-time models and of models with neural controllers is also studied [10,59,61].

As previously mentioned, safety verification of nonlinear continuous-time dynamical models is a mature field. Analysis of such models can be performed using Taylor models [16–19], abstractions [2,7,46,52,53], simulations [25], reduction to first-order logic [40], and via Koopman linearisation [11]. In this work we focus on the challenging setup of verification of nonlinear models with possibly non-Lipschitz vector fields, which violate the working assumptions of all existing verification tools.

In this work we observe that neural ODEs with piecewise constant or affine dynamics induce a partitioning of the state space, which casts them as hybrid automata. The approximation of a (nonlinear) vector field as a hybrid automaton is know as *hybridisation*, and has been extensively utilised for the analysis of nonlinear models [6,8,12,14,21,22,34,37,39,41,58], even with dynamics as simple as piecewise constant ones [9,50]. However, typically these approaches rely on a fixed partitioning of the state space (often based on a grid) that is chosen a-priori, which can limit their scalability. Hybridisation schemes commonly rely on error-based over-approximation to ensure soundness, though this idea is not exclusive to methodologies that partition the state space [3], and has been formalised by means of the notions of (bi)simulation relations [43,48,49].

2 Preliminaries

In this section we introduce the necessary preliminary material for this work, beginning with familiar concepts of dynamical models and neural networks before moving on to the recently introduced neural abstractions.

2.1 Dynamical Models and Safety Verification

This work studies continuous-time autonomous dynamical models which consist of coupled ODEs affected by an uncertain disturbance [38,54]. Denote a dynamical model as

$$\mathcal{F}: \quad \dot{x} = f(x) + d, \quad \|d\| \le \delta, \quad x \in \mathcal{X}, \tag{1}$$

where $f : \mathbb{R}^n \to \mathbb{R}^n$ is a nonlinear vector field, and the state vector x lies within some Euclidean domain of interest $\mathcal{X} \subset \mathbb{R}^n$. More formally, we denote the dynamical model in Eq. (1) as $\mathcal{F}$. The disturbance d is a signal with range $\delta > 0$, where $\| \cdot \|$ denotes a norm operator (unless explicitly stated, we assume all norms are given the same semantics across the paper). The symbol d is interpreted as a non-deterministic disturbance, which at any time can take any possible value within the disturbance bound provided by δ.

We study the following formal verification question: whether a continuous-time, autonomous dynamical model with uncertain disturbances is safe within a region of interest and a given time horizon, and with respect to a region of initial states $\mathcal{X}_0 \subset \mathcal{X}$ and a region of bad states $\mathcal{X}_B \subset \mathcal{X}$. Verifying safety consists of determining that no trajectory initialised in $\mathcal{X}_0$ enters $\mathcal{X}_B$ over the given time horizon.

We tackle safety verification by means of formal abstractions, followed by reachability analysis via flowpipe (i.e., reach sets) propagation. We construct an abstract dynamical system that captures all behaviours (trajectories, executions) of the original system. We then calculate (sound over-approximations of) the flowpipes of these abstractions: i.e., we find (sound over-approximations of) the reachable states from the initial set $\mathcal{X}_0$. If these states do not intersect with the bad states, we can conclude the abstraction is safe. Since we deal with over-approximation, if the abstract model is safe, then the concrete model is necessarily safe too [8]. Since the final check involving set intersection is quite straightforward, the crux of the safety verification problem is the computation of first, the abstract model, and second, the reachable states via flow-pipe propagation - both steps will be the focus of our experimental benchmarks.

2.2 Linear Hybrid Automata

Hybrid automata are useful models of systems comprising both continuous and discrete dynamics [32,55]. They consist of a collection of (discrete) modes, each with their own dynamics over continuous states, rules for transitioning amongst discrete modes, and consequently to re-initialise the continuous dynamics into a new mode. Different semantics exists for hybrid automata; in this work we use the same semantics as SpaceEx [4,27], which is a state-of-the-art tool used for analysis of linear hybrid automata.

The present work considers linear hybrid automata induced by a state space partitioning. Define a hybrid automaton $\mathcal{H} = (Mod, Var, Lab, Inv, Flow, Trans)$, consisting of a labeled graph encoding the evolution of a finite set of continuous variables $Var \subset \mathbb{R}^n$, in which the state variables x take their values. Each vertex $m_i \in Mod$ of the graph is referred to as a mode, and corresponds to a partition within the state space. The state of the automaton is thus given by the pair (m_i, x), whose evolution in time is described by $Flow(m_i, x)$. We describe this evolution as the flow, and in each mode the local flow given by $\dot{x}(t) = f_{m_i}(x) + d$, where $f_{m_i}(x) \in \mathbb{R}^n$ is affine in x and d is a non-deterministic disturbance such that $||d|| \leq \delta$. The edges of the graph describing $\mathcal{H}$ are known as transitions. A transition $(m, \alpha, Guard, m') \in Trans$ has label $\alpha \in Lab$ and allows the system state to jump from (m, x) to the successor state (m', x) instantaneously. We note that only the mode changes during a transition, with x being mapped via the identity function. Transitions are governed by $Guard \subset \mathbb{R}^n$, with a transition being enabled when $x \in Guard$. Moreover, the state may only remain in a mode m if the continuous state is within the invariant $Inv(m) \subset \mathbb{R}^n$. For the purposes of hybrid automata induced by state-space partitioning in this work, both $Guard$ and $Inv(m)$ are closed convex polyhedra $\mathcal{P} = \{x| \bigwedge_i a_i \cdot x \leq b_i\}$, where $a_i \in \mathbb{R}^n$ and $b_i \in \mathbb{R}$. We denote by $\mathcal{P}_i$ the polyhedron that defines the invariant for mode m_i.

2.3 Neural Networks

Let a feed-forward neural network $\mathcal{N}(x)$ consist of an n-dimensional input layer y_0, k hidden layers $y_1, \ldots, y_k$ with dimensions $h_1, \ldots, h_k$ respectively, and an n-dimensional output layer y_{k+1} [42]. Each hidden or output layer with index i is respectively associated to matrices of weights $W_i \in \mathbb{R}^{h_i \times h_{i-1}}$, a vector of biases $b_i \in \mathbb{R}^{h_i}$ and a nonlinear activation function σ_i. The value of every hidden layer and the final output y_{k+1} is given by the following equations:

$$y_i = \sigma_i(W_i y_{i-1} + b_i), \quad y_{k+1} = W_{k+1} y_k + b_{k+1}. \tag{2}$$

We remark that the expression used for the activation functions clearly determines the features of a neural network: later we shall discuss three alternatives, and how these affect abstract models built on these "neural templates".

2.4 Neural Abstractions

The concept of a neural abstraction has been recently introduced in [1] and is recalled here as follows.

Consider a feed-forward neural network $\mathcal{N} \colon \mathbb{R}^n \to \mathbb{R}^n$ together with a user-specified error bound $\epsilon > 0$. Together, these define a neural abstraction $\mathcal{A}$ for the dynamical model $\mathcal{F}$, which is given by nonlinear function f and disturbance radius δ, over a region of interest $\mathcal{X}$, if it holds true that

$$\forall x \in \mathcal{X} \colon \|f(x) - \mathcal{N}(x)\| \leq \epsilon - \delta. \tag{3}$$

Then, the neural abstraction $\mathcal{A}$ consists of the following dynamical system, defined by a neural ODE with bounded disturbance:

$$\dot{x} = \mathcal{N}(x) + d, \quad \|d\| \leq \epsilon, \quad x \in \mathcal{X}. \tag{4}$$

Altogether, we define the neural abstraction $\mathcal{A}$ of a non-linear dynamical system $\mathcal{F}$ as a neural ODE with an additive disturbance, and note that it approximates the dynamics of $\mathcal{F}$, while also accounting for the approximation error. Notably, no assumption is placed on the vector field f: in particular, f is not required to be Lipschitz continuous. This is because the precision of a neural abstraction relies on the condition described by (3), whose certification is performed by an SMT solver (see Sect. 3).

The choice of activation function σ here is significant, as it determines the structure of the final abstraction, which in turn impacts its quality and utility. The work in [1] focuses on neural abstractions consisting of a neural network with ReLU activations, and presents an abstraction workflow and experimental results that depend on this specific choice. The ReLU activation is a suitable choice in neural nets. Such functions originate a polyhedral partitioning of the input domain, since each ReLU function can be 'on' or 'off', corresponding to two half-spaces – one of which has a linear output and the other zero output,

$$\text{ReLU}(x) = \begin{cases} x, & \text{if } x > 0 \\ 0, & \text{otherwise.} \end{cases} \tag{5}$$

Any fixed configuration of ReLU functions as 'on' or 'off' can be therefore be interpreted as the intersection of a finite number of halfspaces, which itself is a convex polyhedron. Furthermore, the output layer of the neural net induces functions that are piecewise affine. In light of this, [1] shows that a feed-forward neural ODE with ReLU activations induces a linear hybrid automaton. Furthermore, it shows how this linear hybrid automaton can be cast from the ReLU network, by constructing the labelled graph for $\mathcal{H}$. This involves calculating the appropriate modes (vertices)—which requires the computation of the convex polyhedral invariants and locally affine flows—and transitions (edges)—whose guards are equivalent to the invariant of the mode being transitioned to. Due to space limitations, details of this cast are not discussed and readers are directed to [1] instead.

In this work, we consider additional instances of neural abstractions which induce a special kind of affine hybrid automaton, in which the local flows are 'constant', as well as 'nonlinear' neural abstractions that do not induce any piecewise state partitioning but just 'regularise' the original dynamics.

2.5 Activation Functions Determine the Shape of the Abstract Neural Models

Approximations of nonlinear vector fields do not need to be piecewise affine. In the present work we consider two additional templates for neural abstractions, resulting in abstract dynamical models of different shapes, namely piecewise constant and nonlinear. We expect the former alternative to be simpler to generate than the piecewise affine one, but arguably less precise (at least if the spatial partitioning is the same). Conversely, nonlinear neural abstractions will be seen as "regularised" alternatives of the original nonlinear vector fields. This aligns with our intuition about the trade-off between simplicity and precision, as elaborated in the introduction of this work. We require a specific activation function for each of the two classes, as discussed next.

Piecewise Constant Templates. Neural piecewise constant templates can be constructed using the unit-step function, also known as the Heaviside function, namely

$$\text{H}(x) = \begin{cases} 1, & \text{if } x > 0 \\ 0, & \text{otherwise.} \end{cases} \tag{6}$$

It is only necessary for the final hidden layer to use piecewise constant activation functions to obtain a piecewise constant output, with all other hidden layers using a ReLU activation function. In other words, to obtain a piecewise constant template we set $\sigma_i(x) = \text{ReLU}(x)$, $i = 1, \ldots, k - 1$ and $\sigma_k = H(x)$. A neural network with piecewise constant output, rather than a piecewise affine output, will also be equivalent to a specific linear hybrid automaton, namely one with constant flows in each mode (rather than affine flows).

Nonlinear Templates. For neural nonlinear templates we utilise the Sigmoid function – resulting in 'sigmoidal' nonlinear templates. Sigmoid serves as a smooth approximation to a step function,

$$\text{Sigmoid}(x) = \frac{1}{1 + e^{-x}}. \tag{7}$$

3 Formal Synthesis of Template-Dependent Neural Abstractions

In this section we discuss a framework for constructing neural abstractions of different expressivities using inductive training procedures coupled with symbolic reasoning. We then describe how these abstractions can be cast from neural ODEs coupled with an error bound to equivalent models (with identical behaviour) that are well studied, alongside mature tools for the analysis of each object. This information is summarised in Fig. 1.

We leverage an iterative synthesis procedure known as counterexample-guided inductive synthesis (CEGIS, [57]), which has been shown to be useful for inductively constructing functions that satisfy desired specifications. It consists of two phases: a training phase which seeks to generate a candidate using data, and a sound certification phase which seeks to (dis)prove this candidate using symbolic reasoning. We now consider each phase in turn.

3.1 Training Procedures for Neural Abstractions

Gradient-based Training. Training neural abstractions for the piecewise affine and the nonlinear templates can be done using gradient descent algorithms (for which we use PyTorch, [47]), which depend on the loss functions used. We seek to minimise the maximum error over the domain; however, this would result in a non-differentiable loss function, which is undesirable for gradient descent. Therefore, we train via a proxy: the mean squared error,

$$\mathcal{L}(s) = \frac{1}{|S|} \sum_{s \in S} \|f(s) - \mathcal{N}(s)\|_2. \tag{8}$$

Here $\|\cdot\|_2$ represents the l^2-norm of its input, and S represents a finite set of data points s that are sampled over $\mathcal{X}$.

	Piecewise constant	Piecewise affine	Nonlinear
Concrete models	Nonlinear, non-Lipschitz	Nonlinear, non-Lipschitz	Nonlinear, non-Lipschitz
Activation function	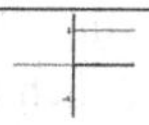 H	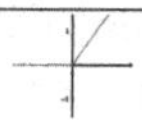 ReLU	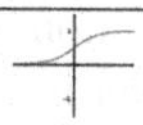 Sigmoid
Training procedure	Particle swarm	Gradient descent	Gradient descent
Loss function	$\max_{s \in S} l^{\infty}(s)$	$\frac{1}{\|S\|} \sum_{s \in S} l^2(s)$	$\frac{1}{\|S\|} \sum_{s \in S} l^2(s)$
Equivalent abstract model	LHA I with disturbance	LHA II with disturbance	Nonlinear ODE with disturbance
Safety verification technology	Symbolic model checking	STC algorithm	Taylor model based flowpipe propagation
Safety verification tool	PHAVer	SpaceEx	Flow*

Fig. 1. Overview of neural abstraction templates presented in this work, alongside details of their synthesis procedure, interpretation, and corresponding state-of-the-art verification technologies. Here, LHA I refers to *linear hybrid automata* with polyhedral guards and invariants and constant flows, whereas LHA II refers to the same object but now with affine flows.

Gradient-free Training. The step-function $H(x)$ (cf. Eq. (6)) has zero gradient almost everywhere: this makes training with gradient descent-based approaches unsuitable. We therefore rely on gradient-free methods for training neural networks, which notably scale less well and converge less quickly. Several options exist as suitable choices: here we use a particle swarm optimisation approach, specifically with PySwarms [45].

Gradient-based training approaches benefit from differentiable loss functions as they ensure gradient calculations are always possible. However, in the case of neural abstraction synthesis this means we do not train based our true objective, but via a proxy. With a non-gradient based approach there are no longer advantages to a differentiable loss function, meaning we can minimise a loss function that better represents our desired specification (cf. Eq. (3)) based on the maximum error over samples. In line with this, we also utilise the l^{∞}-norm, namely

$$\mathcal{L}'(s) = \max_{s \in S} \|f(s) - \mathcal{N}(s)\|_{\infty}. \tag{9}$$

3.2   Certification of the Quality of a Neural Abstraction

Regardless of the template used to generate the abstraction, the procedure for the certification of its quality is identical. It involves using an SMT solver to check that at no point in the entire domain of interest $\mathcal{X}$ is the maximum error between the neural network and the concrete model greater than some given ϵ. This upper bound ϵ on the abstraction error is first estimated empirically using the finite data set S. Next, the SMT solver is asked to find an assignment cex of x, such that the following formula is satisfiable:

$$\phi = \exists x \in \mathcal{X}: \ \|f(x) - \mathcal{N}(x)\| > \epsilon - \delta. \tag{10}$$

If any such assignment is found then the error bound ϵ does not hold over the entire domain $\mathcal{X}$, and a valid abstraction has not been constructed. Instead, the assignment is treated as a "counterexample" and is added to the data set S, as further training of the network continues. On the other hand, if no assignment is found, then the specification in Eq. (3) is valid and the synthesis is complete, returning a sound neural abstraction. Many suitable choices exist for the SMT solver: we use Z3 [23] and DReal [29]. We note that feed-forward neural networks with Sigmoid activation functions are simply elementary functions, and networks with ReLU or step-function activations can be interpreted as linear combinations of formulae in first-order logic. These insights enable the encoding of such networks in SMT-formulae by means of symbolic evaluation.

3.3   Refining the Precision of the Abstraction

The abstraction error given by Eq. (10) is global, that is, it holds across the entire domain $\mathcal{X}$. In reality, the true abstraction error is likely smaller than ϵ throughout much of the domain.

As previously stated, piecewise constant and -affine neural networks induce a hybrid automaton over $\mathcal{X}$; this automaton is realised as part of the proposed framework, after a valid abstraction has been synthesised. Since this automaton induces a polyhedral partitioning of the state-space, we can consider these polyhedra in turn to determine a local abstraction error for each mode. Consider a given mode m_i, with flow f_{m_i} and polyhedral invariant $\mathcal{P}_i$. As described in Sect. 2.2, the local flow f_{m_i} describes the evolution of x while it remains within the invariant given by $\mathcal{P}_i$. We can therefore rewrite Eq. (10) as

$$\phi_i = \exists x \in \mathcal{P}_i: \ \|f_{m_i}(x) - \mathcal{N}(x)\| > \epsilon_i - \delta, \tag{11}$$

where $\epsilon_i \leq \epsilon$ is a candidate upper bound on the abstraction error estimated empirically over the set $S_i := \{s \in S | s \in \mathcal{P}_i\}$.

We provide ϕ_i to an SMT solver as before: if no satisfying assignment is found then the candidate error bound holds for the mode. Note that this does not effect the overall correctness of our abstractions: if a satisfying assignment is found, the original upper bound on abstraction error still holds. We present results on the efficacy of this procedure in Appendix B.

4   Safety Verification Using Neural Abstractions

Performing safety verification using a neural abstractions amounts to verifying a neural ODE with additive bounded disturbance, as in Eq. (4). A positive outcome of safety verification on the abstract model can, in view of (3), be directly claimed to also hold for the concrete model in (1). The bottleneck of this step of the safety verification problem is the computation of the reachable states via flow-pipe propagation, which is in general hard for non-linear dynamics but is intended to be mitigated by the 'simpler' abstract model: this will indeed be the focus of our experimental benchmarks.

While recent literature has made advancements in performing reachability analysis on neural ODEs [30,31,44], to the best of our knowledge no such method exists for neural ODEs with an additive bounded disturbance – a *neural abstraction*. Therefore, rather than handling the neural network directly, we must first construct equivalent models that are amenable to computation. By equivalent, we mean that the trajectories generated by both models are identical. We summarise this information in the final three rows of Fig. 1.

For the piecewise affine neural abstractions, this involves constructing a hybrid automaton with affine dynamics and invariants defined by polyhedra. The reader is referred to [1] for details on this cast, and how this construction can be performed efficiently. The obtained models can be analysed more efficiently using the space-time clustering approximation (STC) algorithm [28], which is implemented in the verification tool SpaceEx [27].

We now turn to the newly-introduced piecewise constant models. Noting the similarity between $H(x)$ and $ReLU(x)$, neural abstractions with piecewise constant activation functions also induce a linear hybrid automaton, except that now the flows are given by a constant term. Verification of such models is the specialty of the tool PHAVer [26], which itself implements a bespoke version of the symbolic model checking algorithm introduced by [5,33].

Finally, we consider nonlinear neural abstractions with Sigmoid activation functions. We observe that the output of these networks can simply be interpreted as 'regularised' nonlinear dynamical models consisting of elementary functions, hence these abstractions are nonlinear ODEs with bounded additive non-determinism. Casting a neural ODE as a nonlinear ODE involves evaluating the network symbolically. Reachability analysis of these kinds of models can be performed using the mature tool Flow* [17], which performs flowpipe (i.e., reach sets) propagation of nonlinear models using Taylor approximations [16], and hence is dependent on local Lipschitz continuity. This means that by constructing a neural abstraction with nonlinear templates of a concrete model that is not locally Lipschitz continuous, we enable safety verification (via Flow*) of that otherwise intractable concrete model.

5   Experimental Results

The experiments presented in this section investigate the trade-off between efficiency and precision of neural abstractions. We consider a number of benchmarks and study how the expressivity of neural abstractions, which depends on their activation functions, varies across their already-demonstrated niche, namely models that are not locally Lipschitz continuous. We provide the first examination of an additional for which neural abstraction is suitable: that of neural ODEs. We demonstrate the efficacy of our abstraction/refinement scheme (cf. Sect. 3.3), which enables us to consider higher dimensional benchmarks, in Appendix B.

5.1 Non-Lipschitz Models

Table 1. Table showing the properties of different abstractions with different templates over a series of benchmarks, and the total time spent for synthesis and flowpipe propagation. Here, 'PWC' denotes piecewise constant, 'PWA' denotes piecewise affine, 'Sig.' denotes nonlinear sigmoidal; W: network architecture (nr. of neurons per layer); ϵ: error bound; $||.||$ denotes the 1-norm, M: number of modes in abstraction; T: total computation time; μ denotes average over 10 repeated runs.

Benchmark	Template	W	$||\epsilon||_1$ min	μ	max	M min	μ	max	T min	μ	max
Water Tank	PWC	[15]	0.16	0.22	0.27	4	5.30	6	6.32	7.32	7.91
	PWA	[12]	0.08	0.09	0.10	6	7.00	8	14.55	70.37	390.90
	Sig.	[4]	0.07	0.07	0.07	1	1.00	1	15.53	18.31	21.53
Non-Lipschitz1	PWC	[20]	0.97	1.21	1.47	14	26.50	45	13.84	16.07	20.56
	PWA	[10]	0.10	0.12	0.14	6	14.00	20	24.58	51.94	92.40
	Sig.	[4]	0.05	0.08	0.10	1	1.00	1	76.24	97.20	105.60
Non-Lipschitz2	PWC	[24]	1.58	1.97	2.25	55	73.40	94	25.81	32.97	41.99
	PWA	[12 10]	0.11	0.12	0.13	11	23.80	62	55.94	86.47	163.16
	Sig.	[10]	0.06	0.08	0.10	1	1.00	1	309.82	368.36	451.04
Water Tank 4D	PWC	[25]	2.47	2.57	2.69	1	1.50	4	173.09	232.33	400.26
	PWA	[12]	0.80	0.80	0.80	4	8.10	18	9.14	21.49	50.70
	Sig.	[7]	0.35	0.35	0.35	1	1.00	1	100.08	133.00	317.80
Water Tank 6D	PWA	[16]	1.30	1.30	1.30	11	16.00	25	76.86	426.16	1659.32
NODE1	PWC	[20]	1.59	1.83	2.12	33	63.10	86	36.23	70.03	104.95
	PWA	[5]	0.20	0.20	0.20	4	5.80	7	17.20	17.92	18.65
	Sig.	[3]	0.20	0.20	0.20	1	1.00	1	120.38	126.51	137.63

The first set of benchmarks we consider *do not* exhibit local Lipschitz continuity, meaning they violate the working assumptions of state of the art safety verification tools such as Flow* and GoTube [31], and are in general challenging to analyse computationally. Three of the benchmarks are taken from [1], and are shown in the first three rows of Table 1. We introduce two new higher dimensional benchmarks, *Water Tank 4D* and *Water Tank 6D*, to demonstrate the ability of our approach to scale to higher-dimensional models with up to six continuous variables. These two additional models are also not locally-Lipschitz continuous. The equations of the dynamics of these models are in Appendix A.

Table 2. Breakdown of the total computation time shown in Table 1. Here, learning time T_L; T_C: certification time; T_f: time to compute flowpipe (i.e., reach sets) over-approximation over horizon of 1 sec, using corresponding verification tool (cf. Fig. 1).

Benchmark	Template	T_L min	μ	max	T_C min	μ	max	T_f min	μ	max
Water-tank	PWC	6.16	7.20	7.78	0.02	0.03	0.03	0.01	0.01	0.02
	PWA	14.43	70.18	390.64	0.02	0.05	0.08	0.05	0.08	0.12
	Sig.	10.99	13.90	16.79	0.00	0.01	0.02	3.90	4.40	5.13
Non-Lipschitz1	PWC	12.94	14.42	16.91	0.19	0.44	0.95	0.16	0.74	3.22
	PWA	22.65	43.64	79.17	0.56	7.21	20.23	0.10	0.99	4.97
	Sig.	19.47	26.63	36.88	0.61	1.60	5.17	46.01	68.97	83.91
Non-Lipschitz2	PWC	11.25	13.75	15.22	1.27	1.86	2.58	8.51	15.97	22.27
	PWA	28.48	54.20	136.79	12.93	23.03	38.76	0.45	4.41	8.49
	Sig.	30.73	65.35	127.13	12.76	37.17	58.26	257.82	265.84	275.55
Water-tank-4d	PWC	172.06	215.69	306.95	0.38	0.58	0.94	0.20	15.84	92.13
	PWA	6.37	12.63	21.94	2.07	8.61	40.64	0.09	0.13	0.17
	Sig.	42.67	46.00	51.54	25.98	57.09	245.52	29.56	29.91	30.92
Water-tank-6d	PWA	38.92	101.84	360.93	16.64	323.45	1619.26	0.18	0.26	0.34
NODE1	PWC	6.04	7.61	8.26	28.97	60.21	95.12	0.11	1.64	7.15
	PWA	3.12	4.24	4.89	12.67	13.60	14.49	0.03	0.04	0.04
	Sig.	18.38	26.35	33.16	20.82	23.33	26.79	72.75	76.83	82.47

For each benchmark, we synthesise an abstraction using one of the proposed templates: piecewise constant, piecewise affine and sigmoidal, and perform flowpipe (i.e., reach sets) propagation over a 1-second time horizon using the appropriate tool[1]. We leverage our procedure's dependence on random seeding and its low computational cost: for each experiment we initialise multiple runs (namely, four) in parallel. When the first of these returns successfully, we discard the remaining and record the result. For statistical and reproducibility reasons, we repeat each experiment 10 times, and present the mean (μ), max and min over these runs.

We present the salient features of the abstractions in Table 1. In particular, we present the architecture of the neural network (the number of neurons in the hidden layers); the size of the resulting abstraction in terms of the number of modes M (non-linear abstractions consist of a single mode) and the 1-norm of the error bound ϵ – a vector representing the error bound in each dimension – for each abstraction. We note that the ϵ presented is the *global* upper bound to the abstraction error, and does not account for any error refinement achieved by the procedure detailed in Sect. 3.3.

---

[1] For the *Water Tank 6D*, we only use the piecewise affine template, as this abstraction performs best in the similar but smaller *Water Tank 4D* experiments.

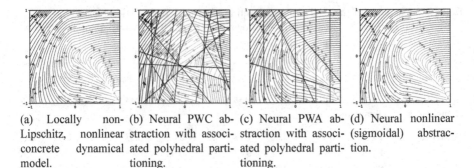

(a) Locally non-Lipschitz, concrete model.
(b) Neural PWC abstraction with associated polyhedral partitioning.
(c) Neural PWA abstraction with associated polyhedral partitioning.
(d) Neural nonlinear (sigmoidal) abstraction.

Fig. 2. Visualisation of Neural Abstractions (underlying flow and relevant partitioning) from piecewise constant (PWC, b), piecewise affine (PWA, c) and sigmoidal (d) templates of a concrete model (a) that does not exhibit local Lipschitz continuity (Non-Lipschitz 2 model).

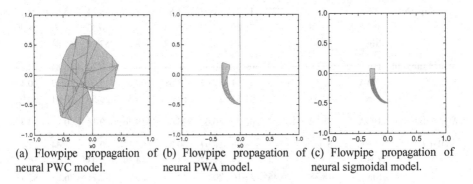

(a) Flowpipe propagation of neural PWC model.
(b) Flowpipe propagation of neural PWA model.
(c) Flowpipe propagation of neural sigmoidal model.

Fig. 3. Flowpipe (i.e., reach sets) propagation for the Non-Lipschitz 2 model using neural abstraction templates and corresponding verification tools. Flowpipe propagation is performed over the same initial set over the same time horizon (1 s).

In practice the mode-wise error is often much lower than the global upper bound (see Appendix B for results on this), which enables us to study higher dimensional models and employ the less expressive piecewise constant templates successfully. The notable exception to this is for *Water Tank 4D*, for which piecewise constant abstractions regularly consist of a very small number of modes, despite a relatively high error: this indicates that the gradient-free learning procedure performs less well for this higher dimensional model, and suggests that neural-based piecewise constant templates are more suitable to smaller dimensional models.

Piecewise constant abstractions perform least well in terms of achieved error bound and in general require larger networks (cf. column W in Table 1) and more modes (col. M) in the resulting hybrid automaton: this is unsurprising, particularly when compared to the "higher-order" piecewise affine abstractions. Meanwhile, piecewise affine and simgoidal templates perform more comparably to each other, though sigmoidal templates do achieve slightly better error bounds while using significantly smaller neural networks.

We also present a breakdown on the time spent within each stage of synthesis – training (T_L) and certification (T_C)) – in Table 2. We prioritised learning speed for the piecewise constant abstractions rather than comprehensive learning, seeking to ensure learning times that were comparable to the other two templates. This is also because we expect safety verification for piecewise constant dynamics to be quite fast. This means that better errors might be achievable for piecewise constant abstractions, however with significantly greater learning times.

We do not perform explicit safety verification using abstract models here, as selecting regions of bad states is arbitrary given that all the abstractions depend on an error bound. Instead, as anticipated earlier, we do perform flowpipe (i.e., reach sets) propagation for each abstraction over a time horizon of 1 s, and present the obtained computation time (T_f) in Table 2. This column highlights the aforementioned trade-off between abstraction templates: flowpipes can be calculated for piecewise constant abstractions (in general) significantly faster than for the other two in lower dimensions. However, for higher-dimensional or more complex models, the flowpipe computation is much slower: this is due to the abstraction error (ϵ) being greater, making flowpipe computation more difficult. Meanwhile, despite the improved accuracy, the sigmoidal abstractions require more computation for flowpipe propagation due to the increase in model complexity (non-linearity).

The flowpipe propagation is run with a 500 s timeout, and may not terminate successfully - e.g., Flow* is sometimes unable to compute the whole flowpipe. Across all experiments (including repetitions), the flowpipe propagation is unsuccessful *only three times*: twice for the sigmoidal templates for *Non-Lipschitz 2*, and once for the piecewise constant templates for *WaterTank 4D*. We also set an overall timeout of 1800 s seconds on the whole procedure (including abstraction synthesis and flowpipe propagation). *Only four* experiments failed to complete before this timeout: once for the piecewise constant template for *Water Tank 4D*, and three times for the *Water Tank 6D* experiments. We emphasise that these outcomes highlight the robustness and 'practical completeness' of the end-to-end procedure. The numerical outcomes shown in the tables exclude those few experiments that time out.

The abstractions for the *Non-Lipschitz 2* model are illustrated in Fig. 2, with the corresponding polyhedral partitioning (for piecewise constant and -affine abstractions) and obtained (locally) approximated vector fields. The results of the corresponding flowpipe (i.e., reach sets) propagation are then depicted in Fig. 3. This figure illustrates further the sorts of safety verification tasks that can be completed using each abstraction type: challenging verification tasks, requiring enhanced precision (ϵ), should be attempted using piecewise affine or nonlinear templates, whereas simpler tasks might be formally verified more efficiently through safe piecewise constant neural templates.

5.2   Abstraction of Neural ODEs

The results presented in Table 1 include a model *(NODE1)* that is Lipschitz continuous, but whose dynamics are described by a neural network. These kinds of models are known as neural ODEs (NODEs), and have become a widely studied tool across machine learning and verification [15]. Here, we discuss neural ODEs as an additional use-case for neural abstractions.

Neural ODEs are commonly trained to approximate real physical models based on sample data. However, it is likely that these models are over-parameterised, making their reachability analysis difficult. For the experiment, we train a neural ODE consisting of a feed-forward network with three hidden layers of hyperbolic tangent activations, based on data from an underlying two-dimensional model. Then, as with the previous experiments, we construct abstractions of this neural ODE with three templates and perform flowpipe propagation. The longest computation time is for the sigmoidal template at 138 s; in contrast, we provide the concrete neural ODE to Flow* with the same settings, which takes 205 s. It is clear that neural abstractions can be beneficial in improving computational efficiency for flowpipe propagation of neural ODEs at the cost of the error bound ϵ. We consider this to be an alternative approach to reducing the Taylor model with Flow*, but instead results in a model which is permanently easier to analyse.

Finally, we note that we do not use state-of-the-art tools for neural ODE reachability analysis here, such as GoTube [31] or NNV [44]. These tools cannot perform reachability analysis on neural abstractions, as they do not account for the non-determinism introduced by the abstraction error. Thus, in order to make a fair comparison we use Flow*, since to the best of our knowledge, no tool specialised to neural ODEs can also handle nonlinear neural abstractions directly. Our results are promising in light of research interest in neural ODEs, and warrant further interest in the reachability analysis of neural abstractions.

6   Conclusions

This work builds on recent literature that employs neural networks as piecewise affine formal abstractions of dynamical models. We extend these abstractions with new neural-based templates which can be cast as models with different, but well studied, semantics – piecewise constant and nonlinear (sigmoidal). After presenting a workflow for the construction of these abstractions, including templating, training and certification, we study the abstractions via equivalent models that are analysable by existing verification technologies. Using existing tools that implement these technologies, we show the advantages of abstractions of different semantics with regards to their precision and ease to analyse, which indicates their suitability for different verification tasks.

We improve on existing procedures for synthesising neural abstractions, allowing for abstraction of higher dimensional models. In addition, we demonstrate that neural abstractions of any template can be used to abstract complex neural ODEs, enabling more efficient, though coarser, reachability-based analysis. These results are promising in light of growing interest in neural ODEs; future work in this area should consider extending tools specialised in reachability of neural ODEs to neural abstractions by incorporating the appropriate non-determinism.

Acknowledgments. Alec is grateful for the support of the EPSRC Centre for Doctoral Training in Autonomous Intelligent Machines and Systems (EP/S024050/1).

A   Benchmarks of NonLinear Dynamical Systems

For each dynamical model, we report the vector field $f : \mathbb{R}^n \rightarrow \mathbb{R}^n$ and the spatial domain $\mathcal{X}$ over which the abstraction is performed.

Water Tank

$$\begin{cases} \dot{x} = 1.5 - \sqrt{x} \\ \mathcal{X}_0 = [0, 0.01] \\ \mathcal{X} = [0, 2] \end{cases} \tag{12}$$

Non-Lipschitz Vector Field 1 (NL1)

$$\begin{cases} \dot{x} = y \\ \dot{y} = \sqrt{x} \\ \mathcal{X}_0 = [0, 0.01] \times [0, 0.01], \\ \mathcal{X} = [0, 1] \times [-1, 1], \end{cases} \tag{13}$$

Non-Lipschitz Vector Field 2 (NL2)

$$\begin{cases} \dot{x} = x^2 + y \\ \dot{y} = \sqrt[3]{x^2} - x, \\ \mathcal{X} = [-1, 1]^2, \\ \mathcal{X}_0 = [-0.005, 0.005] \times [-0.5, -0.49] \end{cases} \tag{14}$$

Water Tank 4D

$$\begin{cases} \dot{x}_0 = 0.2 - \sqrt{x_0} \\ \dot{x}_1 = -x_1 \\ \dot{x}_2 = -x_2 \\ \dot{x}_3 = -0.25(x_0 + x_1 + x_2 + x_3) \\ \mathcal{X} = [0, 1]^4 \\ \mathcal{X}_0 = [0, 0.01] \times [0.8, 0.81] \times [0.8, 0.81] \times [0.8, 0.81] \end{cases} \tag{15}$$

Water Tank 6D

$$\begin{cases} \dot{x}_0 = 0.2 - \sqrt{x_0} \\ \dot{x}_1 = -x_1 \\ \dot{x}_2 = -x_2 \\ \dot{x}_3 = -x_3 \\ \dot{x}_4 = -x_4 \\ \dot{x}_5 = -\frac{1}{6}(x_0 + x_1 + x_2 + x_3 + x_4 + x_5) \\ \mathcal{X} = [0, 1]^6 \\ \mathcal{X}_0 = [0, 0.01] \times [0.8, 0.81] \times [0.8, 0.81] \times \\ [0.8, 0.81] \times [0.7, 0.71] \times [0.65, 0.66] \end{cases} \tag{16}$$

B   Experiments on Error Refinement Scheme

We consider the effect of the error refinement scheme proposed in Sect. 3.3, over a single benchmark (*Non-Lipschitz 2*) for 10 repeated runs and no parallelism (i.e., each run consists of only a single attempt) with and without the error-refinement scheme. We consider first a piecewise affine template, using a network architecture of two hidden layers with 14 and 12 neurons respectively, and a piecewise constant template with a single hidden layer of 20 neurons. These results are shown in Table 3, where we report the *mean* (μ) certification time $\bar{T}_C$ and flowpipe propagation time $\bar{T}_f$. We consider any time spent in SMT-solving to be certification time. The time spent in the learner, T_L is not reported as this is not impacted by the refinement scheme.

It is clear that for both templates, the error refinement provides a significant decrease in flowpipe propagation time, at the cost of a minor increase in certification time. In addition to this, it significantly increases the usability of piecewise affine abstractions for SpaceEx with twice as many successfully terminating. We note that runs are considered unsuccessful if they do not terminate successfully from SpaceEx, or within a 500 s timeout, as in previous experiments. Finally, we present the average error bound over all modes in an abstraction, averaged over all successful runs.

From the reported outcomes, it is clear that the error refinement is able to significantly refine the error for modes in the abstraction, resulting in a more precise abstractions and thus more accurate flowpipe propagation. This error refinement technique is promising, as it can enable the approach to scale to higher-dimensional and more complex models more easily.

Table 3. The impact of error refinement on certification time T_C, flowpipe propagation time T_f, and mean abstraction error over modes $||\epsilon||_1$, as well as the effect on success rate (out of 10 repeats) of flowpipe propagation for the Non-Lipschitz 2 benchmark for piecewise affine (pwa) and piecewise constant models. The results shown for T_C, T_f and $||\epsilon||_1$ are averaged over all successful runs, which we denote using bar notation (rather than μ as above).

Template	Error Refinement	$\bar{T}_C$	$\bar{T}_f$	$||\bar{\epsilon}||_1$	Success Rate
PWC	Without	0.92	16.45	1.89	1.00
	With	1.27	7.87	1.20	1.00
PWA	Without	14.54	54.17	0.14	0.30
	With	16.46	39.67	0.07	0.60

References

1. Abate, A., Edwards, A., Giacobbe, M.: Neural abstractions. In: Thirty-Sixth Conference on Neural Information Processing Systems (2022)
2. Althoff, M.: Reachability analysis of nonlinear systems using conservative polynomialization and non-convex sets. In: HSCC, pp. 173–182. ACM (2013)
3. Althoff, M., Stursberg, O., Buss, M.: Reachability analysis of nonlinear systems with uncertain parameters using conservative linearization. In: CDC, pp. 4042–4048. IEEE (2008)

4. Alur, R., et al.: The algorithmic analysis of hybrid systems. Theoret. Comput. Sci. **138**(1), 3–34 (1995)
5. Alur, R., Henzinger, T., Ho, P.H.: Automatic symbolic verification of embedded systems. IEEE Trans. Software Eng. **22**(3), 181–201 (1996)
6. Alur, R., Henzinger, T., Lafferriere, G., Pappas, G.: Discrete abstractions of hybrid systems. Proc. IEEE **88**(7), 971–984 (2000)
7. Asarin, E., Dang, T.: Abstraction by projection and application to multi-affine systems. In: Alur, R., Pappas, G.J. (eds.) HSCC 2004. LNCS, vol. 2993, pp. 32–47. Springer, Heidelberg (2004). https://doi.org/10.1007/978-3-540-24743-2_3
8. Asarin, E., Dang, T., Girard, A.: Reachability analysis of nonlinear systems using conservative approximation. In: Maler, O., Pnueli, A. (eds.) HSCC 2003. LNCS, vol. 2623, pp. 20–35. Springer, Heidelberg (2003). https://doi.org/10.1007/3-540-36580-X_5
9. Asarin, E., Dang, T., Girard, A.: Hybridization methods for the analysis of nonlinear systems. Acta Informatica **43**(7), 451–476 (2007)
10. Bacci, E., Giacobbe, M., Parker, D.: Verifying reinforcement learning up to infinity. In: IJCAI, pp. 2154–2160. ijcai.org (2021)
11. Bak, S., Bogomolov, S., Duggirala, P.S., Gerlach, A.R., Potomkin, K.: Reachability of black-box nonlinear systems after Koopman operator linearization. IFAC-PapersOnLine **54**(5), 253–258 (2021). 7th IFAC Conference on Analysis and Design of Hybrid Systems ADHS 2021
12. Bak, S., Bogomolov, S., Henzinger, T.A., Johnson, T.T., Prakash, P.: Scalable static hybridization methods for analysis of nonlinear systems. In: HSCC, pp. 155–164. ACM (2016)
13. Bogomolov, S., Frehse, G., Giacobbe, M., Henzinger, T.A.: Counterexample-guided refinement of template Polyhedra. In: Legay, A., Margaria, T. (eds.) TACAS 2017. LNCS, vol. 10205, pp. 589–606. Springer, Heidelberg (2017). https://doi.org/10.1007/978-3-662-54577-5_34
14. Bogomolov, S., Giacobbe, M., Henzinger, T.A., Kong, H.: Conic abstractions for hybrid systems. In: Abate, A., Geeraerts, G. (eds.) FORMATS 2017. LNCS, vol. 10419, pp. 116–132. Springer, Cham (2017). https://doi.org/10.1007/978-3-319-65765-3_7
15. Chen, T.Q., Rubanova, Y., Bettencourt, J., Duvenaud, D.: Neural ordinary differential equations. In: NeurIPS, pp. 6572–6583 (2018)
16. Chen, X., Ábrahám, E., Sankaranarayanan, S.: Taylor model flowpipe construction for nonlinear hybrid systems. In: RTSS, pp. 183–192. IEEE Computer Society (2012)
17. Chen, X., Ábrahám, E., Sankaranarayanan, S.: Flow*: an analyzer for non-linear hybrid systems. In: Sharygina, N., Veith, H. (eds.) CAV 2013. LNCS, vol. 8044, pp. 258–263. Springer, Heidelberg (2013). https://doi.org/10.1007/978-3-642-39799-8_18
18. Chen, X., Mover, S., Sankaranarayanan, S.: Compositional relational abstraction for nonlinear hybrid systems. ACM Trans. Embed. Comput. Syst. **16**(5s), 187:1–187:19 (2017)
19. Chen, X., Sankaranarayanan, S.: Decomposed reachability analysis for nonlinear systems. In: RTSS, pp. 13–24. IEEE Computer Society (2016)
20. Clarke, E., Grumberg, O., Jha, S., Lu, Y., Veith, H.: Counterexample-guided abstraction refinement. In: Emerson, E.A., Sistla, A.P. (eds.) CAV 2000. LNCS, vol. 1855, pp. 154–169. Springer, Heidelberg (2000). https://doi.org/10.1007/10722167_15
21. Dang, T., Maler, O., Testylier, R.: Accurate hybridization of nonlinear systems. In: HSCC, pp. 11–20. ACM (2010)
22. Dang, T., Testylier, R.: Hybridization domain construction using curvature estimation. In: HSCC, pp. 123–132. ACM (2011)
23. de Moura, L., Bjørner, N.: Z3: an efficient SMT solver. In: Ramakrishnan, C.R., Rehof, J. (eds.) TACAS 2008. LNCS, vol. 4963, pp. 337–340. Springer, Heidelberg (2008). https://doi.org/10.1007/978-3-540-78800-3_24

24. Dutta, S., Chen, X., Sankaranarayanan, S.: Reachability analysis for neural feedback systems using regressive polynomial rule inference. In: HSCC, pp. 157–168. ACM (2019)
25. Fan, C., Qi, B., Mitra, S., Viswanathan, M., Duggirala, P.S.: Automatic reachability analysis for nonlinear hybrid models with C2E2. In: Chaudhuri, S., Farzan, A. (eds.) CAV 2016. LNCS, vol. 9779, pp. 531–538. Springer, Cham (2016). https://doi.org/10.1007/978-3-319-41528-4_29
26. Frehse, G.: PHAVer: algorithmic verification of hybrid systems past HyTech. Int. J. Softw. Tools Technol. Transfer **10**(3), 263–279 (2008)
27. Frehse, G., et al.: SpaceEx: scalable verification of hybrid systems. In: Gopalakrishnan, G., Qadeer, S. (eds.) CAV 2011. LNCS, vol. 6806, pp. 379–395. Springer, Heidelberg (2011). https://doi.org/10.1007/978-3-642-22110-1_30
28. Frehse, G., Kateja, R., Le Guernic, C.: Flowpipe approximation and clustering in space-time. In: Proceedings of the 16th International Conference on Hybrid Systems: Computation and Control - HSCC 2013, p. 203. ACM Press, Philadelphia, Pennsylvania, USA (2013)
29. Gao, S., Kong, S., Clarke, E.M.: dReal: an SMT solver for nonlinear theories over the reals. In: Bonacina, M.P. (ed.) CADE 2013. LNCS (LNAI), vol. 7898, pp. 208–214. Springer, Heidelberg (2013). https://doi.org/10.1007/978-3-642-38574-2_14
30. Gruenbacher, S., Hasani, R.M., Lechner, M., Cyranka, J., Smolka, S.A., Grosu, R.: On the verification of neural odes with stochastic guarantees. In: AAAI, pp. 11525–11535. AAAI Press (2021)
31. Gruenbacher, S., et al.: GoTube: scalable stochastic verification of continuous-depth models. In: AAAI (2022)
32. Henzinger, T.A.: The theory of hybrid automata. In: LICS, pp. 278–292. IEEE Computer Society (1996)
33. Henzinger, T.A., Ho, P.H., Wong-Toi, H.: HYTECH: a model checker for hybrid systems. Int. J. Softw. Tools Technol. Transfer **1**(1–2), 110–122 (1997)
34. Henzinger, T.A., Wong-Toi, H.: Linear phase-portrait approximations for nonlinear hybrid systems. In: Alur, R., Henzinger, T.A., Sontag, E.D. (eds.) HS 1995. LNCS, vol. 1066, pp. 377–388. Springer, Heidelberg (1996). https://doi.org/10.1007/BFb0020961
35. Huang, C., Fan, J., Li, W., Chen, X., Zhu, Q.: ReachNN: reachability analysis of neural-network controlled systems. ACM Trans. Embed. Comput. Syst. **18**(5s), 106:1–106:22 (2019)
36. Ivanov, R., Carpenter, T., Weimer, J., Alur, R., Pappas, G., Lee, I.: Verisig 2.0: verification of neural network controllers using Taylor model preconditioning. In: Silva, A., Leino, K.R.M. (eds.) CAV 2021. LNCS, vol. 12759, pp. 249–262. Springer, Cham (2021). https://doi.org/10.1007/978-3-030-81685-8_11
37. Kekatos, N., Forets, M., Frehse, G.: Constructing verification models of nonlinear Simulink systems via syntactic hybridization. In: CDC, pp. 1788–1795. IEEE (2017)
38. Khalil, H.K.: Nonlinear Systems, 3rd edn. Prentice Hall, Upper Saddle River, N.J. (2002)
39. Kong, H., et al.: Discrete abstraction of multiaffine systems. In: Cinquemani, E., Donzé, A. (eds.) HSB 2016. LNCS, vol. 9957, pp. 128–144. Springer, Cham (2016). https://doi.org/10.1007/978-3-319-47151-8_9
40. Kong, S., Gao, S., Chen, W., Clarke, E.: dReach: δ-reachability analysis for hybrid systems. In: Baier, C., Tinelli, C. (eds.) TACAS 2015. LNCS, vol. 9035, pp. 200–205. Springer, Heidelberg (2015). https://doi.org/10.1007/978-3-662-46681-0_15
41. Li, D., Bak, S., Bogomolov, S.: Reachability analysis of nonlinear systems using hybridization and dynamics scaling. In: Bertrand, N., Jansen, N. (eds.) FORMATS 2020. LNCS, vol. 12288, pp. 265–282. Springer, Cham (2020). https://doi.org/10.1007/978-3-030-57628-8_16
42. MacKay, D.J.C.: Information Theory, Inference, and Learning Algorithms. Cambridge University Press, Cambridge (2003)

43. Majumdar, R., Zamani, M.: Approximately bisimilar symbolic models for digital control systems. In: Madhusudan, P., Seshia, S.A. (eds.) CAV 2012. LNCS, vol. 7358, pp. 362–377. Springer, Heidelberg (2012). https://doi.org/10.1007/978-3-642-31424-7_28

44. Manzanas Lopez, D., Musau, P., Hamilton, N.P., Johnson, T.T.: Reachability analysis of a general class of neural ordinary differential equations. In: Bogomolov, S., Parker, D. (eds.) Formal Modeling and Analysis of Timed Systems, pp. 258–277. Springer International Publishing, Cham (2022). https://doi.org/10.1007/978-3-031-15839-1_15

45. Miranda, L.J.V.: PySwarms, a research-toolkit for Particle Swarm Optimization in Python. J. Open Source Softw. **3** (2018)

46. Mover, S., Cimatti, A., Griggio, A., Irfan, A., Tonetta, S.: Implicit semi-algebraic abstraction for polynomial dynamical systems. In: Silva, A., Leino, K.R.M. (eds.) CAV 2021. LNCS, vol. 12759, pp. 529–551. Springer, Cham (2021). https://doi.org/10.1007/978-3-030-81685-8_25

47. Paszke, A., et al.: PyTorch: an imperative style, high-performance deep learning library. In: NeurIPS, pp. 8024–8035 (2019)

48. Pola, G., Girard, A., Tabuada, P.: Approximately Bisimilar symbolic models for nonlinear control systems. arXiv:0706.0246 [math], January 2008

49. Prabhakar, P., Dullerud, G.E., Viswanathan, M.: Stability preserving simulations and bisimulations for hybrid systems. IEEE Trans. Autom. Control **60**(12), 3210–3225 (2015)

50. Prabhakar, P., Garcia Soto, M.: Abstraction based model-checking of stability of hybrid systems. In: Sharygina, N., Veith, H. (eds.) CAV 2013. LNCS, vol. 8044, pp. 280–295. Springer, Heidelberg (2013). https://doi.org/10.1007/978-3-642-39799-8_20

51. Roohi, N., Prabhakar, P., Viswanathan, M.: Hybridization based CEGAR for hybrid automata with affine dynamics. In: Chechik, M., Raskin, J.-F. (eds.) TACAS 2016. LNCS, vol. 9636, pp. 752–769. Springer, Heidelberg (2016). https://doi.org/10.1007/978-3-662-49674-9_48

52. Sankaranarayanan, S.: Automatic abstraction of non-linear systems using change of bases transformations. In: HSCC, pp. 143–152. ACM (2011)

53. Sankaranarayanan, S., Tiwari, A.: Relational abstractions for continuous and hybrid systems. In: Gopalakrishnan, G., Qadeer, S. (eds.) CAV 2011. LNCS, vol. 6806, pp. 686–702. Springer, Heidelberg (2011). https://doi.org/10.1007/978-3-642-22110-1_56

54. Sastry, S.: Nonlinear Systems, Interdisciplinary Applied Mathematics, vol. 10. Springer, New York (1999). https://doi.org/10.1007/978-1-4757-3108-8

55. van der Schaft, A., Schumacher, H.: An Introduction to Hybrid Dynamical Systems. LNCIS, vol. 251. Springer, London (2000). https://doi.org/10.1007/BFb0109998

56. Schilling, C., Forets, M., Guadalupe, S.: Verification of neural-network control systems by integrating Taylor models and zonotopes. In: AAAI (2022)

57. Solar-Lezama, A., Tancau, L., Bodik, R., Seshia, S., Saraswat, V.: Combinatorial sketching for finite programs. SIGOPS Oper. Syst. Rev. **40**(5), 404–415 (2006)

58. Soto, M.G., Prabhakar, P.: Hybridization for stability verification of nonlinear switched systems. In: RTSS, pp. 244–256. IEEE (2020)

59. Tran, H., Cai, F., Lopez, D.M., Musau, P., Johnson, T.T., Koutsoukos, X.D.: Safety verification of cyber-physical systems with reinforcement learning control. ACM Trans. Embed. Comput. Syst. **18**(5s), 105:1–105:22 (2019)

60. Tran, H.-D., et al.: NNV: the neural network verification tool for deep neural networks and learning-enabled cyber-physical systems. In: Lahiri, S.K., Wang, C. (eds.) CAV 2020. LNCS, vol. 12224, pp. 3–17. Springer, Cham (2020). https://doi.org/10.1007/978-3-030-53288-8_1

61. Xiang, W., Tran, H., Rosenfeld, J.A., Johnson, T.T.: Reachable set estimation and safety verification for piecewise linear systems with neural network controllers. In: ACC, pp. 1574–1579. IEEE (2018)

Equilibrium Analysis of Markov Regenerative Processes

András Horváth[1], Marco Paolieri[2]($\boxtimes$), and Enrico Vicario[3]

[1] Department of Computer Science, University of Turin, Turin, Italy
horvath@di.unito.it
[2] Department of Computer Science, University of Southern California,
Los Angeles, USA
paolieri@usc.edu
[3] Department of Information Engineering, University of Florence, Florence, Italy
enrico.vicario@unifi.it

Abstract. We present a solution to compute equilibrium probability density functions (PDFs) for the continuous component of the state in Markov regenerative processes, a class of non-Markovian processes. Equilibrium PDFs are derived as closed-form analytical expressions by applying the Key Renewal Theorem to stochastic state classes computed between regenerations. The solution, evaluated experimentally through the development of an analysis tool, provides the basis to analyze system properties from the equilibrium.

1 Introduction

Stochastic models of discrete-event systems provide a powerful tool for the evaluation of system designs: concurrent activities with stochastic duration can represent service times, arrivals, server breakdowns, or repair actions, for example. Several high-level modeling formalisms are available, including queueing networks [11], stochastic Petri nets [9], stochastic process algebras [5]. Performance and reliability metrics can be evaluated in these models from the transient or steady-state probabilities of their underlying stochastic processes [4].

Steady-state probabilities are computed for the *discrete* component of the system state, such as the number of customers in each queue, or the number of failed servers. These probabilities provide a complete characterization at equilibrium for continuous-time Markov chains (CTMCs), but not for non-Markovian processes, where future evolution also depends on the distribution of the *continuous* component of the state (e.g., remaining time to a failure or service completion).

In fact, CTMCs are memoryless [12]: at any time instant, the future evolution of the process is completely characterized by the current discrete state, independently of previous states or sojourn times. For example, in an M/M/1 queue with arrival rate λ and service rate μ, the discrete state is the number n of customers and times to the next arrival and to the next service (if $n > 0$) are always independent, exponential random variables with rates λ and μ, respectively. Non-Markovian processes do not enjoy such properties: the process evolution after

© The Author(s), under exclusive license to Springer Nature Switzerland AG 2023
N. Jansen and M. Tribastone (Eds.): QEST 2023, LNCS 14287, pp. 172–187, 2023.
https://doi.org/10.1007/978-3-031-43835-6_13

time t depends not only on the discrete state, but also on the distribution of timers at time t [12]. Timers with general (i.e., non-exponential) distributions can "accumulate memory" of previous events and sojourn times: in an $M/G/1$ queue, the distribution of the remaining service time is not known given the number n of customers, but depends on the time since the last service. Multiple general timers that are concurrently enabled become dependent random variables with a joint distribution [16].

In this paper, we propose a solution to evaluate, in addition to the steady-state probability of discrete states, also the equilibrium distribution of the continuous component of each state, the active timers. Our solution is analytical: using the calculus of stochastic state classes [10,16], we compute closed-form expressions for the joint probability density function (PDF) of active timers immediately after each discrete event of a stochastic model; from these, we derive equilibrium PDFs through the construction of a renewal process and application of the Key Renewal Theorem [12]. Our analysis targets Markov regenerative processes (MRGPs): this class of non-Markovian processes satisfies the Markov property at *regeneration points*, which correspond to time instants where the discrete component of the state provides sufficient information to characterize the PDF of active timers, and thus future evolution [12]. Regeneration points occur when all general timers are reset: in the $M/G/1$ queue example, each service completion corresponds to a regeneration point of the underlying stochastic process, which is an MRGP.

We restrict our analysis to irreducible MRGPs with finite state space and "bounded memory," i.e., such that a new regeneration point is reached w.p.1 after a bounded number of discrete events (in general, MRGPs can produce trajectories without regenerations, if their measure is zero). Semi-Markov processes (SMPs) are a special case of MRGPs where regeneration points are reached after each discrete event. In contrast with MRGPs under enabling restriction [7], we allow multiple general timers to be concurrently enabled [10,13]. We develop an implementation based on the freely available tool ORIS [14]. Our implementation can automatically compute steady-state probabilities and equilibrium PDFs for each stochastic state class of models where timers are deterministic or sampled according to expolynomial PDFs (products of exponentials and polynomials), which include exponential, uniform, triangular, and Erlang distributions.

The equilibrium analysis of MRGPs is an important result, as it characterizes the stochastic process at the time of a random inspection in the long-run. It generalizes the well-known result for the *remaining life* of a renewal process, which has PDF $f_Y(y) = [1 - F_X(y)]/E[X]$ after a random inspection if $F_X(x)$ is the cumulative distribution function (CDF) of inter-event times [12, Eq. (8.40)]. When the inspection represents a catastrophic failure, equilibrium PDFs can be modified to reflect its effects and used in transient analysis to compute survivability metrics, similarly to solutions for CTMCs [6,8,15]. Once the equilibrium PDFs are known for each state, the approach also enables the generation of samples from the equilibrium distribution of the MRGP process without the need to monitor convergence and mixing during a simulation, similarly to *perfect sampling* [1–3] methods for DTMCs and CTMCs.

2   Markov Regenerative Processes

2.1   Stochastic Time Petri Nets

We adopt stochastic time Petri nets (STPNs) to specify discrete-event systems governed by stochastic timers. We refer to Appendix A for a complete definition of STPNs and present only their essential elements. An STPN includes a set P of *places* (graphically drawn as circles) and a set T of *transitions* (drawn as vertical bars): transitions represent concurrent activities that move *tokens* between places.

State. The state $s = (m, \vec{\tau})$ of an STPN includes two components: (1) a *marking* $m \colon P \to \mathbb{N}$ that assigns a *token count* to each place and controls the enabling of transitions, and (2) a *time-to-fire* vector $\vec{\tau} = (\tau_1, \dots, \tau_n)$ that specifies the remaining time $\tau_i \in \mathbb{R}_{\geqslant 0}$ to the *firing* of each transition enabled by m (given a total order on T).

State Update. A transition is enabled by marking m if all input places (connected with incoming arcs) contain at least one token. The enabled transition t^* with minimum remaining time in $\vec{\tau}$ fires and produces a new state $s' = (m', \vec{\tau}')$ where: m' is obtained from m by removing one token from each input place of t^* and adding one token to each output place of t^* (connected with outgoing arcs). Transitions enabled before and after each step, but distinct from t^*, are called *persistent*: after the firing, their remaining times to the fire are reduced by that of t^*, i.e., $\tau_i' = \tau_i - \tau_{t^*}$. Other transitions enabled by m' are called *newly-enabled*: their remaining times τ_i' are sampled independently according to CDFs $F_t(x)$ specified by the STPN for each transition t.

We assume that each CDF F_t admits the representation $F_t(x) = \int_0^x f_t(u)\, du$. We represent the PDF of a transition t with deterministic duration $\overline{x}$ using the Dirac delta function $f_t(x) = \delta(x - \overline{x})$. As usual in stochastic Petri nets, the PDF family of a transition is represented graphically using white rectangles for exponential transitions, gray rectangles for deterministic ones, black rectangles for other distributions.

Example 1 (Parallel Producer-Consumer). Figure 1a presents the STPN model of two producers working in parallel to produce parts that are consumed together by a single consumer; consumption begins only when both parts are available, and production of new parts starts when the previous ones have been consumed. Tokens in places $\mathtt{pIn_1}$ and $\mathtt{pIn_2}$ activate the two producers represented by transitions $\{\mathtt{prod_{11}}, \mathtt{prod_{12}}\}$ and $\mathtt{prod_2}$, respectively. The first producer uses two processing units to increase performance: the first among $\mathtt{prod_{11}}$ and $\mathtt{prod_{12}}$ to complete (respectively, with time to fire PDF $f(x) = 2x - 2$ on $[1, 2]$ and uniform on $[1, 2]$) ends the production of the first part. The second producer is modeled by transition $\mathtt{prod_2}$ with PDF $f(x) = 1 - x/2$ on $[0, 2]$. The consumer, modeled

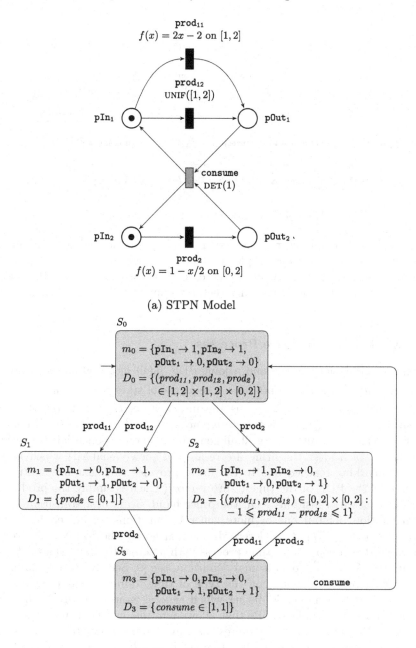

(a) STPN Model

(b) State Class Graph

Fig. 1. Parallel Producer-Consumer Example

by transition consume with deterministic firing time equal to 1, is enabled when
both tokens are moved to places $pOut_1$ and $pOut_2$; after its completion, tokens

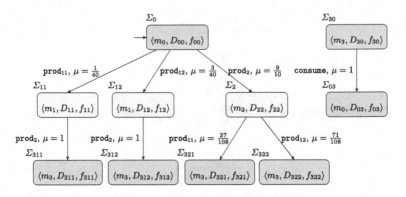

Fig. 2. Trees of Stochastic State Classes from Σ_0 and Σ_{00}

are moved back to places pIn$_1$ and pIn$_2$, and production restarts. Figure 1b presents the *state class graph* [16] for this model, where edges represent possible transitions firings and each node S_i represents the marking m_i and the set D_i of possible values for the time-to-fire vector $\vec{\tau}$ immediately after a firing.

2.2 Markov Regenerative Processes

Given an initial marking m_0 and a joint PDF $f_0(\vec{x})$ of initial times to fire $\vec{\tau}_0$, each execution of the STPN produces a sequence of state changes $s_0 \xrightarrow{t_1} s_1 \xrightarrow{t_2} s_2 \xrightarrow{t_3} \cdots$ where $s_0 = (m_0, \vec{\tau}_0)$ is the initial state, $t_i \in T$ is the ith fired transition, and $s_i = (m_i, \vec{\tau}_i)$ is the state reached after the firing of t_i. Firing of a transition is a *regeneration point* if all general (i.e., non-exponential) transitions that are enabled after the firing are resampled (newly-enabled) which means that the marking reached due to the firing provides sufficient information to reconstruct the PDF of the time-to-fire vector (which is simply the product of PDFs of enabled transitions). On the other hand, firing of a transition is not a regeneration point if one or more general transitions remain enabled. The *marking process* $\{Z(t), t \geqslant 0\}$ records the marking of the STPN as it evolves over time. It is a continuous-time process with a countable state space, the set of markings $M \subseteq \mathbb{N}^P$ (for a formal definition, see [9, Sect. 3.1]). The family of the marking process depends on the type of time-to-fire distributions and on the overlap of time intervals during which general transitions are enabled [4]. We focus on the class of MRGPs that allows multiple general transitions to be enabled at the same time but requires that, in a bounded number of transition firings, the model reaches a regeneration point; we denote the set of markings reached at regeneration points as $R \subseteq M$.

Example 2 (Parallel Producer-Consumer). Regeneration points for the marking process of the STPN in Fig. 1a are highlighted in the state class graph of Fig. 1b by darker backgrounds; they correspond to the firings that lead to marking m_3 (i.e., when production ends) or to marking m_0 (i.e., when consume fires and production restarts), i.e., $R = \{m_3, m_0\}$.

$f_{00}(age, prod_{11}, prod_{12}, prod_2) = (2prod_{11} - prod_{11}\, prod_2 + prod_2 - 2)\, \delta(age)$

$D_{00} = \{(age, prod_{11}, prod_{12}, prod_2) \in [0,0] \times [1,2] \times [1,2] \times [0,2]\}$

$f_{11}(age, prod_2) = -200age^2 - 320age + 40age^2\, prod_2 +$
$\qquad\qquad 120age\, prod_2 - 40age^3 - 160 + 80prod_2$

$D_{11} = \{(age, prod_2) \in [-2,-1] \times [0,1] : -2 \leqslant age - prod_2 \leqslant -1\}$

$f_{12}(age, prod_2) = -\dfrac{80}{3}age^2 - \dfrac{80}{3}age + \dfrac{20}{3}age^2\, prod_2 + \dfrac{40}{3}age\, prod_2 - \dfrac{20}{3}age^3$

$D_{12} = \{(age, prod_2) \in [-2,-1] \times [0,1] : -2 \leqslant age - prod_2 \leqslant -1\}$

$f_{311}(age, consume) = \left(\dfrac{200}{3} + \dfrac{580}{3}age + 200age^2 + \dfrac{260}{3}age^3 + \dfrac{40}{3}age^4\right)\delta(consume - 1)$

$D_{311} = \{(age, consume) \in [-2,-1] \times [1,1]\}$

$f_{312}(age, consume) = \left(-\dfrac{80}{9} - \dfrac{40}{9}age + \dfrac{40}{3}age^2 + \dfrac{100}{9}age^3 + \dfrac{20}{9}age^4\right)\delta(consume - 1)$

$D_{312} = \{(age, consume) \in [-2,-1] \times [1,1]\}$

Fig. 3. PDFs and supports for paths $\Sigma_0 \to \Sigma_{11} \to \Sigma_{311}$ and $\Sigma_0 \to \Sigma_{12} \to \Sigma_{312}$

MRGPs provide a good trade-off between modeling power and complexity of the analysis: concurrent general timers can persist to discrete events, while transient and steady-state probabilities can be computed numerically [10,13].

Transient probabilities $P_{ij}(t) := P(Z(t) = j \mid X_0 = i)$ from all initial regenerations $i \in R$ and for all $j \in M$, $t \geqslant 0$ can be computed from the system of Markov renewal equations [12]

$$\mathbf{P}(t) = \mathbf{L}(t) + \int_0^t d\mathbf{G}(u)\, \mathbf{P}(t - u) \tag{1}$$

where

$$G_{ik}(t) := P(X_1 = k, T_1 \leqslant t \mid X_0 = i)$$

for $i, k \in R$ is the *global kernel* of the MRGP, specifying the joint distribution of the next regeneration X_1 and regeneration point T_1 given that the last regeneration was $X_0 = i$ at time 0, while

$$L_{ij}(t) := P(Z(t) = j, T_1 > t \mid X_0 = i)$$

is the *local kernel* of the MRGP, defined as the probability that, given the initial regeneration $i \in R$ at time 0, no further regeneration has been reached and the marking is $j \in M$ at time t. Informally, the global kernel describes the process of the regeneration points while the local kernel provides the necessary information between two consecutive regeneration points. The system of Eq. (1) is a set of Volterra integral equations that can be solved numerically in the time domain.

The global and local kernels also provide the steady-state probabilities p_j of each marking $j \in M$. If $\vec{\pi}$ is the vector of steady-state probabilities of the discrete-time Markov chain (DTMC) embedded at regeneration points, i.e., $\sum_{k \in R} \pi_k = 1$ and $\vec{\pi} = \mathbf{G}(\infty)\vec{\pi}$, then

$$p_j = \frac{\sum_{i \in R} \pi_i \alpha_{ij}}{\sum_{i \in R, j' \in M} \pi_i \alpha_{ij'}} \tag{2}$$

where $\alpha_{ij} := \int_0^\infty L_{ij}(t)\, dt$ is the expected time spent in j after a regeneration in $i \in R$ and before the next one [13].

2.3   Analysis with Stochastic State Classes

Although Eqs. (1) and (2) provide an elegant solution to compute transient and steady-state probabilities, a major difficulty lies in the evaluation of the global and local kernels for a given model. One approach is to compute the joint PDF of the time-to-fire vector and firing time after each transition firing until a regeneration. This can be accomplished through the calculus of *stochastic state classes* [10,16].

Definition 1 (Stochastic State Class). *A stochastic state class Σ is a tuple $\langle m, D, f \rangle$ where: $m \in M$ is a marking; f is the PDF (immediately after a firing) of the random vector $\langle \tau_{age}, \vec{\tau} \rangle$ including the time-to-fire vector $\vec{\tau}$ of transitions enabled by m and the age variable τ_{age} accumulating previous sojourn times; $D \subseteq \mathbb{R}^{n+1}$ is the support of f.*

The initial stochastic state class has marking m_0 (the initial marking of the STPN) and PDF $f(x_{age}, \vec{x}) = \delta(x_{age}) f_0(\vec{x})$, where f_0 is the PDF of the initial time-to-fire vector $\vec{\tau}_0$ of the STPN and δ is the Dirac delta. Given a class $\Sigma = \langle m, D, f \rangle$ and a transition t enabled by m, the calculus of [16] computes (1) the probability μ that t is the transition that fires in Σ, and (2) the *successor class Σ'*, which includes the marking and time-to-fire PDF after the firing of t in Σ. In the calculus, the age variable τ_{age} is decreased by the sojourn time, to treat it similarly to persistent times to fire [10]; the time of the last firing is thus given by $-\tau_{age}$.

To analyze MRGPs, regeneration points are detected during the computation of successors and new regeneration states are included in the set R. Each regeneration state $i \in R$ uniquely identifies an initial marking and PDF of the time-to-fire vector, which can be used to construct an initial stochastic state class. By computing trees of stochastic state classes from each $i \in R$ to other regeneration points, the MRGP is encoded as a set of *trees* of stochastic state classes.

If INNER(i) and LEAVES(i) are, respectively, the stochastic state classes of inner nodes and leaf nodes in the transient tree enumerated from regeneration $i \in R$, then

$$L_{ij}(t) = \sum_{\substack{\Sigma \in \text{INNER}(i) \text{ s.t.} \\ \Sigma \text{ has marking } j}} p_{in}(\Sigma, t) \quad \text{and} \quad G_{ik}(t) = \sum_{\substack{\Sigma \in \text{LEAVES}(i) \text{ s.t.} \\ \Sigma \text{ has regeneration } k}} p_{reach}(\Sigma, t)$$

for all $i, k \in R$, $j \in M$, and $t \geqslant 0$, where for a class $\Sigma = \langle m, D, f \rangle$ reached through firings with probability $\rho(\Sigma)$,

$$p_{reach}(\Sigma, t) = \rho(\Sigma) \int_{\{(x_{age}, \vec{x}) \in D \,|\, -x_{age} \leqslant t\}} f(x_{age}, \vec{x})\, dx_{age}\, d\vec{x}$$

is the probability that Σ is reached from i within time t and

$$p_{in}(\Sigma, t) = \rho(\Sigma) \int_{\{(x_{age}, \vec{x}) \in D | -x_{age} \leqslant t \text{ and } x_k - x_{age} > t \; \forall k\}} f(x_{age}, \vec{x}) \, dx_{age} \, d\vec{x}$$

is the probability that all and only the transitions leading from i to Σ have fired by time t [10]. When timers are deterministic or with expolynomial PDFs, the integrals in the two equations above (and similar integrals in the rest of the paper) can be evaluated exactly in ORIS by symbolic integration over zones.

Example 3 (Parallel Producer-Consumer). Figures 2 and 3 present the trees of stochastic state classes for the example of Fig. 1a. Regenerations are identified by markings $R = \{m_0, m_3\}$: INNER$(m_0) = \{\Sigma_0, \Sigma_{11}, \Sigma_{12}, \Sigma_2\}$, INNER$(m_3) = \{\Sigma_{30}\}$, while LEAVES$(m_0) = \{\Sigma_{311}, \Sigma_{312}, \Sigma_{321}, \Sigma_{322}\}$, and LEAVES$(m_3) = \{\Sigma_{03}\}$. Edges are labeled with firing probabilities, e.g., the probability $\rho(\Sigma_{321})$ of firing prod$_2$ and then prod$_{11}$ is $\frac{9}{10} \cdot \frac{37}{108}$.

3   Equilibrium Analysis

3.1   Steady-State Probabilities

To compute the probability of observing each stochastic state class in steady state, we follow the strategy of Eq. (2) but consider each class (instead of each marking) as a distinct state j. First, we compute the limit of the global kernel $\mathbf{G}(t)$ as $t \to \infty$. Each entry $G_{ik}(\infty)$ for $i, k \in R$ can be obtained as the product of firing probabilities from regeneration i to all classes in LEAVES(i) that reach regeneration $k \in R$:

$$G_{ik}(\infty) := P(X_1 = k \mid X_0 = i) = \sum_{\substack{\Sigma \in \text{LEAVES}(i) \text{ s.t.} \\ \Sigma \text{ has regeneration } k}} \rho(\Sigma)$$

where $\rho(\Sigma)$ is the product of firing probabilities μ for the sequence of firings that leads from $i \in R$ to Σ.

Next, for each regeneration $i \in R$ and class $j \in$ INNER(i), we compute the expected time α_j spent in j after reaching i and before the next regeneration.

Lemma 1. *Let $j = \langle m, D, f \rangle$ be an inner node in the tree enumerated from regeneration $i \in R$, i.e., $j \in$ INNER(i). Then, the expected sojourn time of the MRGP in j before the next regeneration is*

$$\alpha_j = \rho(j) \sum_{t \in E(j)} \mu^{(t)} \int_{D^{(t)}} x_{k_t} \, f^{(t)}(x_{age}, \vec{x}) \, dx_{age} \, d\vec{x}, \tag{3}$$

where: $\rho(j)$ is the product of firing probabilities of transitions that lead from regeneration i to class j; $E(j) \subseteq T$ is the set of transitions enabled in j; $\mu^{(t)}$ is the probability that $t \in E(j)$ fires in j; $D^{(t)} = \{(x_{age}, \vec{x}) \in D \mid x_k \geqslant x_{k_t} \; \forall k\}$ is *the subset of the support D where τ_{k_t}, the time to fire of t, is minimum (k_t is the index of t in $\vec{\tau}$); and*

$$f^{(t)}(x_{age}, \vec{x}) := f(x_{age}, \vec{x}) \left(\int_{D^{(t)}} f(x_{age}, \vec{x}) \, dx_{age} \, d\vec{x} \right)^{-1} \tag{4}$$

is the PDF of $\langle \tau_{age}, \vec{\tau} \rangle$ conditioned on $\{\tau_{k_t}$ is minimum$\}$.

Proof. Equation (3) follows from the definition of stochastic state class and from the law of total expectation. The events of the firing in j of transitions in $E(j)$ are mutually exclusive and exhaustive, and thus, if S_j is the sojourn time in j, $\alpha_j := \rho(j) E[S_j] = \rho(j) \sum_{t \in E(j)} \mu^{(t)} E[S_j \mid t$ fires in $j]$.

Similarly to Eq. (2), the steady-state probability of class $j \in \text{INNER}(i)$ is

$$p_j = \frac{\pi_i \alpha_j}{\sum_{i' \in R, j' \in \text{INNER}(i')} \pi_{i'} \alpha_{j'}} \tag{5}$$

where $\vec{\pi}$ is such that $\sum_{k \in R} \pi_k = 1$ and $\vec{\pi} = \mathbf{G}(\infty)\vec{\pi}$.

Example 4 (Parallel Producer-Consumer). The MRGP of Fig. 2 has a simple DTMC embedded at regeneration points: the process alternates between regenerations $R = \{m_0, m_3\}$, and thus $\mathbf{G}(\infty) = \left(\begin{smallmatrix} 0 & 1 \\ 1 & 0 \end{smallmatrix} \right)$ and $\vec{\pi} = (0.5, 0.5)$. The sojourn times α_j of inner nodes are: $\left(\frac{1}{40}\frac{4}{3} + \frac{3}{40}\frac{11}{9} + \frac{9}{10}\frac{31}{54} \right) = \frac{77}{120}$ for $j = \Sigma_0$, $\frac{1}{40}\left(\frac{2}{9}\right) = \frac{1}{180}$ for $j = \Sigma_{11}$, $\frac{3}{40}\left(\frac{7}{27}\right) = \frac{7}{360}$ for $j = \Sigma_{12}$, $\frac{9}{10}\left(\frac{37}{108}\frac{34}{37} + \frac{71}{108}\frac{59}{71}\right) = \frac{31}{40}$ for $j = \Sigma_2$, 1 for $j = \Sigma_{30}$. These give steady-state probabilities p_j equal to $\frac{77}{293}, \frac{2}{879}, \frac{7}{879}, \frac{93}{293}, \frac{120}{293}$ for $j = \Sigma_0, \Sigma_{11}, \Sigma_{12}, \Sigma_2, \Sigma_{30}$, respectively.

3.2   Equilibrium PDFs

Marginal PDF of $\vec{\tau}$ in Σ given the firing of t_1. Without loss of generality, we assume that n transitions $t_1, \ldots, t_n$ are enabled in Σ and consider the case where t_1 is the one that fires. Conditioned on this event, the PDF of $\vec{\tau}$ in Σ is

$$f^{(t_1)}(\vec{x}) = \int_{D^{(t_1)}} f(x_{age}, \vec{x}) \, dx_{age} \left(\int_{D^{(t_1)}} f(x_{age}, \vec{x}) \, dx_{age} \, d\vec{x} \right)^{-1} \tag{6}$$

where $D^{(t_1)} = \{(x_{age}, \vec{x}) \in D \mid x_k \geqslant x_1 \, \forall k\}$ is the subset of the support D where τ_1 is minimum. Equation (6) follows by restricting the support to the subset $D^{(t_1)}$, normalizing the PDF $f(x_{age}, \vec{x})$, and then obtaining the marginal PDF of $\vec{\tau}$ by integrating over all possible values for τ_{age} in $D^{(t_1)}$.

Stochastic process $\vec{r}(t)$ of times to fire $\vec{\tau}$ across renewals. Successive visits to the stochastic state class $\Sigma = \langle m, D, f \rangle$ observe copies of $\langle \tau_{age}, \vec{\tau} \rangle$ that are independent and identically distributed (i.i.d.) according to the PDF f, since the MRGP encounters a regeneration point between visits and then performs the same sequence of transition firings. Similarly, visits to Σ that end with the firing of t_1 also observe the same PDF $f^{(t_1)}$ of $\vec{\tau}$ derived in Eq. (6), and their sojourn times are i.i.d. random variables.

We focus our attention on the time intervals of these i.i.d. sojourn times: as time advances, we move from a sojourn in Σ to the next one, always under the hypothesis that t_1 is the transition that fires in Σ. We construct a renewal process $\{N(t), t \geqslant 0\}$ where times between events (i.e., interarrival times) are distributed as a sojourn in Σ that ends with the firing of t_1. We denote interarrival times as $S_1, S_2, \ldots$ and renewal times as $T_k = \sum_{i=1}^{k} S_i$ for $k \geqslant 0$; $N(t) = \max\{k \mid T_k \leqslant t\}$ is the number of sojourns completed by time t and $N(t) = k \Leftrightarrow T_k \leqslant t < T_{k+1}$. The interarrival PDF of this renewal process is given by the marginal PDF of τ_1 given that it is minimum in $\vec{\tau}$, which is

$$g(x_1) = \int_{D^{(t_1)}} f^{(t_1)}(x_1, x_2, \ldots, x_n)\, dx_2 \cdots dx_n \qquad (7)$$

where $D^{(t_1)}$ is the support of $f^{(t_1)}$. Equation (7) integrates over all possible values of $\tau_2, \ldots, \tau_n$ to obtain the marginal PDF of τ_1 when $\tau_1 \leqslant \tau_i$ for all $i = 2, \ldots, n$.

As $N(t)$ evolves across each renewal $T_0, T_1, T_2, \ldots$, a new time-to-fire vector $\vec{\tau}^{(i)}, i = 0, 1, 2, \ldots$ is sampled independently at each T_i according to the same PDF $f^{(t_1)}$. Our goal is to study the evolution of these time-to-fire random vectors over time, subject to the fact that renewal times T_i are also random. We denote by $\{\vec{r}(t),\ t \geqslant 0\}$ the n-dimensional stochastic process describing, for each $t \geqslant 0$, the current value of the time-to-fire vector, i.e., $\vec{r}(t) := \vec{\tau}^{(N(t))} - (t - T_{N(t)})$, and denote its PDF at all $t \geqslant 0$ by $h(t, \vec{x})$, i.e.,

$$P(r_1(t) \leqslant x_1, \ldots, r_n(t) \leqslant x_n) := \int_{-\infty}^{x_1} \cdots \int_{-\infty}^{x_n} h(t, \vec{x})\, dx_1 \cdots dx_n.$$

Equilibrium PDF of $\vec{r}(t)$. Our goal is to compute the equilibrium PDF of $\vec{r}(t)$, i.e., the function $\hat{f}^{(t_1)}(\vec{x}) = \lim_{t \to \infty} h(t, \vec{x})$, which gives the PDF of the times to fire in Σ at equilibrium (given that sojourns end with the firing of t_1). First, we provide the following result, which highlights the fundamental relation between $h(t, \vec{x})$, the object of our analysis, and the PDF $f^{(t_1)}(\vec{x})$, which can be readily computed using Eq. (6).

Lemma 2 (Renewal Equation for h). *If $h(t, \vec{x})$ is the PDF of $\vec{r}(t)$ for each $t \geqslant 0$, $f^{(t_1)}(\vec{x})$ is the PDF of $\vec{\tau}$ at each renewal, and $g(x)$ is the PDF of τ_1 (the interarrival time of the renewal process), the following renewal equation holds:*

$$h(t, \vec{x}) = f^{(t_1)}(\vec{x} + t) + \int_0^t h(t - u, \vec{x})\, g(u)\, du. \qquad (8)$$

Proof. Equation (8) can be derived through a renewal argument: for the first renewal time S_1 we have that either $S_1 > t$ or $S_1 \leqslant t$.

If $S_1 > t$, then the first renewal has not occurred, so that $N(t) = 0$ and $\vec{r}(t) = \vec{\tau}^{(0)} - t$. The PDF of $\vec{r}(t)$ at time t is then given by $f^{(t_1)}(\vec{x} + t)/P(\tau_1 > t)$, i.e., the PDF $f^{(t_1)}$ used to sample $\vec{\tau}^{(0)}$ but conditioned on the event $\{\tau_1 > t\}$ and where each component is shifted by time t (we denote by $\vec{x} + t$ the vector $(x_1 + t, \ldots, x_n + t)$). Then, we have that $h(t, \vec{x} \mid S_1 > t)\, P(S_1 > t) = f^{(t_1)}(\vec{x} + t)$, since $S_1 := \tau_1$.

If $S_1 \leqslant t$, the process $\vec{r}(t)$ "probabilistically restarts" after S_1, when a new time-to-fire vector $\vec{\tau}^{(1)}$ is sampled. Formally, if $S_1 = u$, at least one renewal is encountered by time t, $N(t) = N(t-u)+1$, $T_{N(t-u)+1} = T_{N(t-u)} + u$, and thus

$$\vec{r}(t) = \vec{\tau}^{(N(t-u)+1)} - (t - T_{N(t-u)+1})$$
$$= \vec{\tau}^{(N(t-u)+1)} - [(t - u) - T_{N(t-u)}]$$

for $u \leqslant t$. Given that time-to-fire vectors $\vec{\tau}^{(N(t-u)+1)}$ and $\vec{\tau}^{(N(t-u))}$ have the same PDF $f^{(t_1)}$, it holds that $h(t, \vec{x}) = h(t - u, \vec{x})$ for $u \leqslant t$. By conditioning on all the possible values of $S_1 = u$ and decreasing t accordingly, we have

$$h(t, \vec{x} \mid S_1 \leqslant t)\, P(S_1 \leqslant t) = \int_0^t h(t - u, \vec{x})\, g(u)\, du\,.$$

By putting together the two cases, we obtain Eq. (8).

Lemma 2 establishes a connection between h and $f^{(t_1)}$ and also reveals the recursive structure of h across renewals. This kind of renewal-type equation is well-known for renewal processes and provides a strategy to compute h at the equilibrium through the following result [12, Theorem 8.17].

Theorem 1 (Key Renewal Theorem). *Let $g(x)$ be the PDF of the inter-arrival time, and let h be a solution to the renewal-type equation $h(t) = d(t) + \int_0^t h(t-u)g(u)\, du$. Then, if d is the difference of two non-negative bounded monotone functions and $\int_0^\infty |d(u)|\, du < \infty$,*

$$\lim_{t \to \infty} h(t) = \frac{1}{E[S]} \int_0^\infty d(u)\, du$$

where $E[S] = \int_0^\infty u\, g(u)\, du$ is the mean interarrival time.

Theorem 1 applies to Eq. (8) with $d(t) = f^{(t_1)}(\vec{x} + t)$. Moreover, when the PDFs f_t used to sample newly-enabled transitions are piecewise expolynomials (products of exponentials and polynomials), the joint PDF f of timers, and thus $f^{(t_1)}$, is also piecewise continuous and with bounded variation [16].

By combining Lemma 2 and Theorem 1, we obtain the equilibrium PDF $\hat{f}^{(t_1)}$ of $\vec{\tau}$ in Σ when t_1 is the transition that fires at the end of each sojourn:

$$\hat{f}^{(t_1)}(\vec{x}) := \lim_{t \to \infty} h(t, \vec{x}) = \frac{1}{E[S^{(t_1)}]} \int_0^\infty f^{(t_1)}(\vec{x} + u)\, du \qquad (9)$$

where $E[S^{(t_1)}] = \int_0^\infty u\, g(u)\, du$ is the mean sojourn time in Σ when t_1 fires. The identity of Eq. (9) is a major step for the analysis of the joint PDF of $\vec{\tau}$ at the steady state. Combined with Eq. (6) to obtain $f^{(t_1)}$ from f, and with Eq. (7) to obtain g from $f^{(t_1)}$, it provides a straightforward derivation of the equilibrium PDF under the hypothesis that t_1 is always the transition that fires in Σ.

Equilibrium PDF When Multiple Transitions Can Fire. The equilibrium PDF $\hat{f}^{(t_1)}$ of Eq. (9) assumes that, after each visit to Σ, transition t_1 is always the one that fires among $t_1, \ldots, t_n$. The following theorem removes this hypothesis.

Theorem 2 (Equilibrium PDF). *Let* $\Sigma = \langle m, D, f \rangle$ *be a stochastic state class where transitions* $t_1, \ldots, t_n$ *can fire with probabilities* $\mu^{(t_1)}, \ldots, \mu^{(t_n)}$, *respectively. Then, the equilibrium PDF of* $\vec{\tau} = (\tau_1, \ldots, \tau_n)$ *is given by*

$$\hat{f}(\vec{x}) = \frac{1}{E[S]} \sum_{i=1}^{n} \mu^{(t_i)} \int_0^{\infty} f^{(t_i)}(\vec{x} + u) \, du \qquad (10)$$

where $E[S]$ *is the expected sojourn time in* Σ *and, for all* $i = 1, \ldots, n$, $f^{(t_i)}$ *is the PDF of* $\vec{\tau}$ *conditioned on the firing of* t_i *according to Eq. (6).*

Proof. We focus only on sojourns in class Σ and ignore the rest of the time line. The probability that a sojourn ends with the firing of t_i is $\mu^{(t_i)}$ for $i = 1, \ldots, n$, with $\sum_{i=1}^{n} \mu^{(t_i)} = 1$; conditioned on this event, the expected sojourn time in Σ is $E[S^{(t_i)}]$. Then, the steady-state probability of sojourns in Σ that end with the firing of t_i is given by

$$p_i = \frac{\mu^{(t_i)} E[S^{(t_i)}]}{\sum_{j=1}^{n} \mu^{(t_j)} E[S^{(t_j)}]}$$

which is the mean fraction of time spent in such sojourns. Since $\hat{f}^{(t_i)}$ is the equilibrium PDF when sojourns end with t_i,

$$\hat{f}(\vec{x}) = \sum_{i=1}^{n} p_i \, \hat{f}^{(t_i)}(\vec{x}) = \sum_{i=1}^{n} \left(\frac{\mu^{(t_i)} E[S^{(t_i)}]}{\sum_{j=1}^{n} \mu^{(t_j)} E[S^{(t_j)}]} \right) \hat{f}^{(t_i)}(\vec{x})$$

$$= \frac{1}{\sum_{j=1}^{n} \mu^{(t_j)} E[S^{(t_j)}]} \sum_{i=1}^{n} \mu^{(t_i)} \int_0^{\infty} f^{(t_i)}(\vec{x} + u) \, du$$

which, since $\sum_{j=1}^{n} \mu^{(t_j)} E[S^{(t_j)}] = E[S]$, gives Eq. (10). $\qquad \square$

4   Experimental Evaluation

Steady-state probabilities and equilibrium PDFs represent the equilibrium distribution of the MRGP. When used as an initial distribution for transient analysis, this distribution must result in constant transient probabilities that are equal to the steady-state ones. In this section, we describe how to perform transient analysis from this distribution and validate the correctness of the approach.

In Sect. 3.1, we derived the steady-state probability p_c of each each class $c \in \cup_{i \in R} \mathrm{INNER}(i)$. Given that the MRGP is in class $c = \langle m, D, f \rangle$, the marking is equal to m and the time-to-fire vector $\vec{\tau}$ has equilibrium PDF given by $\hat{f}(\vec{x})$, which is computed from f according to Eq. (10) of Theorem 2. To compute transient probabilities from the equilibrium, we modify the approach of Eq. (1) as follows.

First, for each inner node $c = \langle m, D, f \rangle, c \in \cup_{i \in R} \mathrm{INNER}(i)$, we compute a tree of stochastic state classes until the next regeneration. We construct the initial class $\mathrm{START}(c)$ of this tree using marking m and PDF of $\langle \tau_{age}, \vec{\tau} \rangle$ equal to $g(x_{age}, \vec{x}) = \delta(x_{age})\hat{f}(\vec{x})$, i.e., $\tau_{age} = 0$ and the time-to-fire vector $\vec{\tau}$ has PDF $\hat{f}(\vec{x})$. For each $c \in \cup_{i \in R} \text{INNER}(i)$, we denote the inner nodes and leaves of the tree computed from $\text{START}(c)$ (until the next regeneration) as $\text{STARTINNER}(c)$ and $\text{STARTLEAVES}(c)$, respectively.

Then, we extend the Markov renewal equations of Eq. (1) by introducing an additional regeneration $\hat{r}$ that represents the state of the MRGP at equilibrium. The process starts in $\hat{r}$ at time 0, but never returns to this artificial regeneration: by construction, the next regeneration belongs to R and, afterward, the MRGP cycles through its original trees of stochastic state classes. To achieve this behavior, we set MRGP kernel entries as follows. Let $\hat{R} = R \cup \{\hat{r}\}$ and set, for $i = \hat{r}$,

$$L_{ij}(t) = \sum_{c \in \cup_{i' \in R} \text{INNER}(i')} p_c \left(\sum_{\substack{\Sigma \in \text{STARTINNER}(c) \text{ s.t.} \\ \Sigma \text{ has marking } j}} p_{in}(\Sigma, t) \right) \tag{11}$$

$$G_{ik}(t) = \sum_{c \in \cup_{i' \in R} \text{INNER}(i')} p_c \left(\sum_{\substack{\Sigma \in \text{STARTLEAVES}(c) \text{ s.t.} \\ \Sigma \text{ has regeneration } k}} p_{reach}(\Sigma, t) \right) \tag{12}$$

for all $k \in R$, $j \in M$, and $t \geqslant 0$. Since $\hat{r}$ is never reached again, we set $G_{ik}(t) = 0$ for all $i \in \hat{R}$ when $k = \hat{r}$.

Kernel entries in the additional row $\hat{r}$ model a random choice of the initial stochastic state class c according to the discrete distribution given by p_c for $c \in \cup_{i \in R} \text{INNER}(i)$; for a given class c, the tree computed from $\text{START}(c)$ is used to characterize the system evolution from the equilibrium in c and until the next regeneration. As in Sect. 2.3, measures $p_{in}(\Sigma, t)$ and $p_{reach}(\Sigma, t)$ provide the probability that the MRGP is in the stochastic state class Σ at time t, and that it has reached Σ by time t, respectively.

Example 5. We consider the STPN of Fig. 1a and its underlying MRGP of Fig. 2 with markings $M = \{m_0, m_1, m_2, m_3\}$ (defined in Fig. 1b), regenerations $R = \{m_0, m_3\}$, inner nodes $\text{INNER}(m_0) = \{\Sigma_0, \Sigma_{11}, \Sigma_{12}, \Sigma_2\}$ and $\text{INNER}(m_3) = \{\Sigma_{30}\}$. The steady-state probabilities p_c result in steady-state probabilities of the

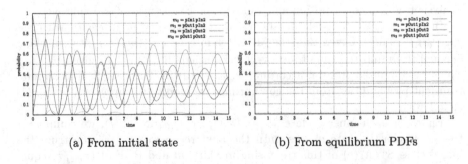

(a) From initial state   (b) From equilibrium PDFs

Fig. 4. Transient Analysis of Parallel Producer-Consumer Example

marking process equal to $\frac{77}{293} \approx 0.263$ for m_0 (steady state probability of Σ_0), $\frac{2}{879} + \frac{7}{879} \approx 0.010$ for m_1 (steady-state probabilities of Σ_{11} and Σ_{12}), $\frac{93}{293} \approx 0.317$ for m_2 (steady-state probability of Σ_2), $\frac{120}{293} \approx 0.410$ for marking m_3 (steady-state probability of Σ_{30}). Figure 4a illustrates the transient probabilities $P_{ij}(t)$ for $0 \leqslant t \leqslant 15$ of the MRGP for $i = m_0$ (i.e., from the initial regeneration) and for each $j \in M$. Figure 4b shows the transient probabilities $P_{\hat{r}j}(t)$ for $0 \leqslant t \leqslant 15$ and each $j \in M$, where the additional kernel row of $\hat{r}$ is computed using Eqs. (11) and (12). As expected, these correspond to the steady-state probabilities.

5   Conclusions

We presented a solution to compute a closed-form expression of the equilibrium distribution of MRGPs. The solution leverages the calculus of stochastic state classes, and it can be applied to a given STPN through the implementation in the ORIS tool. In future work, we plan to apply this solution to compute survivability measures [15] for MRGPs; in particular, equilibrium PDFs can be used to characterize the system after a catastrophic failure at the steady state, providing the initial conditions for the transient analysis of system recovery.

A   Stochastic Time Petri Nets

STPNs are a formal model of concurrent timed systems where: *transitions* (depicted as vertical bars) represent activities; *places* (depicted as circles) represent discrete components of the logical state, with values encoded by a number of *tokens* (depicted as dots); *directed arcs* from *input* places to transitions and from transitions to *output* places represent token moves triggered by the *firing of transitions*. A transition is enabled when all its input places contain at least one token; its firing removes a token from each input place and adds a token to each output place. The time from the enabling to the firing of a transition is a random variable, and the choice between transitions with equal time to fire is solved by a random switch determined by transition *weights*. Moreover, STPNs can: (1) restrict the enabling of a transition using general constraints on token counts (called *enabling functions*); (2) execute additional updates of token counts after a transition firing (specified by *update functions*); (3) restart selected transitions after a firing (using *reset sets*); (4) impose *priorities* among immediate or deterministic transitions.

Definition 2 (Syntax). *An STPN is a tuple $\langle P, T, A^-, A^+, B, U, R, EFT, LFT, F, W, Z \rangle$ where: P and T are disjoint sets of places and transitions, respectively; $A^- \subseteq P \times T$ and $A^+ \subseteq T \times P$ are precondition and post-condition relations, respectively; B, U, and R associate each transition $t \in T$ with an enabling function $B(t) \colon M \to \{\text{TRUE}, \text{FALSE}\}$, an update function $U(t) \colon M \to M$, and a reset set $R(t) \subseteq T$, respectively, where M is the set of reachable markings $m \colon P \to \mathbb{N}$; EFT and LFT associate each transition $t \in T$ with an earliest firing time $EFT(t) \in \mathbb{Q}_{\geqslant 0}$ and a latest firing time $LFT(t) \in \mathbb{Q}_{\geqslant 0} \cup \{\infty\}$ such that*

$EFT(t) \leqslant LFT(t)$; F, W, and Z associate each transition $t \in T$ with a Cumulative Distribution Function (CDF) F_t for its duration $\tau(t) \in [EFT(t), LFT(t)]$ (i.e., $F_t(x) = P\{\tau(t) \leqslant x\}$, with $F_t(x) = 0$ for $x < EFT(t)$, $F_t(x) = 1$ for $x > LFT(t)$), a weight $W(t) \in \mathbb{R}_{>0}$, and a priority $Z(t) \in \mathbb{N}$, respectively.

A place p is said to be an *input* or *output* place for a transition t if $(p, t) \in A^-$ or $(t, p) \in A^+$, respectively. Following the usual terminology of stochastic Petri nets, a transition t is called *immediate* (IMM) if $EFT(t) = LFT(t) = 0$ and *timed* otherwise; a timed transition is called *exponential* (EXP) if $F_t(x) = 1 - \exp(-\lambda x)$ for some rate $\lambda \in \mathbb{R}_{>0}$, or *general* (GEN) if its time to fire has a non-exponential distribution; as a special case, a GEN transition t is *deterministic* (DET) if $EFT(t) = LFT(t) > 0$. For each transition t with $EFT(t) < LFT(t)$, we assume that F_t can be expressed as the integral function of a probability density function (PDF) f_t, i.e., $F_t(x) = \int_0^x f_t(y)\,dy$. The same notation is also adopted for an IMM or DET transition $t \in T$, which is associated with a Dirac impulse function $f_t(y) = \delta(y - \overline{y})$ with $\overline{y} = EFT(t) = LFT(t)$.

A marking $m \in M$ assigns a natural number of tokens to each place of an STPN. A transition t is *enabled* by m if m assigns at least one token to each of its input places and the enabling function $B(t)(m)$ evaluates to TRUE. The set of transitions enabled by m is denoted as $E(m)$.

Definition 3 (State). *The state of an STPN is a pair* $\langle m, \vec{\tau} \rangle$ *where* $m \in M$ *is a marking and vector* $\vec{\tau} \in \mathbb{R}_{\geqslant 0}^{|E(m)|}$ *assigns a* time to fire $\vec{\tau}(t) \in \mathbb{R}_{\geqslant 0}$ *to each enabled transition* $t \in E(m)$.

Definition 4 (Semantics). *Given an initial marking* m_0, *an execution of the STPN is a (finite or infinite) path* $\omega = s_0 \xrightarrow{\gamma_1} s_1 \xrightarrow{\gamma_2} s_2 \xrightarrow{\gamma_3} \cdots$ *such that:* $s_0 = \langle m_0, \vec{\tau}_0 \rangle$ *is the initial state, where the time to fire* $\vec{\tau}_0(t)$ *of each enabled transition* $t \in E(m_0)$ *is sampled according to the distribution* F_t; $\gamma_i \in T$ *is the ith fired transition;* $s_i = \langle m_i, \vec{\tau}_i \rangle$ *is the state reached after the firing of* γ_i. *In each state* s_i:

- *The next transition* γ_{i+1} *is selected from the set of enabled transitions with minimum time to fire and maximum priority according to the distribution given by weights: if* $E_{min} = \arg\min_{t \in E(m_i)} \vec{\tau}_i(t)$ *and* $E_{prio} = \arg\max_{t \in E_{min}} Z(t)$, *then* $t \in E_{prio}$ *is selected with probability* $p_t = W(t) / \left(\sum_{u \in E_{prio}} W(u) \right)$.
- *After the firing of* γ_{i+1}, *the new marking* m_{i+1} *is derived by (1) removing a token from each input place of* γ_{i+1}, *(2) adding a token to each output place of* γ_{i+1}, *and (3) applying the update function* $U(\gamma_{i+1})$ *to the resulting marking. A transition* t *enabled by* m_{i+1} *is termed* persistent *if it is distinct from* γ_{i+1}, *it is not contained in* $R(\gamma_{i+1})$, *and it is enabled also by* m_i *and by the intermediate markings after steps (1) and (2); otherwise,* t *is termed* newly enabled *(thus, transitions in the reset set of* γ_{i+1} *are newly enabled if enabled after the firing).*
- *For each newly enabled transition* t, *the time to fire* $\vec{\tau}_{i+1}(t)$ *is sampled according to the distribution* F_t; *for each persistent transition* t, *the time to fire in* s_{i+1} *is reduced by sojourn time in the previous marking, i.e.,* $\vec{\tau}_{i+1}(t) = \vec{\tau}_i(t) - \vec{\tau}_i(\gamma_{i+1})$.

When features are omitted for a transition $t \in T$, default values are assumed as follows: an always-true enabling function $B(t)(m) = $ TRUE; an identity update function $U(t)(m) = m$ for all $m \in M$; an empty reset set $R(t) = \varnothing$; a weight $W(t) = 1$; and, a priority $Z(t) = 0$.

Arc cardinalities greater than 1 can also be introduced in STPN syntax and semantics, letting the firing of a transition remove an arbitrary number of tokens from each input place or add an arbitrary number of tokens to each output place.

References

1. Balsamo, S., Marin, A., Stojic, I.: Perfect sampling in stochastic Petri nets using decision diagrams. In: MASCOTS, pp. 126–135. IEEE Computer Society (2015).
2. Balsamo, S., Marin, A., Stojic, I.: SPNPS: a tool for perfect sampling in stochastic Petri nets. In: Agha, G., Van Houdt, B. (eds.) QEST 2016. LNCS, vol. 9826, pp. 163–166. Springer, Cham (2016). https://doi.org/10.1007/978-3-319-43425-4_11
3. Casella, G., Lavine, M., Robert, C.P.: Explaining the perfect sampler. Am. Stat. **55**(4), 299–305 (2001)
4. Ciardo, G., German, R., Lindemann, C.: A characterization of the stochastic process underlying a stochastic Petri net. IEEE Trans. Softw. Eng. **20**(7), 506–515 (1994)
5. Clark, A., Gilmore, S., Hillston, J., Tribastone, M.: Stochastic process algebras. In: Bernardo, M., Hillston, J. (eds.) SFM 2007. LNCS, vol. 4486, pp. 132–179. Springer, Heidelberg (2007). https://doi.org/10.1007/978-3-540-72522-0_4
6. Cloth, L., Haverkort, B.R.: Model checking for survivability. In: QEST 2005, pp. 145–154. IEEE Computer Society (2005)
7. German, R.: Iterative analysis of Markov regenerative models. Perform. Eval. **44**(1), 51–72 (2001)
8. Ghasemieh, H., Remke, A., Haverkort, B.R.: Survivability analysis of a sewage treatment facility using hybrid Petri nets. Perform. Eval. **97**, 36–56 (2016)
9. Haas, P.J.: Stochastic Petri Nets. SSOR, Springer, New York (2002). https://doi.org/10.1007/b97265
10. Horváth, A., Paolieri, M., Ridi, L., Vicario, E.: Transient analysis of non-Markovian models using stochastic state classes. Perform. Eval. **69**(7–8), 315–335 (2012)
11. Kleinrock, L.: Queueing Systems. Wiley (1975)
12. Kulkarni, V.: Modeling and Analysis of Stochastic Systems. Chapman & Hall (1995)
13. Martina, S., Paolieri, M., Papini, T., Vicario, E.: Performance evaluation of Fischer's protocol through steady-state analysis of Markov regenerative processes. In: MASCOTS, pp. 355–360. IEEE Computer Society (2016)
14. Paolieri, M., Biagi, M., Carnevali, L., Vicario, E.: The ORIS tool: quantitative evaluation of non-Markovian systems. IEEE Trans. Softw. Eng. **47**(6), 1211–1225 (2021)
15. Trivedi, K.S., Xia, R.: Quantification of system survivability. Telecomm. Syst. **60**(4) (2015)
16. Vicario, E., Sassoli, L., Carnevali, L.: Using stochastic state classes in quantitative evaluation of dense-time reactive systems. IEEE Trans. Softw. Eng. **35**(5), 703–719 (2009)

Max-Entropy Sampling for Deterministic Timed Automata Under Linear Duration Constraints

Benoît Barbot[1(✉)] and Nicolas Basset[2(✉)]

[1] Univ. Paris Est Creteil, LACL, 94010 Creteil, France
benoit.barbot@lacl.fr
[2] Univ. Grenoble Alpes, CNRS, Grenoble INP, VERIMAG,
38000 Grenoble, France
bassetni@univ-grenoble-alpes.fr

Abstract. Adding probabilities to timed automata enables one to carry random simulation of their behaviors and provide answers with statistical guarantees to problems otherwise untractable. Thus, when just a timed language is given, the following natural question arises: *What probability should we consider if we have no a priori knowledge except the given language and the considered length (i.e. number of events) of timed words?* The maximal entropy principle tells us to take the probability measure that maximises the entropy which is the uniform measure on the language restricted to timed word of the given length (with such a uniform measure every timed word has the same chance of being sampled). The uniform sampling method developed in the last decade provides no control on the duration of sampled timed words.

In the present article we consider the problem of finding a probability measure on a timed language maximising the entropy under general linear constraints on duration and for timed words of a given length. The solution we provide generalizes to timed languages a well-known result on probability measure over the real line maximising the Shannon continuous entropy under linear constraints. After giving our general theorem for general linear constraints and for general timed languages, we concentrate to the case when only the mean duration is prescribed (and again when the length is fixed) for timed languages recognised by deterministic timed automata. For this latter case, we provide an efficient sampling algorithm we have implemented and illustrated on several examples.

1 Introduction

Since their introduction in the early 90's, Timed Automata (TA) are extensively used to model and verify the behaviors of real-timed systems and thoroughly explored from a theoretical standpoint. Several lines of research have been developed to add probabilities to these models. The two main motivations for this are

This work was financed by the ANR MAVeriQ (ANR-20-CE25-0012).

(i) modelling systems that both exhibit real-time and probabilistic aspects, for example network protocols where delays are chosen at random to resolve conflicts, as in CSMA-CD (see *e.g.* [9,15]); (ii) statistical model checking of TA with the claim of replacing a prohibitively expensive exhaustive verification of the system by a thorough random simulation of the TA with statistical guarantees (see *e.g.* [11] and reference therein).

For statistical model checking (or any other approach based on statistics) to make sense, the probability distribution defined on the runs of the TA has to be clearly given, which is not always the case as pointed in [2,10]. In the present article we follow a line of research initiated in [2,6] focusing on Deterministic Timed Automata (DTA), where the aim is to give a constructive answer to the following question inspired by Jayne's maximal entropy principle [14]: *What probability should we specify on the runs of a given DTA without any a priori knowledge?* Another aim of this approach, which appears equivalent, is to get a sampler as uniform as possible of runs of the DTA, that is, a sampler that gives to runs of the same length the same (density of) probability. The article [6] was a first theoretical one that dealt with infinite timed words while [2] gave more pragmatic algorithms and their implementation to sample timed words of a given finite length. These two probability measures maximize Shannon continuous entropy adapted to probability measures on finite and infinite timed words respectively. The algorithms in [2] were later implemented in the tool WORDGEN [5] and applied in the context of validation of CPS by generating input signals satisfying a specification given as a DTA [3,4]. With this sampling method, one can choose the length of runs, that is, the number of events occurring in them, but not their duration.

In Sect. 3, we propose, as a main contribution, a maximal entropy theorem for general linear constraints on duration, that is, the expectations (over the timed words of a DTA of a given length) of some arbitrary constraint functions depending only on duration are prescribed. We were inspired by a classical maximal entropy theorem for real-to-real functions and classical Shannon continuous entropy (see the dedicated chapter of the textbook on information theory [12]). As a second main contribution we propose in Sect. 4, a procedure for sampling finite timed words of fixed length with random duration such that the *mean duration is prescribed*. Besides these two main contributions, we introduce, in Sect. 4, the key concept of Laplace Transform of Volumes (LTV) which generalizes the volume functions of [1,2] with an extra parameter. As we will discuss in the conclusion (Sect. 6), we think that replacing these previous volume functions by LTVs can enable us to extend several results by adding a focus on duration. Last but not least, we implemented our algorithms by extending the tool WORD-GEN and in this article we provide several experiments in Sect. 5, one of which features a DTA with thousands of control states.

Before exposing our contributions, we recall preliminaries on timed languages and previous work on max-entropy sampling for DTA in Sect. 2.

2   Preliminaries

In this section, we recall definitions on timed languages, max-entropy probability measure on them and how to sample them.

2.1   General Timed Languages and Their Measure

Let Σ be a finite alphabet of *events*. A *timed letter* (t, a) is a couple of $\mathbb{R}^+ \times \Sigma$ where t represents the *delay* before the *event* a happens. A *timed word* $\mathfrak{w}$ is a sequence of timed letters. We denote by ε the empty timed word. For the sake of conciseness we often remove commas and parentheses and write $t_1 a_1 \cdots t_n a_n$ for $(t_1, a_1) \cdots (t_n, a_n)$. Such timed word is called of length n and of duration $\sum_{i=1}^{n} t_i$ denoted by $\theta(\mathfrak{w})$. A *timed language* is a set of timed words. The universal language is the set of all timed words denoted by $(\mathbb{R}^+ \times \Sigma)^*$. We are mainly interested in timed languages recognized by timed automata described in Sect. 2.2 below but a large part of our theory can be stated for more general timed language.

For a timed language L and a timed letter ta, we define $(ta)^{-1}L$ the left derivative of L wrt ta as

$$(ta)^{-1}L = \{w \mid taw \in L\}.$$

Languages can be defined recursively via defining sub-languages $[L]_n$, $n \in \mathbb{N}$ which are the restriction to L to words of length n.

$$[L]_n = \bigcup_t \bigcup_a (ta).[(ta)^{-1}L]_{n-1} \text{ where the base case is } [L]_0 = \{\epsilon\} \text{ or } [L]_0 = \emptyset.$$

(1)

Example 1. Consider the timed language $L = \{t_1 a t_2 a \ldots t_n a \mid \forall i \geq 1, t_i + t_{i+1} < 1\}$. This language satisfy $(ta)^{-1}L = \{t_2 a \ldots t_n a \mid t_2 < 1 - t \wedge t_i + t_{i+1} < 1\}$ if $t < 1$ and is empty otherwise.

Timed Language as a Collection of Sets of Real-Vectors. Given a timed language L and $n \in \mathbb{N}$, every untimed-word $w = a_1 \cdots a_n \in \Sigma^n$ can be seen as the label for the set $\mathcal{P}_L(w) = \{(t_1, \ldots, t_n) \in \mathbb{R}^n \mid t_1 a_1 \cdots t_n a_n \in L\}$ which is possibly empty.

In the following we only consider timed language for which every $\mathcal{P}_L(w)$ is a measurable set for the Lebesgue measure. In particular we mostly work with timed language for which every $\mathcal{P}_L(w)$ is a union of polytopes but our general theorem applies for the general case of measurable sets.

A polytope in dimension n is a subset of $\mathbb{R}^n$ defined as the set of points whose coordinates satisfy a finite conjunction of linear inequalities, that is, of the form $\sum_{i=1}^{n} \alpha_i t_i \leq \beta$. Note we consider non-necessarily bounded polytope e.g. $\{t_1, t_2 \mid t_1 \leq 3 \wedge t_2 \geq 1 \wedge t_1 + t_2 \geq 3\}$.

Example 2. The polytopes corresponding to the language of Example 1 are depicted below with their respective volumes. Polytopes up to dimension 4 are displayed. For dimension 4 the projection of the polytope with $t_1 = 1$ is displayed.

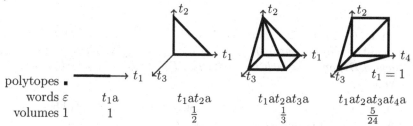

polytopes	■				
words ε		t_1a	t_1at_2a	$t_1at_2at_3a$	$t_1at_2at_3at_4a$
volumes 1		1	$\frac{1}{2}$	$\frac{1}{3}$	$\frac{5}{24}$

Integrating a Function over a Language, Volume and Entropy. Given a timed language L, $n \in \mathbb{N}$ and $f : L_n \to \mathbb{R}$, a real-valued function defined on L_n, we use the following notation for the integral of f over L_n.

$$\int_{L_n} f(\mathfrak{w})d\mathfrak{w} = \sum_{w=a_1\cdots a_n \in \Sigma^n} \int_{(t_1,\ldots,t_n)\in \mathcal{P}_L(w)} f(t_1a_1 \cdots t_na_n)dt_1 \cdot dt_n.$$

When f is non-negative this value is either a non-negative real or $+\infty$. We say that two functions are equal almost everywhere if the set where they differ is of null measure, that is, $\int_{L_n} 1_{f(\mathfrak{w})\neq g(\mathfrak{w})}d\mathfrak{w} = 0$. In the rest of the paper we will just see as equals functions that are almost everywhere equals. Similarly when we say that a function f is a unique solution of a maximisation problem, it means that all functions almost-everywhere equal to f, and only them, are also a solution of the problem. These conventions will also apply to functions from $\mathbb{R}$ to $\mathbb{R}$.

Volumes. For a given n, L_n is the formal union of subsets $\mathcal{P}_L(w)$ of $\mathbb{R}^n$. The volume of L_n denoted $\text{Vol}(L_n)$ is the sum of volumes of these sets. When one is infinite then so is $\text{Vol}(L_n)$. Equivalently it is just the integral of 1 over L_n: $\text{Vol}(L_n) = \int_{L_n} 1d\mathfrak{w}$. Volumes can also be characterised recursively by

$$\text{Vol}(\emptyset) = 0, \text{Vol}(\{\varepsilon\}) = 1, \quad \text{Vol}(L_n) = \sum_{a\in\Sigma} \int_0^\infty \text{Vol}([(ta)^{-1}L]_{n-1})dt. \quad (2)$$

PDF on Timed Languages. A *probability density function* (PDF) over L_n is a function f non-negative for which $\int_{L_n} f(\mathfrak{w})d\mathfrak{w} = 1$. Given a PDF f, and a function g defined on L_n, the expected value of g is denoted by $E_f(g)$ and defined by $\int_{L_n} g(\mathfrak{w})f(\mathfrak{w})d\mathfrak{w}$. We will be mostly interested with the expected duration of timed words of a given language L and PDF f: $E_f(\theta) = \int_{L_n} \theta(\mathfrak{w})f(\mathfrak{w})d\mathfrak{w}$.

Entropy of a PDF. The *entropy* of a PDF f on L_n is[1]

$$H(f) = E_f(-\log_2(f)) = -\int_{L_n} f(\mathfrak{w})\log_2 f(\mathfrak{w})d\mathfrak{w}, \quad (3)$$

---

[1] In this definition the usual convention that $0\log_2 0 = 1$ applies.

This definition is the natural generalisation to timed languages of the Shannon continuous entropy (aka. Shannon differential entropy) for functions from $\mathbb{R}$ to $\mathbb{R}$. The following max-entropy theorem, tells us that the uniform PDF (used in [2] for uniform random sampling) is the unique PDF that maximises the entropy.

Theorem 1. *Given a timed language L and $n \in \mathbb{N}$ such that $\mathrm{Vol}(L_n) < +\infty$, the maximal value of the entropy of a PDF on L_n is $\log_2(\mathrm{Vol}(L_n))$. It is reached by a unique PDF on L_n which the uniform PDF on L_n: the constant function $\mathfrak{w} \mapsto 1/\mathrm{Vol}(L_n)$.*

This theorem will be a particular case of our main theorem (Theorem 2 of Sect. 3) where no constraint is given on duration.

Maximal Entropy Measure and Uniform Sampling for General Timed Languages. Uniform sampling can be defined over general timed languages based on Eq. (2). Exact sampling in L of a word of length n is performed iteratively. When n is equal to 0 the sampling stops otherwise the next letter $a \in \Sigma$ is chosen with probability $\frac{\int_0^\infty \mathrm{Vol}([(ta)^{-1}L]_{n-1})dt}{\mathrm{Vol}(L_n)}$. Next a delay t is sampled[2] from the continuous distribution defined by the Probability Density function $t \mapsto \frac{\mathrm{Vol}([(ta)^{-1}L]_{n-1})}{\int_0^\infty \mathrm{Vol}([(ta)^{-1}L]_{n-1})}$. Then a timed word of length $n-1$ is sampled from the language $(ta)^{-1}L$. This scheme needs a practical way of computing PDF which will be the case for timed languages recognized by deterministic timed automata.

2.2   Languages Recognized by DTA and Their Measures

In this part we restrict our attention to languages recognised by a DTA. For these languages, we can describe computation of the volume of the language with recursive equations on the structure of the DTA recognising the language.

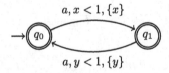

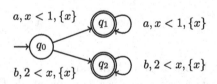

(a) A TA with 2 clocks x and y, 2 states and 2 transitions. The transitions are guarded by an upper-bound of 1. This automaton recognises the language of Example 1

(b) A TA with a single clock recognising either words in a^n with delays bounded by 1 or words in b^n with delays of at least 2.

Fig. 1. Examples of Deterministic Timed Automata

---

[2] One could also sample the delay before the action, this would lead at the end to the same probability distribution on timed words.

Timed Automata. A *timed automaton* (TA) $\mathcal{A}$ is defined as a tuple $(\Sigma, X, Q, q_0, \mathcal{F}, \Delta)$ where Σ is a finite set of events; X is a finite set of *clocks*; Q is a finite set of *locations*; q_0 is the initial location; $\mathcal{F} \subseteq Q$ is a set of final locations; and Δ is a finite set of *transitions*. The finite set of clocks X is a finite set of non-negative real-valued variables. A *guard* is a finite conjunction of clock constraints. For a clock vector $\boldsymbol{x} \in [0, \infty[^X$ and a non-negative real t, we denote by $\boldsymbol{x} + t$ the vector $\boldsymbol{x} + (t, \ldots, t)$. A transition $\delta = (\delta^-, a_\delta, \mathfrak{g}_\delta, \mathfrak{r}_\delta, \delta^+) \in \Delta$ has an *origin* $\delta^- \in Q$, a *destination* $\delta^+ \in Q$, a label $a_\delta \in \Sigma$, a *guard* $\mathfrak{g}_\delta$ and a *reset* function $\mathfrak{r}_\delta$ determined by a subset of clocks $B \subseteq X$. To simplify notation $\mathfrak{r}$ are overloaded to be both a subset of the clocks set and the function assigning 0 to clocks in the subset. Figure 1a depicts a TA with two clocks x and y, two locations q_0, q_1 which are final and two transitions. The transition from q_0 to q_1 is labelled by a is guarded by $x < 1$ and reset x while the transition from q_1 to q_0 is labelled by a is guarded by $y < 1$ and reset y. Other examples of TAs are given in Fig. 1b and Fig. 4.

A *state* $s = (q, \boldsymbol{x}) \in Q \times [0, \infty[^X$ is a pair of location and a clock vector. The initial state of $\mathcal{A}$ is $(q_0, \boldsymbol{0})$. A *timed transition* is a pair (t, δ) of a time *delay* $t \in [0, \infty[$ followed by a discrete transition $\delta \in \Delta$. The delay t represents the time before firing the transition δ.

A run is an alternating sequence $(q_0, \boldsymbol{x}_0) \xrightarrow{t_1, \delta_1} (q_1, \boldsymbol{x}_1) \ldots \xrightarrow{t_n, \delta_n} (q_n, \boldsymbol{x}_n)$ of states satisfying that q_i is the successor of q_{i-1} by δ_i, the vector $\boldsymbol{x}_{i-1} + t$ satisfies the guard $\mathfrak{g}_\delta$ and $\boldsymbol{x}_i = \mathfrak{r}_\delta(\boldsymbol{x}_{i-1} + t)$, $(q_0, \boldsymbol{x}_0)$ is the initial state and q_n is a final location. This run is *labelled* by the *timed word* $(t_1, a_1) \cdots (t_n, a_n)$ where for every $i \leq n$, a_i is the label of δ_i. The set of timed words that label all the runs is called the *timed language* of $\mathcal{A}$. The TA of Fig. 1a recognises the language of Example 1.

A *deterministic* timed automaton (DTA) is a TA where for every location q, transitions starting from q have pairwise disjoint guards or different labels. All TA shown in this paper are DTA.

Max-Entropy Sampling for DTA. In [2–5] algorithms were developed to compute the volume for languages of DTA with some restrictions. Given a DTA, (2) can be instantiated over its zone graph. The first step is to compute the forward-reachability zone graph. The second operation splits the zone graph such that, for any vertex of the graph, for each available transition δ, there exist two functions $\mathbf{lb}_\delta$ and $\mathbf{ub}_\delta$ of the clock vector $\boldsymbol{x}$ such that, for all $t \in \mathbb{R}^+$, $\boldsymbol{x} + t \models \mathfrak{g}_\delta$ is equivalent to $t \in [\mathbf{lb}_\delta(\boldsymbol{x}), \mathbf{ub}_\delta(\boldsymbol{x})]$. Moreover the bounds $\mathbf{lb}_\delta(\boldsymbol{x})$ and $\mathbf{ub}_\delta(\boldsymbol{x})$ are of the form $c - x$ where $c \in \mathbb{N}$ and x is a clock or zero. Details on the computation of the split operation can be found in [2]. In the split zone graph Eq. (2) becomes:

$$v_0(l, \boldsymbol{x}) = 1_{l \in F} \text{ and } v_n(l, \boldsymbol{x}) = \sum_{\delta^- = l} \int_{\mathbf{lb}_\delta(\boldsymbol{x})}^{\mathbf{ub}_\delta(\boldsymbol{x})} v_{n-1}\left(\delta^+, \mathfrak{r}_\delta(\boldsymbol{x} + t)\right) dt \quad (4)$$

Given n, $v_n(l, \boldsymbol{x})$ can be computed in polynomial time with respect to n and the split zone graph size which in the worst case can be exponentially large in the number of clocks. Given n and location l the functions $\boldsymbol{x} \mapsto v_n(l, \boldsymbol{x})$ is a polynomial in $\boldsymbol{x}$. This can be seen by a straightforward recursion since v_0 is a constant and the bound of the integrals to define v_{n+1} from v_n are polynomials.

Since closed-form formulae for v_n are efficiently computable, this provides an effective way of sampling using the procedure presented in Sect. 2.1.

More precisely the kth letter is randomly chosen in state $(q, \boldsymbol{x})$ by choosing the transition $\delta = (q, a, \mathfrak{g}_\delta, \mathfrak{r}_\delta, q')$ labelling a with the discrete probability distribution

$$p_k(\delta \mid q, \boldsymbol{x}) = \frac{\int_{\mathrm{lb}_\delta(\boldsymbol{x})}^{\mathrm{ub}_\delta(\boldsymbol{x})} v_{n-k-1}(q', \mathfrak{r}_\delta(\boldsymbol{x}+t))dt}{v_{n-k}(q, \boldsymbol{x})} \tag{5}$$

and the delay with the PDF:

$$p_k(t \mid a, q, \boldsymbol{x}) = \frac{1_{t \in (\mathrm{lb}_\delta, \mathrm{ub}_\delta)} v_{n-k-1}(q', \mathfrak{r}_\delta(\boldsymbol{x}+t))}{\int_{\mathrm{lb}_\delta(\boldsymbol{x})}^{\mathrm{ub}_\delta(\boldsymbol{x})} v_{n-k-1}(q', \mathfrak{r}_\delta(\boldsymbol{x}+t))dt} \tag{6}$$

Approximate uniform sampling method are presented in [2] to sample words of large length. One of such method is the *receding sampling* where to avoid computing the p_k in (5) and (6) when k is close to a too large sampling length m, one use instead $p_{min(n,k)}$ for a n large enough but small compared to m (see [2] for the details and formal statements).

3 Max-Entropy Theorem Under Linear Duration Constraints

In previous work on uniform sampling, constraint on duration could not be expressed. Here we are interested to express for instance that the mean and the variance of duration are equals to μ and ν, we write $\int_{L_n} \theta(\mathfrak{w})p(\mathfrak{w})d\mathfrak{w} = \mu$ and $\int_{L_n} \theta^2(\mathfrak{w})p(\mathfrak{w})d\mathfrak{w} = \nu + \mu^2$.

More generally we are given m measurable functions $f_i : \mathbb{R}_{\geq 0} \to \mathbb{R}$ and constants $a_i \in \mathbb{R}$, $i = 1..m$, and we address the following max-entropy problem:

Maximise $H(p)$ under constraints $\int_{L_n} f_i(\theta(\mathfrak{w}))p(\mathfrak{w})d\mathfrak{w} = a_i$ for $i \in \{0, \dots, m\}$, \tag{7}

where f_0 is the constant 1 and $a_0 = 1$, to add the constraints that p is a PDF.

Our solution given in the max-entropy theorem below (Theorem 2) is based on expressing this max-entropy problem on L_n as a max-entropy problem for functions of the reals (that take the duration as argument). Before stating this theorem, we first need few definitions and Lemmas.

We say that a function f defined on L_n *depends only on duration* if $f = \tilde{f} \circ \theta$ for some functions on the reals $\tilde{f}$, that is, for every $\mathfrak{w} \in L_n$, $f(\mathfrak{w}) = \tilde{f}(\theta(\mathfrak{w}))$. For a PDF f, "depending only on duration" can be rephrased as "being uniform at fixed duration": for every T, all the timed words of duration T have the same density of probability.

With Lemma 1 we show that integrating a function that depends only on duration, i.e. of the form $\tilde{f} \circ \theta$, can be expressed as integrating $\tilde{f}(T)$ with the duration T as the variable of integration multiplied by a weight denoted $V_n^L(T)$. This weight $V_n^L(T)$ is the $n - 1$-dimensional[3] volume of timed-word of length n restricted to duration T defined by

$$V_n^L(T) = \sum_{w_1 \cdots w_n \in \Sigma^n} \int_{(\mathbb{R}_{\geq 0})^{n-1}} 1_{t_1 w_1 \cdots t_{n-1} w_{n-1} (T - t_1 - \cdots - t_{n-1}) w_n \in L} dt_1 \cdots dt_{n-1}. \quad (8)$$

We will denote $V_n^L(T)$ by $V_n(T)$ when L is clear from the context. We will also use the following notation: $L_{n,T} = \{w \in L_n \mid \theta(w) = T\}$ and consider $(n-1)$-dimensional integral on it:

$$\int_{L_{n,T}} f(w) dw = \sum_{w_1 \cdots w_n \in \Sigma^n} \int_{(\mathbb{R}_{\geq 0})^{n-1}} 1_{w \in L_{n,T}} f(w) dt_1 \cdots dt_{n-1}.$$

In particular $V_n^L(T) = \int_{L_{n,T}} 1 dw$ and hence $w \mapsto 1/V_n^L(T)$ is the uniform PDF on $L_{n,T}$

Lemma 1. *If $f = \tilde{f} \circ \theta$ then $\int_{L_n} f(w) dw = \int_0^{+\infty} \tilde{f}(T) V_n^L(T) dT$*

Remark 1. We give few cases of interest for this Lemma.

1. If $f(w) = 1$, then we get $\text{Vol}(L_n) = \int_0^{+\infty} V_n(T) dT$
2. If $f(w) = \frac{1}{\text{Vol}(L_n)}$, then $T \mapsto \frac{V_n(T)}{\text{Vol}(L_n)}$ is a PDF on reals.
3. If $f(w) = \frac{\theta(w)}{\text{Vol}(L_n)}$ then $\int_{L_n} \frac{\theta(w)}{\text{Vol}(L_n)} dw = \int_0^{+\infty} T \frac{V_n(T)}{\text{Vol}(L_n)} dT$ is the mean duration for the uniform distribution on L_n.

A straightforward consequence of Lemma 1, called Lemma 2 below, is that the max-entropy problem is equivalent to a max-entropy problem for functions defined on the reals. In this Lemma $p \ll V_n$ means that $\forall T \in \mathbb{R}_{\geq 0}$, $V_n(T) = 0 \Rightarrow p(T) = 0$, and

$$H_{V_n}(q) = - \int_0^{+\infty} q(T) \log_2 \frac{q(T)}{V_n(T)} dT.$$

Lemma 2. *For every PDF of the form $\tilde{p} \circ \theta$, the function $p^{\mathbb{R}} : T \mapsto V_n(T)\tilde{p}(T)$ is a PDF such that $H_{V_n}(p^{\mathbb{R}}) = H(p \circ \theta)$. Moreover $\tilde{p} \circ \theta$ is a solution of (7) iff $p^{\mathbb{R}}$ is a solution of*

Maximise $H_{V_n}(p)$ s.t. $\int_0^{+\infty} f_i(T) p(T) dT = a_i$ for $i \in \{0, \ldots, m\}$ and $p \ll V_n$.

$$(9)$$

---

[3] The timed words of duration T have their timed vector $(t_1, \ldots, t_n)$ belonging to the hyperplane $t_1 + \ldots + t_n = T$ which have a null volume. That is why we do not integrate over the last delay which is fixed and equal to $T - t_1 - \cdots - t_{n-1}$.

With Lemma 3 we show that given any PDF, one can always find another PDF with higher or equal entropy that depends only on duration.

Lemma 3. *Let p be a PDF on L_n and[4] $\tilde{p} : T \mapsto \frac{1}{V_n(T)} \int_{L_{n,T}} p(\mathfrak{w})d\mathfrak{w}$ then $H(p) \leq H(\tilde{p} \circ \theta)$ with equality if and only if $p = \tilde{p} \circ \theta$.*

Theorem 2 (Maximal entropy theorem on L_n with integral constraints on duration). *Given the max-entropy problem (7) with its constraints functions f_i and constants a_i. If one find constants $\lambda_i, (0 \leq i \leq m)$ such that the function $p^*(\mathfrak{w}) = \exp\left(\lambda_0 + \sum_{i=1}^m \lambda_i f_i(\theta(\mathfrak{w}))\right)$ is a PDF that satisfies the constraints, then the problem admits a unique solution which is p^*.*

Proof. Let p^* and λ_i as above so that p^* satisfies the constraints, then we have to prove that $H(p^*)$ is maximal and unique for this property. First, Lemma 3 implies that we only have to consider PDF q of the form $q = \tilde{q} \circ \theta$. We show that such PDF q satisfies $H(q) \leq H(p^*)$ with equality iff $q = p^*$. This is equivalent to show that $H_{V_n}(q^{\mathbb{R}}) \leq H_{V_n}(p^{\mathbb{R}})$ with equality iff $q^{\mathbb{R}} = p^{\mathbb{R}}$, with $p^{\mathbb{R}} = \tilde{p}.V_n$ and $q^{\mathbb{R}} = \tilde{q}.V_n$ since by Lemma 2, $H_{V_n}(p^{\mathbb{R}}) = H(p^*)$ and $H_{V_n}(q^{\mathbb{R}}) = H(q)$.

We adapt a classical proof based on Kulback-Lebler divergence that can be found in [12]. The Kullback-Leibler divergence from $p^{\mathbb{R}}$ to $q^{\mathbb{R}}$ is the quantity $D(q^{\mathbb{R}}||p^{\mathbb{R}}) = \int_0^{+\infty} q^{\mathbb{R}}(T) \log \frac{q^{\mathbb{R}}(T)}{p^{\mathbb{R}}(T)} dT$ which is always non-negative, and null iff $p^{\mathbb{R}} = q^{\mathbb{R}}$. We will show that $D(q^{\mathbb{R}}||p^{\mathbb{R}}) = H_{V_n}(p^{\mathbb{R}}) - H_{V_n}(q^{\mathbb{R}})$ and thus $H_{V_n}(p^{\mathbb{R}}) = H_{V_n}(q^{\mathbb{R}})$ will be equivalent to $p^{\mathbb{R}} = q^{\mathbb{R}}$.

We introduce $V_n(T)$ using the fact that $\log_2(q^{\mathbb{R}}/p^{\mathbb{R}}) = \log_2 \frac{q^{\mathbb{R}}(T)}{V_n(T)} - \log_2 \frac{p^{\mathbb{R}}}{V_n(T)}$:

$$D(q^{\mathbb{R}}||p^{\mathbb{R}}) = \int_0^{+\infty} q^{\mathbb{R}}(T) \log_2 \frac{q^{\mathbb{R}}(T)}{V_n(T)} dT - \int_0^{+\infty} q^{\mathbb{R}}(T) \log_2 \frac{p^{\mathbb{R}}(T)}{V_n(T)} dT$$

The first integral is $-H_{V_n}(q^{\mathbb{R}})$, hence it remains to prove that $H_{V_n}(p^{\mathbb{R}}) = -\int q^{\mathbb{R}}(T) \log_2\left(\frac{p^{\mathbb{R}}(T)}{V_n(T)}\right)$. We recall that $\frac{p^{\mathbb{R}}(T)}{V_n(T)} = \tilde{p}(T) = \exp\left(\sum_{i=0}^m \lambda_i f_i(T)\right)$, thus

$$\int_0^{+\infty} q^{\mathbb{R}}(T) \log_2\left(\frac{p^{\mathbb{R}}(T)}{V_n(T)}\right) = \frac{1}{\ln 2} \int_0^{+\infty} q^{\mathbb{R}}(T) \left(\sum_{i=0}^m \lambda_i f_i(T)\right) dT$$

$$= \frac{1}{\ln 2} \sum_{i=0}^m \lambda_i \int_0^{+\infty} q^{\mathbb{R}}(T) f_i(T) dT.$$

Since $q^{\mathbb{R}}$ satisfies the constraints this quantity is equal to $\frac{1}{\ln 2} \sum_{i=0}^m \lambda_i a_i$ The same reasoning holds for p, so we can conclude the proof with the equality:

$$H_{V_n}(p^{\mathbb{R}}) = -\frac{1}{\ln(2)} \sum_{i=0}^m \lambda_i a_i = -\int_0^{+\infty} q^{\mathbb{R}}(T) \log_2\left(\frac{p^{\mathbb{R}}(T)}{V_n(T)}\right) dT.$$

---

[4] Here $\tilde{p}$ generalises the previous concept it was used for: when p depends only on duration then $p = \tilde{p} \circ \theta$.

Theorem 1 is a specialisation of Theorem 2 where there is no constraint on duration. Indeed the function has to be searched as a constant e^{λ_0} with the unique constraint that $\int_{L_n} e^{\lambda_0} d\mathfrak{w} = 1$, that is, $e^{\lambda_0} = 1/\mathrm{Vol}(L_n)$. As there is no constraint on duration the volume must be finite for this theorem to apply.

4   Max-Entropy Theorem with Prescribed Mean Duration

In this section we describe and see how to compute the max-entropy PDF when the mean duration is prescribed. So we will use Theorem 2 with f_1 being the identity function and explain the normalising constant (link between λ_0 and λ_1).

Theorem 3 (Maximal entropy theorem for measures on L_n with prescribed mean duration). *Given a timed language L, length n, duration T_{mean}. If there exist probability measures p that satisfy the constraint $E_p(\theta(w)) = T_{mean}$, then there is a unique one p^* satisfying this constraint that maximizes the entropy. It is given by*

$$p^*(w) = \frac{e^{-s\theta(w)}}{v_{n,s}^L}$$

where the normalising constant $v_{n,s}^L$ is defined by

$$v_{n,s}^L = \int_{L_n} e^{-s\theta(\mathfrak{w})} d\mathfrak{w} = \int_0^{+\infty} e^{-sT} V_n^L(T) dT \qquad (10)$$

with s the unique real such that $\frac{1}{v_{n,s}^L} \frac{\partial v_{n,s}^L}{\partial s} = T_{mean}$.

The normalising constant $v_{n,s}^L$ defined in Eq. (10) can be interpreted as a Laplace transform of the function V_n and hence we call it the *Laplace Transform of Volumes* (LTV) and s is called the *Laplace parameter*. In the next section we will propose efficient computation of it which will be the base of a sampling algorithm. As shown above by tuning the Laplace parameter s we can control the mean duration which is defined wrt. the LTV and its derivative. Controlling the variance could be done by adding an extra parameter. Efficient sampling for this latter case is left for future work, and we focus on fixing only the mean duration so working only with the LTV. Once this done, we can characterise the variance (without prescribing it) with the LTV and its derivatives.

Proposition 1 (Characterization of the variance wrt. LTV and its derivatives). *Let p^* be the PDF given in Theorem 3 and let $Var_{p^*}(\theta(\mathfrak{w})) = E_{p^*}(\theta(\mathfrak{w})^2) - E_{p^*}(\theta(\mathfrak{w}))^2$ the variance of the duration of timed words sampled with this probability. Then*

$$Var_{p^*}(\theta(\mathfrak{w})) = \frac{\partial^2 v_{n,s}^L}{\partial s^2} - \left(\frac{1}{v_{n,s}^L} \frac{\partial v_{n,s}^L}{\partial s} \right)^2.$$

Remark 2. Theorem 1 is again a special case. It suffices to take $s = 0$. We get in addition, a characterisation of the mean of the duration and with Proposition 1 its variance.

The LTV v_n can be characterized via a recursive definition:

Proposition 2 (Recursive definition of v_n).

$$v_{n,s}^L = \sum_{a \in \Sigma} \left(\int_0^{+\infty} e^{-st} v_{n-1,s}^{(ta)^{-1}L} dt \right) \qquad (11)$$

This recursive definition is a step towards dynamic programming computation that will be fully possible now when we focus on DTA.

4.1   Computing LTV and Max-Entropy Sampling for Timed Automata

Theorem 3 tells us the form of the maximal entropy PDF when a mean duration constraint is imposed. In this section we focus on how to sample this PDF for language recognized by a DTA using the LTVs for languages starting on states of this DTA. This PDF is then used to sample its language.

Computation of LTV as Exponential Polynomials for DTA. In this section, we describe how LTV can be computed effectively from a timed automaton. This is done by computing LTV recursively and by showing that all computations are performed over exponential polynomials (Definition. 1) for which the computation of integral is effective. Here we find a suitable generalisation of the volumes functions used in [1,2,6] and recall in Sect. 2. In a nutshell, we multiply by e^{-st} the function which is integrated over the next transition with the delay as integration variable.

In the following we assume that a DTA in split form is given and for every length n, parameter s and state $(l, \boldsymbol{x})$ we denote by $v_{n,s}(l, \boldsymbol{x})$ the LTV of the language starting from $(l, \boldsymbol{x})$ that is $v_{n,s}(q, \boldsymbol{x}) = v_{n,s}^{L_{(l,\boldsymbol{x})}}$.

Definition 1 (Exponential Polynomials). *Given a variable s and a sequence of variable $\boldsymbol{X} = (X_i)_{i=1}^n$, we call exponential polynomials, expressions $EP(s, \boldsymbol{X})$ such that : there exists a finite set $S \subset \{0, 1, \ldots, n\} \times \mathbb{N}$ and a polynomial $P_{i,k}$ for each element of S with $P_{i,k}$ a polynomials over $\boldsymbol{X} \cup \{\frac{1}{s}\}$ with the the convention that $X_0 = 0$; and $EP(s, \boldsymbol{X})$ is written in the form*

$$EP(s, \boldsymbol{X}) = \sum_{(i,k) \in S} P_{i,k} \left(\frac{1}{s}, \boldsymbol{X} \right) e^{s(X_i - k)}$$

Proposition 3. *For every location q and length n, $\boldsymbol{x} \mapsto v_{n,s}(q, \boldsymbol{x})$ is an exponential polynomials. It can be computed in polynomial time using dynamic programming from the following recursive equations:*

$$v_{0,s}(l, \boldsymbol{x}) = 1_{l \in F} \text{ and } v_{n,s}(l, \boldsymbol{x}) = \sum_{\delta^- = l} \int_{\mathrm{lb}_\delta(\boldsymbol{x})}^{\mathrm{ub}_\delta(\boldsymbol{x})} v_{n-1,s} \left(\delta^+, \mathfrak{r}_\delta(\boldsymbol{x} + t) \right) e^{-st} dt. \qquad (12)$$

Random Sampling Using the LTV

The sampling follows the same line as in for the unconstrained case (Sect. 2.1) where the probability distributions for discrete and continuous choice (5) and (6) are replaced by

$$p_{k,s}(a \mid q, \boldsymbol{x}) = \frac{\int_{\text{lb}_\delta(\boldsymbol{x})}^{\text{ub}_\delta(\boldsymbol{x})} v_{n-k-1,s}(q', \mathfrak{r}_\delta(\boldsymbol{x}+t)) e^{-st} dt}{v_{n-k,s}(q, \boldsymbol{x})}; \text{ and} \qquad (13)$$

$$p_{k,s}(t \mid a, q, \boldsymbol{x}) = \frac{1_{(\text{lb}_\delta, \text{ub}_\delta)} v_{n-k-1,s}(q', \mathfrak{r}_\delta(\boldsymbol{x}+t)) e^{-st}}{\int_{\text{lb}_\delta(\boldsymbol{x})}^{\text{ub}_\delta(\boldsymbol{x})} v_{n-k-1,s}(q', \mathfrak{r}_\delta(\boldsymbol{x}+t)) e^{-st} dt}. \qquad (14)$$

5   Experiments

We have implemented our approach in the tool WORDGEN [5], which required developing a data structure for exponential polynomials, implementing the LTV computation and the estimation of parameters s. These developments are freely available with the GPLv3 licence. This section contains experiments on small timed automata (Example 3 and 4) and a case study on a larger automaton (Example 5).

Example 3. We compute the LTVs of the language defined in Example 1 recognized by the timed automaton depicted in Fig. 1a. After computing the split zone graph, the number of locations is still 2. We show the LTVs computed for small n in both of these locations and in the initial one in the following table.

$(q, \boldsymbol{x})\backslash n$	0	1	2
$(q_0, (x,y))$	1	$\frac{1-e^{-s(1-x)}}{s}$	$\frac{1-e^{-s(1-x)}}{s^2} + \frac{(x-1)}{s} e^{-s}$
$(q_1, (x,y))$	1	$\frac{1-e^{-s(1-y)}}{s}$	$\frac{1-e^{-s(1-y)}}{s^2} + \frac{(y-1)}{s} e^{-s}$

If we restrict to timed words of length 2, sampling with the parameter $s = 0$ (using Taylor expansion) provides a uniform distribution in the triangle. The expected duration is $\frac{1}{v_{2,s}(q_0,(0,0))} \frac{\partial v_{2,s}}{\partial s}(q_0, (0,0))$. For $s \neq 0$ it is $\frac{2+2s+s^2-2e^s}{s+s^2-se^s}$. Using Taylor expansions in $s = 0$ of both the numerator and the denominator one can show that the limit when $s \to 0$ is $\frac{2}{3}$.

Figure 2 shows plots of the sampling from this language with different mean durations. We observe a shift of the concentration of points along the axis $y = x$. Note that the time duration of 0.9 is obtained with a negative value of s. This is allowed when all clocks are bounded and thus $V_n(T)$.

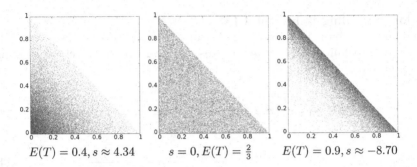

$$E(T) = 0.4, s \approx 4.34 \qquad s = 0, E(T) = \tfrac{2}{3} \qquad E(T) = 0.9, s \approx -8.70$$

Fig. 2. Sample 50000 timed words $t_1 a t_2 a$ from the language of Example 1 (the DTA is in Fig. 1a) with t_1 in the abscissa and t_2 in the ordinate.

Example 4. In this example, we are interested in the language recognized by the automaton of Fig. 1b. We want to generate words of length $3(n = 3)$ with a mean time duration of 5.5. There are only two possible untimed words which are *aaa* and *bbb*. One can see that their corresponding polytopes are $[0, 1)^3$ and $(2, +\infty)^3$. In particular, timed words are either of duration < 3 or > 6, so there are no timed words of the target duration 5.5. However, a mean duration of 5.5 can be reached by computing and taking the appropriate value for the parameter s, that is, 0.52. By sampling 100 000 words with this parameter, we obtain the histogram depicted in Fig. 3 with an average duration of 5.50116.

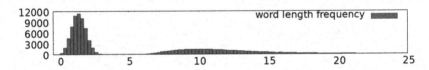

Fig. 3. Histogram of duration of 100000 words sampled from the automaton in Fig. 1b.

Example 5 (Train Gate). We illustrate the scalability of our method using the well-known Train Gate example described extensively in the Uppaal Tutorial [8]. This example does not involve a single automaton but a network of timed automata synchronized by their transitions. The semantics of such a network is a single timed automaton whose state space is the cross product of each automaton. This synchronisation semantics is explained in [8] and implemented in the tool Uppaal, thus we will use the same notation as in this tool.

In this model, N trains want to cross a one-way bridge repeatedly. Each train is modeled as a timed automaton with one clock. The bridge is modeled as a FIFO scheduler, which stops trains or clears them for crossing. The timed automaton for each train is depicted in Fig. 4(left) where *id* is instantiated as an identifier for the train. This automaton is replicated N times with an independent copy of clock x. The clock x is reset on every edge (omitted in the figure).

Note that the states Safe and Stop are unbounded. Figure 4(right) depicts the scheduler whose size depends on the number of trains. The FIFO queue is implemented by an array of identifiers of size N and four functions pop, push, front and last that modify it.

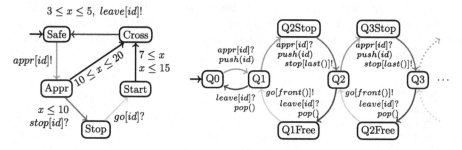

Fig. 4. The automaton on the left depicts the timed automaton for each train where the clock x is reset on every edge. The automaton on the right depicts the scheduler.

In Table 1 we show the performance of the tool WORDGEN in sampling a thousand runs of this automaton of expected duration 50. The number N of trains and parameter n are shown in the first two columns. As there is one clock per train, the number of zones grows exponentially with N, while the splitting only double the number of zones. The computation time is dominated by the computation of $v_{n,s}(l, \boldsymbol{x})$. The sampling time is proportional to the number of trajectories and requires the evaluation of large polynomials for the sampling of timed distributions. As expected, the computation of the zone graph and split zone graph is small, compared to the other computations. The memory appears to be a computational bottleneck as the number of terms in each $v_{n,s}(l, \boldsymbol{x})$ grows exponentially with the number of clocks and they need to be computed on each location.

Table 1. Each experiment samples 1000 timed words of length n and expected duration 50

N	n	number of locations		time (sec)			memory
		reachability	split	reach+split	distribution	sampling	
1	10	9	9	0	0	0.07	6.2 MB
2	10	51	70	0.01	0	0.13	34.8 MB
3	10	1081	1826	0.12	9.04	2.6	1.7 GB
4	10	28353	53534	7.2	647	466	51.9 GB
5	10	88473	178414	35	Out of Memory > 64 GB		
5	7	88473	178414	35	408	205	52.4 GB

By changing the expected duration of timed words, we observe very different behaviors of the system. Figure 5 depicts two timed words sampled from the

Train Gate example with 3 trains and $n = 10$ and of length of length $m = 40$ using receding sampling. Figure 5(a) depicts a timed word sampled with expected duration 80 while Fig. 5(b) depicts a timed word sampled with expected time duration 240. We can see that in Fig. 5(b) the system is saturated since there is almost always a train in the locations Start or Cross which have a lower bound on their waiting time, and the synchronisation blocks other events. In Fig. 5(a) we observe that the system stay in a state where all trains are in location Safe for some time where there are no constraints (no lower nor upper bound) thus trains evolve more independently.

(a) Timed words with an expected duration of 240 time units and 40 events ($s \approx 0.1098$).

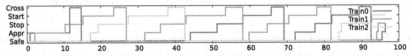

(b) Timed words with an expected duration of 80 time units and 40 events ($s \approx 0.9070$).

Fig. 5. Timed words sampled from the Train Gate example. The state of trains are in ordinates.

6   Conclusion and Future Work

In this article, we address the problem of characterizing probability measures of maximal entropy for timed words of a fixed length of timed languages under duration constraints. We focus our attention on constraint on mean duration, which leads us to define and propose a method for efficiently computing Laplace Transform of Volumes, a key theoretical tool to sample timed word wrt. the maximal entropy measure. Several experiments are provided to illustrate our sampling algorithm.

Ongoing and Future Work. The LTV we compute are Laplace transforms of functions $V_n(T)$ which can be called the volume of timed words with fixed duration T. These functions are crucial in our theoretical development of Sect. 3 but are not computed in this current article. An ongoing work is to write recursive characterization of such functions that could be turned into a sampler of timed words of exact duration T. The computation of closed form formulae for $V_n(T)$ could be used to sample the PDF of the form $p^{\mathbb{R}}(T) = \tilde{p}(T)V_n(T)$, which, coupled with a uniform sampler of exact duration T, would provide a max-entropy sampler for the general linear constraints of Theorem 2. Another promising research direction is to revisit the results of [1,6] with the operator underlying the definition of the LTV. With this approach we aim at defining maximal entropy measure for infinite timed words with a prescribed frequency of event, *e.g.* 0.7 events per time unit. Finally, another possible extension is to consider max-entropy measures when the length is also random. For this we would like to adapt Boltzmann sampling algorithms [13] to our settings. A first adaptation of such a sampling for a very particular subclass of timed languages (used for random sampling of permutations) was proposed in [7]. This latter work though was not concerned at all with duration of timed words (nor entropy).

References

1. Asarin, E., Basset, N., Degorre, A.: Entropy of regular timed languages. Inf. Comput. **241**, 142–176 (2015)
2. Barbot, B., Basset, N., Beunardeau, M., Kwiatkowska, M.: Uniform sampling for timed automata with application to language inclusion measurement. In: Agha, G., Van Houdt, B. (eds.) QEST 2016. LNCS, vol. 9826, pp. 175–190. Springer, Cham (2016). https://doi.org/10.1007/978-3-319-43425-4_13
3. Barbot, B., Basset, N., Dang, T.: Generation of signals under temporal constraints for CPS testing. In: Badger, J.M., Rozier, K.Y. (eds.) NFM 2019. LNCS, vol. 11460, pp. 54–70. Springer, Cham (2019). https://doi.org/10.1007/978-3-030-20652-9_4
4. Barbot, B., Basset, N., Dang, T., Donzé, A., Kapinski, J., Yamaguchi, T.: Falsification of cyber-physical systems with constrained signal spaces. In: Lee, R., Jha, S., Mavridou, A., Giannakopoulou, D. (eds.) NFM 2020. LNCS, vol. 12229, pp. 420–439. Springer, Cham (2020). https://doi.org/10.1007/978-3-030-55754-6_25
5. Barbot, B., Basset, N., Donze, A.: Wordgen: a timed word generation tool. In: Proceedings of the 26th ACM International Conference on Hybrid Systems: Computation and Control, HSCC 2023. Association for Computing Machinery, New York, NY, USA (2023). https://doi.org/10.1145/3575870.3587116
6. Basset, N.: A maximal entropy stochastic process for a timed automaton. Inf. Comput. **243**, 50–74 (2015). https://doi.org/10.1016/j.ic.2014.12.006
7. Basset, N.: Counting and generating permutations in regular classes. Algorithmica **76**(4), 989–1034 (2016). https://doi.org/10.1007/s00453-016-0136-9
8. Behrmann, G., David, A., Larsen, K.G.: A tutorial on UPPAAL. In: Bernardo, M., Corradini, F. (eds.) SFM-RT 2004. LNCS, vol. 3185, pp. 200–236. Springer, Heidelberg (2004). https://doi.org/10.1007/978-3-540-30080-9_7
9. Bertrand, N., et al.: Stochastic timed automata. Log. Methods Comput. Sci. **10**(4) (2014). https://doi.org/10.2168/LMCS-10(4:6)2014
10. Bohlender, D., Bruintjes, H., Junges, S., Katelaan, J., Nguyen, V.Y., Noll, T.: A review of statistical model checking pitfalls on real-time stochastic models. In: Margaria, T., Steffen, B. (eds.) ISoLA 2014. LNCS, vol. 8803, pp. 177–192. Springer, Heidelberg (2014). https://doi.org/10.1007/978-3-662-45231-8_13
11. Budde, C.E., D'Argenio, P.R., Hartmanns, A., Sedwards, S.: A statistical model checker for nondeterminism and rare events. In: Beyer, D., Huisman, M. (eds.) TACAS 2018. LNCS, vol. 10806, pp. 340–358. Springer, Cham (2018). https://doi.org/10.1007/978-3-319-89963-3_20
12. Cover, T.M., Thomas, J.A.: Information theory and statistics. Elem. Inf. Theory **1**(1), 279–335 (1991)

13. Duchon, P., Flajolet, P., Louchard, G., Schaeffer, G.: Boltzmann samplers for the random generation of combinatorial structures. Comb. Probab. Comput. **13**(4), 577–625 (2004). https://doi.org/10.1017/S0963548304006315
14. Jaynes, E.T.: Information theory and statistical mechanics. II. Phys. Rev. Online Arch. (PROLA) **108**(2), 171–190 (1957)
15. Norman, G., Parker, D., Sproston, J.: Model checking for probabilistic timed automata. Formal Methods Syst. Des. **43**(2), 164–190 (2013)

Quasi-Deterministic Burstiness Bound for Aggregate of Independent, Periodic Flows

Seyed Mohammadhossein Tabatabaee[1]([✉]), Anne Bouillard[2]([✉]),
and Jean-Yves Le Boudec[1]([✉])

[1] EPFL, Lausanne, Switzerland
{hossein.tabatabaee,jean-yves.leboudec}@epfl.ch
[2] Huawei Technologies France, Paris, France
anne.bouillard@huawei.com

Abstract. Time-sensitive networks require timely and accurate monitoring of the status of the network. To achieve this, many devices send packets periodically, which are then aggregated and forwarded to the controller. Bounding the aggregate burstiness of the traffic is then crucial for effective resource management. In this paper, we are interested in bounding this aggregate burstiness for independent and periodic flows. A deterministic bound is tight only when flows are perfectly synchronized, which is highly unlikely in practice and would be overly pessimistic. We compute the probability that the aggregate burstiness exceeds some value. When all flows have the same period and packet size, we obtain a closed-form bound using the Dvoretzky-Kiefer-Wolfowitz inequality. In the heterogeneous case, we group flows and combine the bounds obtained for each group using the convolution bound. Our bounds are numerically close to simulations and thus fairly tight. The resulting aggregate burstiness estimated for a non-zero violation probability is considerably smaller than the deterministic one: it grows in $\sqrt{n \log n}$, instead of n, where n is the number of flows.

1 Introduction

The development of industrial automation requires timely and accurate monitoring of the status of the network. In time-sensitive networks, a common assumption for critical types of traffic is that devices send packets periodically. These packets are aggregated and forwarded to the controller. Characterizing this aggregate traffic is then crucial for effective resource management.

Among the analytic tools providing analysis for real-time systems is deterministic network calculus [1,13]. From the characterization of the flows, the description of the switches (offered bandwidth and scheduling policy), it can derive worst-case performance bounds, such as end-to-end delay or buffer occupancy. These performances can grow linearly with the burstiness of the flows [3]. Hence, accurately bounding the burstiness is key for performance evaluation and resource management. However, deterministic network calculus takes into account the worst-case scenario for aggregation of flows, which happens when flows are perfectly synchronized, and this is very unlikely to happen.

To overcome this issue, probabilistic versions of network calculus (known as *Stochastic Network Calculus*) have emerged, and their aim is to compute performances when a small *violation probability* is allowed. Using probabilistic tools such as moment-generating functions [8] or martingales [15], recent works mainly focus on ergodic systems and on the performances at an arbitrary point in time. This does not imply that the probability that the delay bound is never violated during a period of interest is small. Moreover, results are very limited in terms of topology and service policies, and become inaccurate for multiple servers [2]. Methods that compute probabilistic bounds on the burstiness have been discarded as they do not provide as good results for ergodic systems [6].

In this paper, we focus on the burstiness of the aggregation of periodic and independent flows. In other words, each flow sends packets periodically with a fixed packet size and period. The first packet is sent at a random time (the phase) within that period. We assume phases are mutually independent. Our aim is to find a probabilistic bound on the burstiness of the aggregation of flows, that is, finding a burst that is valid *at all times* with large probability. That way, combining these probabilistic burstiness bounds with results of deterministic network calculus lead to delay and backlog bounds that are valid with large probability *at all times*, hence the name *quasi-deterministic*. To our knowledge, this is the first method to obtain quasi-deterministic bounds for independent, periodic flows.

Our contributions are the following:

1. First, in the homogeneous setting, where all flows have the same period and packet size, we provide two probabilistic bounds for the aggregate burstiness. Both of them are based on bounding the probability of some event E relying on the order statistics of the phases. The former (Theorem 1) has a closed form; it uses the Dvoretzky-Kiefer-Wolfowitz (DKW) inequality [14] to bound the probability of E, and when a small positive violation probability is allowed, the burstiness grows in $O(\sqrt{n \log n})$, where n is the number of flows, instead of $O(n)$ for a deterministic bound. The latter (Theorem 2) directly computes the probability of E, which can be implemented iteratively.
2. Second, we focus on two types of heterogeneity: either flows can be grouped into several homogeneous sub-sets, and we use a bounding technique based on convolution (Theorem 3), or flows have the same period but different packet sizes, and the bounds can be adapted from the homogeneous case (Theorem 4).
3. Last, we numerically show that our bounds are close to simulations. The quasi-deterministic aggregate burstiness we obtain with a small, non-zero violation tolerance is considerably smaller than the deterministic one. For the heterogeneous case, we show that our convolution bounding technique provides bounds significantly smaller than those obtained by the union bound.

The rest of the paper is organized as follows: We present our model in Sect. 2, and then in Sect. 3 some results from state of the art. Our contributions are detailed in Sect. 4 for the homogeneous case and Sect. 5 for the heterogeneous case. Finally, we provide some simulation results in Sect. 6 to demonstrate the tightness of the bounds.

2   Assumptions and Problem Statement

We use the notation $\mathbb{N} = \{0, 1, \ldots\}$ and $\mathbb{N}_n = \{1, \ldots, n\}$.

Assumptions. We consider n periodic flows of packets. Each flow $f \in \mathbb{N}_n$ is periodic with period τ_f and phase $\phi_f \in [0, \tau_f)$, and sends packets of size ℓ_f: the number of bits of flow f arriving in the time interval $[0, t)$ is $\ell_f \lceil [t - \phi_f]^+ / \tau_f \rceil$ where we use the notation $[x]^+ = \max(0, x)$ and $\lceil \rceil$ denotes the ceiling.

For every flow f, we assume that ϕ_f is random, uniformly distributed in $[0, \tau_f]$, and that the different $(\phi_f)_{f \in \mathbb{N}_n}$ are independent random variables.

Problem Statement. We consider the aggregation of the n flows and let $A[s, t)$ denote the number of bits observed in time interval $[s, t)$. Our goal is to find a token-bucket arrival curve constraining this aggregate, that is, a rate r and a burst b such that $\forall s \leq t, \; A[s, t) \leq r(t - s) + b$. It follows from the assumptions that each individual flow $f \in \mathbb{N}_n$ is constrained by a token-bucket arrival curve with rate $r_f = \ell_f / \tau_f$ and burst ℓ_f. Therefore, the aggregate flow is constrained by a token-bucket arrival curve with rate $r^{\text{tot}} = \sum_{f=1}^n r_f$ and burst $\ell^{\text{tot}} = \sum_{f=1}^n \ell_f$.

However, due to the randomness of the phases, ℓ^{tot} might be larger than what is observed, and we are rather interested in token-bucket arrival curves with rate r^{tot} and a burst b valid with some probability; specifically, we want to find a bound on the tail probability of the aggregate burstiness, which is defined as the smallest value of B such that the aggregate flow is constrained by a token-bucket arrival curve with rate r^{tot} and burst B, for the entire network lifetime. The aggregate burstiness is given by

$$B = \sup_{t \geq 0} \bar{B}(t). \tag{1}$$

where $\bar{B}(t)$ is the token-bucket content at time t for a token-bucket that is initially empty, and is given by

$$\bar{B}(t) = \sup_{s \leq t} \{A[s, t) - r^{\text{tot}}(t - s)\}. \tag{2}$$

Note that B is a function of the random phases of the flows, therefore, is also random. Assume that $\mathbb{P}(B > b) = \varepsilon$; this means that, with probability $1 - \varepsilon$, after periodic flows started, the aggregate burstiness is $\leq b$. Conversely, with probability ε, the aggregate burstiness is $> b$.

Observe that $\mathbb{P}(B > b) = 0$ for all $b \geq \ell^{\text{tot}}$, as ℓ^{tot} is a deterministic bound on the aggregate burstiness. Then, for some pre-specified value $0 \leq b < \ell^{\text{tot}}$, our problem is equivalent to finding $\epsilon(b)$ that bounds the tail probability of the aggregate burstiness B, i.e.,

$$\mathbb{P}(B > b) \leq \epsilon(b). \tag{3}$$

3   Background and Related Works

Bounding the burstiness of flows in Network Calculus is an important problem since it has a strong influence on the delay and backlog bounds. The deterministic aggregate burstiness can be improved (compared with summing burstiness of all flows) when the phases of the flows are known exactly [7].

Regarding the *stochastically bounded burstiness* [6,11], three models have been proposed, depending on how quantifiers are used,

$$\text{SBB} : \forall 0 \le s \le t, \ \mathbb{P}\left(A[s,t) - r(t-s) > b\right) \le \epsilon(b), \tag{4}$$

$$\text{S}^2\text{BB} : \forall t \ge 0, \ \mathbb{P}(\sup_{0 \le s \le t} \{A[s,t) - r(t-s)\} > b) \le \epsilon(b), \tag{5}$$

$$\text{S}^3\text{BB} : \mathbb{P}(\sup_{t \ge 0} \{\sup_{0 \le s \le t} \{A[s,t) - r(t-s)\}\} > b) \le \epsilon(b). \tag{6}$$

First, notice that $\text{S}^3\text{BB} \implies \text{S}^2\text{BB} \implies \text{SBB}$. Indeed, SBB is a probability upper bound that the arrival curve constraint is invalid for a fixed pair of times $s \le t$. In contrast, S^2BB is the probability that token-bucket content at time t, $\bar{B}(t)$ exceeds b, hence the "$\forall s$" appearing inside the probability. Last, S^3BB represents the violation probability of the aggregate burstiness B of the whole process. A deterministic arrival curve is a special case of S^3BB, with $\epsilon(b) = 0$, which is why, for a non-zero violation probability $\epsilon(b)$, b is called a *quasi-deterministic* bound on the burstiness.

The first model SBB is the weakest, but also the easiest to handle: bounding the arrivals during a given interval of time can be done for many stochastic models. It was also used for the study of aggregated independent flows with periodic patterns [4,6,8,10,12,16]. All the approaches can be summarized as follows: a) defining an event E_s of interest related to some time interval $[s,t)$ and aggregation of the flows; b) combining the events $(E_s)_{s \le t}$ together to obtain a violation probability of the burstiness or of the backlog bound at time t.

The second model S^2BB seems at first more adapted to network calculus analysis, as performance bounds can be directly derived from the formulation. However, the probability bound of S^2BB is usually deduced from SBB, which leads to pessimistic bounds for a single server. Nevertheless, this framework may become necessary for more complex cases [5].

In time-sensitive networks, we are interested in the probability that a delay or backlog bound is not violated during some interval (e.g., the network's lifetime), not just one arbitrary point in time, so the two models SBB and S^2BB are not adapted, as they do not provide the violation probability of a delay bound during a whole period of interest. In contrast, when using S^3BB, we can guarantee, with some probability, that delay and backlog bounds derived by deterministic network calculus are never violated during the network's lifetime, which is why we choose this formulation in our model.

As pointed out in [6, Section 4.4], when arrival processes are stationary and ergodic, S^3BB is always trivial and the bounding function $\epsilon(b)$ in (6) is either zero or one. This is perhaps why the literature was discouraged from studying

S^3BB characterizations. However, it has been overlooked that there is interest in some non-ergodic arrival processes, as in our case. Indeed, with our model, phases ϕ_f are drawn randomly but remain the same during the entire period of interest; thus, our arrival processes are not ergodic.

4   Homogeneous Case

In this section, we consider the case where flows have the same packet size and same period.

More precisely, we assume

(H) There exist $\tau, \ell > 0$ such that $\forall f \in \mathbb{N}_n$, $\ell_f = \ell, \tau_f = \tau$ and $(\phi_f)_{f \in \mathbb{N}_n}$ is a family of independent and identically distributed (iid) uniform random variables (rv) on $[0, \tau)$.

We present two bounds for the aggregate burstiness; the former gives a closed form, unlike the latter, which might be slightly more accurate when the number of flows is small.

Let us first prove a useful result when the period τ is equal to 1; it shows that if the time origin is shifted to the arrival time of the first packet of flow i, the phases of the $n-1$ other flows remain uniformly distributed on $[0,1)$ and mutually independent. For this, we define the function h as $\forall x, y \in [0,1)$,

$$h(x,y) = (x-y)\mathbb{1}x \geq y + (1+x-y)\mathbb{1}x < y. \tag{7}$$

Intuitively, if $x = \phi_j$ and $y = \phi_i$, $h(x,y)$ is the time until the arrival of the first packet of flow j, counted from the arrival time of the first packet of flow i.

Lemma 1. *Let $U_1, \ldots, U_n$ be a sequence of n iid uniform rv on $[0,1)$. Let $i \in \mathbb{N}_n$ and define W_j for $j \in \mathbb{N}_n \setminus \{i\}$ by $W_j = h(U_j, U_i)$. Then, $(W_j)_{j \neq i}$ is a family of $n-1$ iid uniform rv on $[0,1)$.*

Proof. Let us first do a preliminary computation for all $u_i \in [0,1]$ and all bounded measurable function g_j:

$$\mathbb{E}[g_j(h(U_j,u_i))] = \int_{u_j=0}^{1} g_j(h(u_j,u_i))du_j = \int_{u_j=u_i}^{1} g_j(u_j-u_i)du_j + \int_{u_j=0}^{u_i} g_j(1+u_j-u_i)du_j$$

$$= \int_{u_j=0}^{1-u_i} g_j(w_j)dw_j + \int_{u_j=1-u_i}^{1} g_j(w_j)dw_j = \int_{w_j=0}^{1} g_j(w_j)dw_j.$$

Then, consider a collection of bounded measurable functions $(g_j)_{j \neq i}$: we can compute $\mathbb{E}[\prod_{j \neq i} g_j(W_j)] = \mathbb{E}[\prod_{j \neq i} g_j(h(U_j, U_i))] = \int_{u_i=0}^{1} \mathbb{E}[\prod_{j \neq i} g_j(h(U_j, u_i))]du_i = \int_0^1 \prod_{j \neq i} \mathbb{E}[g_j(h(U_j, u_i))]du_i = \int_0^1 \prod_{j \neq i} \mathbb{E}[g_j(V_j)]du_i = \prod_{j \neq i} \mathbb{E}[g_j(V_j)] = \mathbb{E}[\prod_{j \neq i} g_j(V_j)]$, where $(V_j)_{j \neq j}$ is a collection of $n-1$ iid uniformly rv on $[0,1)$. $\square$

The bounds we are to present are based on the order statistics: consider $n-1$ rv $U_1, \ldots, U_{n-1}$ and its order statistics is $U_{(1)} \leq \cdots \leq U_{(n-1)}$, defined by sorting $U_1, \ldots, U_{n-1}$ in non-decreasing order. It is well-known [9, Equation 1.145] that if (U_i) is an iid family of uniform rv on $[0,1]$, the density function of the joint distribution of $U_{(1)}, \ldots, U_{(n-1)}$ is

$$f_{U_{(1)}, \ldots, U_{(n-1)}}(y_1, \ldots, y_{n-1}) = (n-1)! \mathbb{1} 0 \leq y_1 \leq y_2 \leq \ldots \leq y_{n-1} \leq 1. \quad (8)$$

The next proposition connects the order statistics of the phases with the aggregate burstiness, and is key for Theorems 1 and 2.

Proposition 1. *Assume model* (**H**). *For all* $0 \leq b < n\ell$,

$$\mathbb{P}(B > b) \leq n\mathbb{P}(E), \quad (9)$$

with

$$E \stackrel{def}{=} \bigcup_{k=\lfloor b/\ell \rfloor}^{n-1} \left\{ U_{(k)} < \frac{(k+1) - b/\ell}{n} \right\}, \quad (10)$$

where $U_{(1)}, \ldots, U_{(n-1)}$ *is the order statistic of* $n-1$ *iid uniform rv on* $[0,1]$.

Proof. Note that the normalized process $\tilde{A}[s,t] = \frac{1}{\ell}A[\tau s, \tau t)$ follows model (**H**) with $\tau = \ell = 1$, and

$$\mathbb{P}(B > b) = \mathbb{P}(\tilde{B} > b/\ell).$$

We then assume in this proof (and that of Theorem 1) that $\tau = \ell = 1$, and the final result is obtained by replacing b by b/ℓ. One can also remark that the bound is independent of τ.

Let T_j, $j \geq 1$ be the arrival time of the j-th packet in the aggregate. With probability 1, T_j is strictly increasing as we assume all phases are different. First, for all $i \leq j$, for all $(t_i, t_j) \in (T_{i-1}, T_i] \times (T_j, T_{j+1}] \stackrel{def}{=} C_{i,j}$, $A[t_i, t_j) = j - i + 1$, and $H_{i,j} \stackrel{def}{=} \sup_{t_i, t_j \in C_{i,j}} A[t_i, t_j) - n(t_i - t_j) = j - i + 1 - n(T_j - T_i)$. Then, we can rewrite the aggregate burstiness as

$$B = \sup_{1 \leq i \leq j} \sup_{t_i, t_j \in C_{i,j}} A[t_i, t_j) - n(t_i - t_j) = \sup_{1 \leq i \leq j} H_{i,j}. \quad (11)$$

As our model is the aggregation of n flows of period 1, $T_{j+n} = T_j + 1$ for $j \geq 1$, and $H_{i,j+n} = j + n - i + 1 - n(T_j - 1 - T_i) = H_{i,j}$ for all $j \geq i$. Similarly, $H_{i+n,j} = H_{i,j}$ for all $j \geq i + n$. Combine this with (11) and obtain

$$B = \max_{j \geq 1} \max_{i \in \mathbb{N}_j} H_{i,j} = \max_{i \in \mathbb{N}_n} \underbrace{\max_{i \leq j \leq n-1} H_{i,j}}_{B_i}. \quad (12)$$

We now prove that $\forall i \in \mathbb{N}_n$, $\mathbb{P}(B_i > b) = \mathbb{P}(E)$.

Observe that for all $j \geq i$, we have the equality of events $\{H_{i,j} > b\} = \{T_j - T_i < (j - i + 1 - b)/n\}$, so for all $i \in \mathbb{N}_n$, $\{B_i > b\} = \bigcup_{j=i}^{i+n-1} \{T_j - T_i < (j - i + 1 - b)/n\}$.

We can also notice that the sequence $(T_j - T_i)_{j=i+1}^{n+i-1}$ is the ordered sequence of phases starting from time origin T_i. Conditionally to $T_i = \phi_f$, or equivalently $\phi_{(i)} = f$, $(T_j - T_i)_{j=i+1}^{n+i-1}$ is the order statistics of $(\phi_j - \phi_f)_{j \neq f}$, which is, from Lemma 1, iid and uniformly distributed on $[0,1)$. If follows that

$$\mathbb{P}(B_i > b \mid \phi_{(i)} = f) = \mathbb{P}(\cup_{k=1}^{n-1}\{U_{(k)} < \frac{k-b}{n}\}) = \mathbb{P}(\cup_{k=\lfloor b \rfloor}^{n-1}\{U_{(k)} < \frac{k-b}{n}\}) = \mathbb{P}(E),$$

since $U_{(k)} \geq 0$. Then, using the law of total probabilities, $\mathbb{P}(B_i > b) = \sum_{f=1}^{n} \mathbb{P}(B_i > b \mid \phi_{(i)} = f)\mathbb{P}(\phi_{(i)} = f) = \mathbb{P}(E)$.

Lastly, we conclude by using the union bound: $\mathbb{P}(B > b) = \mathbb{P}(\cup_{i=1}^{n} B_i > b) \leq \sum_{i=1}^{n} \mathbb{P}(B_i > b) = n\mathbb{P}(E)$.   □

We now present the first bound on the tail probability of the aggregate burstiness B.

Theorem 1 (Homogeneous case, DKW bound). *Assume model* (**H**) *with* $n > 1$. *For all* $b < n\ell$, *a bound on the tail probability of the aggregate burstiness* B *is given by*

$$\mathbb{P}(B > b) \leq n \, \exp\left(-2(n-1)\left(\frac{\lfloor b/\ell \rfloor}{n-1} - \frac{1}{n}\right)^2\right) \stackrel{def}{=} \varepsilon^{dkw}(n, \ell, b). \tag{13}$$

Proof. Let us assume that $\tau = \ell = 1$ in the proof, as in the proof of Proposition 1. Observe that when $\lfloor b \rfloor < 1 - \frac{1}{n} + \sqrt{\frac{(n-1)\log 2}{2}}$, we have $\varepsilon^{dkw}(n, 1, b) \geq \frac{n}{2}$, hence (13) holds. Therefore we now proceed to prove (13) when $\lfloor b \rfloor \geq 1 - \frac{1}{n} + \sqrt{\frac{(n-1)\log 2}{2}}$.

Step 1: Consider $n - 1$ iid, rv $U_1, \ldots, U_{n-1}$ and its order statistics is $U_{(1)} \leq \cdots \leq U_{(n-1)}$, defined by sorting $U_1, \ldots, U_{n-1}$ in non-decreasing order. For $\varepsilon > 0$, define $E'(\varepsilon)$ by

$$E'(\varepsilon) \stackrel{def}{=} \bigcup_{k=1}^{n-1}\left\{U_{(k)} < \frac{k}{n-1} - \varepsilon\right\}. \tag{14}$$

We now show that if $\varepsilon \geq \sqrt{\frac{\log 2}{2(n-1)}}$,

$$\mathbb{P}(E'(\varepsilon)) \leq e^{-2(n-1)\varepsilon^2}. \tag{15}$$

Let F_{n-1} be the (random) empirical cumulative distribution function of $U_1, \ldots, U_{n-1}$, defined $\forall x \in [0,1]$ by

$$F_{n-1}(x) = \frac{1}{n-1}\sum_{i=1}^{n-1} \mathbb{1}U_{(i)} \leq x. \tag{16}$$

The Dvoretzky-Kiefer-Wolfowitz inequality [14] states that if $\varepsilon \geq \sqrt{\frac{\log 2}{2(n-1)}}$, then

$$\mathbb{P}\left(\sup_{x \in [0,1]}(F_{n-1}(x) - x) > \varepsilon\right) \leq e^{-2(n-1)\varepsilon^2}. \tag{17}$$

We can apply this to find the bound of interest. First, we prove that

$$\sup_{x \in [0,\, 1]} (F_{n-1}(x) - x) > \varepsilon \Leftrightarrow \exists k \in \mathbb{N}_{n-1},\; U_{(k)} < \frac{k}{n-1} - \varepsilon. \qquad (18)$$

Proof of $\Leftarrow$: First, observe that $F_{n-1}(U_{(k)}) = k/(n-1)$, so if $\frac{k}{n-1} - U_{(k)} > \varepsilon$ for some k, then $F_{n-1}\left(U_{(k)}\right) - U_{(k)} > \varepsilon$, and the left-hand side holds.

Proof of $\Rightarrow$: Set $U_{(0)} = 0$ and $U_{(n)} = 1$. Observe that for all $k \in \{0,\dots,n-1\}$, and all $U_{(k)} \le x < U_{(k+1)}$, $F_{n-1}(x) = F_{n-1}\left(U_{(k)}\right) = \frac{k}{n-1}$. Hence, $F_{n-1}(x) - x = \frac{k}{n-1} - x$ is decreasing on each segment $[U_{(k)}, U_{(k+1)})$. Then, the supremum in the left-hand side of (18) is obtained for some $x = U_{(k)}$, i.e., $\sup_{x \in [0,\, 1]}(F_{n-1}(x) - x) = \sup_{k \in \{0,\dots,n-1\}}(F_{n-1}(U_{(k)}) - U_{(k)}) = \sup_{k \in \{0,\dots,n-1\}}(\frac{k}{n-1} - U_{(k)})$, which implies the right-hand side $(F_{n-1}(0) - 0 = 0 < \varepsilon)$.

This proves (18), and Step 1 is concluded by combining it with (17).

Step 2: We now proceed to show that if

$$\varepsilon = \frac{\lfloor b \rfloor}{n-1} - \frac{1}{n}, \qquad (19)$$

then, $E \subseteq E'(\varepsilon)$, where event E is defined in Proposition 1.

It is enough to show that for all $k \in \{\lfloor b \rfloor,\dots,n-1\}$, $\frac{k+1-b}{n} \le \frac{k-\lfloor b \rfloor}{n-1} + \frac{1}{n}$, which can be deduced from the following implications:

$$\frac{k+1-b}{n} \le \frac{k - \lfloor b \rfloor}{n-1} + \frac{1}{n} \Leftrightarrow \frac{k-b}{n} \le \frac{k - \lfloor b \rfloor}{n-1} \Leftarrow \frac{k - \lfloor b \rfloor}{n} \le \frac{k - \lfloor b \rfloor}{n-1} \Leftrightarrow \frac{1}{n} \le \frac{1}{n-1}.$$

Step 3: By Step 2, we have $\mathbb{P}(E) \le \mathbb{P}(E'(\varepsilon))$. Also, observe that $\lfloor \frac{b}{l} \rfloor \ge 1 - \frac{1}{n} + \sqrt{\frac{(n-1)\log 2}{2}}$ implies $\varepsilon \ge \sqrt{\frac{\log 2}{2(n-1)}}$. Thus, combine it with Step 1 to obtain

$$\mathbb{P}(E) \le \mathbb{P}(E'(\varepsilon)) \le \exp\left(-2(n-1)\left(\frac{\lfloor b \rfloor}{n-1} - \frac{1}{n}\right)^2\right). \qquad (20)$$

Combine (20) with Proposition 1 to conclude the theorem. $\qquad \square$

Note that the bound of Theorem 1 is only less than one and is non-trivial when $\lfloor \frac{b}{l} \rfloor \ge 1 - \frac{1}{n} + \sqrt{\frac{(n-1)\log 2}{2}}$.

The following corollary provides a closed-form formulation for the minimum value for the aggregate burstiness with a violation probability of at most ε. It is obtained by setting the right-hand side of (13) in Theorem 1 to ε.

Corollary 1 (Quasi-deterministic burstiness bound). *Assume model* (**H**) *with $n > 1$. Consider some $0 < \varepsilon < 1$, and define*

$$b(n, \ell, \varepsilon) \stackrel{def}{=} \ell \left[1 - \frac{1}{n} + \sqrt{\frac{(n-1)(\log n - \log \varepsilon)}{2}} \right]. \qquad (21)$$

Then, $b(n, \ell, \varepsilon)$ is a quasi-deterministic burstiness bound for the aggregate with the violation probability of at most ε, i.e., $\mathbb{P}(B > b(n, \ell, \varepsilon)) \le \varepsilon$.

Observe that $b(n, \ell, \varepsilon)$ grows in $\sqrt{n \log n}$ as opposed to the deterministic bound ($\ell^{\mathrm{tot}} = n\ell$) that grows in linearly (see Fig. 1b).

Proposition 1 introduces the event E such that an upper bound of $\mathbb{P}(E)$ is used to derive an upper bound on the tail probability of the aggregate burstiness. Theorem 1 is derived from the DKW upper bound of $\mathbb{P}(E)$, which is tight when the number of flows n is large. In Theorem 2, we compute the exact value of $\mathbb{P}(E)$; thus, it provides a slightly better bound when the number of flows is small but at the expense of not having a closed-form expression.

Theorem 2 (Refinement of Theorem 1 for small groups). *Assume model* **(H)** *with* $n > 1$. *For all* $b \geq 0$. *Then, a bound on the tail probability of the aggregate burstiness* B *is*

$$\mathbb{P}\left(B > b\right) \leq n(1 - p(n, \ell, b)) \overset{\text{def}}{=} \varepsilon^{thm2}(n, \ell, b), \tag{22}$$

with

$$p(n, \ell, b) = (n-1)! \int_{y_{n-1}=u_{n-1}}^{1} \int_{y_{n-2}=u_{n-2}}^{y_{n-1}} \cdots \int_{y_i=u_i}^{y_{i+1}} \cdots \int_{y_1=u_1}^{y_2} 1 \, dy_1 \ldots dy_{n-1}, \tag{23}$$

and $u_k = \frac{[(k+1)-b/\ell]^+}{n}$, *for all* $k \in \mathbb{N}_{n-1}$ *and* $[x]^+ = \max(0, x)$.

Note that the computation of the bound of Theorem 2 requires computing $p(n, \ell, b)$ in (23), which is a series of polynomial integrations, and finding a general closed-form formula might be challenging. However, computing the bound can be done iteratively as in Algorithm 1: The integrals are computed from the inner sign to the outer (incorporation factor i from the factorial in the i-th integral). Polynoms are computed at each step and variable q_j^m represents the coefficient of degree j of the m-th integral. Note that we always have $q_m^m = 1$, so the monomial of degree $n - 1$ cancels in (22).

All computations involve exact representations of the integrals (no numerical integration) and use exact arithmetic with rational numbers; therefore, the results are exact with infinite precision.

Algorithm 1: Computation of $\varepsilon^{thm2}(n, \ell, b)$ from Theorem 2

    Inputs   : number of flows n, a burst b, and a packet size ℓ.
    Output   : $\varepsilon^{thm2}(n, \ell, b)$ such that $\mathbb{P}(B > b) \leq \varepsilon^{thm2}(n, \ell, b)$.

1 $m \leftarrow \lfloor b/\ell \rfloor - 1$;

2 $(q_0^m, q_1^m, \ldots, q_m^m) \leftarrow (0, 0, \ldots, 0, 1)$;

3 for $m \leftarrow \lfloor b/\ell \rfloor$ **to** $n - 1$ **do**

4 $u_m \leftarrow (m + 1 - b/\ell)/n$;

5 $q_0^m \leftarrow -\sum_{j=0}^{m-1} \frac{m q_j^{m-1}}{j+1} u_m^{j+1}$;

6   **for** $i \leftarrow 1$ **to** m **do** $q_i^m \leftarrow \frac{m q_{i-1}^{m-1}}{i}$;

7 return $n \sum_{i=0}^{n-2} q_i^{n-1}$

214   S. M. Tabatabaee et al.

Proof. Let $\bar{E}$ be the complementary event of E defined in Proposition 1.

$$\bar{E} = \bigcap_{k=1}^{n-1} \left\{ U_{(k)} \geq \frac{[k+1-b/\ell]^+}{n} \right\}. \tag{24}$$

Let $f_{U_{(1)},\dots,U_{(n-1)}}$ be the density function of the joint distribution of $U_{(1)},\dots,U_{(n-1)}$, given in (8). Then

$$\mathbb{P}\left(\bar{E}\right) = \int_{y_{n-1}=u_{n-1}}^{1} \cdots \int_{y_i=u_i}^{1} \cdots \int_{y_1=u_1}^{1} f_{U_{(1)},\dots,U_{(n-1)}}(y_1,\dots,y_{n-1})\, dy_1 \dots dy_{n-1} \tag{25}$$

$$= \int_{y_{n-1}=u_{n-1}}^{1} \cdots \int_{y_i=u_i}^{1} \cdots \int_{y_1=u_1}^{1} (n-1)!\, \mathbb{1}\,0 \leq y_1 \leq y_2 \leq \dots \leq y_{n-1} \leq 1\, dy_1 \dots dy_{n-1} \tag{26}$$

$$= (n-1)! \int_{y_{n-1}=u_{n-1}}^{1} \cdots \int_{y_i=u_i}^{y_{i+1}} \cdots \int_{y_1=u_1}^{y_2} 1\, dy_1 \dots dy_{n-1} = p(n,\ell,b). \tag{27}$$

Combine it with $\mathbb{P}\left(\bar{E}\right) = 1 - \mathbb{P}\left(E\right)$ and Proposition 1 to conclude the theorem. $\square$

Note that since Theorem 2 computes the exact probability of event E, we have $\varepsilon^{\mathrm{dkw}}(n,\ell,b) \geq \varepsilon^{\mathrm{thm2}}(n,\ell,b)$.

5   Heterogeneous Case

In this section, we consider the case where flows have different periods and packet sizes. We present burstiness bounds in two different settings: First, when flows can be grouped into homogeneous flows; second, when all packets have the same period but with different packet sizes.

Let us first focus on the model where flows are grouped according to their characteristics:

(G) There exists a partition $I_1,\dots,I_g$ of $\mathbb{N}_n$ such that I_i is a group of n_i flows satisfying model **(H)** with packet size ℓ_i and period τ_i. All phases are mutually independent.

Proposition 2 (Convolution Bound). *Let $X_1, X_2, \dots, X_g$ be $g \geq 1$ mutually independent rv on $\mathbb{N}$. Assume that for all $i \in \mathbb{N}_n$, Ψ_i is wide-sense increasing and is a lower bound on the CDF of X_i, namely, $\forall b \in \mathbb{N}$, $\mathbb{P}(X_i \leq b) \geq \Psi_i(b)$. Define ψ_i by $\psi_i(0) = \Psi_i(0)$ and $\psi_i(b) = \Psi_i(b) - \Psi_i(b-1)$ for $b \in \mathbb{N} \setminus \{0\}$. Then, a lower bound on the CDF of $\sum_{i=1}^g X_i$ is given by: $\forall b \in \mathbb{N}$,*

$$\mathbb{P}\left(\sum_{i=1}^g X_i \leq b\right) \geq (\psi_1 * \psi_2 * \dots * \psi_{g-1} * \Psi_g)(b), \tag{28}$$

where, the symbol $*$ denotes the discrete convolution, defined for arbitrary functions $f_1, f_2 : \mathbb{N} \to \mathbb{R}$ by

$$\forall b \in \mathbb{N}, \quad (f_1 * f_2)(b) = \sum_{j=0}^{b} f_1(j) f_2(b-j). \tag{29}$$

Proof. We prove it by induction on g.

Base Case $g = 1$: There is nothing to prove: for all $b \in \mathbb{N}$, $\mathbb{P}(X_1 \leq b) \geq \Psi_1(b)$.

Induction Case: We now assume that Eq. (28) holds for g variables, and we show that it also holds for $g + 1$ variables.

We can apply Eq. (28) to variables $X_2, X_3, \ldots, X_{g+1}$, and let us denote $Y = X_2 + \cdots + X_{g+1}$ and $\Psi = \psi_2 * \cdots * \psi_g * \Psi_{g+1}$. We need to show that for all $b \in \mathbb{N}$,

$$\mathbb{P}(X_1 + Y \leq b) \geq (\psi_1 * \Psi)(b). \tag{30}$$

Let $F(b) = \mathbb{P}(Y \leq b)$ and observe that $\mathbb{P}(Y = 0) = F(0)$ and $\mathbb{P}(Y = b) = F(b) - F(b-1)$ for $b \in \mathbb{N} \setminus \{0\}$. Then, since X_1 and Y are independent,

$$\mathbb{P}(X_1 + Y \leq b) = \sum_{j=0}^{b} \mathbb{P}(X_1 + j \leq b | Y = j)\mathbb{P}(Y = j) = \sum_{j=0}^{b} \mathbb{P}(X_1 + j \leq b)\mathbb{P}(Y = j) \tag{31}$$

$$\geq \sum_{j=0}^{b} \Psi_1(b-j)\mathbb{P}(Y = j) \tag{32}$$

$$\geq \Psi_1(b)F(0) + \sum_{j=1}^{b} \Psi_1(b-j)(F(j) - F(j-1)). \tag{33}$$

We now use Abel's summation by parts in (33) and obtain

$$\mathbb{P}(X_1 + Y \leq b) \geq \Psi_1(b)F(0) + \sum_{j=1}^{b} \Psi_1(b-j)F(j) - \sum_{j=1}^{b} \Psi_1(b-j)F(j-1) \tag{34}$$

$$= \Psi_1(b)F(0) + \sum_{j=1}^{b} \Psi_1(b-j)F(j) - \sum_{j=0}^{b-1} \Psi_1(b-j-1)F(j) \tag{35}$$

$$= \sum_{j=0}^{b} \Psi_1(b-j)F(j) - \sum_{j=0}^{b-1} \Psi_1(b-j-1)F(j) \tag{36}$$

$$= \Psi_1(0)F(b) + \sum_{j=0}^{b-1} (\Psi_1(b-j) - \Psi_1(b-j-1))F(j) \tag{37}$$

$$= \psi_1(0)F(b) + \sum_{j=0}^{b-1} \psi_1(b-j)F(j) = \sum_{j=0}^{b} \psi_1(b-j)F(j) \tag{38}$$

$$\geq \sum_{j=0}^{b} \psi_1(b-j)\Psi(j) = (\psi_1 * \Psi)(b). \tag{39}$$

We can conclude by using the associativity of the discrete convolution: $\psi_1 * \Psi = \psi_1 * \cdots * \psi_g * \Psi_{g+1}$.   □

Remarks 1. 1. Note that $(\psi_1 * \Psi_2)(b) = \sum_{i+j\leq b} \psi_1(i) + \psi_2(j) = (\psi_2 * \Psi_1)(b)$, so the convolution bound is independent of the order of $X_1, \ldots, X_g$.
2. An alternative to Proposition 2 is to use then union bound rather than the convolution bound: for all $(b_1, \ldots, b_g) \in \mathbb{N}^g$ such that $\sum_{i=1}^{g} b_i = b$, we have $\{\sum_{i=1}^{g} X_i > b\} \subseteq \bigcup_{i=1}^{g} \{X_i > b_i\}$, so $\mathbb{P}(X > b) \leq \sum_{i=1}^{g} \mathbb{P}(X_i > b_i) \leq \sum_{i=1}^{g}(1 - \Psi_i(b_i))$. We can choose $(b_i)_{i=1}^{g}$ so as to minimize this latter term, and take the complement to obtain

$$\mathbb{P}(\sum_{i=1}^{g} X_i \leq b) \geq 1 - \min_{b_1+\cdots+b_g=b} \sum_{i=1}^{g}(1 - \Psi_i(b_i)). \tag{40}$$

This bound is also valid when rvs X_i are not independent, but it can be shown that the convolution bound always dominates the union bound. In our numerical evaluations, we find that the convolution bound provides significantly better results than the union bound.

Theorem 3 (Flows with different periods and different packet-sizes).
Assume model (**G**). *Let ε_i be a wide-sense decreasing function that bounds the tail probability of aggregate burstiness B_i of each group $i \in \mathbb{N}_g$: for all $b \in \mathbb{N}$, $\mathbb{P}(B_i > b) \leq \varepsilon_i(b)$ for all $b \in \mathbb{N}$. Define $\Psi_i(b) = 1 - \varepsilon_i(b)$ for $b \in \mathbb{N}$ and define ψ_i by $\psi_i(0) = \Psi_i(0)$ and $\psi_i(b) = \varepsilon_i(b-1) - \varepsilon_i(b)$ for $b \in \mathbb{N} \setminus \{0\}$.*
   Then, a bound on the tail probability of the aggregate burstiness of all flows B is given by $\forall b \in \mathbb{N}_{\ell tot}$,

$$\mathbb{P}(B > b) \leq 1 - (\psi_1 * \psi_2 * \cdots * \psi_{g-1} * \Psi_g)(b). \tag{41}$$

Proof. For all group $i \in \mathbb{N}_g$, let $A^i[s,t)$ be the aggregate of flows of group i during the interval $[s,t)$, r^i, its aggregate arrival rate, and B_i its aggregate burstiness. Observe that for all $s \leq t$, $A(s,t] = \sum_{i=1}^{g} A^i[s,t)$ and $r^{tot} = \sum_{i=1}^{g} r^i$. We then obtain

$$B = \sup_{0\leq s\leq t} \{A(s,t] - r^{tot}(t-s)\} = \sup_{0\leq s\leq t} \left\{ \sum_{i=1}^{g} (A_i(s,t] - r_i(t-s)) \right\} \tag{42}$$

$$\leq \sum_{i=1}^{g} \sup_{0\leq s\leq t} \{A_i(s,t] - r_i(t-s)\} = \sum_{i=1}^{g} B_i \leq \sum_{i=1}^{g} \lceil B_i \rceil. \tag{43}$$

Hence, it follows that $\mathbb{P}(B \leq b) \geq \mathbb{P}(\sum_{i=1}^{g} \lceil B_i \rceil \leq b)$, $b \in \mathbb{N}$.
   We now apply Proposition 2 with $X_i = \lceil B_i \rceil$ and Ψ_i as defined in the theorem: it suffices to observe that $(\lceil B_i \rceil)_{i \in \mathbb{N}_g}$ are mutually independent rv on $\mathbb{N}$; as ε_i is wide-sense decreasing, Ψ_i is wide-sense increasing; Hence, by Proposition 2, we obtain that for all $b \in \mathbb{N}$, $\mathbb{P}(\sum_{i=1}^{g} \lceil B_i \rceil \leq b) \geq (\psi_1 * \psi_2 * \ldots * \psi_{g-1} * \Psi_g)(b)$, which concludes the proof.   □

We now turn to our second heterogeneous model: when all flows have the same period but different packet sizes.

(P) There exists $\tau > 0$ such that $\forall f \in \mathbb{N}_n$, $\tau_f = \tau$; $\ell_1 \geq \ell_2 \geq \cdots \geq \ell_n > 0$ and $(\phi_f)_{f \in \mathbb{N}_n}$ is a family of iid uniform rv on $[0, \tau)$.

Theorem 4 (Flows with the same period but different packet sizes).
Assume model **(P)**. *For all* $0 \leq b < \ell^{tot}$, *set* $\eta \overset{def}{=} \min\left\{ \frac{k}{n-1} - \frac{\sum_{j=1}^{k+1} \ell_j}{\ell^{tot}}, \ k \in \mathbb{N}_{n-1}, \sum_{j=1}^{k+1} \ell_j > b \right\}$. *Then*

1. *A bound on the tail probability of the aggregate burstiness of all flows B is*

$$\mathbb{P}(B > b) \leq n \, \exp\left(-2(n-1)\left(\eta + \frac{b}{\ell^{tot}}\right)^2 \right). \tag{44}$$

2. *For all* $\varepsilon \in (0, 1)$, *for all* $n \geq 2$, *the violation probability of at most ε, i.e.,*
$\mathbb{P}\left(B > b(n, \ell_1, \ldots, \ell_n, \varepsilon) \right) \leq \varepsilon$ *with*

$$b(n, \ell_1, \ldots, \ell_n, \varepsilon) \overset{def}{=} \ell^{tot}\left\lceil \sqrt{\frac{\log n - \log \varepsilon}{2(n-1)}} - \eta \right\rceil. \tag{45}$$

3. *A bound on the tail probability of the aggregate burstiness of all groups B is given by* $\mathbb{P}\left(B > b \right) \leq n(1 - \bar{p}(n, \ell_1, \ldots, \ell_n, b))$, *where* $\bar{p}(n, \ell_1, \ldots, \ell_n, b)$ *is computed as in Equation* (23), *where for all* $k \in \mathbb{N}_{n-1}$, $u_k = \frac{[\sum_{j=1}^{k+1} \ell_j - b]^+}{\ell^{tot}}$.

When all flows have the same packet-sizes ℓ, this is model **(H)** and the bounds provided are exactly the same as in Sect. 4. Algorithm 1 can also be used to compute the bound of item 3 if a) line 1 is replaced by $m \leftarrow \max\{k \geq 0 \mid \sum_{j=1}^{k+1} \ell_j \leq b\}$ and b) the values of u_m are adapted in line 4.

Proof. The proof is done by adapting Proposition 1. Then the proofs of each item follow exactly the steps of Theorems 1, Corollary 1 and Theorem 2. The key difference in Proposition 1 is the computation of $H_{i,j}$: $H_{i,j} \leq \sum_{k=1}^{j-i+1} \ell_k - \ell^{tot}(T_j - T_i)$: we bound this value as if the packets arrived in this arrival where the $j - i + 1$ longest ones. $\qquad\square$

6   Numerical Evaluation

In this section, we numerically illustrate our bounds in Fig. 1 and Fig. 2.

6.1   Homogeneous Case

In Fig. 1a, we consider 250 flows with the same packet size (with respect to a unit, is assumed to be 1) and the same period. We then compute bounds on the tail probability of their aggregate burstiness using Theorems 1 and 2. We also compute the bound using simulations: For each flow, we independently pick a

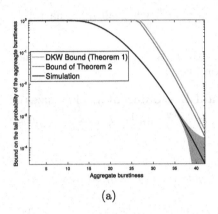
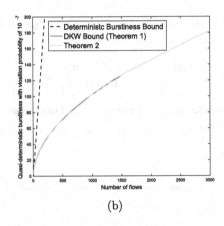

(a)   (b)

Fig. 1. (a): Bound on the tail probability of the aggregate burstiness obtained by Theorems 1 2, and simulations. (b): The obtained quasi-deterministic burstiness with violation probability of 10^{-7} by Theorem 1 and Theorem 2, as the number of flows grows; the deterministic bound (dashed plot) grows linearly with the number of flows.

phase uniformly at random, and we then compute the aggregate burstiness as in (1); we repeat this 10^8 times. We then compute bounds on the tail probability of their aggregate burstiness and its 99% Kolmogorov-Smirnov confidence band. The bound of Theorem 2 is slightly better than that of Theorem 1. Also, compared to simulations, our bounds are fairly tight.

In Fig. 1b, we consider $n \in \{2, \ldots, 3000\}$ flows with the packet size 1 and same period. We then compute a quasi-deterministic burstiness bound with violation probability of 10^{-7} once using Corollary 1 and once using Theorem 2; they are almost equal and as n grows are exactly equal, as Theorem 1 is as tight as Theorem 2 for large n. Also, our quasi-deterministic burstiness bound is considerably less than the deterministic one (i.e., n) and grows in $\sqrt{n} \log n$.

6.2   Heterogeneous Case

To assess the efficiency of the bound in the heterogeneous case, we consider in Fig. 2a 10000 homogeneous flows with period and packet length 1, and divide them into g groups of $10000/g$ flows, for $g \in \{1, 2, 4, 5, 8\}$. We compute a bound for each group by Theorem 1, and combine them once with the convolution bound of Theorem 3 and once by the union bound (as explained after Proposition 2). Our convolution bound is significantly better than the union bound, and the differences increases fast with the number of sets.

In Fig. 2b, we consider 10 (resp. 5) homogeneous groups of 10 (resp. 20) flows, flows of each set $g \in \mathbb{N}_{10}$ (resp. $g \in \mathbb{N}_5$), have a packet-size equal to g, and all flows have the same period. We then compute the bound on the tail probability of the aggregate burstiness once with Theorem 3 and once with Theorem 4. When groups are small (here of 10 flows), Theorem 4 provides better bounds than Theorem 3, but when groups are larger (here of 20 flows), Theorem 3 dominates Theorem 4.

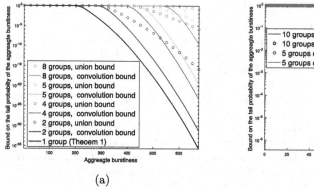
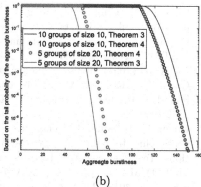

(a)   (b)

Fig. 2. (a): Comparison of the convolution bound of Theorem 3 to the union bound when combining bound obtained for homogeneous sets of flows. (b): Slight improvement of Theorem 4 compared to Theorem 3 when the number of flows per packet-size is small.

7   Conclusion

In this paper, we provided quasi-deterministic bounds on the aggregate bursti-ness for independent, periodic flows. When a small violation tolerance, is allowed, the bounds are considerably better compared to the deterministic bounds. We obtained a closed-form expression for the homogeneous case, and for the het-erogeneous case, we combined bounds obtained for homogeneous sets using the convolution bounding technique.

We on purpose limited our study to the burstiness. Quasi-deterministic delay and backlog bounds can be obtained by applying any method from deterministic network calculus, and combining, either by mean of the union bound or (in case of independence) convolution-like manipulations of the burstiness violation events defined for this paper for all groups of flows. Our results can for example be directly applied to [3, Theorem 5], where the model S^3BB was used to compute probabilistic delay bounds in tandem networks.

References

1. Bouillard, A., Boyer, M., Le Corronc, E.: Deterministic Network Calculus: From Theory to Practical Implementation. Wiley-ISTE (2018)
2. Bouillard, A., Nikolaus, P., Schmitt, J.B.: Unleashing the power of paying multi-plexing only once in stochastic network calculus. Proc. ACM Meas. Anal. Comput. Syst. **6**(2), 31:1–31:27 (2022). https://doi.org/10.1145/3530897
3. Bouillard, A., Nowak, T.: Fast symbolic computation of the worst-case delay in tandem networks and applications. Perform. Eval. **91**, 270–285 (2015). https://doi.org/10.1016/j.peva.2015.06.016

4. Chang, C.S., Chiu, Y.M., Song, W.T.: On the performance of multiplexing independent regulated inputs. In: Proceedings of the 2001 ACM SIGMETRICS International Conference on Measurement and Modeling of Computer Systems, SIGMETRICS 2001, pp. 184–193. Association for Computing Machinery, New York, NY, USA (2001). https://doi.org/10.1145/378420.378782

5. Ciucu, F., Burchard, A., Liebeherr, J.: Scaling properties of statistical end-to-end bounds in the network calculus. IEEE/ACM Trans. Network. (ToN) **14**(6), 2300–2312 (2006)

6. Ciucu, F., Schmitt, J.: Perspectives on network calculus: no free lunch, but still good value. In: Proceedings of the ACM SIGCOMM 2012 Conference on Applications, Technologies, Architectures, and Protocols for Computer Communication, SIGCOMM 2012, pp. 311–322. Association for Computing Machinery, New York, NY, USA (2012). https://doi.org/10.1145/2342356.2342426

7. Daigmorte, H., Boyer, M.: Traversal time for weakly synchronized can bus. In: Proceedings of the 24th International Conference on Real-Time Networks and Systems, RTNS 2016, pp. 35–44. Association for Computing Machinery, New York, NY, USA (2016). https://doi.org/10.1145/2997465.2997477, https://doi.org/10.1145/2997465.2997477

8. Fidler, M., Rizk, A.: A guide to the stochastic network calculus. IEEE Commun. Surv. Tutorials **17**(1), 92–105 (2015). https://doi.org/10.1109/COMST.2014.2337060

9. Gentle, J.: Computational Statistics. Statistics and Computing, Springer New York (2009). https://doi.org/10.1007/978-0-387-98144-4. https://books.google.ch/books?id=mQ5KAAAAQBAJ

10. Guillemin, F.M., Mazumdar, R.R., Rosenberg, C.P., Ying, Y.: A stochastic ordering property for leaky bucket regulated flows in packet networks. J. Appl. Probab. **44**(2), 332–348 (2007). http://www.jstor.org/stable/27595845

11. Jiang, Y.: A basic stochastic network calculus. In: Proceedings of the 2006 Conference on Applications, Technologies, Architectures, and Protocols for Computer Communications, SIGCOMM 2006, pp. 123–134. Association for Computing Machinery, New York, NY, USA (2006). https://doi.org/10.1145/1159913.1159929

12. Kesidis, G., Konstantopoulos, T.: Worst-case performance of a buffer with independent shaped arrival processes. IEEE Commun. Lett. **4**(1), 26–28 (2000). https://doi.org/10.1109/4234.823539

13. Le Boudec, J.-Y., Thiran, P. (eds.): Network Calculus. LNCS, vol. 2050. Springer, Heidelberg (2001). https://doi.org/10.1007/3-540-45318-0

14. Massart, P.: The tight constant in the Dvoretzky-Kiefer-Wolfowitz inequality. Ann. Probab. **18**(3), 1269–1283 (1990). https://doi.org/10.1214/aop/1176990746

15. Poloczek, F., Ciucu, F.: Scheduling analysis with martingales. Perform. Eval. **79**, 56–72 (2014). https://doi.org/10.1016/j.peva.2014.07.004. http://www.sciencedirect.com/science/article/pii/S0166531614000674. Special Issue: Performance 2014

16. Vojnovic, M., Le Boudec, J.Y.: Bounds for independent regulated inputs multiplexed in a service curve network element. In: IEEE Global Telecommunications Conference (Cat. No.01CH37270), GLOBECOM 2001, vol. 3, pp. 1857–1861 (2001). https://doi.org/10.1109/GLOCOM.2001.965896

Matching Distributions Under Structural Constraints

Aaron Bies[(✉)][iD], Holger Hermanns[iD], Maximilian A. Köhl[iD],
and Andreas Schmidt[iD]

Saarland University, Saarland Informatics Campus, Saarbrücken, Germany
{bies,hermanns,koehl,andreas.schmidt}@cs.uni-saarland.de

Abstract. Phase-type distributions, the probability distributions generated by the time to absorption in a continuous-time Markov chain, are a popular tool for modeling time-dependent system behaviour. They are well understood mathematically, and so is the problem of identifying a matching distribution if given information about its moments, as well as fitting to a given distribution or a set of distribution samples. This paper looks at the problem of finding distributions from a structural perspective, namely where system behaviour is known to have a specific structure comprising parallelism, sequencing, and first-to-finish races. We present a general method that, given the coarse system structure with annotations regarding the moments of some fragments, finds a concrete phase-type distribution that fulfils the specification, if one exists. We develop the foundational underpinning in terms of constraint solving with satisfiability modulo theories, spell out the algorithmic details of a divide-and-conquer solution approach, and provide empirical evidence of feasibility, presenting a prototypical solution engine for structural distribution matching.

1 Introduction

Continuous probability distributions are an important topic in statistics, stochastics, and data science because they allow to model and analyze data that are influenced by continuously varying quantities. This is particularly important for real-world phenomena where the continuous variable of interest is time.

Phase-type distributions are a class of continuous probability distributions that are often used in this context. A phase-type (PH) distribution can be thought of as describing the time to absorption in an absorbing continuous-time Markov chain (CTMC) if starting in its initial state at time zero. These distributions have been proven effective in modelling many time-dependent systems, since they can, in principle, represent any distributions on the positive axis with arbitrary precision [3], while the representation size is straightforward to control.

The behaviour of PH distributions is well understood. Given a PH distribution, one can compute the corresponding probability density function, cumulative distribution function as well as any statistical moment using closed formulae.

N. Jansen and M. Tribastone (Eds.): QEST 2023, LNCS 14287, pp. 221–237, 2023.
https://doi.org/10.1007/978-3-031-43835-6_16

```
1   #[async_std::main]
2   async fn main() {
3       let d1 = download_bytes("https://.../assets/orig.svg");        // X1
4       let d2 = download_bytes("https://.../assets/orig.png");        // X2
5       let d3 = download_bytes("https://.../assets/flat.png");        // X3
6
7       let d4 = async {                                               // X4.X5
8           let text = download_text("https://.../assets/style.css").await;  // X4
9           async_std::fs::write("style.css", &text).await.unwrap();   // X5
10          text
11      };
12
13      let image = d1.race(d2).race(d3);                              // X1 + X2 + X3
14      let (image_bytes, css_text) = image.join(d4).await;           // ... || X4.X5
15
16      dbg!(image_bytes.len(), css_text.len());
17  }
```

Listing 1: An example program written in Rust

They are also composable, as they are closed under typical operations on distributions such as minimum, maximum, convolution, and mixture.

Finding a phase-type distribution for certain data involves estimating the parameters of the absorbing CTMC including its size, shape, and parameters. There are a number of sophisticated algorithms for doing so, which come in two flavors: They either fit to data points [1,4,9,10,14,21,23] available from sampling, or aim at matching the moments [2,11–13,22] of the distribution. They usually aim at finding a PH distribution that resembles the observed behaviour as accurately as possible, but is otherwise as small and simple as possible.

Now, when considering real-world applications of finding PH distributions, we might have more than only some measured characteristics at hand – and this is the twist we are considering worthwhile to investigate: In particular, we may also have insights about the coarse structure of the system. This may refer to the parallel or sequential nature of the program execution studied, together with measurements that relate to fragments of this structure.

Motivating Example. Consider the program in Lst. 1, which is written in Rust using the `async-std` library. Here, the functions `download_bytes` and `download_text` return futures (promises in other languages) that can be composed with each other to create structural concurrency. `race` runs two futures in parallel and outputs the result of the future that finished first. `join` runs two futures in parallel and returns both results, i.e., only finishes once both futures finish. After downloading the stylesheet, we also write its content to a local file.

Now suppose we have benchmarked the program and determined that the program takes 0.15 s to terminate on average. Additionally, we measured that the `style.css` file only takes about 0.1 s to download and save. Using these sparse measurements and our knowledge of the program structure, we would like to construct a complete model for the execution time of this program. This program can be abstractly expressed as

$$\overline{(X_1+X_2+X_3) \parallel \overline{X_4.X_5}|_{E=0.1}}|_{E=0.15}$$

Here, we use some common process calculi operators $\{\,.\,,+,\parallel\}$ for sequence, choice, and parallelism, and $\overline{\Psi}|_{E=c}$ to constrain the expected value for the over-lined process Ψ to c. Choice and parallelism echo the Rust constructs race and join, the variables X_1 to X_4 represent the individual download durations, and X_5 represents the time it takes to store the file in the example.

We are not aware of methods that allow for such structural constraints to be taken into account. This paper pioneers the consideration of structural constraints in finding phase-type distributions, and it does so focusing on moment matching techniques. The paper develops the theory and algorithmic ingredients to fill in small PH distributions for X_1 to X_5 so that altogether the constraints present are satisfied. Using our prototype solver STRUMPH, we find that the exponential distribution $Exp(3.25)$ for X_1 to X_3, approximately $Exp(10.965)$ for X_4, and approximately $Exp(113.655)$ for X_5 is a valid solution for this example.

Contribution. In this paper, we discuss the problem of finding a PH distribution subject to a set of structural and moment constraints. We restrict to the first two moments for the sake of simplicity, and need to only work with acyclic PH distributions (APH). To do so, we assume that structural knowledge determining the problem at hand is given in a formal language which we call CCcSAT—a variant of the CCC process calculus [20] that exploits the closure properties discussed above. We discuss properties and transformations of CCcSAT problems that are leveraged by a general semi-decision procedure. This procedure decomposes the problem at hand in a divide-and-conquer manner and is guaranteed to find the smallest APH solution, if one exists.

In short, the contribution of this paper is fourfold: a) We formalize the problem of moment matching under structural constraints as a CCcSAT problem. b) We propose an approach for determining an APH distribution of smallest possible size solving a CCcSAT problem. c) We present a prototypical implementation in the tool STRUMPH that embeds SMT solving into a problem-specific iterative procedure. d) We report on empirical evidence regarding the practicality of the approach.

Organization of the Paper. In Sec. 2, we give related work and the mathematical background, after which we introduce our CCcSAT calculus in Sec. 3. Techniques to find constraint-satisfying variable assignments for CCcSAT expressions are discussed in Sec. 4. Our implementation STRUMPH is used for evaluation in Sec. 5. Finally, Sec. 6 concludes the paper and provides an outlook.

2   Preliminaries

Moments. For a (continuous) random variable X, characterized by probability density function f_X, the *expected value of X* (also called mean) is $E(X) =$

$\int_{\mathbb{R}} x\, f_X(x)\, dx$, and its k-*th moment* is $E(X^k)$. The *variance* of X is $V(X) = E((X - E(X))^2) = E(X^2) - E(X)^2$, and the *coefficient of variation* c_X is defined as $\sqrt{V(X)}\,/\,E(X)$.

Furthermore, if D is a probability distribution and X is a random variable distributed according to D, then $E[D]$ refers to $E(X)$, and $V[D]$ refers to $V(X)$. We say a distribution is *Dirac* if it assigns full probability to a single outcome.

Phase-Type Distributions. A *Phase-Type Distribution* [17], is a probability distribution modeled by the time from start to absorption in an absorbing *continuous-time Markov chain* (CTMC). In an absorbing CTMC, every state is either *absorbing*, meaning it has no outgoing transitions, or *transient*, in which case it can reach an absorbing state. Note that an *acyclic* CTMC is always absorbing.

Definition 1. *The* (continuous) n-phase phase-type distribution $PH(\vec{\alpha}, \boldsymbol{A})$ *is the distribution of the time-until-absorption in an* $(n+1)$-*state CTMC* $(S, \boldsymbol{R}, \vec{\pi})$ *with parameters*

$$\vec{\pi} = (\vec{\alpha}, 0)\,, \qquad \boldsymbol{R} = \begin{pmatrix} \boldsymbol{A} & \vec{A} \\ \vec{0} & 0 \end{pmatrix}$$

where all but the last state are transient.

The tuple $(\vec{\alpha}, \boldsymbol{A})$ is commonly referred to as the *phase-type representation* of the associated PH distribution. The parameter n is called the *order* of the representation. A PH distribution is *acyclic* if it is represented by an acyclic CTMC.

These distributions are mathematically well-understood and easy to analyze. Given an n-phase PH distribution $PH(\vec{\alpha}, \boldsymbol{A})$, we can compute its

- cumulative distribution function (CDF) with $F(t) = 1 - \vec{\alpha}\exp(t\,\boldsymbol{A})\,\vec{1}$,
- probability density function (PDF) with $f(t) = \vec{\alpha}\exp(t\,\boldsymbol{A})\,\vec{A}$,
- Laplace-Stieltjes transform (LST) with $\tilde{f}(s) = \vec{\alpha}_{n+1} + \vec{\alpha}(s\boldsymbol{I} - \boldsymbol{A})^{-1}\vec{A}$, and
- k-th moment with $E(X^k) = (-1)^k k!\,\vec{\alpha}\,\boldsymbol{A}^{-k}\vec{1}$ (see [18]).

Here $\exp(\cdot)$ refers to the matrix exponential. We remark that the column vector $\vec{A}$ can be computed from the $n \times n$-matrix $\boldsymbol{A}$ since every row in $\boldsymbol{R}$ must sum to zero. The set of (acyclic) phase-type distributions is closed under convolution, minimum, maximum, and mixture (aka convex combination) [18].

Cox & Convenience Calculus. To conveniently describe acyclic phase-type distributions (APH) in a structured manner, we recall the *Cox & Convenience Calculus* (Ccc) [20] – a simple stochastic process calculus, spanned by all expressions that adhere to the following grammar[1]

$$P, Q \in \mathcal{L} \quad ::= \quad (\lambda) \quad | \quad (\mu) \triangleleft (\lambda)P \quad | \quad P.Q \quad | \quad P+Q \quad | \quad P \parallel Q$$

---

[1] Listed here in order of operator precedence, from highest to lowest. The three binary operators are left-associative.

with rates $\lambda \in \mathbb{R}_+$ and $\mu \in \mathbb{R}_{\geq 0}$. Any CCC process P represents an acyclic PH-distribution, formally denoted by $PH(P)$. We refer to [20] for details of the semantic mapping $PH(\cdot)$. The three binary operators sequence (.), choice (+), and parallel ($\parallel$) correspond to the convolution, minimum, and maximum on PH distribution respectively, exploiting the closure properties discussed above. Intuitively, given processes P and Q,

- $P.Q$ terminates once Q terminates which is started only after P terminates.
- $P \parallel Q$ terminates once the *last* of P and Q terminates, with both P and Q started at the same time.
- $P + Q$ terminates once the *first* of P and Q terminates, with both P and Q started at the same time.

The remaining two operators are used to model more primitive distributions, where (λ) stands for a simple exponential distribution, while the operator $(\mu) \lhd (\lambda)P$ (called the "Cox"-operator) assures completeness. It allows one to express any possible Cox^2 distribution [6], and hence any possible APH distribution [20]. Practically speaking, the following distribution families can be encoded into CCC inductively, for $n > 0$. We will make use of their encodings in the sequel.

Exponential distribution: $Exp(\lambda) = (\lambda)$;
Erlang distribution: $Erl(n + 1, \lambda) = (\lambda).Erl(n, \lambda)$, and $Erl(1, \lambda) = (\lambda)$;
Hypoexponential distribution: $Hypo(\lambda_1, R) = (\lambda_1).Hypo(R)$ where R is of shape $\lambda_2, \ldots, \lambda_n$, and $Hypo(\lambda) = (\lambda)$;
Cox distribution: $Cx([\lambda_1, p_1], R) = ((1 - p_1)\lambda_1) \lhd (p_1\lambda_1).Cx(R)$ where R is of shape $[\lambda_2, p_2], \ldots, [\lambda_{n-1}, p_{n-1}], \lambda_n$, and $Cx(\lambda) = (\lambda)$.

Notably, choice (+) and parallel ($\parallel$) do not occur in the above encodings. They are added for modeling convenience as they echo typical usage scenarios.

Related Work. Moment matching, the problem of finding a PH distribution with a given set of moments, is well-explored. There have been too many contributions to list here, but some we would like to highlight.

Johnson and Taaffe investigated the use of mixtures of Erlang distributions of common order in the context of moment matching [13], showing among others that, except for corner cases, this class of phase-type distributions can be used to match the first k finite moments, for any choice of k. Asmussen et. al. developed an extended version of the expectation-maximization algorithm to approximate sampled data by phase-type distributions [1]. The paper [22] discusses minimal representations and moment matching methods for Markovian Arrival Processes (MAP), and [5] looks into similarity transformations and investigates representations maximizing the first joint moment of MAP. Building on those results, [4]

---

[2] A Cox distribution $PH(\vec{\alpha}, A)$ corresponds to an acyclic absorbing CTMC with bidiagonal A, Dirac α, and ascending (but negative) diagonal. Pictorially it can be seen as a sequence of exponential distributions (like hypo-exponential distributions) of decreasing rate with escape options to the absorbing state at each state. Any APH distribution can be represented as a Cox distribution of equal order.

presents a variation of the EM algorithm, for online estimation of the parameters of PH distributions. Finally, for a summary of existing phase-type approximation techniques, we would like to highlight the Bachelor's thesis of Komárková [15].

The above methods let us construct PH distributions that fulfill given moment constraints, but they do not allow for structural constraints to be taken into account. We believe to be the first to explore PH distributions with structural and moment constraints in this manner.

3   The Ccc Satisfiability Problem (CccSat)

In this section, we will formally define the problem statement of this paper. To do so, we will first introduce CccSat, a derivative of the Ccc process calculus, and define its semantics as well as a notion of satisfiability.

Syntax. Based on the Ccc language, we define the following syntax.

Definition 2. *Let $\mathcal{P}$ be a language defined by the following grammar*

$$\Psi, \Phi \in \mathcal{P} \quad ::= \quad X \quad | \quad \overline{\Psi}|_{E=c} \quad | \quad \overline{\Psi}|_{V=c} \quad | \quad \Psi.\Phi \quad | \quad \Psi + \Phi \quad | \quad \Psi \parallel \Phi$$

where $X \in \mathrm{Var}$ is a variable for some Ccc process, and $c \in \mathbb{R}_{\geq 0}$ is a constant.

We will refer to an element of $\mathcal{P}$ as a CccSat *problem instance* or a *problem* for short. While $\overline{\Psi}|_{E=c}$ will constrain the expected value of Ψ (thus $E[\Psi]$) to be c, $\overline{\Psi}|_{V=c}$ will constrain the variance $V[\Psi]$ to be c.[3]

Semantics. We define a constraint-oblivious semantics for CccSat problems, assuming an assignment Γ assigning to each variable a specific Ccc process.

Definition 3. *Let $\Gamma \in \mathrm{Var} \to \mathcal{L}$ be an assignment. For problem instance $\Psi \in \mathcal{P}$, we define $[\![\Psi]\!]_\Gamma$ as the Ccc process acquired by the following substitutions.*

- *Variables are substituted, i.e., $[\![X]\!]_\Gamma = \Gamma(X)$ for any $X \in \mathrm{Var}$.*
- *Constraint operators are dropped, i.e., $[\![\, \overline{\Psi}|_{...} \,]\!]_\Gamma = [\![\Psi]\!]_\Gamma$ for any $\Psi \in \mathcal{P}$.*
- *$[\![\Psi \circ \Phi]\!]_\Gamma = [\![\Psi]\!]_\Gamma \circ [\![\Phi]\!]_\Gamma$ for any $\Psi, \Phi \in \mathcal{P}$ and $\circ \in \{\,.\,, +, \parallel\,\}$.*

By defining the semantics of $\mathcal{P}$ using the function $[\![\cdot]\!]_\Gamma \in \mathcal{P} \to \mathcal{L}$, many statements that have been shown for Ccc processes and PH distributions are immediately applicable to $\mathcal{P}$. In particular, the semantics of the three binary operators sequence (.), choice (+), and parallel ($\parallel$) are inherited from Ccc.

---

[3] In all that follows, there will be no need to distinguish $\overline{\Psi}|_{E=c}|_{V=d}$ from $\overline{\Psi}|_{V=d}|_{E=c}$.

Satisfiability. The significance of the constraint operators that have been discarded so far manifests in the following notion of satisfiability.

Definition 4. *For a given problem $\Psi \in \mathcal{P}$ and assignment $\Gamma \in \mathrm{Var} \to \mathcal{L}$, we say Γ satisfies Ψ, written as $\Gamma \vDash \Psi$, if the following hold*

$$\frac{}{\Gamma \vDash X} \qquad \frac{\Gamma \vDash \Psi \quad E[\,PH([\![\Psi]\!]_\Gamma)\,] = c}{\Gamma \vDash \overline{\Psi}\big|_{E=c}}$$

$$\frac{\Gamma \vDash \Psi \quad \Gamma \vDash \Phi}{\Gamma \vDash \Psi \circ \Phi} \qquad \frac{\Gamma \vDash \Psi \quad V[\,PH([\![\Psi]\!]_\Gamma)\,] = c}{\Gamma \vDash \overline{\Psi}\big|_{V=c}}$$

with $X \in \mathrm{Var}$, problem instances $\Psi, \Phi \in \mathcal{P}$, and operators $\circ \in \{\,.\,, +, \parallel\,\}$.

Example 1. Consider the problem instance $\Psi \in \mathcal{P}$ defined as follows.

$$\Psi = \overline{X_1 + \overline{X_2}\big|_{E=2}\big|_{V=3}}\big|_{E=5} \parallel \overline{X_3}\big|_{E=4}.X_4$$

By the definition above, the statement $\Gamma \vDash \Psi$ expands to the following.

$$E[\,\Gamma(X_1) + \Gamma(X_2)\,] = 5 \quad \wedge \quad V[\,\Gamma(X_1) + \Gamma(X_2)\,] = 3$$
$$\wedge \quad E[\,\Gamma(X_2)\,] = 2 \quad \wedge \quad E[\,\Gamma(X_3)\,] = 4$$

Note that some variables, such as X_1 and X_2, are constrained multiple times, which will make solving for those variables harder. On the other hand, X_3 can be solved for independently, as it appears in just one constraint by itself. It is easy to see that X_4 may be chosen arbitrarily, as it is completely unconstrained.

The goal of this paper is to develop techniques for finding an assignment Γ for a given problem instance Ψ, such that $\Gamma \vDash \Psi$, if one exists. We will see later that our methods tend to find the simplest solution for a given problem instance. We call a problem $\Psi \in \mathcal{P}$ *satisfiable*, denoted by SAT(Ψ), if there is an assignment Γ such that $\Gamma \vDash \Psi$. Otherwise we call Ψ *unsatisfiable*, denoted by UNSAT(Ψ).

4   Semi-Algorithm for Solving CccSat

Now that we have arrived at a formal problem statement, let us walk through how we go about its solution, i.e., given a problem instance $\Psi \in \mathcal{P}$, we want to find $\Gamma \in \mathrm{Var} \to \mathcal{L}$, such that $\Gamma \vDash \Psi$. Our approach is split up into three stages: a) Immediately handle cases that can be solved analytically (Sec. 4.1). b) Divide the problem (recursively) into smaller problems that can be solved independently, if possible (Sec. 4.2). c) Solve the residual cases using an SMT solver (Sec. 4.3).

4.1 Satisfying Simple Cases

First, we will cover some cases that can be solved analytically. In what follows, we abuse notation and abbreviate the assignment that maps each variable on the same constant process P just as P. We call a problem instance $\Psi \in \mathcal{P}$ *unconstrained*, if Ψ does not contain any constraint operator ($\overline{\,\rceil}...$).

Theorem 1. *Let $\Psi \in \mathcal{P}$ be unconstrained. This implies*

1. $\Gamma \vDash \Psi$ *where Γ is arbitrary,*
2. $\Gamma \vDash \overline{\Psi}|_{E=c}$ *where $\Gamma = (\hat{c}/c)$ and $\hat{c} = E\big[[\![\Psi]\!]_{(1)}\big]$, and*
3. $\Gamma \vDash \overline{\Psi}|_{V=d}$ *where $\Gamma = (\sqrt{\hat{d}/d})$ and $\hat{d} = V\big[[\![\Psi]\!]_{(1)}\big]$.*

The first statement is obvious in light of how $\vDash$ is defined. For the remaining statements, it suffices to show that scaling all rates in a PH-representation scales the moments accordingly. For the first case, our implementation defaults to the constant assignment (1).

Remark 1. Special cases of Theorem 1 are the problems $\overline{X}|_{E=c}$ and $\overline{X}|_{V=d}$ respectively, where $X \in \text{Var}$. These are well known to be solvable using exponential distributions with rates $\lambda = 1/c$ and $\lambda = 1/\sqrt{d}$, which aligns with the above theorems (since $Exp(1)$ has a mean and variance of one).

The case $\overline{X}|_{E=c}|_{V=d}$ also has a well-known solution, which however we need to spend some thought on. Recall the squared coefficient of variation $c_X^2 = d\,/\,c^2$. For $c_X^2 \geq 0.5$, we can solve it with a 2-phase Coxian distribution $\text{Cx}([\lambda_1, p_1], \lambda_2)$ where $\lambda_1 = 2/c$, $p_1 = 1\,/\,2c_X^2$ and $\lambda_2 = \lambda_1 p_1$, due to [16]. This gives us the CCC process mapping $\Gamma(X) = ((1 - p_1)\lambda_1) \lhd (p_1\lambda_1)(\lambda_2)$. For $c_X^2 < 0.5$, we would obtain $p_1 > 1$ however, which would result in negative rates, so an alternative method is required.

A popular approach to create PH distributions with $0 < c_X \leq 1$ is using a mixture of two Erlang distributions with identical rates and sizes differing by one [24]. Yet, due to the absence of a mixture operator and of non-Dirac initial distributions in CCC, this solution cannot be expressed directly.

One way to handle this case is to instead prefix a 2-phase Coxian distribution with an Erlang distribution, i.e. using $\text{Erl}(k, \lambda).\text{Cx}([\lambda_1, p_1], \lambda_2)$. The idea is to use an Erlang distribution to offset the coefficient of variation required for the Coxian distribution, such that we can find it with the method outlined above. For the parameters k and λ of the Erlang distribution, we require that

$$\frac{d - \frac{k}{\lambda^2}}{\left(c - \frac{k}{\lambda}\right)^2} > \frac{1}{2}, \qquad d > \frac{k}{\lambda^2}, \qquad c > \frac{k}{\lambda}, \qquad 0 < \frac{d}{c^2} < \frac{1}{2}.$$

Due to the use of inequalities, there are infinitely many solutions. One possible solution is to fix $k = \left\lfloor \frac{c^2}{d} \right\rfloor$ and $\lambda = \frac{kc}{c^2 - 2d}$. The 2-phase Coxian distribution is constructed as described above with mean $c - k/\lambda$ and variance $d - k/\lambda^2$. These observations can be harvested by the following CCC constructions, which are notable for being fully described by their size k and up to 4 rate parameters.

Definition 5. *We define the class of* Tailweight *distributions as follows.*

$$Tw(\lambda_1) = (\lambda_1)$$
$$Tw([\lambda_3, \lambda_2], \lambda_1) = (\lambda_3) \lhd (\lambda_2)(\lambda_1)$$
$$Tw([k, \lambda_4], [\lambda_3, \lambda_2], \lambda_1) = Erl(k, \lambda_4).(\lambda_3) \lhd (\lambda_2)(\lambda_1)$$

The above recipe for problems of the form $\overline{X}|_{E=c}|_{V=d}$ (where $X \in$ Var and $c, d \in \mathbb{R}_+$) indeed always constructs a Tailweight distribution.

Corollary 1. $\forall c, d \in \mathbb{R}_+. \exists P \in Tailweight. E[P] = c \wedge V[P] = d.$

4.2   Problem Decomposition

Next, we discuss how we can break large problem instances down into multiple smaller problems that can be solved independently. Let $\circ$ denote any binary operator of $\mathcal{P}$, i.e., $\circ \in \{ \,.\, , +, \| \,\}$. For $\Psi \in \mathcal{P}$, we define $\mathrm{Vars}(\Psi)$ as the set of variables in Ψ. Further, we say $\Psi, \Phi \in \mathcal{P}$ are *disjoint*, if $\mathrm{Vars}(\Psi) \cap \mathrm{Vars}(\Phi) = \emptyset$.

Disjointness seems like a strong assumption to make, yet many problem instances we care about in practice use any variable only once. When modeling for instance a production chain, we want to represent each step in the process by a unique variable. The reuse of a variable asserts that two steps have exactly the same distribution, which is hard to guarantee in practice. Therefore, it is not uncommon for sub-expressions of a problem instance to be pairwise disjoint.

Theorem 2. $\forall \Psi, \Phi \in \mathcal{P}. \ disjoint(\Psi, \Phi) \wedge SAT(\Psi) \wedge SAT(\Phi) \to SAT(\Psi \circ \Phi)$

This holds since the function $\Gamma_\Psi \cup \Gamma_\Phi$ is well-formed due to disjointness, and solves $\Psi \circ \Phi$. Using the above theorem, we can decompose problem instances where the topmost operation is a binary operator and work on both operands independently, as long as they are disjoint. Note that this theorem can be applied recursively, and thus truly is the basis for a divide-and-conquer approach.

Theorem 3. *Using only Theorems 1 and 2, we can solve all* CccSat *problems containing neither nested constraints nor repeated variables.*

The above can be shown by structural induction on $\mathcal{P}$. Yet, Theorem 2 only allows us to decompose until we reach the first constraint operator. To deal with constraints, additional techniques are required.

Remark 2. To show that $\Psi \in \mathcal{P}$ is UNSAT, it suffices to show that any of its sub-expressions is UNSAT. This means while refuting Ψ, we are permitted to check all sub-expressions independently.

- $\forall \Psi, \Phi \in \mathcal{P}.\ \mathrm{UNSAT}(\Psi) \vee \mathrm{UNSAT}(\Phi) \to \mathrm{UNSAT}(\Psi \circ \Phi)$
- $\forall \Psi \in \mathcal{P}.\ \mathrm{UNSAT}(\Psi) \to \mathrm{UNSAT}(\overline{\Psi}|_{...})$

Constraint propagation through sequences. The following theorem allows us to rewrite specific problems involving the sequence operator between four forms.

Theorem 4. *For problems $\Psi, \Phi \in \mathcal{P}$, assignment Γ and property $p \in \{E, V\}$ the following statements are pairwise equivalent:*

$$\Gamma \vDash \overline{\overline{\Psi}|_{p=c}.\overline{\Phi}|_{p=d}}|_{p=c+d} \qquad \Gamma \vDash \overline{\overline{\Psi}|_{p=c}.\Phi}|_{p=c+d}$$

$$\Gamma \vDash \overline{\Psi.\overline{\Phi}|_{p=d}}|_{p=c+d} \qquad \Gamma \vDash \overline{\Psi}|_{p=c}.\overline{\Phi}|_{p=d}$$

This follows from the fact that both the mean and variance are cumulative. The last form gives us an unconstrained sequence operator at the top of the expression, which potentially allows us to decompose the problem further using Theorem 2. Once again, this theorem can be applied recursively to push constraints further into the problem.

Remark 3. Theorem 4 also yields some rejection criteria for CccSat problems. In cases where all three constraints around a sequence operator are given but do not add up as shown, the problem admits no solutions.

  If only two constraints are given, we may still reject the problem if we infer that the third constraint must be negative, since the mean and variance of PH distributions are always positive.

Example 2. Consider $\Psi = \overline{\overline{X_1.X_2}|_{E=5}}|_{E=3}$. Due to Theorem 4, $\Gamma \vDash \Psi$ is equivalent to $\Gamma \vDash \overline{X_1}|_{E=-2}.\overline{X_2}|_{E=5}$, which has no solution.

Choice and parallel are less well-behaved. If the remaining two operators behaved like the sequence operator (.), i.e., if a theorem similar to Theorem 4 could be shown for operators choice (+) and parallel ($\parallel$), we would be able to push all constraints to the innermost sub-expressions of a given CccSat problem. This means problems could be broken down recursively until only variable expressions with constraints remain, which can be easily solved using the methods discussed in Sec. 4.1. Alas, no such rule exists. We will outline why this is the case here.

Theorem 5. *For unknown $P, Q \in \mathcal{L}$, $E[P + Q]$, $V[P + Q]$, $E[P \parallel Q]$ and $V[P \parallel Q]$ are not fully determined by $E[P]$, $V[P]$, $E[Q]$ and $V[Q]$.*

Counterexamples are easy to find, for instance by setting Q to (2) and comparing $P = (1).(1/3)$ and $P' = (1/10) \lhd (2/5)(2/5)$. Here $E[P] = E[P']$, and $V[P] = V[P']$ hold, but the compound expressions appearing in the theorem give different values if P is replaced by P'. As a result, we will need to treat the moments of choice (+) and parallel ($\parallel$) as completely opaque in this paper.

Remark 4. The above is equivalent to showing that equality of the first two moments does not yield a congruence relation on Ccc.

  PH-equivalence ($\approx_{PH}$) on the other hand is a congruence relation for all operators of Ccc [19] and defined as $\forall P, Q \in \mathcal{L}. PH(P) = PH(Q) \to P \approx_{PH} Q$. Using [25], two Ccc processes are PH-equivalent if and only if their first $2n$ moments agree, where n is the order of the larger PH representation.

4.3   Reduction to SMT Instances

We now present a general procedure for solving a CCCSAT problem $\Psi \in \mathcal{P}$ based on iteration over *template assignments* $\Gamma_\Lambda \in \text{Var} \to \mathcal{L}[\Lambda]$, i.e., assignments with rate parameters $\Lambda \in (\mathbb{R}_{\geq 0})^d$ for a template with d parameters to fill in. For each such template assignment, we present an encoding of the statement $\Gamma_\Lambda \vDash \Psi$ as a system of equations to be solved for Λ by an SMT solver. We show that this procedure is exhaustive, i.e., guaranteed to find a solution Γ for Ψ, if one exists, and that it finds the smallest solution for each variable.

Enumerating assignments. To iterate over template assignments, we need a function $\gamma \in \mathbb{N} \to \text{Var} \to \mathcal{L}[\Lambda]$ producing a template assignment $\Gamma_\Lambda \in \text{Var} \to \mathcal{L}[\Lambda]$ in each step $i \in \mathbb{N}$. We first construct *template generators* $\tau \in \mathbb{N}_+ \to \mathcal{L}[\Lambda]$ and then discuss how to use them to iterate over template assignments.

For templates, we make use of Coxian distributions because they are canonical forms for general APH distributions, i.e., every APH distribution can be transformed into a Coxian distribution while preserving its CDF [7]. This allows us to restrict our search to the set of Coxian distributions without missing out on possible solutions which guarantees exhaustiveness.

Definition 6. *We define the Coxian template generator* $\tau_{\text{Cx}} \in \mathbb{N}_+ \to \mathcal{L}[\Lambda]$ *by*

$$\tau_{\text{Cx}}(1) = (\lambda_1), \qquad \tau_{\text{Cx}}(j+1) = (\lambda_{2j+1}) \lhd (\lambda_{2j}) \, \tau_{\text{Cx}}(j).$$

While τ_{Cx} covers all APH distributions, the resulting templates $\tau_{\text{Cx}}(j)$ have $2j-1$ rate parameters λ_1 to λ_{2j-1}, so here $\Lambda \in (\mathbb{R}_{\geq 0})^{2j-1}$. Our experiments (Sec. 5) show that the number of rate parameters has a significant impact on SMT solving time. Hence, we also define the following alternative template generators with fewer parameters. In contrast to the Coxian generator, using them does, however, not guarantee exhaustiveness.

Definition 7. *We define the Erlang* (τ_{Erl}), *the Hypoexponential* (τ_{Hypo}), *and the Tailweight* (τ_{Tw}) *template generators as follows.*

$$\tau_{\text{Erl}}(1) = (\lambda_1), \quad \tau_{\text{Erl}}(j+1) = (\lambda_1).\tau_{\text{Erl}}(j)$$
$$\tau_{\text{Hypo}}(1) = (\lambda_1), \quad \tau_{\text{Hypo}}(j+1) = (\lambda_{j+1}).\tau_{\text{Hypo}}(j)$$
$$\tau_{\text{Tw}}(1) = (\lambda_1), \quad \tau_{\text{Tw}}(2) = (\lambda_3) \lhd (\lambda_2)(\lambda_1), \quad \tau_{\text{Tw}}(j+2) = (\lambda_4).\tau_{\text{Tw}}(j+1)$$

To lift a template generator τ to a function $\gamma \in \mathbb{N} \to \text{Var} \to \mathcal{L}[\Lambda]$ for generating template assignments, we use linear search thereby starting with the smallest assignment and working our way upward. Note that for some of the template generators defined above, there are distributions that can only be represented using a specific order, i.e. parameter j. Thus, we must consider all possible combinations of template orders for each variable. Further, we want to choose the sequence of template assignments such that we find the smallest solution for each variable first. To this end, we define the following auxiliary function f.

$$f(1, s) = \{(s)\}, \qquad f(N, s) = \bigcup_{x=0}^{s} \{(x, t) \mid t \in f(N-1, s-x)\}$$

Here, $f(N, s)$ is the set of all tuples $\mathbb{N}^N$ whose elements sum up to $s \in \mathbb{N}$. Now, we take the tuples from $f(|\text{Vars}(\Psi)|, s)$ for increasing values of s, obtaining a sequence of tuples with monotonically increasing sums. Finally, by mapping every number x in each of the tuple to $\tau(x + 1)$, we get our enumeration of template assignments $\gamma \in \mathbb{N} \rightarrow \text{Var} \rightarrow \mathcal{L}[\Lambda]$.

This algorithm is guaranteed to find the smallest APH distribution for each variable, but as the set of templates we need to search is countably infinite, the algorithm diverges if Ψ is UNSAT.

Remark 5. Rather than minimizing the order of templates, i.e., the order of the CCC processes substituted for the variables, one may be interested in the assignment that produces the smallest process once all variables are substituted. Using [18], the order of CCC processes can be computed recursively.

$$\text{ord}(P.Q) = \text{ord}(P) + \text{ord}(Q), \qquad \text{ord}(P + Q) = \text{ord}(P) \cdot \text{ord}(Q),$$
$$\text{ord}(P \parallel Q) = \text{ord}(P) \cdot \text{ord}(Q) + \text{ord}(P) + \text{ord}(Q)$$

Finding the next tuple in the sequence then amounts to minimizing a multivariate polynomial over the natural numbers while not repeating solutions.

SMT encoding. So far, we defined a linear search function $\gamma \in \mathbb{N} \rightarrow \text{Var} \rightarrow \mathcal{L}[\Lambda]$ which gives us a template assignment $\Gamma_\Lambda = \gamma(i)$ for every step i in our search. Now, assuming $\Gamma_\Lambda \in \text{Var} \rightarrow \mathcal{L}[\Lambda]$ is given, we need to encode $\Gamma_\Lambda \models \Psi$ into a system of equations, which can be solved using an SMT solver. As described in Sec. 3, this can be done by generating an equation for every occurrence of the constraint operator ($\overline{\cdot|}...$) where the left-hand side is the computed moment of $[\![\Psi]\!]_{\Gamma_\Lambda}$ and the right-hand side is a constant given by the constraint. The final SMT encoding is the conjunction of all generated equations.

To compute the k-th moment of a random variable $X \sim PH(\vec{\alpha}, A)$, we can use the equation $E(X^k) = (-1)^k k! \vec{\alpha} A^{-k} \vec{1}$ mentioned in Sec. 2. As remarked in [25], we can avoid the matrix inverse by computing the first k moments iteratively. First we solve for $\vec{\beta}_1$ and continue by inductively solving for $\vec{\beta}_k$ in

$$\vec{\beta}_1 A = -\vec{\alpha}, \qquad \vec{\beta}_{k+1} A = -(k+1)\vec{\beta}_k.$$

In every step, we can compute the k-th moment $E(X^k) = \vec{\beta}_k \vec{1}$. Since the PH-representations constructed from CCC processes are acyclic, A is an upper-triangular matrix. This means if A is of order n, each $\vec{\beta}_k$ can be computed using backward substitution in $\mathcal{O}(n^2)$.

5 Empirical Evaluation

We have implemented the methods discussed in Sec. 4 in a prototypical solver for CCCSAT problems, called STRUMPH (Structural Matching of PH-distributions). The problems we consider are more general than those that can be attacked by moment matching or phase-type fitting methods (unless constraints are present on the outermost level only). Therefore no meaningful comparison with existing tools is possible. This section provides empirical observations of the solution of STRUMPH on a selection of example problems.

Overview. STRUMPH is a CLI application written in Rust, which takes a problem instance either in the CCCSAT syntax or a simplified JSON format and returns a valid assignment to stdout, and stores it as a JSON dictionary. We use the rsmt2 library to interface with any solver that conforms to the SMT-LIB v2 standard.

The benchmarks in this section have been performed on a desktop PC with an Intel Core i7-6700K CPU at 4 GHz and 24 GB of RAM running Windows 10. We use the SMT solver z3 [8] as a backend. To measure execution times of our solver, we use the Measure-Command utility that comes with Powershell. Executions taking longer than 1 h were considered as timeout.

Decomposable Cases. The easiest CCCSAT problems for our solver are those where constraints do not overlap, i.e., every subexpression is part of at most one constraint. In this case, our implementation can always decompose the problem until each subproblem only has one constraint, which makes it possible to use moment matching for each of them.

Example 3. The following problem can be solved by decomposing it into three smaller problems, which are solved independently using moment matching (see Theorem 1). Our implementation solves this in about 15 ms.

$$\overline{X_1.X_2}\big|_{E=5} \parallel \overline{X_3}\big|_{V=3}.X_4$$

If a problem does not have any constraints, our implementation outputs the assignment $\Gamma = (1)$ immediately.

Large sequential problems. In the subset of CCCSAT problems constructed without choice $(+)$ and parallel $(\parallel)$, the solver performs well even on larger problem instances. For problems with many moment constraints, STRUMPH quickly manages to propagate constraints to the innermost subexpressions which are then solved by moment matching.

Example 4. Our implementation can solve problem Ψ below only using constraint propagation and moment matching in about 103 ms.

$$\Psi_1 = \overline{\overline{\overline{X_1}\big|_{E=2}\big|_{V=3/2}.X_2}\big|_{E=4}\big|_{V=3}}$$

$$\Psi_2 = \overline{\overline{\overline{X_3}\big|_{E=1/2}\big|_{V=1}.X_4}\big|_{E=5}\big|_{V=2}.\overline{X_5}\big|_{E=2}\big|_{V=2}.X_6}\big|_{E=10}\big|_{V=5}$$

$$\Psi_3 = \overline{X_7}\big|_{E=1/2}\big|_{V=1}.X_8$$

$$\Psi = \overline{\Psi_1.\Psi_2.\Psi_3}\big|_{E=16}\big|_{V=10}$$

If constraints are too sparse, STRUMPH must resort to SMT solving, but the SMT instances generated in these cases are easily solved.

Example 5. Consider the following CCCSAT problem.

$$\Psi = \overline{X_1.\overline{X_2.X_3.X_4.X_5.X_6.X_7}\big|_{V=3}.X_8}\big|_{E=5}$$

When we lower $\Gamma \vDash \Psi$ into an SMT problem with $\Gamma(X_k) = (\lambda_k)$, we get the following system of equations.

$$\frac{1}{\lambda_1} + \frac{1}{\lambda_2} + \frac{1}{\lambda_3} + \frac{1}{\lambda_4} + \frac{1}{\lambda_5} + \frac{1}{\lambda_6} + \frac{1}{\lambda_7} + \frac{1}{\lambda_8} = 5,$$

$$\frac{1}{\lambda_2^2} + \frac{1}{\lambda_3^2} + \frac{1}{\lambda_4^2} + \frac{1}{\lambda_5^2} + \frac{1}{\lambda_6^2} + \frac{1}{\lambda_7^2} = 3,$$

Using z3, we find the following solution (consistently in roughly 58 ms).

$$\lambda_1 = \lambda_4 = \lambda_5 = \lambda_6 = \lambda_7 = \lambda_8 = 2, \qquad \lambda_2 = \lambda_3 = 1$$

Challenging constraints in sequences. We now turn to synthetic cases that are meant to pose challenges to the solution engine. We consider the following example parametric in n.

$$\Psi_n = \overline{X_2. \overline{X_1}\big|_{E=n}\big|_{V=n}.X_2}\big|_{E=n+1}$$

By construction we know that the smallest solution for Ψ_n is given by $\Gamma(X_1) = Erl(n,1)$ and $\Gamma(X_2) = (2)$. The outer constraint and variable X_2 however prevent STRUMPH from decomposing the problem[4] and solving it within a few milliseconds. Table 1 shows solution times for Ψ_n using different templates. As

Table 1. Solution times for varying sizes n measured in *ms* or timeout (-).

	Ψ_n									Ψ'_n					Ψ''_n		
n	1	2	3	4	5	6	7	8	9	1	2	3	4	5	1	2	3
Erl	39	75	117	185	372	534	639	812	962	39	87	130	4545	-	42	148	-
Hypo	41	74	138	268	291	690	2400	-	-	45	72	121	4436	-	40	271	-
Tail	42	66	147	323	462	-	-	-	-	37	89	126	-	-	41	158	-
Cox	54	73	344	-	-	-	-	-	-	38	74	464	-	-	47	169	-

we can see, the number of parameters as well as the size of the templates has a significant impact on the execution times of our solver.

We also notice that our solver gets stuck for over an hour on certain problems, even though it can solve just slightly smaller problems within a few seconds. After further investigation, we find the z3 SMT solver is quick to reject incorrect template assignments, but stalls while trying to compute the exact rates for the correct solution. We believe this is due to the SMT solver internally choosing the wrong strategy for those cases, yet we had little success figuring out what causes this behavior or how to prevent it.

---

[4] Specifically, decomposition fails here because the occurrences of variable X_2 makes the subexpressions of the sequence non-disjoint.

Choice & Parallel. To investigate the performance of the other operators, we repeat the above experiment with two slightly modified versions of Ψ_n:

$$\Psi'_n = \overline{\overline{X_1}\big|_{E=n}\big|_{V=n} + X_2}\Big|_{E=n/2} \qquad \Psi''_n = \overline{\overline{X_1}\big|_{E=n}\big|_{V=n} \parallel X_2}\Big|_{E=n+1}$$

As above, the problem is deliberately constructed as a challenge for the solver. It must be solved using templates of size at least n, since choice ($+$) and parallel ($\parallel$) do not allow for constraints to be propagated. The smallest solution for X_1 is $Erl(n,1)$. Table 1 displays the solution times for both Ψ'_n and Ψ''_n. In both cases, our implementation reaches the timeout a lot sooner than in the previous example for most templates. Since choice and parallel quickly result in CTMCs of higher order, the equations for both mean and variance quickly become more complex and harder for the SMT solver to handle. We expect that a variety of tailored strategies can be devised to overcome this. Notably the issue is entirely absent if parallel processes are decomposable, enabling divide-and-conquer.

6   Conclusion

In this paper, we discussed the problem of finding a PH distribution that fulfils structural and moment constraints. To formalize the problem, we introduced CCCSAT, a derivative of the CCC process calculus. We discussed how to decompose CCCSAT problems into smaller problems, how to transform problems into one another, and how to apply existing moment matching methods to solve base cases. This discussion has culminated in the creation of a general semi-decision procedure, which is guaranteed to find the smallest acyclic solution to any given problem, if one exists. We presented STRUMPH, a prototypical implementation of this procedure, and studied its performance on challenging cases.

Future Work. Programs with structural concurrency, such as the one shown in Lst. 1, can be translated into CCCSAT in a very literal and direct way. We believe it is possible for this translation to be performed completely automatically, and this could also interface with benchmarking utilities to infer moment constraints. This is especially interesting in the context of Rust, due to its rich type system and macro support.

The methods for solving CCCSAT problems discussed in this paper assume moment constraints must be matched exactly. Yet in reality, these constraints are likely acquired via real-world measurements, which are notoriously inexact. This leaves room for follow-up work, for instance on a CCCSAT solver which emphasizes solution size and speed over solution accuracy.

Acknowledgements. This work has received support by the Deutsche Forschungsgemeinschaft (DFG, German Research Foundation) – project number 389792660 – TRR 248 – CPEC, see https://perspicuous-computing.science.

References

1. Asmussen, S., Nerman, O., Olsson, M.: Fitting phase-type distributions via the em algorithm. Scand. J. Stat. **23**(4), 419–441 (1996). http://www.jstor.org/stable/4616418

2. Bobbio, A., Horváth, A., Telek, M.: Matching three moments with minimal acyclic phase type distributions. Stoch. Model. **21**(2–3), 303–326 (2005). https://doi.org/10.1081/STM-200056210

3. Bolch, G., Greiner, S., de Meer, H., Trivedi, K.S.: Steady-state solutions of markov chains, chap. 3, pp. 103–151. John Wiley and Sons, Ltd (1998). https://doi.org/10.1002/0471200581.ch3

4. Buchholz, P., Dohndorf, I., Kriege, J.: An online approach to estimate parameters of phase-type distributions. In: 2019 49th Annual IEEE/IFIP International Conference on Dependable Systems and Networks (DSN), pp. 100–111 (2019). https://doi.org/10.1109/DSN.2019.00024

5. Buchholz, P., Felko, I., Kriege, J.: Transformation of acyclic phase type distributions for correlation fitting. In: Dudin, A., De Turck, K. (eds.) ASMTA 2013. LNCS, vol. 7984, pp. 96–111. Springer, Heidelberg (2013). https://doi.org/10.1007/978-3-642-39408-9_8

6. Cox, D.R.: A use of complex probabilities in the theory of stochastic processes. Math. Proc. Cambridge Philos. Soc. **51**(2), 313–319 (1955). https://doi.org/10.1017/S0305004100030231

7. Cumani, A.: On the canonical representation of homogeneous Markov processes modelling failure-time distributions. Microelectron. Reliab. **22**(3), 583–602 (1982). https://doi.org/10.1016/0026-2714(82)90033-6

8. de Moura, L., Bjørner, N.: Z3: an efficient SMT solver. In: Ramakrishnan, C.R., Rehof, J. (eds.) TACAS 2008. LNCS, vol. 4963, pp. 337–340. Springer, Heidelberg (2008). https://doi.org/10.1007/978-3-540-78800-3_24

9. Feldmann, A., Whitt, W.: Fitting mixtures of exponentials to long-tail distributions to analyze network. Perform. Eval. **31**(3–4), 245–279 (1998). https://doi.org/10.1016/S0166-5316(97)00003-5

10. Horváth, A., Telek, M.: PhFit: a general phase-type fitting tool. In: Field, T., Harrison, P.G., Bradley, J., Harder, U. (eds.) TOOLS 2002. LNCS, vol. 2324, pp. 82–91. Springer, Heidelberg (2002). https://doi.org/10.1007/3-540-46029-2_5

11. Horváth, A., Telek, M.: Matching more than three moments with acyclic phase type distributions. Stoch. Model. **23**(2), 167–194 (2007). https://doi.org/10.1080/15326340701300712

12. Horváth, G.: Moment matching-based distribution fitting with generalized hyper-erlang distributions. In: Dudin, A., De Turck, K. (eds.) ASMTA 2013. LNCS, vol. 7984, pp. 232–246. Springer, Heidelberg (2013). https://doi.org/10.1007/978-3-642-39408-9_17

13. Johnson, M.A., Taaffe, M.R.: Matching moments to phase distributions: mixtures of erlang distributions of common order. Commun. Stat. Stoch. Models **5**(4), 711–743 (1989). https://doi.org/10.1080/15326348908807131

14. Khayari, R.E.A., Sadre, R., Haverkort, B.R.: Fitting world-wide web request traces with the EM-algorithm. Perform. Eval. **52**(2–3), 175–191 (2003). https://doi.org/10.1016/S0166-5316(02)00179-7

15. Komárková, Z.: Phase-type approximation techniques (2012). https://is.muni.cz/th/ysfsq/thesis.pdf. Bachelor Thesis, Masaryk University, Faculty of Informatics

16. Marie, R.A.: Calculating equilibrium probabilities for $\lambda(n)/ck/1/n$ queues. In: Proceedings of the 1980 International Symposium on Computer Performance Modelling, Measurement and Evaluation, PERFORMANCE '80, pp. 117–125. Association for Computing Machinery, New York (1980). https://doi.org/10.1145/800199. 806155

17. Neuts, M.F.: Probability distributions of phase type. Liber Amicorum Prof. Emeritus H. Florin (1975)

18. Neuts, M.F.: Matrix-Geometric Solutions in Stochastic Models: An Algorithmic Approach. The Johns Hopkins University Press, Baltimore (1981). https://doi. org/10.1002/net.3230130219

19. Pulungan, M.R.: Reduction of acyclic phase-type representations. Ph.D. thesis, Saarland University (2009). https://doi.org/10.22028/D291-25951

20. Pulungan, R., Hermanns, H.: A construction and minimization service for continuous probability distributions. Int. J. Softw. Tools Technol. Transf. **17**(1), 77–90 (2015). https://doi.org/10.1007/s10009-013-0296-8

21. Reinecke, P., Krauß, T., Wolter, K.: Phase-type fitting using HyperStar. In: Balsamo, M.S., Knottenbelt, W.J., Marin, A. (eds.) EPEW 2013. LNCS, vol. 8168, pp. 164–175. Springer, Heidelberg (2013). https://doi.org/10.1007/978-3-642-40725-3_13

22. Telek, M., Horváth, G.: A minimal representation of Markov arrival processes and a moments matching method. Perform. Eval. **64**(9–12), 1153–1168 (2007). https:// doi.org/10.1016/j.peva.2007.06.001

23. Thümmler, A., Buchholz, P., Telek, M.: A novel approach for phase-type fitting with the EM algorithm. IEEE Trans. Dependable Secure Comput. **3**(3), 245–258 (2006). https://doi.org/10.1109/TDSC.2006.27

24. Tijms, H.C.: Stochastic Models?: An Algorithmic Approach. Wiley, New York (1994)

25. Wolf, V.: Equivalences on phase type processes. Ph.D. thesis, University of Mannheim (2008). https://madoc.bib.uni-mannheim.de/1911/

Comparing Two Approaches to Include Stochasticity in Hybrid Automata

Lisa Willemsen[1]([✉]) [ID], Anne Remke[1,2] [ID], and Erika Ábrahám[3] [ID]

[1] University of Twente, Enschede, The Netherlands
l.c.willemsen@utwente.nl
[2] University of Münster, Münster, Germany
anne.remke@uni-muenster.de
[3] RWTH Aachen University, Aachen, Germany
abraham@cs.rwth-aachen.de

Abstract. Different stochastic extensions of hybrid automata have been proposed in the past, with unclear expressivity relations between them. To structure and relate these modeling languages, in this paper we formalize two alternative approaches to extend hybrid automata with stochastic choices of discrete events and their time points. The first approach, which we call decomposed scheduling, adds stochasticity via stochastic races, choosing random time points for the possible discrete events and executing a winner with an earliest time. In contrast, composed scheduling first samples the time point of the next event and then the event to be executed at the sampled time point. We relate the two approaches regarding their expressivity and categorize available stochastic extensions of hybrid automata from the literature.

Keywords: Formal Modelling · Stochastic Hybrid Models · Classification · Expressivity

1 Introduction

Hybrid automata (HA) [11] are well-suited to model the interplay of continuous and discrete behavior. Hybrid automata naturally exhibit non-determinism, e.g., *discrete non-determinism* via multiple simultaneously enabled discrete events (so-called jumps), or *time non-determinism* via time evolution with non-deterministic duration. In this paper we focus on these aspects and assume that the initial state, successor states of jumps, as well as the continuous evolution of the system state during time elapse are deterministic.

During the execution of a hybrid automaton, every non-deterministic choice has to be resolved by a scheduler. Hybrid automata have been extended with stochastic choices in multiple ways, leading to *stochastic hybrid models (SHM)* [3,15] with different features and expressivity. In most existing formalisms on SHM, all decisions are made randomly and they completely replace the non-determinism present in the underlying HA.

This work is supported by the DFG grant 471367371.

For example discrete probability distributions have been added to decidable subclasses of hybrid automata [21,22], which can be analyzed e.g. with the SISAT tool using abstraction [23]. Another possible extension are stochastic delays via stochastic resets or random clocks [6,16], which are sampled from a continuous distribution to determine how much time should pass between consecutive discrete jumps. In the PROHVER [7] framework, stochastic resets are abstracted to non-deterministic probabilistic resets. Another over-approximating approach [9] discretizes the support of the random variables and then abstracts to Markov decision processes. Some of these formalisms model stochastic choices by stochastic kernels [3,15], each of them responsible for random decisions of a certain kind; we refer to this technique as *composed scheduling*. Others use several stochastically independent random decisions that are in "race" [6,16]; we call this approach *decomposed scheduling*. Due to these differences and diverging mathematical notation, it is often hard to compare existing formalisms with respect to their expressivity.

Contribution. (i) In this paper, we formalize two stochastic HA extensions, that implement composed respectively decomposed scheduling. (ii) We define the stochastic processes induced by each of the approaches. (iii) We relate the expressivity of the two approaches and show that composed scheduling is more expressive than decomposed scheduling. (iv) We discuss how existing formalisms implement such stochastic choices and relate different lines of work.

Outline. Section 2 introduces HA and the necessary stochastic notation. Section 3 formalizes and relates the two HA extensions. Section 4 discusses related work and classifies existing formalisms. Section 5 concludes the paper.

2   Fundamentals

Let $\mathbb{R}$ denote the set of all real numbers, $\mathbb{R}_{\geq 0}$ the nonnegative reals, $\mathbb{N}$ the natural numbers (including zero) and $\mathbb{Z}$ the integers. For a set S, 2^S is the set of all subsets of S. We start with introducing hybrid automata in Sect. 2.1 and recall some basic definitions from probability theory in Sect. 2.2.

2.1   Hybrid Automata

Hybrid automata extend discrete transition systems with the notion of time and continuous evolution. In the below standard definition [10] we omit modeling constructs for parallel composition, as they are not central for this work.

Definition 1 (Hybrid automata: Syntax). *A hybrid automaton (HA) is a tuple $\mathcal{H} = (Loc, Var, Flow, Inv, Edge, Init)$ with the following components:*

- *Loc is a non-empty finite set of* locations *or* control modes.
- *$Var = \{x_1, \ldots, x_d\}$ is a finite ordered set of* variables. *We call d the* dimension, *$\nu \in \mathbb{R}^d$ a* valuation, *and $\sigma = (\ell, \nu) \in Loc \times \mathbb{R}^d = \Sigma$ a* state *of $\mathcal{H}$.*

- *Flow : Loc → ($\mathbb{R}^d$ → $\mathbb{R}^d$) specifies for each location its* flow *or* dynamics.
- *Inv : Loc → $2^{\mathbb{R}^d}$ specifies an* invariant *for each location. We define $\Sigma_{Inv} = \{(\ell, \nu) \in \Sigma \,|\, \nu \in Inv(\ell)\}$.*
- *Edge ⊆ Loc × $2^{\mathbb{R}^d}$ × ($\mathbb{R}^d$ → $\mathbb{R}^d$) × Loc is a finite set of discrete transitions or jumps. For a jump $(\ell_1, g, r, \ell_2) \in Edge$, ℓ_1 and ℓ_2 are its source resp. target locations, g its* guard, *and r its* reset.
- *Init : Loc → $2^{\mathbb{R}^d}$ defines initial valuations. We call a state $(\ell, \nu) \in \Sigma$* initial *if $\nu \in Inv(\ell) \cap Init(\ell)$.*

Executions of a hybrid automaton evolve an initial state by time steps and discrete steps. *Time steps (flows)* model continuous evolution of the variable values according to the flow condition of the current location, while satisfying the current location's invariant. When flows define constant derivatives for all variables then we talk about *linear behavior,* for linear predicates (i.e., linear differential equations) about *linear dynamics,* and in the case of more expressive predicates (involving e.g. polynomials or trigonometric functions) about *nonlinear dynamics. Discrete steps (jumps)* $(\ell, g, r, \ell') \in Edge$ can move the control from location ℓ to ℓ' and change the valuation from $\nu \in \mathbb{R}^d$ to $r(\nu) \in \mathbb{R}^d$, assuming that the jump is *enabled* in (ℓ, ν), i.e. $\nu \in g$ and $r(\nu) \in Inv(\ell')$.

Definition 2 (Hybrid automata: Semantics). *For a hybrid automaton $\mathcal{H} = (Loc, Var, Flow, Inv, Edge, Init)$ of dimension d, its operational semantics is given by the following rules:*

$$\frac{\begin{array}{c} \ell \in Loc \quad \nu, \nu' \in \mathbb{R}^d \quad t \in \mathbb{R}_{\geq 0} \quad f : [0, t] \to \mathbb{R}^d \quad df/dt = \dot{f} : (0, t) \to \mathbb{R}^d \\ f(0) = \nu \quad f(t) = \nu' \quad \forall t' \in (0, t). \; \dot{f}(t') = Flow(\ell)(f(t')) \\ \forall t' \in [0, t]. \; f(t') \in Inv(\ell) \end{array}}{(\ell, \nu) \xrightarrow{t} (\ell, \nu')} \; Flow$$

$$\frac{e = (\ell, g, r, \ell') \in Edge \quad \nu, \nu' \in \mathbb{R}^d \quad \nu \in g \quad \nu' = r(\nu) \quad \nu' \in Inv(\ell')}{(\ell, \nu) \xrightarrow{e} (\ell', \nu')} \; Jump$$

Let $\mathcal{H} = (Loc, Var, Flow, Inv, Edge, Init)$ be a HA of dimension d. A *path* of $\mathcal{H}$ is a finite or infinite sequence $\pi = \sigma_0 \xrightarrow{t_0} \sigma_0' \xrightarrow{e_0} \sigma_1 \xrightarrow{t_1} \ldots$ of alternating time steps and jumps with $\sigma_0 \in \Sigma_{Inv}$; π is said to be *initial* if σ_0 is initial, we define its *length* $len(\pi)$ as the number of jumps in it, and its *duration* $dur(\pi)$ as the sum of the durations of all of its time steps. Let $\Pi(\sigma)$ and $\Pi_{fin}(\sigma)$ be the set of all infinite resp. finite paths starting in $\sigma \in \Sigma_{Inv}$. A state $\sigma \in \Sigma$ of $\mathcal{H}$ is *reachable* iff there is an initial path leading to it.

An infinite path is *time-convergent* if its duration is finite, and *time-divergent* otherwise. A state of $\mathcal{H}$ has a *timelock* iff all infinite paths starting in it are time-convergent; $\mathcal{H}$ has a *timelock* if any of its reachable states has a timelock, and is *timelock-free* otherwise. An infinite path is *Zeno* iff it is time-convergent and contains infinitely many jumps. $\mathcal{H}$ is *Zeno-free* if it has no initial Zeno path.

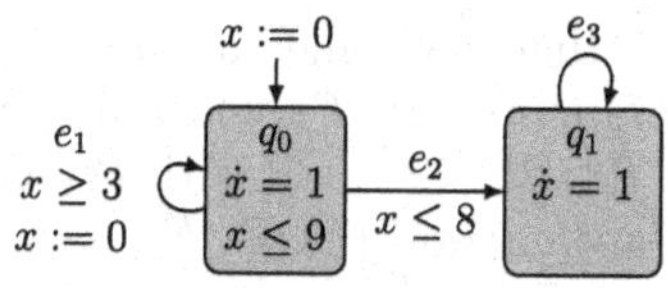

$$T(\sigma, e_1) = [3,9] \quad T(\sigma, e_2) = [0,8] \quad T(\sigma, e_3) = \emptyset$$
$$T(\sigma) = [0,9] \qquad t_{max}(\sigma) = 9$$
$$E_\Sigma(\sigma) = \{e_2\} \qquad E_{Loc}(\sigma) = E_{Loc}(q_0) = \{e_1, e_2\}.$$

Fig. 1. A hybrid automaton and characteristics of its state $\sigma = (q_0, \nu)$, $\nu(x) = 0$.

We discuss how choices over jumps and the length of time steps are made stochastically in Sect. 3, where we use the following notions, partly generalized to hybrid automata from [3] and illustrated in Fig. 1. For $e \in Edge$ and $\sigma \in \Sigma_{Inv}$, we define

$$T(\sigma, e) = \{t \in \mathbb{R}_{\geq 0} \mid \exists \sigma', \sigma'' \in \Sigma_{Inv}.\ \sigma \xrightarrow{t} \sigma' \wedge \sigma' \xrightarrow{e} \sigma''\}, \quad T(\sigma) = \bigcup_{e \in Edge} T(\sigma, e),$$
$$t_{max}(\sigma) = \sup\{t \in \mathbb{R} \mid \exists \sigma' \in \Sigma_{Inv}.\ \sigma \xrightarrow{t} \sigma'\},$$
$$E_\Sigma(\sigma) = \{e \in Edge \mid \exists \sigma' \in \Sigma_{Inv}.\ \sigma \xrightarrow{e} \sigma'\},$$
$$E_{Loc}(\ell) = \bigcup_{\nu \in V} E_\Sigma((\ell, \nu)), \qquad\qquad\qquad E_{Loc}((\ell, \nu)) = E_{Loc}(\ell).$$

We call σ *jump-enabled*[1] iff $T(\sigma) \neq \emptyset$, and *immediate-jump-enabled* iff $E_\Sigma(\sigma) \neq \emptyset$.

2.2 Basic Stochastic Notions

A random *experiment* has an uncertain outcome from a *sample space* Ω, whose subsets are called *events*. A σ-*algebra* $\mathcal{F}$ is a set of events containing the maximal event Ω and being closed under complement and countable union. The standard Borel σ-algebra $\mathcal{B}(\Omega)$ is the smallest σ-algebra containing all open events. An event is $\mathcal{F}$-*measurable* if it is in $\mathcal{F}$. The pair $(\Omega, \mathcal{F})$ is called a *measurable space*.

Given $(\Omega, \mathcal{F})$, a *probability measure* is a function $Pr : \mathcal{F} \rightarrow [0,1] \subseteq \mathbb{R}$ with (i) $Pr(\Omega) = 1$, (ii) $Pr(E) = 1 - Pr(\bar{E})$ for all $E \in \mathcal{F}$ and (iii) $Pr(\bigcup_{i=0}^{\infty} E_i) = \Sigma_{i=0}^{\infty} Pr(E_i)$ for any $E_i \in \mathcal{F}$ with $E_i \cap E_j = \emptyset$ for all $i, j \in \mathbb{N}$, $i \neq j$.

A *probability space* is a triple $(\Omega, \mathcal{F}, Pr)$ with $(\Omega, \mathcal{F})$ a measurable space and Pr a probability measure for $(\Omega, \mathcal{F})$.

Let $(\Omega, \mathcal{F})$ and (S, Σ) be measurable spaces, $X : \Omega \rightarrow S$, $s \in S$ and $\sigma \in \Sigma$. We define $X \sim s$ to be $\{\omega \in \Omega \mid X(\omega) \sim s\}$ and $X^{-1}(\sigma) = \bigcup_{s \in \sigma}(X \sim s)$ with $\sim \in \{\leq, <, =, >, \geq\}$. X is *measurable* (wrt. $(\Omega, \mathcal{F})$ and (S, Σ)) if $X^{-1}(\sigma) \in \mathcal{F}$ for all $\sigma \in \Sigma$. A *random variable* is a measurable function $X : \Omega \rightarrow S$; we call $X(\omega)$ the *realization* of X for $\omega \in \Omega$.

For the following we instantiate $\Omega = S = \mathbb{R}_{\geq 0}$, $\mathcal{F} = 2^{\mathbb{R}_{\geq 0}}$ and X the identity. For $f : \mathbb{R}_{\geq 0} \rightarrow \mathbb{R}_{\geq 0}$ we define its *support* as $supp(f) = \{\omega \in \mathbb{R} \mid f(\omega) > 0\}$, with:

- If $supp(f)$ is countable and $\sum_{\omega \in supp(f)} f(\omega) = 1$, f is called a *discrete probability distribution*, which induces the unique probability measure $Pr : 2^{\mathbb{R}_{\geq 0}} \rightarrow [0,1]$ with $Pr(E) = \sum_{\omega \in E \cap supp(f)} f(\omega)$ for all $E \subseteq \mathbb{R}_{\geq 0}$.

---

[1] In the original definition, this is called *non-blocking*.

- If f is absolute continuous with $\int_0^\infty f(\omega)\,d\omega = 1$ then f is called a *continuous probability distribution* or a *probability density function (PDF)*, which induces for all $a \in \mathbb{R}_{\geq 0}$ the unique probability measure $Pr : 2^{\mathbb{R}_{\geq 0}} \to [0,1]$ with $Pr(X \leq a) = \int_0^a f(\omega)\,d\omega$, and the *cumulative distribution function* (CDF) $F : \mathbb{R}_{\geq 0} \to [0,1]$ with $F(a) = Pr(X \leq a)$.

We denote the set of all discrete resp. continuous probability distributions by $Dist_d$ resp. $Dist_c$ and call elements from $Dist = Dist_d \cup Dist_c$ *probability distributions*. A random variable is *discrete* if its underlying probability measure Pr is induced by a discrete probability distribution, and *continuous* otherwise. By $\mathbb{X}_d$ and $\mathbb{X}_c$ we denote the set of all discrete resp. continuous random variables.

Given two measurable spaces $(\Omega_1, \mathcal{F}_1)$ and $(\Omega_2, \mathcal{F}_2)$, a *stochastic kernel from* $(\Omega_1, \mathcal{F}_1)$ *to* $(\Omega_2, \mathcal{F}_2)$ [14] is a function $\kappa : \mathcal{F}_2 \times \Omega_1 \to [0,1]$ with:

- For each $E_2 \in \mathcal{F}_2$, the function $f_{E_2}^\kappa : \Omega_1 \to [0,1]$ with $f_{E_2}^\kappa(\omega_1) = \kappa(E_2, \omega_1)$ is measurable w.r.t. $(\Omega_1, \mathcal{F}_1)$ and $([0,1], \mathcal{B}([0,1]))$.
- For each $\omega_1 \in \Omega_1$, the function $Pr_{\omega_1}^\kappa : \mathcal{F}_2 \to [0,1]$ with $Pr_{\omega_1}^\kappa(E_2) = \kappa(E_2, \omega_1)$ is a probability measure on $(\Omega_2, \mathcal{F}_2)$.

Stochastic kernels are used to express the state-dependent probability $\kappa(E_2, \omega_1)$ of event $E_2 \in \mathcal{F}_2$ in system state $\omega_1 \in \Omega_1$. κ is *discrete* if each $Pr_{\omega_1}^\kappa$ can be induced by a discrete probability distribution, and *continuous* otherwise.

A *stochastic process* over an index set T is a family of random variables $\{X(t) \mid t \in T\}$, defined over a common probability space and taking values in the same measurable space. In this work we use continuous-time stochastic processes with $T \subseteq \mathbb{R}_{\geq 0}$. A stochastic process has the *Markov property* if its future is independent of its past evolution [14].

3   Extending Hybrid Automata with Stochasticity

In this section we formalize two stochastic HA extensions: *Composed scheduling*, introduced in Sect. 3.1, randomly chooses first a delay and then a jump to be taken after the delay. Conversely, *decomposed scheduling*, introduced in Sect. 3.2, chooses a delay for each jump separately, where the jump with the minimal delay is taken. Our approach chooses delays from $\mathbb{R}_{\geq 0}$, which might be unrealizable due to invariants, or after which no jump might be enabled, so we need to introduce mechanisms to manage these cases. We mention that other formalisms like [3] assume an "oracle" and can therefore sample only over realizable time delays after which there is an enabled jump.

To put a clear focus on the differences between composed and decomposed scheduling, our HA definition assumes deterministic flows and jump resets. In addition, for simplicity we assume in the following a unique initial state. However, our languages and results can easily be extended to relax these restrictions.

In Sects. 3.1 and 3.2 we assume that all invariants evaluate to true; we use $\top$ to denote the trivial invariant, i.e. $\top(\ell) = \mathbb{R}^d$ for all $\ell \in Loc$. In Sect. 3.5 we discuss which adaptions in the definitions are required to apply (de-)composed scheduling to HA with invariants.

3.1   Composed Scheduling

In *composed scheduling*, the durations of time steps and the jumps to be taken are sampled according to two stochastic kernels Ψ_c resp. Ψ_d. For a state σ, the probability distributions induced by Ψ_c and Ψ_d are denoted $\mathrm{Dist}_\sigma^{\Psi_c}$ resp. $\mathrm{Dist}_\sigma^{\Psi_d}$.

To schedule time delays, approaches like e.g. [3] sample only durations after which there exists an enabled jump. This alternative is meaningful for decidable subclasses only (as one needs to determine valid time durations) and for them it would result in the same expressivity as our approach, which is as follows.

Assume a d-dimensional hybrid automaton $\mathcal{H}_{in} = (Loc, Var, Edge_{in}, Act, \top, Init)$. Without an oracle, it might happen that Ψ_c samples a time duration after which there are no enabled jumps. To handle such cases, we introduce for each location $\ell \in Loc$ a unique *(composed) resampling jump* $\epsilon_\ell = (\ell, g, r, \ell)$ with guard $g = \mathbb{R}\setminus(\cup_{(\ell,g',r',\ell')\in Edge_{in}} g')$ and reset $r(\nu) = \nu$ for all $\nu \in \mathbb{R}^d$. Let $Edge_r^{comp} = \{\epsilon_\ell^r \mid \ell \in Loc\}$. We call $\mathcal{H} = (Loc, Var, Edge_{in} \cup Edge_r^{comp}, Act, \top, Init)$ the *composed resampling extension* of $\mathcal{H}_{in}$.

Definition 3 (Composed Syntax). *A hybrid automaton with composed scheduling is a tuple $\mathcal{C} = (\mathcal{H}, \Psi_c, \Psi_d)$, where:*

- *$\mathcal{H} = (Loc, Var, Edge, Act, \top, Init)$ with states Σ is the composed resampling extension of a HA $\mathcal{H}_{in}$ with trivial invariants and deterministic initial state.*
- *$\Psi_c : \mathcal{B}(\mathbb{R}_{\geq 0}) \times \Sigma \to [0,1]$ is a continuous stochastic kernel from $(\Sigma, \mathcal{B}(\Sigma))$ to $(\mathbb{R}_{\geq 0}, \mathcal{B}(\mathbb{R}_{\geq 0}))$.*
- *$\Psi_d : \mathcal{B}(Edge) \times \Sigma \to [0,1]$ is a discrete stochastic kernel from $(\Sigma, \mathcal{B}(\Sigma))$ to $(Edge, \mathcal{B}(Edge))$ such that $supp(\mathrm{Dist}_\sigma^{\Psi_d}) \subseteq E_\Sigma(\sigma)$ for all $\sigma \in \Sigma$.*

After each time step, if any non-resampling jump of $\mathcal{H}_{in}$ is enabled then resampling is scheduled with probability 0, and otherwise with probability 1. In each jump successor state σ, a fresh delay is sampled according to $\mathrm{Dist}_\sigma^{\Psi_c}$.

Definition 4 (Composed Semantics). *Assume a HA with composed scheduling $\mathcal{C} = (\mathcal{H}, \Psi_c, \Psi_d)$. A path of $\mathcal{C}$ is a path $\pi = \sigma_0 \xrightarrow{t_0} \sigma_0' \xrightarrow{e_0} \sigma_1 \xrightarrow{t_1} \dots$ of $\mathcal{H}$ with $t_j \in supp(\mathrm{Dist}_{\sigma_j}^{\Psi_c})$ and $e_j \in supp(\mathrm{Dist}_{\sigma_j'}^{\Psi_d})$ for all $j \geq 0$.*

3.2   Decomposed Scheduling

Let $\mathcal{H}_{in} = (Loc, Var, Edge_{in}, Act, \top, Init)$ be a d-dimensional HA. Instead of centralized decisions via stochastic kernels, *decomposed scheduling* chooses jumps and the length of time steps by associating each jump with a continuous random variable from a non-empty set $\mathbb{X} = \{X_1, \dots, X_k\}$ via a function $a_{in} : Edge_{in} \to \{1, \dots, k\}$, such that two jumps with the same random variable are never enabled simultaneously. We call such a pair $(\mathcal{H}_{in}, a_{in})$ an $\mathbb{X}$-*labeled HA*.

The realisations of the random variables indicate the delay after which a jump with the given random variable should be taken; if no such jump is enabled then we again need a mechanism for resampling.

For each location $\ell \in Loc$ and random variable $X_i \in \mathbb{X}$ we introduce a *(decomposed) resampling jump* $\epsilon^r_{\ell,i} = (\ell, g_{\ell,i}, r, \ell)$ with reset $r(\nu) = \nu$ for all $\nu \in \mathbb{R}^d$ and guard $g_{\ell,i} = \mathbb{R}^d \setminus (\cup_{e \in \{e' = (\ell,g,r,\ell') \in Edge_{in} \mid a_{in}(e')=i\}} g)$. We extend the edge set to $Edge = Edge_{in} \cup Edge^{decomp}_r$ with the resampling edges $Edge^{decomp}_r = \{\epsilon^r_{\ell,i} \mid \ell \in Loc \wedge 1 \leq i \leq k\}$. Let $\mathcal{H} = (Loc, Var, Edge, Act, \top, Init)$.

We extend also a_{in} to cover the resampling edges, defining a : $Edge \to \{1, \ldots, k\}$ with $a(e) = a_{in}(e)$ for $e \in Edge_{in}$ and $a(\epsilon^r_{\ell,i}) = i$ for $\epsilon^r_{\ell,i} \in Edge^{decomp}_r$. Let $a^{-1} : \{1, \ldots, k\} \to 2^{Edge}$ with $a^{-1}(i) = \{e \in Edge \mid a(e) = i\}$ for $i = 1, \ldots, k$.

We call $(\mathcal{H}, a)$ the *decomposed resampling extension* of the $\mathbb{X}$-labeled HA $(\mathcal{H}_{in}, a_{in})$. Note that $(\mathcal{H}, a)$ itself is an $\mathbb{X}$-labeled HA.

Definition 5 (Decomposed Syntax). *A hybrid automaton with decomposed scheduling is a tuple $\mathcal{D} = (\mathcal{H}, \mathbb{X}, a)$ where:*

- $\mathbb{X} = \{X_1, \ldots, X_k\}$ *is a finite non-empty ordered set of continuous random variables.*
- $(\mathcal{H}, a)$ *with $\mathcal{H} = (Loc, Var, Edge, Act, \top, Init)$ is the decomposed resampling extension of some $\mathbb{X}$-labeled HA $(\mathcal{H}_{in}, a_{in})$, where $\mathcal{H}_{in}$ has trivial invariants and a deterministic initial state.*

We store the current realizations of the random variables in a sequence $\mathcal{R} = (x_1, \ldots, x_k) \in \mathbb{R}^k_{\geq 0}$, and use $\mathcal{R}[j]$ to refer to x_j. The stochastic race between the random variables is "won" by the random variable which *expires* first as it has a smallest current realisation. The presence of resampling jumps ensures, that a jump can be scheduled, even if there is no enabled edge associated with the winning random variable. Note that, since all random variables follow a continuous probability distribution, the probability that two random variables expire at the same time is 0 and in this case it is irrelevant which jump is taken.

Definition 6 (Decomposed Semantics). *Assume a HA with decomposed scheduling $\mathcal{D} = (\mathcal{H}, \mathbb{X}, a)$ with $\mathbb{X} = \{X_1, \ldots, X_k\}$. A path of $\mathcal{D}$ has the form*

$$\pi = (\sigma_0, \mathcal{R}_0) \xrightarrow{t_0} (\sigma'_0, \mathcal{R}'_0) \xrightarrow{e_0} (\sigma_1, \mathcal{R}_1) \xrightarrow{t_1} \ldots \text{ with } \sigma_i = (\ell_i, \nu_i), \sigma'_i = (\ell_i, \nu'_i)$$

such that $\sigma_0 \xrightarrow{t_0} \sigma'_0 \xrightarrow{e_0} \sigma_1 \xrightarrow{t_1} \ldots$ is a path of $\mathcal{H}$ and such that for all $i \in \mathbb{N}$:

- $\mathcal{R}_i, \mathcal{R}'_i \in \mathbb{R}^k_{\geq 0}$ *and for all $j \in \{1, \ldots, k\}$, $\mathcal{R}_0[j]$ is sampled according to X_j's probability distribution in σ_0.*
- $t_i = \min_{j \in \{1,\ldots,k\}} \mathcal{R}_i[j]$ *and $\mathcal{R}'_i[j] = \mathcal{R}_i[j] - t_i$ for all $j \in \{1, \ldots, k\}$.*
- $\mathcal{R}'_i[m_i] = 0$ *for $m_i = a(e_i)$.*
- *The value $\mathcal{R}_{i+1}[m_i]$ is sampled according to X_{m_i}'s probability distribution in σ_{i+1}, and $\mathcal{R}_{i+1}[j] = \mathcal{R}'_i[j]$ for all $j \in \{1, \ldots, k\} \setminus \{m_i\}$.*

Example 1 ((De-)composed scheduling). We illustrate the application of composed and decomposed scheduling on the example HA depicted in Fig. 2(a). Note that the resulting HA with (de-)composed scheduling in this example have different underlying stochastic processes (c.f. Sect. 3.3).

In the composed scheduling in Fig. 2(b), choices over the time steps' durations are governed by a kernel Ψ_c, which characterises for each state with location ℓ_0

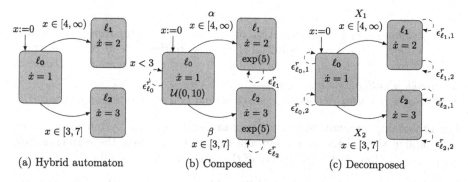

(a) Hybrid automaton   (b) Composed   (c) Decomposed

Fig. 2. Illustration to Example 1. Hybrid automaton shown in (a) is extended with either composed scheduling in (b) or decomposed scheduling in (c).

a uniform distribution $\mathrm{Dist}^{\Psi_c}_{(\ell_0,\nu)} = \mathcal{U}(0,10)$, and an exponential distribution $\mathrm{Dist}^{\Psi_c}_{(\ell_i,\nu)} = \exp(5)$ otherwise. The kernel Ψ_d characterises a discrete probability distribution over enabled jumps. Hence, $\Psi_d(e_1,(\ell_0,\nu)) = \alpha$, $\Psi_d(e_2,(\ell_0,\nu)) = \beta$ and $\Psi_d(\epsilon^r_{\ell_0},(\ell_0,\nu)) = 1 - \alpha - \beta$, where e_1 denotes the jump to location ℓ_1 and e_2 the jump to location ℓ_2. The values of α and β depend on the enabling status of e_1 and e_2. Hence, $\alpha = 1$ and $\beta = 0$, if e_1 is the only jump enabled. Similarly, $\alpha = 0$, $\beta = 1$ in case e_2 is enabled but e_1 is not. If both jumps are enabled then we set $\alpha = 0.7$ and $\beta = 0.3$. If neither e_1 nor e_2 is enabled, then $\alpha = \beta = 0$ and the resampling edge is scheduled with probability 1.

In the decomposed scheduling in Fig. 2(c), there is a race between two competing random variables X_1 and X_2, and the "winner" decides on the delay as well as on the scheduled jump. For our example, in the initial state $\sigma_0 = (\ell_0,\nu)$ with $\nu(x) = 0$ we sample the values x_1 and x_2 for X_1 resp. X_2 from the exponential distribution with parameter 0.2. After a delay of $t = \min\{x_1,x_2\}$ a jump takes place. If $t = x_1$ then it is the unique enabled jump associated with X_1 (i.e. either the jump to ℓ_1 or the resampling jump $\epsilon_{\ell_0,1}$), otherwise if $t = x_2$ then it is the unique enabled jump associated with X_2 (i.e. either the jump to ℓ_2 or the resampling jump $\epsilon_{\ell_0,2}$).

3.3 Induced Stochastic Process

The semantics of a hybrid automaton with unique initial state and composed or decomposed scheduling is fully stochastic. The execution corresponds to a continuous-time stochastic process $\{X(t) \mid t \in \mathbb{R}_{\geq 0}\}$, where the random variable $X(t)$ takes values from the measurable space $(\Sigma, \mathcal{B}(\Sigma))$.

The corresponding probability space for finite paths of length $len(\pi) = n$ is given by $(\Omega, \mathcal{F}, Pr)$, where $\Omega = (\Sigma)^n$ and $\mathcal{F} = \mathcal{B}(\Omega)$. The probability measure Pr also depends on the chosen scheduling method. Since the probability of a single path is in most cases zero, we define measurable probabilities for *traces*.

Definition 7 (Trace). *A trace of a HA $\mathcal{H}$ is a finite sequence $\tau = (\sigma_0, e_0, e_1, \ldots, e_n)$ composed of a state $\sigma_0 \in \Sigma_{Inv}$ and a (possibly empty) sequence of jumps $e_0, \ldots, e_n$ of $\mathcal{H}$. The trace τ is initial if σ_0 is an initial state of $\mathcal{H}$. A sub-trace of τ is a trace $\tau' = (\sigma_i, e_i, \ldots, e_n)$ for some $1 \leq i \leq n$, where σ_i is reachable in $\mathcal{H}$ through a path $\sigma_0 \xrightarrow{t_0} \sigma_0' \xrightarrow{e_0} \sigma_1 \xrightarrow{t_1} \ldots \xrightarrow{e_{i-1}} \sigma_i$.*
A trace of a HA with (de-)composed scheduling is a trace of its underlying HA.

The probability measure of a trace is obtained recursively by integrating over the enabling time of the first jump and taking into account the corresponding jump probabilities. Note that traces of $\mathcal{C}$ might include resampling jumps.

Definition 8 (Composed Probability Measure). *For a HA with composed scheduling $\mathcal{C} = (\mathcal{H}, \Psi_c, \Psi_d)$ and a trace $\tau = (\sigma_0, e_0, e_1 \ldots, e_n)$ of $\mathcal{C}$ we define $Pr_{\mathcal{C}}(\tau)$ to be 1 if $\tau = (\sigma_0)$, and otherwise*

$$Pr_{\mathcal{C}}(\tau) = \int_{t \in T(\sigma_0, e_0)} \Psi_c(\sigma_0)(t) \cdot \Psi_d(\sigma_0')(e_0) \cdot Pr_{\mathcal{C}}(\sigma_1, e_1, \ldots, e_n) dt \qquad (1)$$

where σ_0' and σ_1 are the unique states of $\mathcal{H}$ with $\sigma_0 \xrightarrow{t} \sigma_0'$ and $\sigma_0' \xrightarrow{e_0} \sigma_1$.

Definition 9 (Decomposed Probability Measure). *Assume a HA with decomposed scheduling $\mathcal{D} = (\mathcal{H}, \mathbb{X}, a)$, $\mathbb{X} = \{X_1, X_2, \ldots, X_k\}$. For any state σ of $\mathcal{H}$ and any $t \in \mathbb{R}_{\geq 0}$, let σ^t and be the unique state of $\mathcal{H}$ with $\sigma \xrightarrow{t} \sigma^t$, and for any $e \in E_\Sigma(\sigma)$ let σ^e and be the unique state of $\mathcal{H}$ with $\sigma \xrightarrow{e} \sigma^e$.*
For a trace $\tau = (\sigma_0, e_0, e_1, \ldots, e_n)$ of $\mathcal{D}$ we define

$$Pr_{\mathcal{D}}(\tau) = \qquad (2)$$

$$\int_0^\infty Dist_{X_1}(\sigma_0, t_1) \ldots \int_0^\infty Dist_{X_k}(\sigma_0, t_k) \; P(\sigma_0^{\delta_0}, \mathcal{R}, e_0, e_1 \ldots, e_n) \; dt_k \ldots dt_1$$

where $\delta_0 = min\{t_m \mid 1 \leq m \leq k\}$, $\mathcal{R} \in \mathbb{R}_{\geq 0}^k$ with $\mathcal{R}[m] = t_m - \delta_0$ for all $1 \leq m \leq k$, and where P is defined as follows.
For any a trace $\tau = (\sigma_0, e_0, e_1, \ldots, e_n)$ of $\mathcal{D}$ and any $\mathcal{R} \in \mathbb{R}_{\geq 0}^k$ with $\sigma_0 = (\ell_0, \nu_0)$ and $m_0 = a(e_0)$ we set $P(\sigma_0, \mathcal{R}) = 1$ and for a non-empty sequence $e_0, \ldots, e_n$:

$$P(\sigma_0, \mathcal{R}, e_0, \ldots, e_n) = \qquad (3)$$

$$\begin{cases} 0 & \text{if } e_0 \notin E_\Sigma(\sigma_0) \text{ or } \mathcal{R}[m_0] \neq 0 \\ \int_0^\infty Dist_{X_{m_0}}(\sigma_0^{e_0}, t_{m_0}) \cdot P((\sigma_0^{e_0})^{\delta_1}, \mathcal{R}', e_1, \ldots, e_n) dt_{m_0} & \text{otherwise,} \end{cases}$$

where $\delta_1 = min(\{t_{m_0}\} \cup \{\mathcal{R}[m] \mid 1 \leq m \leq k \wedge m \neq m_0\})$, $\mathcal{R}'[m_0] = t_{m_0} - \delta_1$ and $\mathcal{R}'[m] = \mathcal{R}[m] - \delta_1$ for all $m \in \{1, \ldots, k\} \setminus \{m_0\}$.

When decomposed scheduling is applied, the above-defined probability measure over traces (which might also include resampling jumps) is obtained by sampling all random variables from their corresponding probability distributions. Afterwards, it recursively computes the probability measure P of the trace for the sampled durations, after letting time elapse by the minimum realisation δ_0

under all random variables. In the definition of P in Eq. 3, the first case applies if the jump e_0 is either disabled in the trace's starting state or the realisation of its random variable is not yet expired (i.e. not 0); in this case, the probability of the trace is 0. Otherwise, we take the jump e_0 with successor state $\sigma_0^{e_0}$, resample the random variable X_{m_0} of e_0, let again time elapse with the minimum realisation δ_1 over all random variables, and apply the definition recursively. Note that e_0 might be a resampling jump and that though several realisations can expire simultaneouly, it might happen only with probability 0.

Remark 1. Hybrid automata with composed scheduling C and HA with decomposed scheduling D both extend the jump set of their underlying hybrid automaton $\mathcal{H}$. Therefore, C and D have more paths than $\mathcal{H}$. This means that Zeno paths are inherited from $\mathcal{H}$ to C and D, however, Zeno-freedom of $\mathcal{H}$ does not imply Zeno-freedom of D and C.

Consider for example the hybrid automaton $\mathcal{H}$ from Fig. 2(a) and its composed scheduling extension C from Fig. 2(b). The automaton $\mathcal{H}$ is Zeno-free but C does have Zeno paths, e.g. all paths that take the resampling jump ϵ_{ℓ_0} of ℓ_0 infinitely often. Even though the stochastic kernel of C almost surely excludes such paths (i.e. if τ_k is the trace with k repeated ϵ_{ℓ_0}-jumps from the initial state σ_0 then $\lim_{k \to \inf} P(\tau_k) = 0$), changing the distribution in ℓ_0 from $\mathcal{U}(0, 10)$ to $\mathcal{U}(0, \frac{1-x}{2})$ would increase their probability to 1. Hence, modelers should carefully choose stochastic distributions in order to ensure that additional Zeno behavior is almost surely excluded.

3.4   Relation of Composed and Decomposed Scheduling

Previously, we discussed two different approaches on how hybrid automata can be extended with stochasticity. In this section we show that composed scheduling is *more expressive* than decomposed scheduling w.r.t. *Pr-equivalence*.

Definition 10 (Trace Probability Equivalence). *Let D be a HA with decomposed scheduling, C a HA with composed scheduling, and τ a common trace of D and C. The trace τ is Pr-equivalent in D and C iff $Pr_D(\tau) = Pr_C(\tau)$ and for each σ being either the first state of τ or the first state of a sub-trace of τ if holds for all $t \in \mathbb{R}_{\geq 0}$ that $Pr_D^\sigma(X \leq t) = Pr_C^\sigma(X \leq t)$, where X models the duration of a time step starting in σ.*

D and C are Pr-equivalent if the sets of their initial traces are equal and each of their initial traces is Pr-equivalent.

Theorem 1 (Expressivity Composed vs Decomposed Scheduling)

1. *Let D be a HA with decomposed scheduling. Then there is a HA with composed scheduling C such that D and C are Pr-equivalent.*
2. *There exists a HA with composed scheduling C' such that there is no HA with decomposed scheduling D' such that D' and C' are Pr-equivalent.*

Proof (Theorem 1) Statement 1. Assume a HA with decomposed scheduling $\mathcal{D}=(\mathcal{H},\mathbb{X},\mathsf{a})$ with $\mathcal{H} = (Loc, Var, Flow, \top, Edge, Init)$, $Loc = \{\ell_1,\ldots,\ell_n\}$, $Var = \{v_1,\ldots,v_d\}$ and $\mathbb{X} = \{X_1,\ldots,X_k\}$. We construct a HA with composed scheduling $\mathcal{C} = (\mathcal{H}', \Psi_c, \Psi_d)$ with $\mathcal{H}' = (Loc, Var', Flow', \top, Edge', Init')$ as follows.

We encode into the state of $\mathcal{H}'$ for each random variable (i) the state in which it was sampled the last time and (ii) the time which has evolved since then. We also encode (iii) the time spent in the current location since last entering.

To account for (i), we introduce fresh variables $D = \{d_i \mid 1 \le i \le k\}$ to store the location components and $C = \{c_{i,j} \mid 1 \le i \le k, 1 \le j \le d\}$ to store the variable values before the last sampling of X_i. Hence, for the i-th random variable $X_i \in \mathbb{X}$, we encode the state of its last (re)sampling by the values of $(d_i, (c_{i,1},\ldots,c_{i,d}))$.

To encode (ii), we introduce k variables $R = \{r_i \mid 1 \le i \le k\}$ which capture the time since the last (re)sampling of each random variable $X_i \in \mathbb{X}$ in r_i.

Finally, for (iii) we use t_{jump} to store the time duration since the last jump.

Thus our encoding uses $d' = d + k \cdot (d + 2) + 1$ variables ordered as $Var' = \{v_1,\ldots,v_d,d_1,\ldots,d_k,c_{1,1},\ldots,c_{k,d},r_1,\ldots,r_k,t_{jump}\}$. For $\nu \in \mathbb{R}^{d'}$, we use $\nu\downarrow_d$ to denote the first d components $(\nu_1,\ldots,\nu_d)$ of $\nu = (\nu_1,\ldots,\nu_{d'}) \in \mathbb{R}^{d'}$. Furthermore, for any $\nu \in \mathbb{R}^{d'}$, $a \in Var'$ and $b \in \mathbb{R}$, by $\nu[a \mapsto b]$ we denote ν after changing the entry at the position of variable a (as defined in Var') to b.

The above encoding is implemented by extending $Flow$ to $Flow'$ in each location $\ell \in Loc$ with derivative 0 for each variable in $C \cup D$ and derivative 1 for each variable in $R \cup \{t_{jump}\}$. I.e., for each $\ell \in Loc$ and $\nu \in \mathbb{R}^{d'}$, $Flow'(\nu) = (Flow(\nu\downarrow_d), \mathbf{0}, \mathbf{1})$ with $\mathbf{0}$ is a sequence of $k(d+1)$ zeros and $\mathbf{1}$ of $k+1$ ones.

Further, $Edge'=\{e' \mid e \in Edge\}$ contains for each $e = (\ell_{j_1},g,r,\ell_{j_2}) \in Edge$ with $\mathsf{a}(e) = i$ the jump $e' = (\ell_{j_1},g',r',\ell_{j_2})$ which extends e to handle the new variables; formally, $g' = \{\nu \in \mathbb{R}^{d'} \mid \nu\downarrow_d \in g\}$ and for all $\nu \in \mathbb{R}^{d'}$, $r'(\nu) = \nu[v_1 \mapsto r(\nu)[1]]\ldots[v_d \mapsto r(\nu)[d]][d_i \mapsto j_2][c_{i,1} \mapsto r(\nu)[1]]\ldots[c_{i,d} \mapsto r(\nu)[d]][r_i \mapsto 0][t_{jump} \mapsto 0]$.

For each $\ell \in Loc$, $Init'(\ell)$ consist of all $\nu \in \mathbb{R}^{d'}$ for which $\nu\downarrow_d \in Init(\ell)$, and such that $\nu(d_i) = \ell$, $\nu(c_{i,j}) = \nu(v_j)$ and $\nu(r_i) = \nu(t_{jump}) = 0$ for all $i = 1,\ldots,k$ and $j = 1,\ldots,d$.

Now we define the kernel Ψ_c. With decomposed scheduling, the duration between two samplings of a random variable (defined by its realisation) might cover consecutive stays in different locations. However, in the composed setting, we are forced to sample a new duration upon entering a new location.

For each state $\sigma = (\ell,\nu) \in \Sigma_\mathcal{C}$ let $\sigma_{X_i} = (\ell_{X_i},\nu_{X_i}) \in \Sigma_\mathcal{D}$ denote the state in which X_i was (re)sampled the last time, as encoded in the values of the auxiliary variables: $\ell_{X_i} = \ell_{\nu(d_i)}$ and $\nu_{X_i}(v_j) = \nu(c_{i,j})$ for $j = 1,\ldots,d$. For each random variable $X_i \in \mathbb{X}$ with CDF $F_{X_i}^{\sigma_{X_i}}$ and density function $f_{X_i}^{\sigma_{X_i}}$ in state σ_{X_i}, we first define another random variable X_i' whose probability distribution is first conditioned in that samples are at least as large as the time $\nu(r_i)$ passed since

the last (re)sampling of X_i, and then shifted by $\nu(r_i)$ to the left: $F^\sigma_{X_i'}(x) =$
$Pr(X_i \leq \nu(r_i) + x \mid X_i > \nu(r_i)) = \frac{\int_{\nu(r_i)}^{\nu(r_i)+x} f^{\sigma X_i}_{X_i}(t)dt}{1-F^{\sigma X_i}_{X_i}(\nu(r_i))}$. Let $\mathbb{X}' = \{X_i' \mid X_i \in \mathbb{X}\}$.

For each first state in a (sub)-trace of $\mathcal{D}$ and $\mathcal{C}$, the probability distribution over the delay must be the same in both automata. We let Ψ_c specify for each state $\sigma \in \Sigma_\mathcal{C}$ a probability distribution over the time delay until the next random variable $X_i \in \mathbb{X}$ expires, i.e. $\text{Dist}^{\Psi_c}_\sigma = f_M$, where the random variable $M=min(\mathbb{X}')$ is the minimum over all shifted random variables, as defined in [24].

For the kernel Ψ_d, we observe that for each random variable X_i, in each state $\sigma_\mathcal{D}$ exactly one X_i-labeled jump is enabled in $\mathcal{D}$. Our construction of $Edge'$ maintains this characteristics in $\mathcal{C}$. To formalize the probability that an enabled jump is taken, we define for each random variable X_i another random variable X_i'', $\mathbf{X}'' = \{X_1'', \dots, X_k''\}$, which is defined as X_i' but its CDF $F^\sigma_{X_i''}(x)$ is shifted with $\nu(r_i) - \nu(t_{jump})$, instead of $\nu(r_i)$, to model the probabilities of samples beyond the time point of the last jump (i.e. in the definition of $F^\sigma_{X_i'}(x)$ we replace $\nu(r_i)$ by $\nu(r_i) - \nu(t_{jump})$). We let the discrete kernel define for each state $\sigma_\mathcal{C}$ and each edge $e' \in Edge'$ with $a(e) = i$ the probability

$$\Psi_d(\sigma_\mathcal{C})(e') = \begin{cases} Pr^{\sigma_\mathcal{C}}_\mathcal{C}(X_i'' \leq min(\mathbf{X}'' \setminus \{X_i''\})) & \text{if } e \text{ is enabled in } \sigma_\mathcal{C}, \\ 0 & \text{otherwise.} \end{cases} \quad (4)$$

Hence, given an arbitrary HA $\mathcal{D}$ with decomposed scheduling, we can construct a HA $\mathcal{C}$ with composed scheduling such that $\mathcal{D}$ and $\mathcal{C}$ have the same trace set and the same initial trace set, and such that each common trace τ is Pr-equivalent in $\mathcal{D}$ and $\mathcal{C}$. As furthermore Ψ_c is specified such that it mimics the distribution over the duration until the next random variable expires, it is assured that $Pr^\sigma_\mathcal{D}(X \leq t) = Pr^\sigma_\mathcal{C}(X \leq t)$.

Statement 2. To show statement 2 of Theorem 1 we consider Fig. 3 as a counterexample. The depicted HA with composed scheduling is constructed such that in the initial state a delay distributed according to the uniform distribution $\mathcal{U}(0,10)$ is sampled, before a jump to location ℓ_1 is taken with probability 0.25 and a jump to location ℓ_2 with probability 0.75.

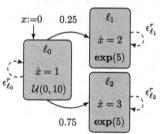

For decomposed scheduling, we associate each edge with an independent random variable X_1 resp. X_2, where $min(X_1, X_2)$ should be uniformly distributed over the intersection of the support of X_1 and X_2 in location ℓ_0 to achieve Pr-equivalence. However, the minimum of two continuous random variables with overlapping support can never be uniformly distributed, see [24]. □

Fig. 3. Counterexample.

3.5 Extending (De-)composed Scheduling with Invariants

If we allow non-trivial invariants, unrealizable time durations might have a positive probability. To manage such cases, we use the concept of *forced jumps* for both, composed and decomposed scheduling. Forced jumps ensure that no time step larger than $t_{max}(\sigma)$ is executed. For both, composed and decomposed scheduling, the HA is adapted such that each location has a forced jump which is used to leave a location before its invariant is violated. Furthermore, the semantics for composed and decomposed scheduling (Definitions 4 and 6) are altered such that for each state σ, the time delay is capped by $t'_i = \min(t_i, t_{max}(\sigma))$.

As the probability mass of sampling a delay larger than $t_{max}(\sigma)$ in state σ is shifted to the forced jumps, this has to be considered in the probability measures of Sect. 3.3 by integrating over $(t_{max}(\sigma), \infty)$ in case of a forced jump.

4 Classification of Existing Approaches

This paper allows for a broad classification covering multiple formalisms. Representatives from literature can be found for both variants, as indicated in Table 1. We discuss one representative formalism for composed and one for decomposed scheduling in more depth. Furthermore, we discuss which formalisms do not fit into our classification and relate our comparison to the work in [17] which focuses on Markovian stochastic hybrid models.

Composed Scheduling. Lygeros and Prandini [15] introduce a general class of *continuous-time stochastic hybrid automata* (CTSHA). This approach has been abstracted to discrete-time e.g., in [2,20]. CTSHA implement composed scheduling, as the stochastic information over the delay is attached to the location and a stochastic kernel chooses the jump-successor state randomly. Technically, the actual jump times are the stopping time of the inhomogeneous Poisson process with state-dependent rate $\lambda(t) = \lambda(q(t), \mathbf{x}(t))$. This results in delays which are sampled according to an inhomogeneous exponential distribution which can explicitly be expressed by the stochastic kernel Ψ_c used for composed scheduling by: $Pr(X > t \mid (\ell(s), \mathbf{x}(s))) = e^{-\left(\int_s^{s+t} \lambda(u)\, du\right)}$. The integral can be computed, if for each location, the continuous state evolves with a deterministic rate. Thus, $(\ell(s+t'), \mathbf{x}(s+t'))$ is well defined for any $t' \leq t \in \mathbb{R}_{\geq 0}$ and given $(\ell(s), \mathbf{x}(s))$.

The inhomogeneous Poisson process used in CTSHA can define a phase-type distribution which can approximate any continuous probability distribution. Hence, just as for composed scheduling, the sampled delay can follow any probability distribution in CTSHA. Moreover, CTSHA directly extend our proposed HA with composed scheduling by including stochastic differential equations (SDEs) to describe the continuous state's evolution. Furthermore, the initial state is sampled according to a probability measure and the kernel Ψ_d is extended to a continuous stochastic kernel, enabling random resets of the continuous state.

Table 1. Classification of existing formalisms.

	[2]	[3]	[4,5]	[6]	[8]	[9]	[15]	[16,19]	[20]
composed	✓	✓	✓			✓	✓		✓
decomposed				✓	✓	✓		✓	

Decomposed Scheduling. Pilch et al. [16,19] introduce singular automata with random clocks (SARC) which basically apply decomposed scheduling. They implement this approach by adding *random clocks* which induce random variables characterizing the delay for each jump. Upon expiration, a new random variable is induced which follows the predefined probability distribution with a constant parameter of the corresponding random clock. Additionally, SARC allow for transitions which are not associated with the expiration of a random variable which adds non-determinism to the model. The concept of random clocks can be extended to other sub-classes of hybrid automata, e.g., rectangular automata [6] with random clocks, which again implicitly assign random variables to jumps via random clocks and restrict the syntax to ensure a sound probability measure. In contrast, decomposed scheduling allows to directly attach random variables to jumps and the semantics ensures that the resulting probability measure is sound. This simplifies modeling.

Markovian Stochastic Hybrid Models. The formalism of *Continuous-Time Markov Chains* (CTMCs) which are discrete-state models without variables, has been extended in the past to stochastic hybrid models [4,5,8,12], which correspond to (different kinds of) Markov processes. *Piecewise Deterministic Markov Processes* (PDMPs) [4,5] implement composed scheduling where the evolution of a continuous state is piece-wise deterministic and can be restricted by invariants. In PDMPs the jump times are ruled by an inhomogenous Poisson process and jumps and their effects are chosen probabilistically by a transition kernel. This stochastic kernel allows to encode guards implicitly.

In contrast, *Switching Diffusion Processes* (SDPs) [8] describe the continuous variables' evolution via stochastic differential equations. They do not allow for invariants or resets of the continuous state and the discrete state evolves according to a *controlled Markov chain* [13], whose transition matrix depends on the continuous state. This allows the user to encode guards into the model. Due to the underlying Markov chain, which can be characterised via a generator matrix describing competing random variables, SDPs can be seen as an approach with decomposed scheduling. *Stochastic hybrid systems* (SHSs) [12] simplify CTSHA as discussed in [15] by relaxing the inhomogenous Poisson process which determines the random delays.

The formalisms mentioned above coincide under certain assumptions, as discussed in [17]. For example, the authors state that the formalism SDP, which is decomposed in our classification, and the formalism SHS, which we classify as composed, coincide iff invariants and guards of the SHS evaluate to true.

Clearly, restricting to exponentially distributed delays renders the counterexample of Theorem 1 invalid, as the minimum of two exponentially distributed random variables is again exponentially distributed. However, additional restrictions are necessary to ensure that the probability spaces of (de-)composed scheduling coincide in the presence of guards and invariants. Specifying such restrictions (and proving their correctness) is out of scope for this paper.

We refer to [17] for a more detailed comparison on the expressivity of Markovian stochastic hybrid models.

Other Formalisms. Several existing formalisms for stochastic hybrid models do not fit into the proposed classification, as they focus e.g., solely on randomly distributed initial states [18] or on a non-deterministic choice over discrete probability distributions for choosing a successor state [21].

The formalism presented in [3] defines a fully stochastic semantics for timed automata by randomly choosing delays and jumps. It applies a composed semantics, however without resampling jumps. Timelock-freedom is ensured by restricting the support of the probability distributions to executable samples.

The formalism presented in [9] proposes networks of stochastic hybrid automata which fit in both, the composed as well as the decomposed approach. Such stochastic hybrid automata allow to reset continuous variables to realizations of continuous random variables. Thus, at each jump we can either sample a randomly distributed delay for the location like in composed scheduling, or associate the samples as delays with the jumps as in decomposed scheduling.

5   Conclusion

In this paper we formalized two approaches to extend hybrid automata with stochasticity. The first approach applies *composed scheduling*, where two stochastic kernels are used to sample the lengths of time steps and the successor states of jumps. In contrast, the second approach yields *decomposed scheduling* via competing random variables. As the realisations of the random variables specify the delay after which the corresponding jump is taken, a race-condition is induced. The minimum realisation of the random variables then fully characterises the next execution step. We formalized the syntax and semantics for (de-)composed scheduling and the stochastic processes underlying the different resulting models. We defined *trace probability equivalence* and showed that it is possible to construct for every given HA with decomposed scheduling, an equivalent HA with composed scheduling. Via a simple counterexample, we showed that a HA with composed scheduling exists, for which no equivalent HA with decomposed scheduling can be constructed.

To connect the theoretical constructs developed in this paper to existing formalisms, we classified several existing formalisms according to their semantics and pointed to approaches which we cannot capture yet. To include them in our classification, future work will consider more expressive systems, e.g., including stochastic resets and stochastic noise. Furthermore, we plan to investigate the relation of the presented classes to approaches without resampling.

Acknowledgements. We thank the ARCH competition 2023 for fruitful discussions on expressing example models from the ARCH report [1] in the formalism of Lygeros et al. [15].

References

1. Abate, A., et al.: Arch-comp22 category report: stochastic models. In: Proceedings of 9th International Workshop on Applied Verification of Continuous and Hybrid Systems (ARCH22), EPiC Series in Computing, vol. 90, pp. 113–141. EasyChair (2022). https://doi.org/10.29007/lsvc

2. Abate, A., Prandini, M., Lygeros, J., Sastry, S.: Probabilistic reachability and safety for controlled discrete time stochastic hybrid systems. Automatica **44**(11), 2724–2734 (2008). https://doi.org/10.1016/j.automatica.2008.03.027

3. Bertrand, N., et al.: Stochastic timed automata. Logical Methods Comput. Sci. **10**(4) (2014). https://doi.org/10.2168/LMCS-10(4:6)2014

4. Bujorianu, M.L., Lygeros, J.: Reachability questions in piecewise deterministic markov processes. In: Maler, O., Pnueli, A. (eds.) HSCC 2003. LNCS, vol. 2623, pp. 126–140. Springer, Heidelberg (2003). https://doi.org/10.1007/3-540-36580-X_12

5. Davis, M.H.: Markov Models & Optimization, 1st edn. Routledge, Milton Park (1993). https://doi.org/10.1201/9780203748039

6. Delicaris, J., Schupp, S., Ábrahám, E., Remke, A.: Maximizing reachability probabilities in rectangular automata with random clocks. In: David, C., Sun, M. (eds.) Theoretical Aspects of Software Engineering - 17th International Symposium, TASE 2023. Lecture Notes in Computer Science, vol. 13931, pp. 164–182. Springer, Cham (2023). https://doi.org/10.1007/978-3-031-35257-7_10

7. Fränzle, M., Hahn, E.M., Hermanns, H., Wolovick, N., Zhang, L.: Measurability and safety verification for stochastic hybrid systems. In: Proceedings of the 14th ACM International Conference on Hybrid Systems: Computation and Control (HSCC 2011), pp. 43–52. ACM (2011)

8. Ghosh, M.K., Arapostathis, A., Marcus, S.I.: Ergodic control of switching diffusions. SIAM J. Control Optim. **35**(6), 1952–1988 (1997). https://doi.org/10.1137/S0363012996299302

9. Hahn, E.M., Hartmanns, A., Hermanns, H., Katoen, J.: A compositional modelling and analysis framework for stochastic hybrid systems. Formal Methods Syst. Des. **43**(2), 191–232 (2013). https://doi.org/10.1007/s10703-012-0167-z

10. Henzinger, T.A.: The theory of hybrid automata. In: Inan, M.K., Kurshan, R.P. (eds) Verification of Digital and Hybrid Systems. NATO ASI Series, vol. 170, pp. 265–292. Springer, Berlin (2000). https://doi.org/10.1007/978-3-642-59615-5_13

11. Henzinger, T.A., Kopke, P.W., Puri, A., Varaiya, P.: What's decidable about hybrid automata? J. Comput. Syst. Sci. **57**(1), 94–124 (1998). https://doi.org/10.1006/jcss.1998.1581

12. Hu, J., Lygeros, J., Sastry, S.: Towards a theory of stochastic hybrid systems. In: Lynch, N., Krogh, B.H. (eds.) HSCC 2000. LNCS, vol. 1790, pp. 160–173. Springer, Heidelberg (2000). https://doi.org/10.1007/3-540-46430-1_16

13. Kesten, H., Spitzer, F.: Controlled Markov chains. Ann. Probab. **3**(1), 32–40 (1975). https://doi.org/10.1214/aop/1176996445

14. Klenke, A.: Probability Theory: A Comprehensive Course. Springer, London (2014). https://doi.org/10.1007/978-1-4471-5361-0_1

15. Lygeros, J., Prandini, M.: Stochastic hybrid systems: a powerful framework for complex, large scale applications. Eur. J. Control **16**(6), 583–594 (2010). https://doi.org/10.3166/ejc.16.583-594
16. Pilch, C., Schupp, S., Remke, A.: Optimizing reachability probabilities for a restricted class of stochastic hybrid automata via flowpipe-construction. In: Abate, A., Marin, A. (eds.) QEST 2021. LNCS, vol. 12846, pp. 435–456. Springer, Cham (2021). https://doi.org/10.1007/978-3-030-85172-9_23
17. Pola, G., Bujorianu, M., Lygeros, J., Benedetto, M.D.D.: Stochastic hybrid models: an overview. In: IFAC Conference on Analysis and Design of Hybrid Systems (ADHS 2003). IFAC Proceedings Volumes, vol. 36, pp. 45–50. Elsevier (2003). https://doi.org/10.1016/S1474-6670(17)36405-4
18. Shmarov, F., Zuliani, P.: ProbReach: verified probabilistic δ-reachability for stochastic hybrid systems. In: Proceedings of the 18th ACM International Conference on Hybrid Systems: Computation and Control, (HSCC 2015), HSCC 2015, pp. 134–139. ACM (2015). https://doi.org/10.1145/2728606.2728625
19. da Silva, C., Schupp, S., Remke, A.: Optimizing reachability probabilities for a restricted class of stochastic hybrid automata via flowpipe-construction. Trans. Model. Comput. Simul. (2023). https://doi.org/10.1145/3607197
20. Soudjani, S.E.Z., Abate, A.: Adaptive and sequential gridding procedures for the abstraction and verification of stochastic processes. SIAM J. Appl. Dyn. Syst. **12**(2), 921–956 (2013). https://doi.org/10.1137/120871456
21. Sproston, J.: Decidable model checking of probabilistic hybrid automata. In: Joseph, M. (ed.) FTRTFT 2000. LNCS, vol. 1926, pp. 31–45. Springer, Heidelberg (2000). https://doi.org/10.1007/3-540-45352-0_5
22. Sproston, J.: Verification and control of probabilistic rectangular hybrid automata. In: Sankaranarayanan, S., Vicario, E. (eds.) FORMATS 2015. LNCS, vol. 9268, pp. 1–9. Springer, Cham (2015). https://doi.org/10.1007/978-3-319-22975-1_1
23. Teige, T., Fränzle, M.: Constraint-based analysis of probabilistic hybrid systems. In: 3rd IFAC Conference on Analysis and Design of Hybrid Systems, (ADHS'09), IFAC Proceedings Volumes, vol. 42, pp. 162–167. Elsevier (2009). https://doi.org/10.3182/20090916-3-ES-3003.00029
24. Willemsen, L., Remke, A., Ábrahám, E.: Comparing two approaches to include stochasticity in hybrid automata (2023). https://doi.org/10.48550/arXiv.2307.08052

Analysis of an Epoch Commit Protocol for Distributed Processing Systems

Paul Ezhilchelvan[1(✉)], Isi Mitrani[1], and Jim Webber[2]

[1] School of Computing, Newcastle University, Newcastle upon Tyne NE4 5TG, UK
{paul.ezhilchelvan,isi.mitrani}@ncl.ac.uk
[2] Neo4j UK, Union House, 182-194 Union Street, London SE1 0LH, UK
Jim.Webber@neo4j.com

Abstract. A policy that reduces communication overheads by committing together all transactions completed within an interval of time is examined. A model of the system involving two queues served alternatively with preemptions is analysed in the steady-state under Markovian assumptions. An exact and easily implementable solution is derived and is used in order to determine performance measures such as average occupancy or average latency. The optimal length of the operative interval is evaluated numerically. A non-preemptive policy is simulated and is shown to be considerably less efficient than the preemptive one analysed here. A generalization to non-Markovian operative intervals is outlined.

Keywords: Distributed Transactions · Epoch commit policy · Tandem Queues · Performance evaluation · Optimization

1 Introduction

In distributed processing systems where transactions access data from multiple nodes, a commitment protocol is executed, whose purpose is to guarantee transaction atomicity and durability when nodes are prone to failures. This is typically the 'two-phase commit', or 2PC protocol that aborts a transaction in case of intervening failures or commits it otherwise (e.g. see Bernstein et al. [2]). However, executing 2PC at the end of each transaction is expensive; each operation involves two network round trips and two disk accesses. This has prompted multiple research efforts focusing on reducing the costs of 2PC executions.

Recently, Lu et al., [9] have proposed a simple and easily implementable policy that is effective in reducing the overheads of commit operations. Instead of executing 2PC at the end of every distributed transaction, that policy executes it once for all transactions processed during an interval of time called an 'epoch'. The rationale for such an approach is that with modern hardware, breakdowns are rare events. It is therefore worth paying the price of a few extra aborted transactions, in order to save frequent executions of 2PC. This idea is a generalization of the 'group-commit' policy proposed by DeWitt et al., [6] in order to reduce disk I/O latency for single-node databases.

© The Author(s), under exclusive license to Springer Nature Switzerland AG 2023
N. Jansen and M. Tribastone (Eds.): QEST 2023, LNCS 14287, pp. 255–269, 2023.
https://doi.org/10.1007/978-3-031-43835-6_18

Here we analyse a stochastic model of the epoch commit policy. Transactions are held in two unbounded queues, first waiting and being processed, and then waiting and being committed. Service is provided by a number of (reliable) servers which, from time to time, interrupt processing in order to perform 2PC operations. The resulting two-dimensional Markov process is solved exactly, leading to the precise determination of the stability condition and the computation of performance measures. This has not been done before.

When applying the epoch commit policy in practice, the main problem is how to set the length of the processing interval. There are trade-offs to be addressed. If that interval is too short, the benefits of postponing the 2PC execution are lost; the useful service capacity, and hence the throughput, is reduced. On the other hand, very long intervals can be equally undesirable because they delay the departure of transactions, leading to deteriorating system occupancy and latency. It is thus important to be able to compute the optimal processing interval. Our solution algorithms provide that facility.

1.1 Related Literature

An approximate model of the epoch commit policy was examined by Waudby et al., [16], using both analysis and simulations. The effect of server breakdowns was considered. Upper and lower bounds on performance were obtained, from which the optimal processing interval could be roughly estimated. One of the conclusions was that the rare breakdowns have a minimal effect on performance.

A similar epoch policy was proposed by Tu et al., [14] in the context of databases held in memory. The aim there was to reduce the cost of shared memory writes.

The decision to ignore server breakdowns when modeling the performance of epoch commit is supported by real-life studies of failure events. In Garraghan et al., [7], Birke et al., [3] and Wang et al., [15] it is shown that mean times to failure are several orders of magnitude larger than transaction processing times.

Among other proposed solutions to the 2PC problem are attempts to minimise or eliminate distributed transactions by means of workload or data partitioning (Curino et al., [4], Das et al., [5], Lin et al., [8]), or scheduling based on pre-declared read and write sets (Thomson et al., [13], Ren et al., [12]). Those approaches have had a limited acceptance because they introduce overheads of their own, or require information which is not readily available.

Our model belongs to a class involving queues in tandem. There is quite a large body of literature on such models (see, for example, Balsamo et al., [1] and Perros [11]). However, there do not appear to be any results that would apply to the present case.

The model is described in Sect. 2. Its solution, and the algorithms for evaluating performance measures, are shown in Sect. 3. The results of several numerical experiments, including a comparison between the preemptive and non-preemptive policies, are presented in Sect. 4. A generalization to non-exponential processing intervals is outlined in Sect. 5.

2   The Model

Transactions, or 'jobs', arrive into a processing system in a Poisson stream with rate λ and join an unbounded queue referred to as the 'processing queue'. There are N identical parallel servers, each of which processes jobs one at a time and independently of the others. Service times are i.i.d. random variables distributed exponentially with mean $1/\mu$. Completed jobs leave the processing queue but do not depart from the system immediately. They are kept in a second unbounded queue, called the 'commit queue', until they are committed.

At random points in time, 'commit signals' are received simultaneously by all servers. If at such a point the commit queue is not empty, all servers interrupt whatever they are doing and perform a joint commit operation (i.e., they execute 2PC). The durations of those operations are i.i.d. random variables distributed exponentially with mean $1/\eta$. During a commit operation, jobs continue to arrive and join the processing queue, but the servers are not available for processing. When the commit operation completes, all jobs currently in the commit queue depart from the system simultaneously, while the jobs whose processing was pre-empted by the commit signal resume their service from the point of interruption.

If a commit signal is received when the commit queue is empty, it has no effect: the servers ignore it and carry on with their work. A new commit signal is then scheduled. The intervals between the completion of a commit operation, or an ineffective commit signal, and the next commit signal, will be referred to as 'work periods'. These are the 'epochs' introduced in [9]. The work periods are i.i.d. random variables distributed exponentially with mean $1/\xi$.

The system structure is illustrated in Fig. 1.

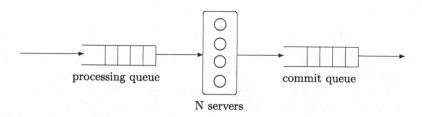

processing queue   commit queue

N servers

Fig. 1. Transaction processing system

Denote by $p(i,j)$ the steady-state probability that there are i jobs in the processing queue, j jobs in the commit queue, and the servers are in a work period. Similarly, let $q(i,j)$ be the steady-state probability that there are i jobs in the processing queue, j jobs in the commit queue, and the servers are in a commit period. These probabilities exist when the system is stable, i.e. when the offered load, $\rho = \lambda/\mu$, is strictly lower than the available service capacity. The latter is not obvious, because the intervals during which the servers process jobs depend on the queue sizes. The stability condition will be established in the course of the analysis.

The probabilities $p(i,j)$ and $q(i,j)$ satisfy a set of balance equations. Note that the rate at which jobs are served during work periods is state-dependent in a limited way: it is $i\mu$ when $i < N$ and $N\mu$ when $i \geq N$. Remember also that following the completion of a commit period, the commit queue becomes empty, regardless of its prior size. Hence, for $j = 0$ we can write

$$(\lambda + i\mu)p(i,0) = \lambda p(i-1,0) + \eta q(i,\cdot) \; ; \quad i = 0,1,\ldots,N-1 , \tag{1}$$

and

$$(\lambda + N\mu)p(i,0) = \lambda p(i-1,0) + \eta q(i,\cdot) \; ; \quad i = N, N+1,\ldots , \tag{2}$$

where $p(-1,0) = 0$ by definition; $q(i,\cdot)$ is the marginal probability that there are i jobs in the processing queue and a commit period is in progress (i.e., it is the sum of $q(i,j)$ over all $j > 0$).

When $j > 0$ the equations become

$$(\lambda + i\mu + \xi)p(i,j) = \lambda p(i-1,j) + (i+1)\mu p(i+1,j-1) \; ; \; i = 0,1,\ldots,N-1 , \tag{3}$$

and

$$(\lambda + N\mu + \xi)p(i,j) = \lambda p(i-1,j) + N\mu p(i+1,j-1) \; ; \; i = N, N+1,\ldots,N-1 , \tag{4}$$

Again, any probability involving a negative queue size is 0 by definition.

Since jobs arrive but are not served during commit periods, the balance equations for the probabilities $q(i,j)$ are

$$(\lambda + \eta)q(i,j) = \lambda q(i-1,j) + \xi p(i,j) \; ; \; i = 0,1,\ldots \; ; \; j = 1,2,\ldots , \tag{5}$$

where $q(-1,j) = 0$ by definition. We have $q(i,0) = 0$ for all i, because commit operations are not initiated when the commit queue is empty. Summing over all $j > 0$, we can write equations for the marginal probabilities $q(i,\cdot)$ that appear in (1):

$$(\lambda + \eta)q(i,\cdot) = \lambda q(i-1,\cdot) + \xi[p(i,\cdot) - p(i,0)] \; ; \; i = 0,1,\ldots , \tag{6}$$

where $p(i,\cdot)$ is the marginal probability that there are i jobs in the processing queue and a work period is in progress.

The balance equations will be solved by first transforming them into relations between one-dimensional and two-dimensional generating functions. The main performance measure that we wish to determine is the average total number of jobs in the system, L or, equivalently, the average time a job spends in the system, $W = L/\lambda$. Other quantities of interest are the stability threshold, the server availability (i.e. the fraction of time the servers are available for processing jobs) and possibly higher moments of the processing and commit queues.

3   Solution

Define by $g_i(z)$ the generating function of the commit queue distribution during work periods, when there are i jobs in the processing queue:

$$g_i(z) = \sum_{j=0}^{\infty} p(i,j)z^j \; ; \; i = 0,1,\ldots . \tag{7}$$

Multiplying Eqs. (1)–(4) by z^j and summing, we transform the relations between probabilities $p(i,j)$ into the relations between functions $g_i(z)$:

$$(\lambda + i\mu + \xi)g_i(z) = \lambda g_{i-1}(z) + (i+1)\mu z g_{i+1}(z) + \xi p(i,0) + \eta q(i,\cdot) \ ; \ i < N \ , \quad (8)$$

and

$$(\lambda + N\mu + \xi)g_i(z) = \lambda g_{i-1}(z) + N\mu z g_{i+1}(z) + \xi p(i,0) + \eta q(i,\cdot) \ ; \ i \geq N \ . \quad (9)$$

Equations for the marginal probabilities, $p(i,\cdot)$, that appear in (6), are obtained by setting $z = 1$ in (8) and (9). Later we shall need those equations for $i < N$:

$$(\lambda + i\mu + \xi)p(i,\cdot) = \lambda p(i-1,\cdot) + (i+1)\mu p(i+1,\cdot) + \xi p(i,0) + \eta q(i,\cdot) \ . \quad (10)$$

Next, define the two-dimensional generating functions, $G(y,z)$ and $H(y,z)$, encapsulating the joint distribution of the processing and commit queues during work and commit periods respectively:

$$G(y,z) = \sum_{i=0}^{\infty}\sum_{j=0}^{\infty} p(i,j)y^i z^j \ ; \ H(y,z) = \sum_{i=0}^{\infty}\sum_{j=0}^{\infty} q(i,j)y^i z^j \ . \quad (11)$$

Multiplying (8) and (9) by y^i and summing over all i we obtain, after some manipulations, a single equation expressing $G(y,z)$ in terms of $H(y,1)$, $G(y,0)$ and the N generating functions $g_i(z)$ where the service rate is state-dependent:

$$a(y,z)G(y,z) = \xi y G(y,0) + \eta y H(y,1) + \mu(y-z)S(y,z) \ , \quad (12)$$

where

$$a(y,z) = \lambda y(1-y) + N\mu(y-z) + \xi y \quad (13)$$

and

$$S(y,z) = \sum_{i=0}^{N-1}(N-i)y^i g_i(z) \quad (14)$$

Similarly, multiplying (5) by $y^i z^j$ and summing over all i and $j > 0$ provides an expression for $H(y,z)$ in terms of $G(y,z)$ and $G(y,0)$:

$$[\lambda(1-y) + \eta]H(y,z) = \xi[G(y,z) - G(y,0)] \ . \quad (15)$$

We also have a normalizing equation, stating that the sum of all probabilities is 1. This can be written as

$$G(1,1) + H(1,1) = 1 \ . \quad (16)$$

3.1   Performance Metrics

Note that $G(y, y)$ is the generating function of the distribution of the sum of the processing and commit queue sizes during work periods. At $y = 1$, $G(1, 1)$ gives the server availability. That is, the unconditional probability that the servers are in a work period. Similarly, $H(y, y)$ is the generating function of the distribution of the sum of the processing and commit queue sizes during commit periods. At $y = 1$, $H(1, 1)$ is the server unavailability; the unconditional probability that the servers are in a commit period.

The total average number of jobs present, L, can be obtained as the sum of the derivatives of $G(y, y)$ and $H(y, y)$ at $y = 1$:

$$L = G'(1, 1) + H'(1, 1) . \tag{17}$$

To derive expressions for these derivatives, start by setting $z = y$ in (12) and (15). This yields

$$[\lambda(1 - y) + \xi]G(y, y) = \xi G(y, 0) + \eta H(y, 1) , \tag{18}$$

and

$$[\lambda(1 - y) + \eta]H(y, y) = \xi[G(y, y) - G(y, 0)] . \tag{19}$$

The function $H(y, 1)$ can be expressed in terms of $G(y, 1)$ and $G(y, 0)$ by setting $z = 1$ in (15):

$$[\lambda(1 - y) + \eta]H(y, 1) = \xi[G(y, 1) - G(y, 0)] . \tag{20}$$

Either of the last three equations at $y = 1$, together with the normalizing condition $G(1, 1) + H(1, 1) = 1$, provide expressions for $G(1, 1)$ and $H(1, 1)$ in terms of $G(1, 0)$:

$$G(1, 1) = \frac{\eta}{\xi + \eta} + \frac{\xi}{\xi + \eta}G(1, 0) , \tag{21}$$

and

$$H(1, 1) = \frac{\xi}{\xi + \eta} - \frac{\xi}{\xi + \eta}G(1, 0) , \tag{22}$$

The value of $G(1, 0)$ is the probability that a work period is in progress and the commit queue is empty.

Differentiating (18) and (20) at $y = 1$, adding the resulting equations and performing cancellations, we obtain

$$G'(1, 1) = \frac{\lambda}{\xi} + \frac{\partial}{\partial y}G(1, 1) . \tag{23}$$

Next, differentiating (19) at $y = 1$ yields

$$H'(1, 1) = \frac{\xi}{\eta}G'(1, 1) + \frac{\lambda}{\eta}H(1, 1) - \frac{\xi}{\eta}\frac{\partial}{\partial y}G(1, 0) . \tag{24}$$

The derivative of $G(y,1)$ at $y = 1$, appearing in the right-hand side of (23), represents the average number of jobs in the processing queue during work periods. Similarly, the derivative of $G(y,0)$ at $y = 1$, appearing in (24), represents the average number of jobs in the processing queue during work periods when the commit queue is empty.

Equation (24), with the addition of $G'(1,1)$ to both sides, allows us to express the total average number of jobs in the system as

$$L = \frac{\xi + \eta}{\eta} G'(1,1) + \frac{\lambda}{\eta} H(1,1) - \frac{\xi}{\eta} \frac{\partial}{\partial y} G(1,0) , \tag{25}$$

where $G'(1,1)$ is given by (23).

Our task is thus reduced to computing the functions $G(y,0)$ and $G(y,1)$, together with their derivatives, at point $y = 1$. In order to determine those quantities, we shall also need to compute the 'boundary' probabilities $p(i,0)$, $p(i,\cdot)$ and $q(i,\cdot)$, for $i = 0, 1, \ldots, N - 1$.

3.2   Determination of $G(y,0)$ and $G(y,1)$

Two equations for the unknown functions $G(y,0)$ and $G(y,1)$ can be obtained from expression (12). Setting $z = 0$ in (12), substituting $H(y,1)$ from (20) and dividing by y, we obtain

$$a(y)G(y,0) + b(y)G(y,1) = \mu S(y,0) , \tag{26}$$

where

$$a(y) = \lambda(1 - y) + N\mu + \frac{\xi\eta}{\lambda(1 - y) + \eta} ,$$

$$b(y) = -\frac{\xi\eta}{\lambda(1 - y) + \eta} ,$$

and, according to (14),

$$S(y,0) = \sum_{i=0}^{N-1} (N - i)y^i p(i,0) .$$

Also, setting $z = 1$ in (12), substituting $H(y,1)$ from (20), and dividing by $(y - 1)$, yields

$$c(y)G(y,0) + d(y)G(y,1) = \mu S(y,1) , \tag{27}$$

where

$$c(y) = \frac{\lambda\xi y}{\lambda(1 - y) + \eta} ,$$

$$d(y) = N\mu - \lambda y - \frac{\lambda\xi y}{\lambda(1 - y) + \eta} ,$$

and

$$S(y,1) = \sum_{i=0}^{N-1} (N - i)y^i p(i,\cdot) .$$

These two equations determine $G(y,0)$ and $G(y,1)$ (and their derivatives) in terms of $S(y,0)$ and $S(y,1)$, which in turn depend on the values of the $2N$ boundary probabilities $p(i,0)$ and $p(i,\cdot)$, $i=0,1,\ldots,N-1$.

We start by deriving expressions for $G(1,1)$ and $G(1,0)$ in terms of the boundary probabilities. Eliminating $G(1,0)$ from (21) and (27) (after setting $y=1$ in the latter), and dividing by μ, we obtain

$$NG(1,1) - \rho = \sum_{i=0}^{N-1}(N-i)p(i,\cdot)\,, \tag{28}$$

where $\rho = \lambda/\mu$ is the offered load.

This equation has a simple intuitive interpretation. The first term in the left-hand side is the average number of servers capable of serving jobs, while the second term, according to Little's result, is the average number of servers actually serving jobs. Hence, the left-hand side represents the steady-state average number of idle servers. The right-hand side represents the same quantity in terms of the processing queue distribution during work periods.

Given $G(1,1)$, we obtain $G(1,0)$ from (21).

Next, differentiating (26) and (27) at $y=1$, we have a set of two equations which determine $\frac{\partial}{\partial y}G(1,0)$ and $\frac{\partial}{\partial y}G(1,1)$ in terms of the boundary probabilities:

$$(N\mu + \xi)\frac{\partial}{\partial y}G(1,0) - \xi\frac{\partial}{\partial y}G(1,1) = r_1\,, \tag{29}$$

and

$$\frac{\lambda\xi}{\eta}\frac{\partial}{\partial y}G(1,0) + (N\mu - \lambda - \frac{\lambda\xi}{\eta})\frac{\partial}{\partial y}G(1,1) = r_2\,, \tag{30}$$

where

$$r_1 = \mu\sum_{i=1}^{N-1}i(N-i)p(i,0) + \lambda\frac{\eta-\xi}{\eta}G(1,0) + \lambda\frac{\xi}{\eta}G(1,1)\,;$$

and

$$r_2 = \mu\sum_{i=1}^{N-1}i(N-i)p(i,\cdot) - \frac{\lambda\xi(\lambda+\eta)}{\eta^2}G(1,0) + \lambda[1 + \frac{\xi(\lambda+\eta)}{\eta^2}]G(1,1)\,.$$

All required quantities are now expressed in terms of the $2N$ probabilities $p(i,0)$ and $p(i,\cdot)$, $i=0,1,\ldots,N-1$. To determine them, we have the N balance Eq. (1) for $i=0,1,\ldots,N-1$, which involve $p(i,0)$ and $q(i,\cdot)$. Another N equations involving $q(i,\cdot)$ and $p(i,\cdot)$ are provided by (6), for $i=0,1,\ldots,N-1$. A further $N-1$ equations involving $p(i,\cdot)$, $q(i,\cdot)$ and $p(i,0)$ are found in (10), for $i=0,1,\ldots,N-2$.

Thus, we have $3N-1$ homogeneous linear equations for the $3N$ unknown probabilities $p(i,0)$, $p(i,\cdot)$ and $q(i,\cdot)$, $i=0,1,\ldots,N-1$. The last, and only non-homogeneous linear equation involving those probabilities is provided by (28).

Eliminating $G(1,0)$ and $G(1,1)$ from (21), (28) and (26) (after setting $y = 1$ in the latter), leads to a normalizing equation expressed only in terms of the unknown probabilities:

$$w \sum_{i=0}^{N-1} (N - i)[p(i, \cdot) - p(i, 0)] + \sum_{i=0}^{N-1} (N - i)p(i, \cdot) = N - \rho(1 + w) , \qquad (31)$$

where

$$w = \frac{N\mu\xi}{\eta(N\mu + \xi)} .$$

Note that, since the left-hand side of the above equation is positive when the probabilities appearing in it exist, a necessary condition for the stability of the two queues is

$$\rho(1 + w) < N . \qquad (32)$$

For sufficiency, one should also add the requirement $\xi > 0$, which ensures the stability of the commit queue. That queue is emptied at every commit instant, regardless of its current size. If $\xi = 0$ (so $w = 0$) and $\rho < N$, the processing queue would be stable but the commit queue would grow without bound.

There are now enough equations to determine all unknown probabilities. Having computed them, one can use (28), (21), (29) and (30) to evaluate $G(y, 0)$ and $G(y, 1)$, and their derivatives, at $y = 1$. Then (25) yields the total average number of jobs, L. In fact, (12) and (15) now provide the full joint distribution of the two queues during work and commit periods. The distributions of the total number of jobs in the system during work and commit periods are given by

$$G(y, y) = \frac{\xi[\lambda(1 - y)G(y, 0) + \eta G(y, 1)]}{[\lambda(1 - y) + \xi][\lambda(1 - y) + \eta]} , \qquad (33)$$

and

$$H(y, y) = \frac{\xi[G(y, y) - G(y, 0)]}{\lambda(1 - y) + \eta} . \qquad (34)$$

3.3 Fast Numerical Algorithm

The computational complexity of the above solution, when the unknown probabilities are computed from a set of $3N$ simultaneous linear equations, is on the order of $O(N^3)$. However, that complexity can be reduced to $O(N)$ by treating the balance equations as recurrences. Starting with $i = 0$, Eqs. (1), (6) and (10) are

$$\lambda p(0, 0) = \eta q(0, \cdot) , \qquad (35)$$

$$(\lambda + \eta)q(0, \cdot) = \xi[p(0, \cdot) - p(0, 0)] , \qquad (36)$$

and, subtracting (1) from (10),

$$(\lambda + \xi)[p(0, \cdot) - p(0, 0)] = \mu p(1, \cdot) . \qquad (37)$$

Set $p(0,0)$ to an arbitrary positive constant, e.g. $p(0,0) = 1$. Then (35), (36) and (37), applied in turn, yield values for $q(0,\cdot)$, $p(0,\cdot)$ and $p(1,\cdot)$ which are proportional to the true probabilities. That constitutes step 0 of the recurrence algorithm.

At step i of the algorithm $(i = 1, 2, \ldots, N-1)$, values for $p(i-1,0)$, $q(i-1,\cdot)$, $p(i-1,\cdot)$ and $p(i,\cdot)$ are already available. The balance equations of step i are

$$(\lambda + i\mu)p(i,0) = \lambda p(i-1,0) + \eta q(i,\cdot) , \tag{38}$$

$$(\lambda + \eta)q(i,\cdot) = \lambda q(i-1,\cdot) + \xi[p(i,\cdot) - p(i,0)] , \tag{39}$$

and, again subtracting (1) from (10),

$$(\lambda + i\mu + \xi)[p(i,\cdot) - p(i,0)] = \lambda[p(i-1,\cdot) - p(i-1,0)] + (i+1)\mu p(i+1,\cdot) . \tag{40}$$

Equations (38) and (39) can be solved for $p(i,0)$ and $q(i,\cdot)$. Then (40) yields $p(i+1,\cdot)$. This completes step i.

After step $N-1$, values proportional to all unknown probabilities $p(i,0)$ and $p(i,\cdot)$ have been computed. The final step is to normalize these values, so that Eq. (31) is satisfied. This is done by computing a normalization constant, C, equal to the ratio between the left-hand side and the right-hand side of (31). Each value is then multiplied by C.

Remark. The stochastic process analysed here could also be formulated as a Quasi-Birth-and-Death proces, and solved by the matrix-geometric method. That would be a numerical solution which, in addition to the boundary probabilities, would require the computation of a matrix R solving a quadratic matrix equation. The relationships revealed by the analysis would not be apparent.

4   Experimental Results

We have implemented the above solution and have carried out several numerical experiments aimed at examining various aspects of the behaviour of the epoch commit policy. A 64-server system is assumed as a representative platform for large-scale processing. The average job execution time is fixed at $1/\mu = 1$, without loss of generality (this simply fixes the unit of time). The other parameters are varied.

The first experiment illustrates the trade-offs between performance and available service capacity. The average total number of jobs in the system, L, is computed as a function of the average work interval, $1/\xi$, for two values of the average commit interval, $1/\eta = 1$ and $1/\eta = 2$. The arrival rate is kept at $\lambda = 30$.

The results are shown in Fig. 2. When the average commit period is 1, the system becomes unstable if the average work interval is shorter than $1/\xi = 0.9$. The service capacity (computed either as $NG(1,1)$ in Eq. (28) or as $N/(1+w)$ as suggested by Eq. (31)), increases from 32 to 50 (rounded to the nearest integer) when $1/\xi$ increases from 1 to 3.4. Consequently, since the offered load is $\rho = 30$, the server utilisation decreases from 94% to 60%.

For these parameters, the optimal average work interval is $1/\xi = 1.9$. The corresponding minimal average number of jobs is $L = 155$ and the average response time is $W = 5.2$. To the left of the optimum, the increase in L is governed by an increasing processing queue, while to the right it is governed by an increasing commit queue.

When the average commit period is twice as long, $1/\eta = 2$, the system would be unstable if the average work period is shorter than about $1/\xi = 1.8$. The service capacity increases from 36 to 46 as $1/\xi$ increases from 2.5 to 4.9. The server utilisation decreases from 83% to 65%. The optimal average work interval is now considerably larger, $1/\xi = 3.7$, and the curve is flatter. This is due to the fact that performance is now dominated by the larger size of the commit queue.

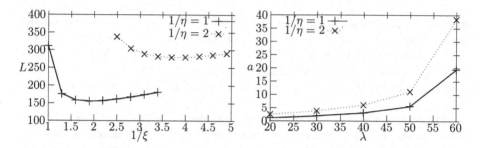

Fig. 2. Varying work interval $N = 64$, $\lambda = 30$, $\mu = 1$

Fig. 3. Optimal average work interval $N = 64$, $\mu = 1$

In the next experiment, the optimal average work interval, $a = 1/\xi^*$, is found numerically by a search and is plotted as a function of the arrival rate. The results are shown in Fig. 3, for the same two values of η. It is to be expected that the heavier the offered load, the larger the work interval should be, and that is indeed what is observed. In addition, we notice that the doubling of the average commit interval leads, almost exactly, to a doubling of the optimal work interval. It seems that what is important here is the ratio η/ξ.

The epoch commit policy that we have analysed is 'preemptive', in the sense that a commit signal interrupts the services in progress and immediately initiates a commit operation. An alternative, 'non-preemptive' policy would treat a commit signal as an order to complete the services in progress, but not start new ones, and then initiate a commit operation. It is interesting to compare the performance of these two policies.

In Fig. 4, the average system occupancy, L, is plotted against the arrival rate, λ, for the preemptive and non-preemptive policies. The latter is evaluated by simulation, since we do not have analytical results for the non-preemptive model. The other parameters are as in the previous example, with $\eta = 1$ and $1/\xi = 1.9$ (that is the optimal value for $\lambda = 30$, from Fig. 2).

The figure shows that, as the offered load increases, the non-preemptive policy performs considerably worse than the preemptive one. This is explained by the

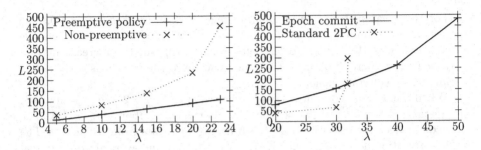

Fig. 4. To preempt or not to preempt? $N = 64$, $1/\xi = 1.9$, $\eta = 1$, $\mu = 1$

Fig. 5. To use epoch commit or not? $N = 64$, $\eta = 1$, $\mu = 1$

following observation. While services are being completed following a commit signal, more and more servers remain idle, unable to start new transactions, thus reducing the available service capacity. The effect of that reduction becomes more significant as the load increases. Also, postponing the start of the commit operation causes the completed transactions to remain in the commit queue for longer.

There is another important question that should be addressed. Is it always advantageous to use the epoch commit policy? Intuitively, it seems that the 'standard 2PC' policy which executes the 2PC protocol at the end of every transaction without stopping the other servers, might perform better than epoch commit when the offered load is light. To illustrate the situation, we have compared the two policies as the arrival rate increases. The standard 2PC policy is modelled by adding a commit operation to each transaction processing time, letting other transactions proceed without interference. Thus, the standard 2PC system behaves like an $M/G/n$ queue where the service times are sums of two exponentially distributed random variables with means $1/\mu$ and $1/\eta$, respectively.

An effective and simple approximation for that model was proposed in Whitt [17]. It works by first evaluating the average number of jobs *waiting* in the $M/M/n$ queue, $Q_{M/M/n}$, and then applying an adjustment to account for the non-exponential distribution of the service times:

$$L_{M/G/n} = \lambda E(T) + \frac{E(T^2)}{2E(T)^2} Q_{M/M/n} , \qquad (41)$$

where $E(T)$ and $E(T^2)$ are the first and second moments of the service time. An expression for $Q_{M/M/n}$ is provided by Erlang's delay formula, also known as the Erlang's C formula (e.g., see [10]).

Figure 5 shows the average occupancy, L, plotted against λ, under the epoch commit and the standard 2PC policies. A 64-server system is again assumed, with $\eta = 1$. The epoch commit policy is assumed to use the optimal work interval, determined by a search for each λ.

We observe that indeed the standard 2PC policy performs better than epoch commit when the offered load is light. However, the system becomes unstable under standard 2PC when $\lambda \geq 32$. The average occupancy and the average response time become infinite. On the other hand, the epoch commit policy remains stable for all $\lambda < 64$ by choosing the work interval appropriately. That policy is clearly preferable if the offered load is likely to become heavy.

5   Non-exponential Work Periods

Since the intervals between commit signals are controlled by the system, they are likely to be nearly constant, rather than being distributed exponentially. It is therefore desirable to analyse a model where those intervals have a low coefficient of variation. This can be done by assuming that they have an Erlang distribution with K phases. In other words, the interval between the start of a work period and the next commit signal is a sum of K independent exponentially distributed random variables, each with parameter ξ. Such an interval starts in phase 1, and a commit signal is issued at the end of phase K. If the commit queue is then empty, a new work interval starts immediately in phase 1; otherwise the new work interval starts in phase 1 after the completion of the commit operation.

The mean of the Erlang distribution is K/ξ. The coefficient of variation (variance divided by the square of the mean) does not depend on ξ and is equal to $1/K$. The larger the number of phases used, the closer to constant would be the work periods, but also themore complex would be the solution.

The system state during work periods is now described by three integers: i, j and k, where i is the size of the processing queue, j is the size of the commit queue and k is the phase of the work period. The corresponding steady-state probabilities are denoted by $p_k(i,j)$, $k = 1, 2, \ldots, K$. The state probabilities during commit periods are $q(i,j)$, as before.

We now need K two-dimensional generating functions, $G_k(y,z)$, describing the joint distribution of the processing and commit queues during work periods:

$$G_k(y,z) = \sum_{i=0}^{\infty} \sum_{j=0}^{\infty} p_k(i,j) y^i z^j \; ; \; k = 1, 2, \ldots, K \; . \tag{42}$$

Writing the steady-state balance equations and transforming them into generating functions, we obtain a set of equations that generalize (12) and depend on the phase k.

$$a(y,z)G_1(y,z) = \xi y G_K(y,0) + \eta y H(y,1) + \mu(y-z)S_1(y,z) \; , \tag{43}$$

and

$$a(y,z)G_k(y,z) = \xi y G_{k-1}(y,z) + \mu(y-z)S_k(y,z) \; ; \; k = 2, 3, \ldots, K \; , \tag{44}$$

where $a(y,z)$ is the same as in (13), and

$$S_k(y,z) = \sum_{i=0}^{N-1} (N-i) y^i \sum_{j=0}^{\infty} p_k(i,j) z^j \; ; \; k = 1, 2, \ldots, K \; . \tag{45}$$

The two-dimensional generating function $H(y, z)$, defined as in (11) now satisfies an equation similar to (15) but involving $G_K(y, z)$:

$$[\lambda(1 - y) + \eta]H(y, z) = \xi[G_K(y, z) - G_K(y, 0)] .\qquad(46)$$

The generating functions for the total number of jobs in the system during phase k of a work period, and during a commit period, are $G_k(y, y)$ and $H(y, y)$, respectively. The average total number of jobs in the system, L, is given by the sum of the derivatives of those functions at $y = 1$, in an expression similar to (17):

$$L = \sum_{k=1}^{K} G'_k(1, 1) + H'(1, 1) .\qquad(47)$$

Setting $z = y$ in (43), (44) and (46) allows us to express the functions $G_k(y, y)$, for all k, and $H(y, y)$, in terms of $G_K(y, 0)$ and $G_K(y, 1)$.

The rest of the analysis will be presented in outline only. Setting $z = 0$ and $z = 1$ in (43) and (44) produces a set of $2K$ equations involving the functions $G_k(y, 0)$ and $G_k(y, 1)$ for $k = 1, 2, \ldots, K$, together with the boundary probabilities $p_k(i, 0)$ and $p_k(i, \cdot)$ for $i < N$. These equations can be reduced explicitly to two equations for $G_K(y, 0)$ and $G_K(y, 1)$, which are generalisations of (26) and (27).

At that point, the unknown functions are expressed in terms of $2KN$ boundary probabilities, $p_k(i, 0)$ and $p_k(i, \cdot)$, for $k = 1, 2, \ldots, K$ and $i = 0, 1, \ldots, N-1$. The latter, plus the N probabilities $q(i, \cdot)$, are determined by solving a set of $(2K + 1)N$ simultaneous linear equations, including balance equations and normalizing equation. Additional $K - 1$ equations are provided by the zeros of a certain polynomial in the interior of the unit disk.

6   Conclusions

We have presented a model of a practical and efficient epoch commit policy for distributed processing systems. The solution obtained is exact. A simple numerical algorithm yields performance measures and can be used to determine the optimal average length of the work interval under different parameter settings. A comparison between the preemptive and non-preemptive versions of the policy has confirmed and quantified the intuition that the former is much more efficient than the latter. The epoch commit policy has also been compared with the standard 2PC policy, which has a lower saturation point.

A generalization to non-exponential work intervals is possible in principle and is described in outline. The solution would be considerably more complex and computationally demanding. The details of the analysis, and the implementation of the solution, are left as a topic for future work.

References

1. Balsamo, S., de Nitto Personé, V., Onvural, R.: Analysis of Queueing Networks with Blocking, Kluwer Academic Publishers, Alphen aan den Rijn (2001)

2. Bernstein, P.A., Hadzilacos, V., Goodman, N.: Concurrency Control and Recovery in Database Systems, Addison-Wesley, Boston (1987)
3. Birke, R., Giurgiu, I., Chen, L.Y., Wiesmann, D., Engbersen, T.: Failure analysis of virtual and physical machines: patterns, causes and characteristics. In: 44th Annual IEEE/IFIP International Conference on Dependable Systems and Networks, pp. 1–12 (2014)
4. Curino, C., Zhang, Y., Jones, E.P.C., Madden, S.: Schism: a workload-driven approach to database replication and partitioning. Procs. VLDB Endowment $3(1)$, 48–57 (2010)
5. Das, S., Agrawal, D., El Abbadi, A.: G-Store: a scalable data store for transactional multi key access in the cloud. In: Proceedings of 1st ACM Symposium on Cloud Computing, pp. 163–174 (2010)
6. DeWitt, D.J., Katz, H.R., Olken, F., Shapiro, L.D., Stonebraker, M., Wood, D.A.: Implementation techniques for main memory database systems. In: Proceedings of SIGMOD'84 Annual Meeting, pp. 1–8 (1984)
7. Garraghan, P., Townend, P., Xu, J.: An empirical failure-analysis of a large-scale cloud computing environment. In: IEEE 15th International Symposium on High-Assurance Systems Engineering, pp. 113–120 (2014)
8. Lin, Q., Chang, P., Chen, G., Ooi, B.C., Tan, K.L., Wang, Z.: Towards a non-2PC transaction management in distributed database systems. In: Proceedings of ACM SIGMOD International Conference on Management of Data, pp. 1659–1674 (2016)
9. Lu, Y., Yu, X., Cao, L., Madden, S.: Epoch-based commit and replication in distributed OLTP databases. Procs. VLDB Endowment $14(5)$, 743–756 (2021)
10. Mitrani, I.: Probabilistic Modelling, Cambridge University Press, Cambridge (1998)
11. Perros, H.G.: Queueing Networks with Blocking. Exact and Approximate Solutions. Oxford University Press, Oxford (1994)
12. Ren, K., Li, D., Abadi, D.J.: SLOG: serializable, low-latency, geo-replicated transactions. Procs. VLDB Endowment $12(11)$, 1747–1761 (2019)
13. Thomson, A., Diamond, T., Weng, S., Ren, K., Shao, P., Abadi, D.J.: Calvin: fast distributed transactions for partitioned database systems. In: Proceedings of the 2012 ACM SIGMOD International Conference on Management of Data, pp. 1–12 (2012)
14. Tu, S., Zheng, W., Kohler, E., Liskov, B., Madden, S.: Speedy transactions in multicore in-memory databases. In: ACM SIGOPS 24th Symposium on Operating Systems Principles, pp. 18–32 (2013)
15. Wang, G., Zhang, L., Xu, W.: What can we learn from four years of data center hardware failures. In: 47th Annual IEEE/IFIP International Conference on Dependable Systems and Networks, pp. 25–36 (2017)
16. Waudby, J., Ezhilchelvan, P., Mitrani, I., Webber, J.: A performance study of epoch-based commit protocols in distributed OLTP databases. In: Proceedings of 41st International Symposium on Reliable Distributed Systems (SRDS), pp. 189–200 (2022)
17. Whitt, W.: Approximations for the GI/G/m queue. Prod. Oper. Manage. $2(2)$, 114–161 (1993)

Causal Reversibility Implies Time Reversibility

Marco Bernardo[1]([✉]), Ivan Lanese[2], Andrea Marin[3], Claudio A. Mezzina[1],
Sabina Rossi[3], and Claudio Sacerdoti Coen[4]

[1] Department of Pure and Applied Sciences,
University of Urbino, Urbino, Italy
`marco.bernardo@uniurb.it`
[2] Focus Team, University of Bologna & INRIA,
Bologna, Italy
[3] Department of Environmental Sciences, Informatics
and Statistics, University Ca' Foscari, Venice, Italy
[4] Department of Informatics – Science and Engineering,
University of Bologna, Bologna, Italy

Abstract. Several notions of reversibility exist in the literature. On the
one hand, causal reversibility establishes that an action can be undone
provided that all of its consequences have been undone already, thereby
making it possible to bring a system back to a past consistent state. On
the other hand, time reversibility stipulates that the stochastic behavior
of a system remains the same when the direction of time is reversed,
which supports efficient performance evaluation. In this paper we show
that causal reversibility is a sufficient condition for time reversibility. The
study is conducted on extended labeled transition systems. Firstly, they
include a forward and a backward transition relations obeying the loop
property. Secondly, their transitions feature an independence relation
as well as rates for their exponentially distributed random durations.
Our result can thus be smoothly applied to concurrent and distributed
models, calculi, and languages that account for performance aspects.

1 Introduction

Reversible computing is a paradigm that allows computational steps to be executed not only in the standard forward direction, but also in the backward
one so as to recover past states. It has attracted an increasing interest due to
its applications in many areas, including low-power computing [16,2], program
debugging [8,27,20], robotics [24], wireless communications [38], fault-tolerant
systems [6,41,17,40], biological modeling [33,34], and parallel discrete-event simulation [30,36]. However, different communities instantiated the idea of reversible
computing in different ways, in order to better fit the intended application area.

In the field of concurrent and distributed systems, a critical aspect of
reversibility is that there may not be a total order over executed actions, hence
the idea of "undoing actions in reverse order" used in other settings does not
apply. This triggered the proposal of *causal reversibility* [5], whereby an action

N. Jansen and M. Tribastone (Eds.): QEST 2023, LNCS 14287, pp. 270–287, 2023.
https://doi.org/10.1007/978-3-031-43835-6_19

can be undone provided that its consequences, if any, have been undone beforehand. Notably, in this setting the concept of causality is used in place of the
concept of time to decide whether an action can be undone or not.

In the field of performance evaluation, instead, the notion of *time reversibility*
is considered [13]. It studies the conditions under which the stochastic behavior
of a system remains the same when the direction of time is reversed. In addition to its theoretical interest, this property allows one to tackle the problems
related to state space explosion and numerical stability often encountered when
solving performance models, especially those based on continuous-time Markov
chains [14,39], where rates of exponential distributions govern state changes.

The two notions evolved independently until very recently. As far as we know,
they were jointly investigated only in the setting of the reversible Markovian
process calculus of [3]. After defining its syntax and semantics by following the
method for reversing process calculi of [32], the calculus was shown to satisfy
causal reversibility by construction through the application of the axiomatic
technique of [22] after importing the notion of concurrent transitions from [5].
On the other hand, two sufficient conditions for time reversibility were provided.
The former requires that every backward rate is equal to the corresponding
forward rate regardless of the syntactical structure of process terms, whilst the
latter requires that parallel composition does not occur within the scope of action
prefix or alternative composition regardless of the values of backward rates.

In [3] it was conjectured that the entire calculus should be time reversible by
construction, which would imply the full robustness of the method of [32] with
respect to both forms of reversibility. In this paper we show that the conjecture
is indeed true by proving that time reversibility follows from causal reversibility.

To be precise, the latter alone is not enough, as it accounts for actions but
not for rates at which actions take place. Causal reversibility stipulates that it is
possible to backtrack correctly, i.e., without encountering previously inaccessible
states, and flexibly, i.e., along any causally equivalent path, which is a path
where independent actions are undone in an order possibly different from the
one in which they were executed when going forward. The simplest case is given
by a commuting square, which is formed by four transitions $t : s \xmapsto{a_1} s'_1$, $u :$
$s \xmapsto{a_2} s'_2$, $u' : s'_1 \xmapsto{a_2} s''$, $t' : s'_2 \xmapsto{a_1} s''$ where t and u are independent of
each other. To ensure time reversibility, in addition to causal reversibility we
need to require that $rate(t) \cdot rate(u') = rate(u) \cdot rate(t')$ and $rate(\underline{u'}) \cdot rate(\underline{t}) = rate(\underline{t'}) \cdot rate(\underline{u})$ where underlines denote reverse transitions. Requiring identical
rate products along a commuting square is more general than requiring that
opposite transitions in the commuting square have the same rate, where the
latter is the case when doing or undoing two exponentially timed actions that
are causally independent.

Our study is conducted on labeled transition systems [12] featuring a forward
and a backward transition relations obeying the loop property [5]: between any
two states there can only be pairs of identically labeled transitions, of which
one is forward and the other is backward. Moreover, transitions are enriched
with an independence relation [35] as well as rates of exponentially distributed

random durations. Thus our result can be smoothly applied to concurrent and distributed models, calculi, and languages including performance aspects. In particular, those already shown to meet causal reversibility would automatically satisfy time reversibility too, as identical rate products usually holds for them.

This paper is organized as follows. In Sect. 2 we recall some background notions and results about causal and time reversibilities. In Sect. 3 we present our investigation of the relationships between the two forms of reversibility on the suitably extended model of labeled transition system. In Sect. 4 we provide an illustrating example based on dining philosophers. In Sect. 5 we conclude with some final remarks and directions for future work.

2   Background on Reversibility

2.1   Causal Reversibility of Concurrent Systems

The behavior of computing systems can be represented through state-transition graphs in which transitions are labeled with actions whose execution causes state changes. Reversibility in a computing system has to do with the possibility of reverting the last performed action. In a sequential system this is very simple as there is just one last performed action, hence the only challenge is how to store the information needed to reverse that action.

In a concurrent system the situation is much more complex, because the last performed action may not be uniquely identifiable. Indeed, there might be several concurrent last actions. A good approximation is to consider as last action every action that has not caused any other action yet. This is at the basis of the notion of *causal reversibility*[1], which combines reversibility with causality [5].

Intuitively, an executed action can be undone provided that all of its consequences, if any, have been undone beforehand, so that no previously inaccessible state can be encountered. On the other hand, some flexibility is allowed while backtracking in the sense that one is not required to preserve history, i.e., to follow the same path undertaken in the forward direction. Indeed, one can take any path causally equivalent to the forward one, i.e., any path in which independent actions are performed in an order possibly different from the forward one.

Following [22] we introduce a behavioral model consisting of a labeled transition system [12] extended with an independence relation over transitions [35]. Since we are interested in reversible systems, the model is equipped with two transition relations – a forward one and a backward one – such that, for all actions, between any pair of states either there are no transitions labeled with that action, or there are two such transitions with one being forward and the other being backward. This is the so-called *loop property*, which establishes that any executed action can be undone and any undone action can be redone [5].

Definition 1 (reversible LTS with independence). *A* reversible labeled transition system with independence (RLTSI) *is a tuple* $(S, A, \rightarrow, \dashrightarrow, \iota)$ *where:*

---

[1] It is often called causal-consistent reversibility in the literature after [18].

- $S \neq \emptyset$ is an at most countable set of states.
- $A \neq \emptyset$ is a countable set of actions.
- $\rightarrow \subseteq S \times A \times S$ is a forward transition relation.
- $\dashrightarrow \subseteq S \times A \times S$ is a backward transition relation.
- Loop property: for all $s, s' \in S$ and $a \in A$, $(s, a, s') \in \rightarrow$ iff $(s', a, s) \in \dashrightarrow$.
- $\iota \subseteq \mapsto \times \mapsto$ is an irreflexive and symmetric independence relation over transitions, where $\mapsto = \rightarrow \dot{\cup} \dashrightarrow$ (disjoint union). ∎

We use s and r to range over states, t and u to range over transitions, and a and b to range over actions. A forward transition from s to s' labeled with a (a-transition for short) is denoted by $s \xrightarrow{a} s'$ instead of $(s, a, s') \in \rightarrow$. The notation is similar, i.e., $s' \dashrightarrow^{a} s$, for a backward transition. Given a forward (resp. backward) transition t, we denote by $\underline{t}$ the corresponding backward (resp. forward) transition, whose existence is guaranteed by the loop property. If t is a transition from s to s', we call s the source and s' the target of t. Two transitions are said to be coinitial if they have the same source and cofinal if they have the same target. Two transitions are composable when the target of the first transition coincides with the source of the second transition.

A possibly empty, finite sequence of pairwise composable transitions is called a path. We use ω to range over paths, ε to denote the empty path, and $|\omega|$ to indicate the length of ω expressed as the number of transitions constituting ω. The notions of source, target, coinitiality, cofinality, and composability naturally extend to paths. We call cycle a nonempty path whose source and target coincide. If ω is a forward (resp. backward) path, then we denote by $\underline{\omega}$ the corresponding backward (resp. forward) path – in which the actions appear in the reverse order – whose existence is again guaranteed by the loop property.

According to [22] we formalize causal reversibility over RLTSI models through the property of causal consistency. We start with the notion of commuting square, a fragment of RLTSI consisting of two pairs of identically labeled transitions, say $t : s \xmapsto{a_1} s'_1$ and $t' : s'_2 \xmapsto{a_1} s''$ together with $u : s \xmapsto{a_2} s'_2$ and $u' : s'_1 \xmapsto{a_2} s''$. Across the two pairs we have two coinitial and independent transitions (t and u) as well as two cofinal transitions (t' and u'), such that each coinitial transition is composable with only one of the two cofinal transitions and along the two resulting paths ($t\,u'$ and $u\,t'$) the actions turn out to commute ($a_1\,a_2$ and $a_2\,a_1$). This can be viewed as the effect of doing or undoing two actions (a_1 and a_2) that are causally independent of each other.

Definition 2 (commuting square). Transitions t, u, u', t' form a commuting square iff:

- $t : s \xmapsto{a_1} s'_1$, $u : s \xmapsto{a_2} s'_2$, $u' : s'_1 \xmapsto{a_2} s''$, $t' : s'_2 \xmapsto{a_1} s''$.
- $(t, u) \in \iota$. ∎

As observed in [22], if four forward transitions form a commuting square, it is reasonable to expect that the four corresponding backward transitions form a commuting square too, and vice versa. This is achieved by applying several times the propagation of coinitial independence[2] property. It allows independence to propagate from the initial corner of a commuting square to one of the two adjacent corners by exploiting reverse transitions so as to ensure coinitiality.

Definition 3 (propagation of coinitial independence). *An RLTSI meets* propagation of coinitial independence (PCI) *iff it holds that* $(\underline{t}, u') \in \iota$ *for all commuting squares* t, u, u', t'. ∎

Then we introduce the notion of *causal equivalence* over paths [5], which relies on two operations. Swap has to do with commuting squares and identifies the two coinitial and cofinal paths of length 2 constituting each such square. Cancellation identifies with the empty path ε any path of length 2 formed by a transition and its reverse. Causal equivalence is closed with respect to path composition by definition, i.e., when each of two causally equivalent paths is composed with a third path, then the two resulting paths are causally equivalent too. Unlike [5], this is formalized below by explicitly plugging the swap and cancellation operations into the context of two arbitrary paths, as it will turn out to be useful for a suitable treatment of cycles in the main result of our paper.

Definition 4 (causal equivalence). Causal equivalence *is the smallest equivalence relation* $\asymp$ *over paths that satisfies the following for all paths* $\omega_1, \omega_2, \omega_3, \omega_4$ *that are composable with the transitions mentioned in the two operations below:*

(swap) *For all* t, u, u', t' *forming a commuting square,* $\omega_1\, t\, u'\, \omega_2 \asymp \omega_1\, u\, t'\, \omega_2$.
(cancellation) *For all transitions* t, $\omega_1\, t\, \underline{t}\, \omega_2 \asymp \omega_1\, \omega_2$ *and* $\omega_3\, \underline{t}\, t\, \omega_4 \asymp \omega_3\, \omega_4$. ∎

As noted in [22, Lemma 4.9], there is a relationship between causally equivalent paths in terms of the number of transitions labeled with a certain action.

Proposition 1. *For all paths* ω_1, ω_2 *and actions* a, *if* $\omega_1 \asymp \omega_2$ *then the number of* a-transitions in ω_1 *is equal to the number of* a-transitions in ω_2, *where each such forward (resp. backward) transition is counted* +1 *(resp.* −1*). In particular, if* ω_1 *and* ω_2 *are both forward or both backward, then* $|\omega_1| = |\omega_2|$. ∎

We are finally in a position of defining causal consistency, which guarantees the correctness and flexibility of rollbacks.

Definition 5 (causal consistency). *An RLTSI meets* causal consistency (CC) *iff every two coinitial and cofinal paths* ω_1, ω_2 *satisfy* $\omega_1 \asymp \omega_2$. ∎

As shown in [22, Corollary 3.8], when CC holds there can be at most one pair[3] of identically labeled transitions – of which one is forward and the other is backward – between any two distinct states. Moreover, we observe that under CC there cannot be self-loops, i.e., transitions whose source and target coincide. In both cases – relevant for Sect. 3 – the reason is that, according to Definition 4, a path formed by a single transition is causally equivalent only to itself.

---

[2] It was called coinitial propagation of independence (CPI) in [22].
[3] A property called unique transitions (UT) in [22].

Definition 6 (uniqueness of pairs). *An RLTSI meets* uniqueness of pairs
(UP) *iff, for all states* s, s', *whenever* $s \xmapsto{a_1} s'$ *and* $s \xmapsto{a_2} s'$, *then* $a_1 = a_2$.   ∎

Proposition 2. *If an RLTSI meets CC, then it meets UP too.*   ∎

Definition 7 (absence of self-loops). *An RLTSI meets* absence of self-loops
(AS) *iff, for all states* s, *there is no action* a *such that* $s \xmapsto{a} s$.   ∎

Proposition 3. *If an RLTSI meets CC, then it meets AS too.*   ∎

2.2   Time Reversibility of Continuous-Time Markov Chains

A different notion of reversibility is considered in the performance evaluation
field. Given a stochastic process, which describes the evolution of some random
phenomenon over time through a set of random variables, one for each time
instant, reversibility has to do with time. We illustrate it in the specific case
of *continuous-time Markov chains*, which are discrete-state stochastic processes
characterized by the *memoryless property* [14]: the probability of moving from
one state to another does not depend on the particular path that has been
followed in the past to reach the current state, hence that path can be forgotten.

Definition 8 (continuous-time Markov chain). *A stochastic process* $X(t)$
taking values from a discrete state space S *for* $t \in \mathbb{R}_{\geq 0}$ *is a* continuous-time
Markov chain *(CTMC) iff* $\Pr\{X(t_{n+1}) = s_{n+1} \mid X(t_i) = s_i, 0 \leq i \leq n\} =$
$\Pr\{X(t_{n+1}) = s_{n+1} \mid X(t_n) = s_n\}$ *for all* $n \in \mathbb{N}$, *time instants* $t_0 < t_1 < \cdots <$
$t_n < t_{n+1} \in \mathbb{R}_{\geq 0}$, *and states* $s_0, s_1, \ldots, s_n, s_{n+1} \in S$.   ∎

A CTMC can be equivalently represented as a labeled transition system or
as a state-indexed matrix. In the first case, each transition is labeled with some
probabilistic information describing the evolution from the source state s to the
target state s' of the transition itself. In the second case, the same information is
stored into an entry, indexed by those two states, of a square matrix. The value
of this probabilistic information is, in general, a function of time.

We restrict ourselves to *time-homogeneous* CTMCs, in which conditional
probabilities of the form $\Pr\{X(t + t') = s' \mid X(t) = s\}$ do not depend on t, so
that the information considered above is given by $\lim_{t' \to 0} \frac{\Pr\{X(t+t')=s' \mid X(t)=s\}}{t'}$.
This limit yields a number called the *rate* at which the CTMC moves from state s
to state s' and characterizes the exponentially distributed random time taken by
the considered move. It can be shown that the sojourn time in any state $s \in S$
is independent from the other states. Moreover, it is exponentially distributed
with rate given by the sum of the rates of the moves from s. The average sojourn
time in s is the inverse of such a sum and the probability of moving from s to s'
is the ratio of the corresponding rate to the aforementioned sum.

A CTMC is *irreducible* iff each of its states is reachable from every other
state with probability greater than 0. A state $s \in S$ is *recurrent* iff the CTMC
will eventually return to s with probability 1, in which case s is called *positive
recurrent* iff the expected number of steps until the CTMC returns to it is finite.

A CTMC is *ergodic* iff it is irreducible and all of its states are positive recurrent. Ergodicity coincides with irreducibility in the case that the CTMC has finitely many states as they form a finite, strongly connected component.

Every time-homogeneous and ergodic CTMC $X(t)$ is *stationary*, which means that $(X(t_i + t'))_{1 \leq i \leq n}$ has the same joint distribution as $(X(t_i))_{1 \leq i \leq n}$ for all $n \in \mathbb{N}_{\geq 1}$ and $t_1 < \cdots < t_n, t' \in \mathbb{R}_{\geq 0}$. In this case, $X(t)$ has a unique *steady-state probability distribution* $\boldsymbol{\pi} = (\pi(s))_{s \in S}$ that fulfills $\pi(s) = \lim_{t \to \infty} \Pr\{X(t) = s \mid X(0) = s'\}$ for any $s' \in S$ because the CTMC has reached equilibrium.

These probabilities are computed by solving the linear system of *global balance equations* $\boldsymbol{\pi} \cdot \mathbf{Q} = \mathbf{0}$ subject to $\sum_{s \in S} \pi(s) = 1$ and $\pi(s) \in \mathbb{R}_{>0}$ for all $s \in S$. The *infinitesimal generator matrix* $\mathbf{Q} = (q_{s,s'})_{s,s' \in S}$ contains, for each pair of distinct states, the rate of the corresponding move, which is 0 in the absence of a direct move between them. In contrast, $q_{s,s} = -\sum_{s' \neq s} q_{s,s'}$ for all $s \in S$, i.e., every diagonal element contains the opposite of the total exit rate of the corresponding state, so that each row of $\mathbf{Q}$ sums up to 0. Therefore, $\boldsymbol{\pi} \cdot \mathbf{Q} = \mathbf{0}$ means that, once reached equilibrium, for every state the incoming probability flux equals the outgoing probability flux. Self-loops are not taken into account in the construction of $\mathbf{Q}$, which is akin to the AS property in Definition 7.

Due to state space explosion and numerical stability problems, the calculation of the solution of the global balance equation system is not always feasible [39]. However, it can be tackled in the case that the behavior of the considered CTMC remains the same when the direction of time is reversed [13].

Definition 9 (time reversibility). *A CTMC $X(t)$ is* time reversible (TR) *iff $(X(t_i))_{1 \leq i \leq n}$ has the same joint distribution as $(X(t' - t_i))_{1 \leq i \leq n}$ for all $n \in \mathbb{N}_{\geq 1}$ and $t_1 < \cdots < t_n, t' \in \mathbb{R}_{\geq 0}$.* ∎

In this case $X(t)$ and its reversed version $X^{\mathrm{r}}(t) = X(-t)$, $t \in \mathbb{R}_{\geq 0}$, are stochastically identical. In particular, they are stationary and share the same steady-state probability distribution $\boldsymbol{\pi}$. The following two necessary and sufficient conditions for time reversibility of stationary CTMCs were provided in [13].

Theorem 1. *Let $X(t)$ be a stationary CTMC. Then the following statements are equivalent:*

1. *$X(t)$ is time reversible.*
2. *For all distinct $s, s' \in S$, it holds that $\pi(s) \cdot q_{s,s'} = \pi(s') \cdot q_{s',s}$.*
3. *For all distinct $s_1, \ldots, s_n \in S$, $n \geq 2$, it holds that $q_{s_1,s_2} \cdot \cdots \cdot q_{s_{n-1},s_n} \cdot q_{s_n,s_1} = q_{s_1,s_n} \cdot q_{s_n,s_{n-1}} \cdot \cdots \cdot q_{s_2,s_1}$.* ∎

Condition 2 is based on the *partial balance equations* $\pi(s) \cdot q_{s,s'} = \pi(s') \cdot q_{s',s}$, called detailed balance equations in [13]. In order for them to be satisfied, it is necessary that both $q_{s,s'}$ and $q_{s',s}$ are equal to 0 or different from 0, i.e., between any pair of distinct states s and s' there must be either no transitions, or exactly one transition from s to s' and one transition from s' back to s, which is a variant of the loop property in Definition 1. It is worth observing that the sum of the partial balance equations for $s \in S$ yields the global balance equation

$\pi(s) \cdot |q_{s,s}| = \sum_{s' \neq s} \pi(s') \cdot q_{s',s}$. Condition 3, requiring products of rates along forward and backward cycles to coincide, is the one that we will exploit to prove the main result of our paper. It is trivially satisfied when the considered distinct states do not form a cycle, as in that case at least one of the rates is 0.

A well-known example of time-reversible CTMCs is given by stationary birth-death processes [13]. A *birth-death process* comprises a totally ordered set of states, such that every state different from the final one has a (birth) transition to the next state and every state different from the initial one has a (death) transition to the previous state. Time reversibility extends to tree-like variants of birth-death processes [13], where each such variant comprises a partially ordered set of states such that every non-final state may have several birth transitions, while every non-initial state has one death transition to its only parent state.

3 Relationships Between Causal and Time Reversibilities

3.1 Extending RLTSI with Rates

In this section we introduce a Markovian extension of RLTSI, in which transitions are labeled not only with actions but also with positive real numbers representing rates of exponentially distributed random durations. This model can also be viewed as an action-labeled CTMC equipped with a transition independence relation as well as a generalization of the loop property, called *rate loop property*, in which the rates of corresponding forward and backward transitions are allowed to be different. From now on we use $\lambda, \mu, \xi, \delta, \gamma$ to range over rates.

Definition 10 (reversible Markovian LTS with independence). *A reversible Markovian labeled transition system with independence (RMLTSI) is a tuple $(S, A \times \mathbb{R}_{>0}, \rightarrow, \dashrightarrow, \iota)$ where:*

- *$S \neq \emptyset$ is at most countable and $A \neq \emptyset$ is countable.*
- *$\rightarrow \subseteq S \times (A \times \mathbb{R}_{>0}) \times S$ and $\dashrightarrow \subseteq S \times (A \times \mathbb{R}_{>0}) \times S$.*
- *Rate loop property: for all $s, s' \in S$ and $a \in A$, there exists $\lambda \in \mathbb{R}_{>0}$ such that $(s, a, \lambda, s') \in \rightarrow$ iff there exists $\mu \in \mathbb{R}_{>0}$ such that $(s', a, \mu, s) \in \dashrightarrow$.*
- *$\iota \subseteq \mapsto \times \mapsto$ is irreflexive and symmetric, where $\mapsto = \rightarrow \cup \dashrightarrow$.*   ■

The notions of corresponding forward and backward transitions/paths, commuting square, propagation of coinitial independence (PCI), causal equivalence ($\asymp$), causal consistency (CC), uniqueness of pairs (UP), and absence of self-loops (AS) extend to RMLTSI models along with their related propositions by abstracting from rates, whereas time reversibility (TR) and its necessary and sufficient conditions extend to RMLTSI models by abstracting from actions.

Between any two different states of an RMLTSI there may be several forward transitions (and as many corresponding backward transitions) where the forward (resp. backward) transitions differ for their actions or their rates. In this case, the two entries indexed by those two states in the infinitesimal generator matrix $\mathbf{Q}$ of the *underlying CTMC* would respectively contain the sum of the rates of the forward and backward transitions. Likewise, any state of an RMLTSI may have several self-loops differing for their actions or their rates, but they would be all ignored due to the definition of the diagonal elements of $\mathbf{Q}$. We recall from Propositions 2 and 3 that, under CC, the RMLTSI meets UP and AS too.

3.2 From Causal Reversibility to Time Reversibility

We now prove the main result of our paper, i.e., that CC implies TR, by exploiting the necessary and sufficient condition for TR related to products of rates along corresponding forward and backward cycles, i.e., Theorem 1(3).

To this purpose, denoting by $rate(t)$ the rate labeling transition t, given a path ω formed by the pairwise composable transitions t_i, $1 \leq i \leq |\omega|$, we indicate with $rateprod(\omega) = \prod_{1 \leq i \leq |\omega|} rate(t_i)$ the product of the rates labeling the transitions in ω, where $rateprod(\omega) = 1$ when $|\omega| = 0$.

We observe that positive real numbers with multiplication and unit form a *cancellative monoid*, i.e., for all $\lambda, \mu, \xi \in \mathbb{R}_{>0}$ it holds that $\lambda \cdot \xi = \mu \cdot \xi \implies \lambda = \mu$ and $\xi \cdot \lambda = \xi \cdot \mu \implies \lambda = \mu$. This fact will be exploited in the proof of the forthcoming lemma from which the main result will follow.

CC alone is not sufficient to obtain TR. The reason is that CC relies on $\asymp$ that in turn relies on commuting squares, where these squares now include rates among which we do not know whether specific relationships exist. Given a commuting square t, u, u', t', we consider the property according to which its two coinitial and cofinal paths $t\,u'$ and $u\,t'$ feature the same product of rates, and this holds for $\underline{u'}\,\underline{t}$ and $\underline{t'}\,\underline{u}$ in the reverse commuting square too. Note that this is more general than requiring that opposite transitions in the commuting square have the same rate, which is what we would get when doing or undoing two exponentially timed actions that are causally independent of each other.

Definition 11 (product preservation along squares). *An RMLTSI meets product preservation along squares (PPS) iff, in all commuting squares t, u, u', t', $rate(t) \cdot rate(u') = rate(u) \cdot rate(t')$ and $rate(\underline{u'}) \cdot rate(\underline{t}) = rate(\underline{t'}) \cdot rate(\underline{u})$.* ∎

As a preliminary result, we show that under CC and PPS, whenever we consider two coinitial and cofinal paths, say ω and ω', the product of the rates along ω (resp. ω') is equal to the product of the rates along the corresponding reverse path iff the same holds true for ω' (resp. ω). From this it will follow that, in the case of a cycle, the product of the rates along the cycle is equal to the product of the rates along the corresponding reverse cycle.

Lemma 1. *If an RMLTSI meets CC and PPS, then for all coinitial and cofinal paths ω, ω' we have $rateprod(\omega) = rateprod(\underline{\omega})$ iff $rateprod(\omega') = rateprod(\underline{\omega'})$.* ∎

Theorem 2. *If an RMLTSI meets CC and PPS, then for all cycles ω we have $rateprod(\omega) = rateprod(\underline{\omega})$.* ∎

Corollary 1. *If an RMLTSI meets CC and PPS and its underlying CTMC is stationary, then it meets TR too.* ∎

3.3 Discussion

The result in Corollary 1 supersedes the two sufficient conditions for TR in [3], respectively relying on identical forward and backward rates and on process syntax constraints, and confirms the conjecture at the end of [3], i.e., the reversible Markovian process calculus of [3] is time reversible by construction. As a consequence, the method for reversing process calculi of [32] is robust not only with respect to causal reversibility, but also with respect to time reversibility.

The additional constraint, i.e., PPS, is trivially satisfied in a stochastic process algebraic setting like the one of [3]. Indeed, in the case of two exponentially timed actions that are causally independent of each other, i.e., respectively executed by two process terms composed in parallel, the corresponding transitions would naturally form a commuting square in which opposite transitions are labeled not only with the same actions, but also with the same rates.

Our result has been proven in the newly introduced model of RMLTSI, which relies on the LTSI model studied in [35]. The latter has been considered within a classification of concurrency models such as Petri nets [31], process algebras [29,9], Hoare traces [11], Mazurkievicz traces [26], synchronization trees [43], and event structures [42], which are all based on atomic units of change – transitions, actions, events – from which computations are built and differ for the fact of being behavioral vs. system models, interleaving vs. truly concurrent models, or linear-time vs. branching-time models.

RMLTSI generalizes LTSI in such a way that our result applies to concurrent and distributed models, calculi, and languages including performance aspects. In addition to the reversible Markovian process calculus of [3], we think of Markovian extensions of models like reversible Petri nets [28], calculi like RCCS [5], CCSK [32], and reversible higher-order π [19], and languages like reversible Erlang [21,15]. Since all of them have already been shown to meet causal reversibility, they automatically satisfy time reversibility too as they enjoy PPS.

Even if we started from an RMLTSI instead of a description provided in one of the formalisms above, proving TR via CC and PPS may be simpler than the direct proof of TR based on Theorem 1(3). One reason is that the latter requires identifying all the cycles in the RMLTSI while the former works with commuting squares. The other is that, following [22, Propositions 3.6 and 3.4], CC can be proven by showing that the RMLTSI meets the *square property* – whenever two coinitial transitions are independent then there exist two transitions respectively composable with the previous two such that a commuting square is obtained – *backward transitions independence* – any two coinitial backward transitions are independent – and *well foundedness* – there are no infinite backward paths. Note that if this sufficient condition for CC does not apply to the RMLTSI, it can nevertheless provide some diagnostic information, in the form of which of the three properties fails and for which transitions, that could be useful for TR.

We conclude by mentioning that the proof of our main result has been verified with the assistance of the interactive theorem prover Matita [1]. The proof does not make use of the irreflexivity of the independence relation ι over the transitions of the RMLTSI. As far as the properties considered in the present paper are concerned, this means that it is safe to assume every transition to be independent from itself, which is consistent with the fact that a path formed by a single transition is causally equivalent to itself due to the reflexivity of $\asymp$. We finally observe that the requirement of being stationary does not mean that the state space of the CTMC underlying the RMLTSI has to be finite.

4   An Application to the Dining Philosophers Problem

4.1   The Problem

Dining philosophers [7] is a classical concurrency problem, which is formulated as follows. There are $k \geq 2$ philosophers sitting around a circular table, who alternately think and eat. At the center of the table there is a large plate of noodles. A philosopher needs two chopsticks to take and eat a helping of noodles. Unfortunately, only k chopsticks are available, with a single chopstick being placed between each pair of philosophers and every philosopher being allowed to use only the neighboring chopsticks, i.e., the one on the right and the one on the left. Since each pair of adjacent philosophers is forced to share one chopstick, but two of them are required to eat, it is necessary to set up an appropriate synchronization policy for governing the access of philosophers to chopsticks.

If the philosophers behave the same – i.e., everyone thinks, takes the right chopstick, takes the left chopstick, eats, releases the left chopstick and, finally, releases the right chopstick – then a deadlock may occur. There are several solutions to the dining philosophers problem, such as breaking the symmetry among the philosophers – by letting one of them to get the chopsticks in a different order – or introducing randomization with respect to the order in which chopsticks are taken – by having every philosopher flipping a coin. In particular, the algorithm in which every philosopher takes the right and left chopsticks in one atomic operation is one of the most well known and studied in the literature, even from a quantitative point of view (see, e.g., [44] and the references therein).

4.2   The RMLTSI Model

In our reversible setting, we consider the case in which every philosopher, after taking the right chopstick, may also release it without eating, which happens when the left one is not available within a short amount of time. This can be viewed as the abortion of an operation because of a timeout expiration.

More formally, let $I_k = \{1, \ldots, k\}$ for $k \geq 2$ be the index set for philosophers, with index increments and decrements being circular, i.e., $i + 1 = 1$ when $i = k$ and $i - 1 = k$ when $i = 1$. The RMLTSI DP_k describing the dining philosophers problem with k philosophers and k chopsticks, which is illustrated in Fig. 1 for $k = 2$ (backward actions coincide with the forward ones), is defined as follows:

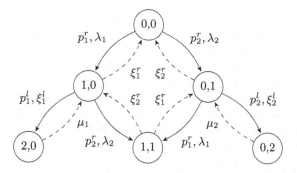

Fig. 1. RMLTSI DP_2: the overall behavior of two dining philosophers

– The set of states is:

$$S_k = \{(n_1, \ldots, n_k) \in \{0,1,2\}^k \mid \sum_{i=1}^{k} n_i \leq k, n_i = 2 \Rightarrow n_{i+1} = 0 \text{ for } i \in I_k\}$$

where $n_i \in \{0,1,2\}$ denotes the number of chopsticks held by philosopher i, who is eating iff $n_i = 2$, and $(0, \ldots, 0)$ is the initial state.
– The set of actions is:

$$A_k = \{p_i^r, p_i^l \mid i \in I_k\}$$

where p_i stands for pick up in the forward direction or put down in the backward direction (taking and releasing chopsticks are inverse operations), r stands for right, and l stands for left.
– The set of rate metavariables is:

$$\{\lambda_i, \mu_i, \xi_i^r, \xi_i^l \mid i \in I_k\}$$

where λ_i accounts for the thinking time and the time to take the right chopstick, μ_i describes the eating time and the time to release the left chopstick, and ξ_i^r (resp. ξ_i^l) models the time to release the right (resp. take the left) chopstick; clearly $\phi \ll \psi$, hence $1/\phi \gg 1/\psi$, for $\phi \in \{\lambda_i, \mu_i\}$ and $\psi \in \{\xi_i^r, \xi_i^l\}$.
– The transition relation $\longrightarrow$ comprises forward transitions of the form:

$$t_i^r : (n_1, \ldots, n_i, \ldots, n_k) \xrightarrow{p_i^r, \lambda_i} (n_1, \ldots, n_i + 1, \ldots, n_k) \text{ if } n_i = 0 \wedge n_{i-1} \neq 2$$

$$t_i^l : (n_1, \ldots, n_i, \ldots, n_k) \xrightarrow{p_i^l, \xi_i^l} (n_1, \ldots, n_i + 1, \ldots, n_k) \text{ if } n_i = 1 \wedge n_{i+1} = 0$$

representing the fact that the i-th philosopher thinks and picks the right (resp. picks the left) chopstick up with action p_i^r (resp. p_i^l) at rate λ_i (resp. ξ_i^l) when the philosopher to the right (resp. left) is not holding it.
– The transition relation $\dashrightarrow$ comprises the backward transitions $\underline{t}_i^l$ and $\underline{t}_i^r$ corresponding to the forward transitions defined above:

$$\underline{t}_i^l : (n_1, \ldots, n_i, \ldots, n_k) \overset{p_i^l, \mu_i}{\dashrightarrow} (n_1, \ldots, n_i - 1, \ldots, n_k) \text{ if } n_i = 2$$

$$\underline{t}_i^r : (n_1, \ldots, n_i, \ldots, n_k) \overset{p_i^r, \xi_r}{\dashrightarrow} (n_1, \ldots, n_i - 1, \ldots, n_k) \text{ if } n_i = 1$$

representing the fact that the i-th philosopher eats and puts the left (resp. puts the right) chopstick down with action p_i^l (resp. p_i^r) at rate μ_i (resp. ξ_i^r) when holding both chopsticks (resp. only the right chopstick).

- The independence relation ι is the smallest irreflexive and symmetric relation over transitions satisfying the PCI property and such that for all squares t, u, u', t' it holds that $(t, u) \in \iota$. The commuting squares have all one of the following forms or rotations thereof – depicted in Fig. 2 – where, for the sake of simplicity, we denote by $(i : v_i, j : v_j)$ any state $(n_1, \ldots, n_i, \ldots, n_j, \ldots, n_k)$ with $n_i = v_i$, $n_j = v_j$, $1 \leq i, j \leq k$, and $i \neq j$:

- $t : (i : 0, j : 0) \xrightarrow{p_i^r, \lambda_i} (i : 1, j : 0)$, $u : (i : 0, j : 0) \xrightarrow{p_j^r, \lambda_j} (i : 0, j : 1)$,
  $u' : (i : 1, j : 0) \xrightarrow{p_j^r, \lambda_j} (i : 1, j : 1)$, $t' : (i : 0, j : 1) \xrightarrow{p_i^r, \lambda_i} (i : 1, j : 1)$.

- $t : (i : 1, j : 1) \xrightarrow{p_i^l, \xi_i^l} (i : 2, j : 1)$, $u : (i : 1, j : 1) \xrightarrow{p_j^l, \xi_j^l} (i : 1, j : 2)$,
  $u' : (i : 2, j : 1) \xrightarrow{p_j^l, \xi_j^l} (i : 2, j : 2)$, $t' : (i : 1, j : 2) \xrightarrow{p_i^l, \xi_i^l} (i : 2, j : 2)$.

- $t : (i : 0, j : 1) \xrightarrow{p_i^r, \lambda_i} (i : 1, j : 1)$, $u : (i : 0, j : 1) \xrightarrow{p_j^l, \xi_j^l} (i : 0, j : 2)$,
  $u' : (i : 1, j : 1) \xrightarrow{p_j^l, \xi_j^l} (i : 1, j : 2)$, $t' : (i : 0, j : 2) \xrightarrow{p_i^r, \lambda_i} (i : 1, j : 2)$.

- $t : (i : 1, j : 0) \xrightarrow{p_i^l, \xi_i^l} (i : 2, j : 0)$, $u : (i : 1, j : 0) \xrightarrow{p_j^r, \lambda_j} (i : 1, j : 1)$,
  $u' : (i : 2, j : 0) \xrightarrow{p_j^r, \lambda_j} (i : 2, j : 1)$, $t' : (i : 1, j : 1) \xrightarrow{p_i^l, \xi_i^l} (i : 2, j : 1)$.

4.3   Time Reversibility and Performance Evaluation

Based on the main result of this paper, we show that DP_k meets TR by proving that it satisfies CC and PPS. As already mentioned in Sect. 3.3, CC can in turn be established by demonstrating that DP_k meets the square property, backward transitions independence, and well foundedness [22]. The proof turns out to be simple, especially compared to the direct proof that DP_k meets TR relying on Theorem 1(3).

Proposition 4. *For all $k \in \mathbb{N}_{\geq 2}$, DP_k meets CC and PPS and hence TR.* ■

Since DP_k meets TR, by exploiting the partial balance equations in Theorem 1(2) we derive a product-form expression for the steady-state probability distribution π_k. This allows us to perform a quantitative analysis of DP_k by computing first the unnormalized steady-state probabilities as in Proposition 5 and then their normalized expressions thanks to Proposition 6. For simplicity, we assume the same rate metavariables $\lambda, \mu, \xi^r, \xi^l$ for all philosophers. The value of the normalizing constant G_k is provided in the case $\xi^r = \xi^l$.

Proposition 5. *For all $\bar{n} = (n_1, \ldots, n_k) \in S_k$, we have $\pi_k(\bar{n}) = \frac{1}{G_k} \cdot \prod_{i=1}^{k} \prod_{j=1}^{n_i} \left(\frac{\delta_j}{\gamma_j} \right)$ where $\lambda_i = \delta_1$, $\xi_i^r = \gamma_1$, $\xi_i^l = \delta_2$, $\mu_i = \gamma_2$ for all $i \in I_k$ and $G_k \in \mathbb{R}_{>0}$.* ■

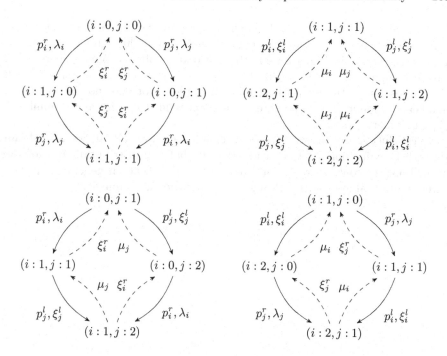

Fig. 2. Commuting squares inside the RMLTSI DP_k

Proposition 6. *For* $\xi^r = \xi^l = \xi$, $G_k = k \cdot \left(\frac{\lambda}{\xi}\right)^{k-1} \cdot {}_2F_1\left(\frac{1}{2} - \frac{k}{2}, 1 - \frac{k}{2}; -k; -\frac{4\xi^2}{\lambda\mu}\right) + \left(\frac{\lambda}{\xi}\right)^k \cdot {}_2F_1\left(\frac{1}{2} - \frac{k}{2}, -\frac{k}{2}; -k; -\frac{4\xi^2}{\lambda\mu}\right) + \left(1 + \frac{\lambda}{\mu}\right) \cdot \sum_{t=0}^{k-2} \left(\frac{\lambda}{\xi}\right)^t \cdot \binom{k}{t} \cdot {}_2F_1\left(\frac{1}{2} - \frac{t}{2}, -\frac{t}{2}; -k; -\frac{4\xi^2}{\lambda\mu}\right)$ *where the hypergeometric function is defined as* ${}_2F_1(u, v; x; z) = \sum_{n=0}^{\infty} \frac{(u)_n \cdot (v)_n}{(x)_n} \cdot \frac{z^n}{n!}$ *with* $(q)_n = q \cdot (q+1) \cdot \ldots \cdot (q+n-1)$ *being the Pochhammer symbol.* ∎

We conclude by showing in Fig. 3 the plots of the total throughput and the normalized throughput of DP_k, where by throughput we mean the average number of philosophers completing the eating phase per unit of time:

$$T_k = \mu \cdot \sum_{\bar{n} \in S_k} \pi_k(\bar{n}) \cdot |\{i \in I_k \mid n_i = 2\}|$$

$$= \frac{1}{G_k} \cdot \mu \cdot \sum_{t=2}^{k} \sum_{e=0}^{\lfloor t/2 \rfloor} \frac{e \cdot (k-e)!}{e! \cdot (t-2 \cdot e)! \cdot (k-t)!} \cdot \left(\frac{\lambda}{\xi}\right)^{t-2 \cdot e} \cdot \left(\frac{\lambda}{\mu}\right)^e +$$

$$\frac{1}{G_k} \cdot \mu \cdot \sum_{t=0}^{k-2} \sum_{e=0}^{\lfloor t/2 \rfloor} \frac{(e+1) \cdot (k-e)!}{e! \cdot (t-2 \cdot e)! \cdot (k-t)!} \cdot \left(\frac{\lambda}{\xi}\right)^{t-2 \cdot e} \cdot \left(\frac{\lambda}{\mu}\right)^{e+1}$$

$$= \frac{1}{G_k} \cdot \frac{\xi^2}{\lambda} \cdot (k-1) \cdot \sum_{t=2}^{k} \left(\frac{\lambda}{\xi}\right)^t \cdot \binom{k-2}{t-2} \cdot {}_2F_1\left(1 - \frac{t}{2}, \frac{3}{2} - \frac{t}{2}; 1 - k; -\frac{4 \cdot \xi^2}{\lambda \cdot \mu}\right)$$

The analysis of the model, obtained by applying the expression of T_k above, illustrates the impact of the concurrency level on the throughput. In particular, the right plot in Fig. 3 reveals that an odd number of philosophers has a negative impact on the normalized throughput. Indeed, when passing from 2 to 3 philosophers, the total throughput in the left plot does not increase by a significant amount. As can be noted, this effect tends to vanish as the number k of philosophers grows.

In conclusion, we observe that the reversible nature of DP_k allows us to obtain an explicit expression for the steady-state probability distribution and for the throughput, in contrast with non-reversible models that, in general, require a purely numerical approach with higher computational complexity.

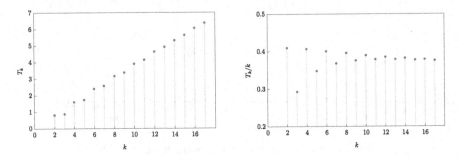

Fig. 3. Numerical evaluation of the throughput of DP_k for $\lambda = 3$, $\mu = 1$, $\xi = 20$: total throughput (left) and normalized throughput (right)

5   Conclusions

In this paper we have shown that CC and PPS ensure TR, thus validating the conjecture according to which the Markovian process calculus of [3], which is causally reversible by construction, is time reversible by construction as well. This witnesses in turn the robustness of the method for reversing process calculi of [32] with respect to both forms of reversibility.

In the future, we plan to investigate whether constraints less demanding than PPS exist under which causal reversibility still implies time reversibility. More interestingly, we would like to study the opposite direction, i.e., under which conditions time reversibility implies causal reversibility.

For example, it is well known from [13], and recalled in [3], that birth-death processes and their tree-like variants are time reversible. Furthermore, our result shows that also concurrent variants satisfying PPS are time reversible. In these stochastic processes there is an order over states such that causality can never be violated when going backward. However, a notion of independence between transitions departing from the same state is lacking, essentially because those transitions are such that the death (backward) transition retracts the premises (causes) for the birth (forward) transition.

Moreover, if we consider a circular variant of birth-death process, which may not be necessarily time reversible if backward rates are not equal to their corresponding forward rates, the number of performed death (backward) transitions may exceed the number of performed birth (forward) transitions. In the terminology of [22], this amounts to a violation of well foundedness, i.e., the absence of infinite backward paths, which is exploited to prove causal reversibility.

Another direction worth investigating is whether equivalence relations such as bisimilarity [29,23,10] and aggregations such as lumpability [14,37,4] play a role in connecting different forms of reversibility, taking inspiration from [25].

From a practical point of view, we would like to exploit our result to analyze naturally reversible systems like biochemical ones, where the direction of computation depends on physical conditions such as temperature or pressure [33,34], as well as reversibility in stochastic phenomena like the Ehrenfest model [13].

Acknowledgments. This work has been supported by the Italian PRIN 2020 project *NiRvAna – Noninterference and Reversibility Analysis in Private Blockchains*, the French ANR-18-CE25-0007 project *DCore – Causal Debugging for Concurrent Systems*, and the INdAM-GNCS E53C22001930001 project *RISICO – Reversibilità in Sistemi Concorrenti: Analisi Quantitative e Funzionali*.

Data Availability Statement. The machine-checkable proof of the result in Sect. 3 has been accepted by the QEST 2023 artifact evaluation committee and is available at https://github.com/sacerdot/Causal2TimedFormalization.

References

1. Asperti, A., Ricciotti, W., Sacerdoti Coen, C.: Matita tutorial. J. Formaliz. Reason. **7**, 91–199 (2014)
2. Bennett, C.H.: Logical reversibility of computation. IBM J. Res. Dev. **17**, 525–532 (1973)
3. Bernardo, M., Mezzina, C.A.: Bridging causal reversibility and time reversibility: a stochastic process algebraic approach. Log. Methods Comput. Sci. **19**(2:6), 1–27 (2023)
4. Buchholz, P.: Exact and ordinary lumpability in finite Markov chains. J. Appl. Probab. **31**, 59–75 (1994)
5. Danos, V., Krivine, J.: Reversible communicating systems. In: Gardner, P., Yoshida, N. (eds.) CONCUR 2004. LNCS, vol. 3170, pp. 292–307. Springer, Heidelberg (2004). https://doi.org/10.1007/978-3-540-28644-8_19
6. Danos, V., Krivine, J.: Transactions in RCCS. In: Abadi, M., de Alfaro, L. (eds.) CONCUR 2005. LNCS, vol. 3653, pp. 398–412. Springer, Heidelberg (2005). https://doi.org/10.1007/11539452_31
7. Dijkstra, E.W.: Hierarchical ordering of sequential processes. Acta Informatica **1**, 115–138 (1971)
8. Giachino, E., Lanese, I., Mezzina, C.A.: Causal-consistent reversible debugging. In: Gnesi, S., Rensink, A. (eds.) FASE 2014. LNCS, vol. 8411, pp. 370–384. Springer, Heidelberg (2014). https://doi.org/10.1007/978-3-642-54804-8_26

9. Hennessy, M.: Algebraic Theory of Processes. MIT Press, Cambridge (1988)
10. Hillston, J.: A Compositional Approach to Performance Modelling. Cambridge University Press, Cambridge (1996)
11. Hoare, C.: Communicating Sequential Processes. Prentice Hall, Hoboken (1985)
12. Keller, R.M.: Formal verification of parallel programs. Commun. ACM **19**, 371–384 (1976)
13. Kelly, F.P.: Reversibility and Stochastic Networks. Wiley, Hoboken (1979)
14. Kemeny, J.G., Snell, J.L.: Finite Markov Chains. Van Nostrand, New York (1960)
15. Lami, P., Lanese, I., Stefani, J.B., Sacerdoti Coen, C., Fabbretti, G.: Reversibility in Erlang: Imperative constructs. In: Mezzina, C.A., Podlaski, K. (eds.) RC 2022. LNCS, vol. 13354, pp. 187–203. Springer, Cham (2022). https://doi.org/10.1007/978-3-031-09005-9_13
16. Landauer, R.: Irreversibility and heat generated in the computing process. IBM J. Res. Dev. **5**, 183–191 (1961)
17. Lanese, I., Lienhardt, M., Mezzina, C.A., Schmitt, A., Stefani, J.-B.: Concurrent flexible reversibility. In: Felleisen, M., Gardner, P. (eds.) ESOP 2013. LNCS, vol. 7792, pp. 370–390. Springer, Heidelberg (2013). https://doi.org/10.1007/978-3-642-37036-6_21
18. Lanese, I., Mezzina, C.A., Stefani, J.-B.: Reversing higher-order Pi. In: Gastin, P., Laroussinie, F. (eds.) CONCUR 2010. LNCS, vol. 6269, pp. 478–493. Springer, Heidelberg (2010). https://doi.org/10.1007/978-3-642-15375-4_33
19. Lanese, I., Mezzina, C.A., Stefani, J.B.: Reversibility in the higher-order π-calculus. Theoret. Comput. Sci. **625**, 25–84 (2016)
20. Lanese, I., Nishida, N., Palacios, A., Vidal, G.: CauDEr: a causal-consistent reversible debugger for Erlang. In: Gallagher, J.P., Sulzmann, M. (eds.) FLOPS 2018. LNCS, vol. 10818, pp. 247–263. Springer, Cham (2018). https://doi.org/10.1007/978-3-319-90686-7_16
21. Lanese, I., Nishida, N., Palacios, A., Vidal, G.: A theory of reversibility for Erlang. J. Log. Algebr. Methods Program. **100**, 71–97 (2018)
22. Lanese, I., Phillips, I., Ulidowski, I.: An axiomatic approach to reversible computation. In: FoSSaCS 2020. LNCS, vol. 12077, pp. 442–461. Springer, Cham (2020). https://doi.org/10.1007/978-3-030-45231-5_23
23. Larsen, K., Skou, A.: Bisimulation through probabilistic testing. Inf. Comput. **94**, 1–28 (1991)
24. Laursen, J.S., Ellekilde, L.P., Schultz, U.P.: Modelling reversible execution of robotic assembly. Robotica **36**, 625–654 (2018)
25. Marin, A., Rossi, S.: On the relations between Markov chain lumpability and reversibility. Acta Informatica **54**, 447–485 (2017)
26. Mazurkiewicz, A.: Basic notions of trace theory. In: de Bakker, J.W., de Roever, W.-P., Rozenberg, G. (eds.) REX 1988. LNCS, vol. 354, pp. 285–363. Springer, Heidelberg (1989). https://doi.org/10.1007/BFb0013025
27. McNellis, J., Mola, J., Sykes, K.: Time travel debugging: root causing bugs in commercial scale software. CppCon talk (2017). https://www.youtube.com/watch?v=l1YJTg_A914
28. Melgratti, H.C., Mezzina, C.A., Ulidowski, I.: Reversing place-transition nets. Log. Methods Comput. Sci. **16**(4:5), 1–28 (2020)
29. Milner, R.: Communication and Concurrency. Prentice Hall, Hoboken (1989)
30. Perumalla, K.S., Park, A.J.: Reverse computation for rollback-based fault tolerance in large parallel systems - evaluating the potential gains and systems effects. Clust. Comput. **17**, 303–313 (2014)

31. Petri, C.A.: Kommunikation mit Automaten. Ph.D. thesis (1962)
32. Phillips, I.C.C., Ulidowski, I.: Reversing algebraic process calculi. J. Logic Algebraic Program. **73**, 70–96 (2007)
33. Phillips, I., Ulidowski, I., Yuen, S.: A reversible process calculus and the modelling of the ERK signalling pathway. In: Glück, R., Yokoyama, T. (eds.) RC 2012. LNCS, vol. 7581, pp. 218–232. Springer, Heidelberg (2013). https://doi.org/10.1007/978-3-642-36315-3_18
34. Pinna, G.M.: Reversing steps in membrane systems computations. In: Gheorghe, M., Rozenberg, G., Salomaa, A., Zandron, C. (eds.) CMC 2017. LNCS, vol. 10725, pp. 245–261. Springer, Cham (2018). https://doi.org/10.1007/978-3-319-73359-3_16
35. Sassone, V., Nielsen, M., Winskel, G.: Models of concurrency: towards a classification. Theoret. Comput. Sci. **170**, 297–348 (1996)
36. Schordan, M., Oppelstrup, T., Jefferson, D.R., Barnes, P.D., Jr.: Generation of reversible C++ code for optimistic parallel discrete event simulation. N. Gener. Comput. **36**, 257–280 (2018)
37. Schweitzer, P.: Aggregation methods for large Markov chains. In: Proceedings of the International Workshop on Computer Performance and Reliability, pp. 275–286. North Holland (1984)
38. Siljak, H., Psara, K., Philippou, A.: Distributed antenna selection for massive MIMO using reversing Petri nets. IEEE Wirel. Commun. Lett. **8**, 1427–1430 (2019)
39. Stewart, W.J.: Introduction to the Numerical Solution of Markov Chains. Princeton University Press, Princeton (1994)
40. Vassor, M., Stefani, J.-B.: Checkpoint/rollback vs causally-consistent reversibility. In: Kari, J., Ulidowski, I. (eds.) RC 2018. LNCS, vol. 11106, pp. 286–303. Springer, Cham (2018). https://doi.org/10.1007/978-3-319-99498-7_20
41. de Vries, E., Koutavas, V., Hennessy, M.: Communicating transactions. In: Gastin, P., Laroussinie, F. (eds.) CONCUR 2010. LNCS, vol. 6269, pp. 569–583. Springer, Heidelberg (2010). https://doi.org/10.1007/978-3-642-15375-4_39
42. Winskel, G.: Events in computation. Ph.D. thesis (1980)
43. Winskel, G.: Synchronisation trees. Theoret. Comput. Sci. **34**, 33–82 (1985)
44. Ycart, B.: The philosophers' process: an ergodic reversible nearest particle system. Ann. Appl. Probab. **3**, 356–363 (1993)

Deductive Controller Synthesis
for Probabilistic Hyperproperties

Roman Andriushchenko[1], Ezio Bartocci[2], Milan Češka[1]([✉]),
Francesco Pontiggia[2]([✉]), and Sarah Sallinger[2]

[1] Brno University of Technology, Brno, Czech Republic
ceskam@fit.vutbr.cz
[2] TU Wien, Wien, Austria
francesco.pontiggia@tuwien.ac.at

Abstract. Probabilistic hyperproperties specify quantitative relations between the probabilities of reaching different target sets of states from different initial sets of states. This class of behavioral properties is suitable for capturing important security, privacy, and system-level requirements. We propose a new approach to solve the controller synthesis problem for Markov decision processes (MDPs) and probabilistic hyperproperties. Our specification language builds on top of the logic HyperPCTL and enhances it with structural constraints over the synthesized controllers. Our approach starts from a family of controllers represented symbolically and defined over the same copy of an MDP. We then introduce an abstraction refinement strategy that can relate multiple computation trees and that we employ to prune the search space deductively. The experimental evaluation demonstrates that the proposed approach considerably outperforms HYPERPROB, a state-of-the-art SMT-based model checking tool for HyperPCTL. Moreover, our approach is the first one that is able to effectively combine probabilistic hyperproperties with additional intra-controller constraints (e.g. partial observability) as well as inter-controller constraints (e.g. agreements on a common action).

1   Introduction

Controller synthesis in a probabilistic environment [10] is a well-known and highly addressed problem. There exist several off-the-shelf tools (e.g. PRISM [33] or Storm [29]) that are able to synthesize optimal policies for large Markov decision processes (MDPs) with respect to probability measures over a given set of paths. However, recent studies [2,3,18,20,21] have brought up the necessity to investigate a new class of properties that relate probability measures over different sets of executions. This class of requirements is called **probabilistic hyperproperties** [2] and standard logics, such as Probabilistic Computation Tree Logic

This work has been supported by the Czech Science Foundation grant GA23-06963S (VESCAA), the Czech Ministry of Education, Youth and Sports project LL1908 of the ERC.CZ programme, the Vienna Science and Technology Fund 10.47379/ICT19018, and by the European Union's ERC CoG ARTIST 101002685 grant and the Horizon 2020 research and innovation programme under grant no. 101034440 ▣.

N. Jansen and M. Tribastone (Eds.): QEST 2023, LNCS 14287, pp. 288–306, 2023.
https://doi.org/10.1007/978-3-031-43835-6_20

(PCTL) [28], are not expressive enough for them. Classical examples include security and privacy requirements, such as quantitative information-flow [32], probabilistic noninterference [27,35] and differential privacy for databases [23]. Probabilistic hyperproperties can be expressed using Probabilistic HyperLogic (PHL) [18] and HyperPCTL [2]. Both are probabilistic temporal logics that extends PCTL with i) quantifiers over schedulers and ii) predicates over functions of probability expressions. Verification of HyperPCTL or PHL formulae can be formalized as the synthesis of multiple, correlated strategies that represent a witness or a counterexample to the specification. Furthermore, pioneering studies have shown that probabilistic hyperproperties have novel applications in robotics and planning [18], and in causality analysis for probabilistic systems [9]. To deal with these new constraints, it is necessary to establish a connection not only between the probabilities of different strategies, but also between the choices they make in corresponding states, i.e. reasoning about the **structural characteristics** of the resulting controllers. Finally, the study of probabilistic hyperproperties in a **partially observable** setting is mostly unexplored.

Motivating Example 1. Consider a planning problem depicted in Fig. 1: a robot in a probabilistic environment attempts to reach the target location T. A controller for the robot is a program that, for each position, selects the next action to take. Consider the synthesis of two distinct controllers for two possible initial states s_a and s_b. In a confidential set-

Fig. 1. Planning under probabilistic noninterference.

ting, we want to combine performance constraints on success probabilities with additional privacy and structural requirements. *Probabilistic noninterference* is a hyperproperty asserting that the success probability must be the same for s_a and s_b, so that an attacker cannot infer any information about the selected initial location by observing the percentage of successful robots. If further the choice made in some state s_χ is assumed to be observable to the attacker, the two strategies must agree on their action choices in s_χ. Finally, the *partial observability* constraint requires a controller to make the same choice in states s_c and s_d having the same observation.

Related Work. In [3] and [1], the authors introduce, respectively, the model-checking problem and the parameter synthesis problem for discrete-time Markov chains satisfying probabilistic hyperproperties. The works in [2,18] have recently studied the verification of probabilistic hyperproperties in the context of MDPs and proposed two suitable logics: HyperPCTL and Probabilistic HyperLogic (PHL), respectively. However, in both cases the model-checking problem is undecidable. The solution proposed for the former logic is to consider only the class of deterministic memoryless controllers, where the problem is shown to be NP-complete by a suitable SMT encoding. Despite recent improvements of HYPER-PROB [19,21], a state-of-the-art tool implementing this approach, its scalability remains very limited: the complexity of the encoding gets out of reach even for modest models. For instance, a simple side-channel analysis introduced in [2] (checking that the probability of certain outcomes is equal) on an MDP with 244 states leads to an SMT problem with over a million variables, which HYPER-PROB fails to solve. Finally, HYPERPROB does not provide support for structural constraints describing non-probabilistic relations between controllers, and the presented techniques cannot principally handle neither partial observability.

On the other hand, PHL is able to express such structural constraints using HyperCTL* [16] formulae in conjunction with probability expressions. Hence, with PHL, it is possible to require different paths to take the same (or a different) choice in a certain state. However, the model checking procedure developed in [18] employs a bounded synthesis algorithm which supports only *qualitative* constraints, and cannot effectively handle the probability expressions of a PHL formula - the synthesis loop has to check them a posteriori. As a result, when the qualitative constraints do not restrict the controllers sufficiently, a large part of the search space has to be explicitly enumerated.

Various works have addressed (reactive) synthesis problems in non probabilistic settings [14,24,25,36]. In particular, decidability for HyperLTL [16] specifications and finite-state plants has been shown in [14].

Our Contribution. To mitigate the limitations of the existing approaches, we propose a novel controller synthesis framework for MDPs and probabilistic hyperproperties. We enrich expressive power of HyperPCTL with **structural constraints** to support additional requirements on the controllers (Sect. 2). We treat the synthesis problem as an exploration problem over a **symbolic representation** of a family of controllers defined over the original MDP (Sect. 3). Inspired by recent advances in the synthesis of probabilistic systems [5], we tailor a deductive exploration method to support hyper-reasoning. In particular, we propose an **abstraction refinement strategy** that relates multiple computation trees and allows us to deductively identify sets of violating controllers that can be safely removed from the search space (Sect. 4). A detailed experimental evaluation (Sect. 5) demonstrates that our approach considerably outperforms the state-of-the-art tool HYPERPROB [19]. In particular, our approach scales to more complicated synthesis problems where HYPERPROB times out, and supports, for the first time, various structural constraints.

2   Problem Formulation

In this section we present the formalisms involved in our synthesis framework as well as the problem statement. We refer to [8,11] for further background.

2.1   Probabilistic Models

A (discrete) *probabilistic distribution* over a finite set X is $\mu : X \to [0,1]$, where $\sum_{x \in X} \mu(x) = 1$. The set of all distributions on X is denoted as $Distr(X)$.

Definition 1 (MDP). *A* Markov Decision Process (MDP) *is a tuple* $(S, Act, \mathcal{P})$ *with a finite set* S *of states, a finite set* Act *of actions and a transition function* $\mathcal{P} : S \times Act \to Distr(S)$. *An MDP with* $|Act| = 1$ *is a* Markov chain (MC) $(S, \mathcal{P})$ *with* $\mathcal{P} : S \to Distr(S)$.

We depart from the standard notion of MDP [11] and do not specify initial states: those will be derived from the context given by the specification. Additionally, to simplify the presentation, we enable all actions in every state. An MDP can be equipped with a state-action reward (or cost) function $rew : S \times Act \to \mathbb{R}_{\geqslant 0}$, where reward $rew(s, \alpha)$ is earned upon executing action α in state s. A (memoryless, deterministic) *controller* $\sigma : S \to Act$ selects in each state s an action $\sigma(s)$; let Σ^M denote the set of all such controllers in M. Imposing controller σ onto MDP $M = (S, Act, \mathcal{P})$ yields MC $M^{\sigma} = (S, \mathcal{P}^{\sigma})$ where $\mathcal{P}^{\sigma}(s) = \mathcal{P}(s, \sigma(s))$. For an MC M, $\mathbb{P}(M, s \models \Diamond T)$ denotes the probability of reaching the set $T \subseteq S$ of target states when starting in state $s \in S$. If $\mathbb{P}(M, s \models \Diamond T) = 1$, then $\mathbb{R}(M, s \models \Diamond T)$ denotes the expected reward accumulated from s before reaching T. Using the standard techniques, one can effectively compute a minimizing controller $\sigma_{\min}$ that minimizes the probability to reach T, i.e. $\forall \sigma \in \Sigma^M$: $\mathbb{P}(M^{\sigma_{\min}}, s \models \Diamond T) \leq \mathbb{P}(M^{\sigma}, s \models \Diamond T)$. Naturally, this can be lifted to the expected rewards.

2.2 HyperPCTL with Structural Constraints

Assume an MDP $M = (S, Act, \mathcal{P})$. Our specification language builds on the syntax of HyperPCTL [2], enriched by structural constraints. For both principal and technical reasons discussed below, we support only the following fragment of the original hyperlogic:

$$\varphi \quad := \exists \sigma_1, \dots, \sigma_n \in \Sigma^M : \varphi^{struc} \wedge \varphi^{state}$$

$$\varphi^{struc} := \varphi^{struc} \wedge \varphi^{struc} \mid \mathcal{X}_s(\Sigma) \mid \mathcal{O}_{S'}(\sigma_i)$$

$$\varphi^{state} := \mathcal{Q}_1 s_1 \in I_1[\sigma^1] \dots \mathcal{Q}_m s_m \in I_m[\sigma^m] : \varphi^{prob}$$

$$\varphi^{prob} := \neg \varphi^{prob} \mid \varphi^{prob} \wedge \varphi^{prob} \mid \varphi^{atom}$$

$$\varphi^{atom} := \mathbb{P}(M^{\sigma^i}, s_i \models \Diamond T) \bowtie \lambda \mid \mathbb{P}(M^{\sigma^i}, s_i \models \Diamond T_i) \bowtie \mathbb{P}(M^{\sigma^j}, s_j \models \Diamond T_j) \mid$$
$$\mathbb{R}(M^{\sigma^i}, s_i \models \Diamond T) \bowtie \rho \mid \mathbb{R}(M^{\sigma^i}, s_i \models \Diamond T_i) \bowtie \mathbb{R}(M^{\sigma^j}, s_j \models \Diamond T_j)$$

where $\mathcal{Q} \in \{\exists, \forall\}$, $\bowtie \in \{<, \leq, =, >, \geq\}$, $k \in \mathbb{N}$, $\lambda \in [0, 1]$ and $\rho \in \mathbb{R}_{\geq 0}$. The individual parts of the syntax and semantics are described below.

Controller quantification (φ): We consider the *synthesis* problem, i.e. only the existential quantification over the controllers σ_i is supported.[1]

---

[1] A *verification* problem defined using universal quantification can be treated as synthesis of a counterexample. Alternation in the controller quantification is not supported as it principally complicates the inductive synthesis strategy. To the best of our knowledge, only very few techniques so far practically support alternation [13,30].

Structural constraints (φ^{struc}): In contrast to HyperPCTL [2], we support conjunctions of structural constraints of two types: i) intra-controller constraint $\mathcal{O}_{S'}(\sigma_i)$ requiring that the controller σ_i takes the same action in all states from the set $S' \subseteq S$ (this can effectively encode partial observability) and ii) inter-controller constraint $\mathcal{X}_s(\Sigma)$ requiring that all controllers in $\Sigma \subseteq \{\sigma_1, \ldots, \sigma_n\}$ take the same action in state $s \in S$.

State quantification (φ^{state}): Allows to quantify over the computational trees rooted in the given initial states. We support an arbitrary quantification $\mathcal{Q}_i s_i \in I_i[\sigma^i]$, where each s_i is associated with its domain $I_i \subseteq S$ and some controller $\sigma^i \in \{\sigma_1, \ldots, \sigma_n\}$. Note that multiple state variables can be associated with the same controller ($n \leq m$).

Probability constraints (φ^{prob}): We consider a Boolean combination of the standard inequalities between the reachability probability (reward) and a threshold as well as the inequalities that correspond to the class of 2-hyperproperties [17], where a relationship between two computation trees is considered[2].

To evaluate whether given n controllers for MDP M satisfy a state formula φ^{state}, we instantiate the initial states on the probabilistic constraint φ^{prob}.

Definition 2 (Instantiation of state quantifiers). *Assume a state formula $\mathcal{Q}_1 s_1 \in I_1[\sigma^1] \ldots \mathcal{Q}_m s_m \in I_m[\sigma^m] : \varphi^{prob}$. Each quantifier $\mathcal{Q}_i s_i \in I_i[\sigma^i]$ is instantiated on φ^{prob} by substituting φ^{prob} with $\nabla_{s \in I_i} \varphi^{prob}[s_i \mapsto s]$, where $\nabla \equiv \bigvee$ if $\mathcal{Q}_i \equiv \exists$ and $\nabla \equiv \bigwedge$ otherwise.*

Upon instantiation, we obtain a Boolean combination of inequalities over probabilities and rewards in the corresponding MCs having the given initial states.

Remark 1. HyperPCTL syntax also allows to reason about stepwise *paired* multiple traces by associating the atomic propositions in the path formulae of probability expressions with state variables (i.e. the initial state of a particular computation). Then, to evaluate a formula with m state quantifications, the semantics requires to build the m-cross-product (also called *self-composition* [2,18], or *trivial coupling* [12]), of the original model M. In this paper, for the sake of simplicity, we avoid such an involved construction and consider only single, uncoupled computation trees, where all atomic propositions share the same state variable specified in the probability expression. However, handling such constraints would require only an additional technical step performing the cross-product when needed [37], and no principle obstacles.

Problem Statement. The key problem statement in this paper is *feasibility*:

---

[2] An extension to nested-free PCTL path formulae is straightforward. Principally, we can extend the probability constraints to allow more complicated expressions (e.g. sum/difference of reachability probabilities from different states) eventually requiring reasoning about more than two computation trees.

> Given an MDP M and a hyperspecification
>
> $$\exists \sigma_1, \ldots, \sigma_n \in \Sigma^M : \varphi^{struc} \wedge \mathcal{Q}_1 s_1 \in I_1[\sigma^1] \ldots \mathcal{Q}_m s_m \in I_m[\sigma^m] : \varphi^{prob}$$
>
> find controllers $\sigma_1, \ldots, \sigma_n$ in Σ^M that satisfy φ^{struc} and the instantiated φ^{prob}.

Remark 2. We follow existing approaches [2, 20, 21] and consider only the class of deterministic memoryless controllers where synthesis is decidable. These controllers are, however, too weak for some synthesis problems considered in the experimental evaluation. Therefore, we enhance the underlying MDPs with a memory to explore a richer class of controllers, as in [18].

Example 2. Consider Example 1 from the introduction. The probabilistic non-interference specification for the synthesis of two controllers can be expressed as

$$\exists \sigma_1, \sigma_2 : \varphi^{struc} \wedge \forall s_1 \in \{s_a\}[\sigma_1] \ \forall s_2 \in \{s_b\}[\sigma_2] : \varphi^{prob}, \text{where}$$

$$\varphi^{struc} \equiv \mathcal{X}_{s_x}(\{\sigma_1, \sigma_2\}) \wedge \mathcal{O}_{\{s_c, s_d\}}(\sigma_1) \wedge \mathcal{O}_{\{s_c, s_d\}}(\sigma_2)$$

$$\varphi^{prob} \equiv \mathbb{P}(M^{\sigma_1}, s_1 \models \Diamond T) = \mathbb{P}(M^{\sigma_2}, s_2 \models \Diamond T) \wedge \mathbb{P}(M^{\sigma_1}, s_1 \models \Diamond T) \geq \lambda$$

Given σ_1 and σ_2, the instantiation of the state quantifiers on φ^{prob} yields the following probability constraint $\mathbb{P}(M^{\sigma_1}, s_a \models \Diamond T) = \mathbb{P}(M^{\sigma_2}, s_b \models \Diamond T) \wedge \mathbb{P}(M^{\sigma_1}, s_a \models \Diamond T) \geq \lambda$.

Remark 3. Our approach also supports various variants of **optimal synthesis** problems, where an additional minimizing/maximizing objective over a reachability probability or reward is considered. For example, we can maximize the difference $\mathbb{P}(M^{\sigma_1}, s_1 \models \Diamond T_1) - \mathbb{P}(M^{\sigma_2}, s_2 \models \Diamond T_2)$. To simplify the technical exposition of the paper, we will omit the precise formalisation of these problems.

3   Family of Controllers

Consider an MDP M for which we wish to synthesize n controllers $\sigma_1, \ldots \sigma_n$, further referred to as an *n-controller* $\overline{\sigma} = (\sigma_1, \ldots, \sigma_n)$, that satisfy a hyperproperty $\varphi^{struc} \wedge \varphi^{state}$. Motivated by the inductive synthesis method presented in [5], we base our approach on a symbolic representation of a *family of n-controllers* over MDP M. The following definition allows us to efficiently represent all n-controllers that satisfy φ^{struc}.

Definition 3 (Parameter space). *Assume an MDP $M = (S, Act, \mathcal{P})$ and a hyperspecification $\exists \sigma_1, \ldots, \sigma_n : \varphi^{struc} \wedge \varphi^{state}$. For each $i \in \{1, \ldots, n\}$ and $s \in S$, we introduce a parameter $k(i, s)$ with domain $\text{dom}(i, s) = Act$. Relation $k \simeq k'$ denotes that parameters k, k' are synonymous, i.e. k and k' are different names for the same parameter. For each $\mathcal{O}_{S'}(\sigma_i) \in \varphi^{struc}$ we require that $\forall s, s' \in S' : k(i, s) \simeq k(i, s')$. Additionally, for each $\mathcal{X}_s(\Sigma) \in \varphi^{struc}$ we require that $\forall \sigma_i, \sigma_j \in \Sigma : k(i, s) \simeq k(j, s)$. Let $\mathcal{K}$ denote the set of all parameters.*

Parameter $k(i,s)$ essentially encodes possible action selections available to σ_i at state s. A realisation then encodes a specific assignment of these parameters.

Definition 4 (Realisation). *A realisation is a mapping* $r\colon \mathcal{K} \to Act$ *such that* $r(k(i,s)) \in \mathrm{dom}(i,s)$. $\mathcal{R}$ *denotes the set of all realisations. Realisation r induces* n-controller $\overline{\sigma}^r := (\sigma_1^r, \ldots, \sigma_n^r)$ *where* $\forall i \in \{1, \ldots, n\}\, \forall s \in S\colon \sigma_i^r(s) = r(k(i,s))$.

Realisation $r \in \mathcal{R}$ specifies for each σ_i a specific action $r(k(i,s))$ to be selected in state s, and thus describes an n-controller $\overline{\sigma}^r \in (\Sigma^M)^n$. Using synonymous parameters implies that r always induces controller $\overline{\sigma}^r$ satisfying structural constraints φ^{struc}. Thus, the set $\mathcal{R}$ of realisations directly corresponds to the family $\mathcal{F}$ of controllers that satisfy φ^{struc}.

Definition 5 (Family of controllers). *A family of i-th controllers is the set* $\mathcal{F}_i = \{\sigma_i^r \mid r \in \mathcal{R}\}$. *A family of n-controllers is the set* $\mathcal{F} := \{\overline{\sigma}^r \mid r \in \mathcal{R}\}$.

From this definition it is clear that $\mathcal{F} \subseteq (\Sigma^M)^n$ and that $\overline{\sigma} \in \mathcal{F} \Leftrightarrow \overline{\sigma} \models \varphi^{struc}$. We say that an n-controller $\overline{\sigma} = (\sigma_1, \ldots, \sigma_n)$ from the set $(\Sigma^M)^n \setminus \mathcal{F}$ is *inconsistent*, where we distinguish between the following two types of inconsistency.

- $\overline{\sigma} \not\models \mathcal{O}_{S'}(\sigma_i)$ is said to be *intra-controller inconsistent*: controller σ_i diverges in its choices $\sigma_i(s_a) \neq \sigma_i(s_b)$ for the action selection in states $s_a, s_b \in S'$.
- $\overline{\sigma} \not\models \mathcal{X}_s(\Sigma)$ is said to be *inter-controller inconsistent*: controllers $\sigma_i, \sigma_j \in \Sigma$ diverge in their choices $\sigma_i(s) \neq \sigma_j(s)$ for the action selection in state s.

Conversely, if $\overline{\sigma} \in \mathcal{F}$, then there exists exactly one realisation $r \in \mathcal{R}$ s.t. $\overline{\sigma}^r = \overline{\sigma}$. Subsequently, if $\sigma_i \in \mathcal{F}_i$, then the following set of realisations:

$$\mathcal{R}[\sigma_i] := \{r \in R \mid \overline{\sigma}^r = (\sigma_1, \ldots, \sigma_i, \ldots, \sigma_n)\}$$

is not empty and contains all realisations that induce σ_i. From $\mathcal{F} \subseteq (\Sigma^M)^n$ it follows that the size of the family $|\mathcal{F}| \leq |Act|^{|S|n}$ is exponential in the number $|S|$ of states and n, rendering enumeration of each n-controller $\overline{\sigma} \in \mathcal{F}$ infeasible.

4   Deductive Exploration

We present a deductive method allowing for an efficient exploration of the given family $\mathcal{F}$ of n-controllers. We adapt the abstraction refinement approach for PCTL specifications presented in [6,15] to the synthesis against probabilistic hyperproperties. We explain the method on a canonical synthesis problem $\varphi_c \equiv \exists \sigma_1, \sigma_2 : \varphi^{struc} \wedge \exists s_1 \in \{s_1\}[\sigma_1], \exists s_2 \in \{s_2\}[\sigma_2] : \varphi^{prob}$, where $\varphi^{prob} \equiv \mathbb{P}(M^{\sigma_1}, s_1 \models \Diamond T_1) \leq \mathbb{P}(M^{\sigma_2}, s_2 \models \Diamond T_2)$. A generalisation to more complicated specifications is discussed in Subsect. 4.3.

4.1   Guided Abstraction Refinement Scheme

MDP M and the corresponding set $(\Sigma^M)^n$ of controllers provide an over-approximation of the set $\mathcal{F}$ of admissible controllers. The key idea is to analyse M using standard techniques to obtain the following four controllers: $\sigma_1^{\min}$ ($\sigma_1^{\max}$) minimising (maximising) the probability of reaching T_1 from s_1, and controllers $\sigma_2^{\min}$ and $\sigma_2^{\max}$ minimising (maximising) the probability of reaching T_2 from s_2. Let $l_1 := \mathbb{P}(M^{\sigma_1^{\min}}, s_1 \models \Diamond T_1)$ denote the minimum probability of reaching T_1 from s_1 via $\sigma_1^{\min}$; since the corresponding $\sigma_1^{\min}$ is not necessarily in $\mathcal{F}_1$, then this probability is a lower bound on $\min_{\sigma \in \mathcal{F}_1} \mathbb{P}(M^\sigma, s_1 \models \Diamond T_1)$. Let u_1 be the corresponding upper bound on the probability of reaching T_1 from s_1, and similarly for l_2 and u_2. It is clear that $l_1 \le u_1$ and $l_2 \le u_2$.

Comparing the corresponding numerical intervals $[l_1, u_1]$ and $[l_2, u_2]$ may allow us to determine whether $(\Sigma^M)^n$ (and, by extension, $\mathcal{F}$) can satisfy φ^{prob}. The procedure is summarised in Table 1. For instance, if the two intervals do not overlap, then either all controllers satisfy φ^{prob} (case (1)) or none do (case (2)).

Table 1. Reasoning about feasibility of family $\mathcal{F}$ using interval logic.

case	condition	feasibility	SAT realisations
(1)	$[l_1, u_1] \le [l_2, u_2]$	all SAT	$\mathcal{R}$
(2)	$[l_2, u_2] < [l_1, u_1]$	all UNSAT	$\emptyset$
(3)	$l_1 \le [l_2, u_2]$	ambiguous	$\mathcal{R}[\sigma_1^{\min}]$
(4)	$[l_1, u_1] \le u_2$	ambiguous	$\mathcal{R}[\sigma_2^{\max}]$
(5)	$l_2 \le l_1 \le u_2 \le u_1$	ambiguous	$\mathcal{R}[\sigma_1^{\min}] \cap \mathcal{R}[\sigma_2^{\max}]$

In the remaining cases (3)–(5), when the intervals do overlap, it is ambiguous whether $\mathcal{F}$ contains feasible solutions, so additional analysis is necessary. For example, if $l_1 \le l_2$ (case (3)) and $\sigma_1^{\min}$ satisfies all intra-controller constraints $\mathcal{O}_{S'}(\sigma_1) \in \varphi^{struc}$, then $\sigma_1^{\min} \in \mathcal{F}_1$ and thus there exists realisation $r \in \mathcal{R}[\sigma_1^{\min}]$ such that $\bar{\sigma}^r \models \varphi_c$. Analogous reasoning can be applied to case (4). Finally, in case (5), an n-controller $(\sigma_1^{\min}, \sigma_2^{\max})$ represents a solution if it is consistent; then, any realisation $r \in \mathcal{R}[\sigma_1^{\min}] \cap \mathcal{R}[\sigma_2^{\max}]$ that induces both of these controllers is accepting. In all of these cases (3)–(5), if the consistency requirements are not met, $\mathcal{R}$ may still contain feasible solutions, but not involving $\sigma_1^{\min}$ or $\sigma_2^{\max}$, or the two of them together. In other terms, our over-approximation M is too coarse and needs to be *refined* to obtain more precise results. Refinement will be the focus of Subsect. 4.2.

Special Cases of φ_c. Note that if $T_1 = T_2$, then $\sigma_1^{\min} = \sigma_2^{\min}$: the controller minimising the reachability probability for state s_1 also minimises the reachability probability from state s_2. Therefore, only two model checking queries are required, as opposed to four. Finally, if $T_1 \ne T_2$ but the property quantifies over a single controller, e.g. $\exists \sigma : \varphi^{struc} \land \exists s_1 \in \{s_1\}[\sigma], \exists s_2 \in \{s_2\}[\sigma] : \varphi^{prob}$, then again four controllers are computed. However, additional care must be taken in case **(5)**. If $\sigma_1^{\min}$ and $\sigma_2^{\max}$ are consistent, note that the property calls for a 1-controller, so we need $\sigma_1^{\min} = \sigma_2^{\max}$. If this is not the case, i.e. $\sigma_1^{\min}(s) \neq \sigma_2^{\max}(s)$ for some $s \in S$, it might happen that for one controller this action selection is not relevant to guarantee feasibility, so we can ensure $\sigma_1^{\min}(s) = \sigma_2^{\max}(s)$. If the two controllers cannot reach an agreement for all action selections, we say that these controllers are *incompatible*.

4.2 Splitting Strategy

Consider controller $\sigma \in \Sigma^M$ provided by the model checker from which a feasible realisation $r \in \mathcal{R}$ could not have been deduced because σ is (intra-controller) inconsistent (cases **(3)**–**(4)**, $\mathcal{R}[\sigma] = \emptyset$)[3]. Let i be the index controller σ is associated with in the set of controllers to be synthesized. σ being inconsistent means there exists a *conflicting parameter* $k \in \mathcal{K}$ associated with multiple states for which σ selected different actions. Formally, let $S_k := \{s' \in S \mid k(i, s') \simeq k\}$ denote the set of states associated with k and let $\mathcal{A}_k := \{\sigma(s) \mid s \in S_k\}$ be the set of actions selected by σ in these states; $|\mathcal{A}_k| > 1$ iff σ is inconsistent.

The goal is to partition $\mathcal{R}$ into subsets of realisations in which this conflict cannot occur, i.e. σ cannot be derived by the model checker. Such a partition can be achieved by partitioning the domain $\text{dom}(i, s)$ of the conflicting parameter. We construct the partition $\mathfrak{X}$ using singletons containing conflicting actions: $\mathfrak{X} = \{\{a\} \mid a \in \mathcal{A}_k\} \cup \{\text{dom}(i, s)\backslash\mathcal{A}_k\}$. Let $\{\mathcal{R}_X\}_{X \in \mathfrak{X}}$ denote the corresponding partition of the set $\mathcal{R}$ of realisations where $\mathcal{R}_X := \{r \in \mathcal{R} \mid r(k(i, s)) \in X\}$. Finally, for each $X \in \mathfrak{X}$, we consider a *restriction* $M[\mathcal{R}_X]$ of MDP M where only actions from the set X are available in states S_k. Note that this restriction will be further applied only in the computation of the i-th controller σ. Also note that, if X is a singleton, inconsistencies and incompatibilities on parameter k are trivially excluded in the restriction. It is easy to see that $\sigma \notin \Sigma^{M[\mathcal{R}_X]}$, indicating that $M[\mathcal{R}_X]$ is a more precise over-approximation. Subsequent model checking queries of $M[\mathcal{R}_X]$ (related to the i-th controller) will generate a different optimal controller having a tighter bound. Eventually, R_X will contain a single realization, for which the lower and upper bounds of the corresponding restriction coincide.

When σ violates multiple structural constraints, there are multiple conflicting parameters. To select a single parameter to be used for splitting, we employ a heuristic inspired by [6]. Let $\mathcal{C} \subseteq \mathcal{K}$ denote the set of conflicting parameters in minimizing or maximizing controller σ. For the purpose of splitting, we wish to identify a single parameter $k^* \in \mathcal{C}$ that will produce sub-families $\{\mathcal{R}_X\}_{X \in \mathfrak{X}}$ that are more likely to fall into cases **(1)** or **(2)**. Since predicting the exact analysis of resulting sub-families is impossible, we propose the following heuristics based on the results of the model checking queries. For a probability constraint, we introduce the *immediate impact* of action a in state s:

---

[3] Case **(5)**, when σ is consistent but incompatible with another controller σ', is handled analogously, as well as when they are inter-controller inconsistent.

$$\gamma(s,a) := exp(s) \cdot \Big[\sum_{s'} \mathcal{P}(s,a,s') \cdot \mathbb{P}(M^\sigma, s' \models \Diamond T) \Big],$$

where $exp(s)$ is the expected number of visits of state s in MC M^σ. For reward-based constraints, the immediate impact is defined in a similar way. Having computed the immediate impact for each state $s \in S$ and action $a \in Act$, the parameter associated with the highest average difference of γ is selected for splitting:

$$k^* \in \arg\max_{k \in \mathcal{C}} avg_{s \in \mathcal{S}_k} \Big\{ \max_{a \in \mathcal{A}_k} \gamma(s,a) - \min_{a \in \mathcal{A}_k} \gamma(s,a) \Big\}.$$

4.3   Synthesis Algorithm (AR Loop)

Algorithm 1 summarises the proposed synthesis approach, further denoted as *abstraction refinement (AR) loop*. We first instantiate the state quantifications of φ^{state} on φ^{prob} (see Definition 2). The result is Φ, a Boolean combination of constraints. Afterwards, we construct the family of realisations $\mathcal{R}$ according to Definition 5 and push $\mathcal{R}$ to the stack $\overline{\mathcal{R}}$ of families to be processed. Then, the main synthesis loop starts: each iteration corresponds to the analysis of one (sub-)family of realisations. Firstly, an iteration analyses the restricted MDP $M[\mathcal{R}]$ with respect to each atomic property ϕ appearing in Φ. This is the most computationally demanding part of the algorithm, where bounds on particular reachability probabilities (rewards) involved in ϕ are computed. The produced

---

Algorithm 1. Deductive syntheses using abstraction refinement (AR loop).

Input: MDP $M = (S, Act, \mathcal{P})$, specification $\exists \sigma_1, \ldots, \sigma_n \in \Sigma^M : \varphi^{struc} \wedge \varphi^{state}$ where $\varphi^{state} = \mathcal{Q}_1 s_1 \in I_1[\sigma^1] \ldots \mathcal{Q}_m s_m \in I_m[\sigma^m] : \varphi^{prob}$

Output: $\overline{\sigma} = (\sigma_1, \ldots \sigma_n)$ that satisfies $\varphi^{struc} \wedge \varphi^{prob}$, or "unfeasible".

1: $\Phi \leftarrow$ instantiateQuantifiers($\varphi^{state}, \varphi^{prob}$)   ▷ Use Def. 2
2: $\mathcal{R} \leftarrow$ constructFamily(M, n, φ^{struc})   ▷ Use Def. 5
3: $\overline{\mathcal{R}}$.push($\mathcal{R}$)   ▷ Create a stack of (sub-)families $\overline{\mathcal{R}}$
4: **while** nonEmpty($\overline{\mathcal{R}}$) **do**
5:    $\mathcal{R} \leftarrow \overline{\mathcal{R}}$.pop()
6:    **For each** $\phi \in$ atoms(Φ) **do**
7:       *bounds, controllers* $\leftarrow$ analyseMDP($M[\mathcal{R}], \phi$)   ▷ Compute bounds wrt. ϕ
8:       *consistency* $\leftarrow$ checkConsistency(*controllers*)
9:       *results*(ϕ) $\leftarrow$ intervalLogic(*bounds, consistency*)   ▷ Apply Tab. 1
10:    $\Phi' =$ replaceAndEvaluate(Φ, *results*)
11:    **if** $\Phi' \equiv \top$ **then return** any $\overline{\sigma}^r$ s.t. $r \in \mathcal{R}$ ▷ $\mathcal{R}$ is feasible
12:    **if** $\Phi' \equiv \bot$ **then continue**   ▷ $\mathcal{R}$ is unfeasible and thus pruned
13:    **if** $(\mathcal{R}' =$ compose(Φ', *results*)) $\neq \varnothing$ **then return** any $\overline{\sigma}^r$ s.t. $r \in \mathcal{R}'$ ▷ SAT
14:    $\phi \leftarrow$ ambiguousAtom(Φ')
15:    $\overline{\mathcal{R}}$.push(split($\mathcal{R}$, *results*(ϕ)))   ▷ Use splitting wrt. ϕ as described in Sec. 4.2
16: **return unfeasible**

---

controllers are checked for consistency, and together with bounds we use reasoning from Sect. 4.1, and in particular Table 1, to mark feasibility of $\mathcal{R}$ with respect to ϕ as one of 'all SAT', 'all UNSAT' or 'ambiguous' (with the, possibly empty, set of satisfying controllers). To decide whether Φ can be satisfied, on Line 10 we replace each (UN)SAT atom with $\top$ ($\bot$). If the resulting formula Φ' is trivially $\top$, then any realisation in $\mathcal{R}$ is a satisfying solution. Conversely, if Φ' is $\bot$, there is no satisfying realisation, hence $\mathcal{R}$ is discarded. Finally, if Φ' is neither $\top$ nor $\bot$, we try to compose a satisfying n-controller from individual controllers (obtained when analysing ambiguous atoms of Φ) by performing intersections or unions of the corresponding sets $\mathcal{R}[\sigma]$ of SAT realisations. If such composition returns an empty set, then $\mathcal{R}$ is partitioned (see Sect. 4.2) into sub-families that are put on top of stack $\overline{\mathcal{R}}$.

Correctness of the synthesis algorithm follows from the interval logic defined in Table 1. Recall that the logic uses safe lower and upper bounds on the probabilities (rewards) attained by controllers in the given sub-family. These bounds together with the consequent consistency analysis ensures that in every iteration either i) a satisfying solution is found, or ii) the entire sub-family is correctly pruned as non-satisfying or iii) the sub-family is decomposed into sub-families that are further analysed. The algorithm terminates as singleton sub-families (i.e. families containing only a single n-controller) are solved in the given iteration.

Complexity of the algorithm is given by the number of the iterations of the main loop (line 4). In the worst case, the algorithm needs to decompose the original family to a set of singleton sub-families and to analyse each n-controller individually. This yields an exponential worst-case complexity with respect to $|S|$ and n. However, in practice, a principally smaller number of iterations is needed, as will be shown in the experimental evaluation.

5   Experiments

Our experimental evaluation focuses on the following key questions:

Q1: *How does the proposed deductive approach perform in comparison to the state-of-the-art SMT based approach implemented in* HYPERPROB [19]*?*

Q2: *Can the deductive approach effectively handle synthesis problems that include structural constraints?*

Implementation and Experimental Setting. We have implemented the proposed AR loop on top of the tool PAYNT [7] – a tool for the synthesis of probabilistic programs. PAYNT takes input models expressed as extended PRISM [33] files, and relies on the probabilistic model checker STORM [29] to analyse the underlying MDPs. The implementation of the synthesis algorithm and all benchmarks are publicly available at https://doi.org/10.5281/zenodo.8116528 in the form of an artifact. All experiments were run on a Linux Ubuntu 20.04.4 LTS server with a 3.60 GHz Intel Xeon CPU and 64 GB of RAM. A timeout of 1 h is enforced. All run-times are in seconds.

Q1: Comparison with HYPERPROB

In this section, we present a detailed performance comparison of the proposed approach with the SMT encoding implemented in HYPERPROB.

We first consider the main case studies presented in [19], namely, **TA** (timing attack), **PW** (password leakage), **TS** (thread scheduling) and **PC** (probabilistic conformance). A short description of the models and considered variants can be found in [4, Appendix B]. Note that the first three case studies are formulated as *verification* problems and thus the dual formulation, i.e., *synthesis of a counter-example*, is considered. Also note that all considered variants are feasible, i.e. a counter-example exists as well as a valid solution for **PC**.

Table 2 reports for each benchmark the size of the underlying MDP and the main performance statistics of the proposed AR loop and HYPERPROB, respectively. For AR loop, it reports the size of the family $\mathcal{R}$, the run-time and the number of AR iterations (the number of investigated sub-families of $\mathcal{R}$). For HYPERPROB, it reports the number of variables (vars) and subformulae (sub) in the SMT encoding, together with the encoding time (e-time) and the time to solve the formula (s-time).

Table 2. Comparison on HYPERPROB benchmarks. * denotes a more complicated variant not considered in [19]. For more details, see [4, Appendix B].

Case Study	MDP	AR loop			HYPERPROB			
	size	family	time	iters	vars	sub	e-time	s-time
PW ($n = 4$)	298	5E21	<1 s	1	357898	597	93	40
PW* ($n = 4$)	298	5E21	<1 s	1	1792172	597	548	TO
TA ($n = 4$)	244	5E21	<1 s	1	240340	489	63	491
TA* ($n = 4$)	244	5E21	<1 s	1	1204628	489	373	TO
TS ($h = (10, 20)$)	45	2E6	<1 s	1	12825	91	4	<1 s
PC ($s = 0 \ldots 4$)	20	3E9	TO(8%)	452k	6780	41	7	<1 s

Table 3. SD on different variants of maze problems.

Maze	Feasible	MDP	AR loop			HYPERPROB			
		size	family	time	iters	vars	sub	e-time	s-time
simple	x	10	16k	1	1.7k	590	21	<1 s	1
splash-1	✓	16	1E7	<1 s	63(10%)	1424	33	<1 s	50
larger-1	✓	25	3E11	194	183k(16%)	3350	51	1	TO
larger-2	x	25	7E10	581	575k	3350	51	1	TO
larger-3	✓	25	7E10	<1 s	120(10%)	3350	51	1	TO
splash-2	x	25	3E11	1524	2E6	3350	51	<1 s	TO
train	x	48	2E10	7	10.2k	11952	97	3	TO

We can observe that the first three case studies are trivial for the AR loop as only a single iteration is required to find a counter-example that disproves the specification. On the other hand, HYPERPROB has to construct and solve non-trivial SMT formulae and, in some cases, fails to do so within the 1-hour timeout. The situation is very different for **PC**. HYPERPROB is able to quickly find the solution due to the manageable size and compact structure of the underlying SMT encoding. On the other hand, the AR loop fails to find a valid solution (within the timeout it explores only 8% or the family members). A deeper investigation indicates that, in this case, the current splitting strategy is not effective.

These results are quite encouraging as they show that the proposed approach can almost trivially solve a number of relevant verification problems considered in the literature. On the other hand, they do not provide insight into the performance of our approach because the very coarse abstraction is already sufficient to find a satisfying solution. In the following section, we consider synthesis problems that are principally much harder for our approach.

Stochastic Domination (SD) [12] is a relation property that has been introduced for random variables. We adapt it to the reachability problem in MDPs. We say that state s_2 *dominates* state s_1 with respect to the target set T if, for all controllers $\sigma_i \in \Sigma^M$, it holds that $\mathbb{P}(M^\sigma, s_1 \models \Diamond T) > \mathbb{P}(M^\sigma, s_2 \models \Diamond T)$. To verify **SD**, we consider synthesis of a counterexample, i.e. a controller proving that s_2 does not dominate s_1 formulated using the following hyperproperty:

$$\exists \sigma : \forall s_1 \in \{s_a\}[\sigma] \, \forall s_2 \in \{s_b\}[\sigma] \; : \; \mathbb{P}(M^\sigma, s_1 \models \Diamond T) \geq \mathbb{P}(M^\sigma, s_2 \models \Diamond T).$$

We verify this specification against several variants of the problem from Example 1, where we consider a robot navigating through a maze [34]. Due to its faulty locomotion, after each step the robot may end up in a different location than the one suggested by the controller. We assume full observability of the maze, hence the controller uses the current state to decide the next movement direction. Individual variants of the problem are described in [4, Appendix B]. Evaluation results are reported in Table 3, where we explicitly report whether the problem is feasible. For feasible instances, the column *iters* also reports the portion of the family members explored before the satisfying solution was found. Observe that HYPERPROB fails to solve a majority of the problems within the 1-hour timeout, despite the fact that the underlying SMT encoding is smaller comparing to the HYPERPROB benchmarks investigated in Table 2. This is probably due to the fact that these **SD** problems are either unfeasible or have only few satisfying solutions. On the other hand, AR loop is able to solve all **SD** variants within the timeout. Note that, although these benchmarks consider MDPs with tens of states, the individual run-times differ significantly. This is because a specific structure of the MDP can have a huge impact on the number of iterations (splittings) required to obtain unambiguous cases.

To further emphasise the capabilities as well as the limitations of the proposed approach, we consider the *complete synthesis* problem: instead of synthesizing a single satisfying controller, we look for **all** controllers that meet the specification, which requires exploration of the entire family. Note that this problem is closely related to a construction of *permissive controllers* or *safety shields* in the context of safe reinforcement learning [22,31]. Table 4 shows the performance of the AR loop as well as the number of feasible instances. As expected, the complete synthesis for feasible instances is significantly harder. Moreover, we observe that abstraction-based pruning is typically less efficient comparing to the unfeasible instances. For the variant larger-1, only 86% of the family members were explored in the 1-hour timeout (the entire exploration took 80 min). Complete synthesis is also very challenging for the SMT approach since it requires multiple calls to the SMT solver, where each subsequent call rules out the previously found solution. HYPERPROB currently does not support complete synthesis.

Table 4. Complete synthesis for the feasible variants of the **SD** problem.

Maze	AR loop		feasible
	time	iters	instances
splash-1	<1 s	1233	1 (~0%)
larger-1	4771	4E6	4(~0%)
larger-3	169	149k	2590 (~0%)

Q2 Synthesis Under Structural Constraints

To investigate the performance of the proposed approach on more complicated synthesis problems that include structural constraints on the resulting controllers, we extend the robotic scenario from Example 1. We consider *probabilistic noninterference* [26] and *opacity* [36], both ensuring location privacy with respect to attacker observability. In both cases we formulate the problem within a planning scenario in a maze. To demonstrate the scalability with respect to the complexity of the underlying MDP, we consider two larger variants of the maze, (see the MDP sizes in Table 5) and a more complicated objective. The maze is equipped with checkpoint locations and the robot earns a reward when repeatedly passing through at least two of these locations. Hence, a good strategy makes the robot move back and forth among two or more checkpoints. There are no traps, but every move has a small chance of unrecoverable error. Additionally, we assume *partial observability* - the robot can only observe the available

Table 5. Probabilistic Noninterference and Opacity specifications for more complex variants of the maze problem.

Maze	MDP size	Noninterference			Opacity			
		family	time	iters	family	time	iters	distance
larger-4	432	221k	1	337	3E9	<1s	96	6/13
larger-5	1260	3E6	<1s	7	5E11	<1s	43	8/15
larger-4+mem	864	8E20	1551	110K	2E38	193	13k	32/32
larger-5+mem	2520	1E25	481	16k	2E46	309	9k	22/38

directions (up, down, left, right). This is encoded using intra-controller constraints $\mathcal{O}_{S'}(\sigma)$.

Probabilistic Noninterference is formulated as a 2-controller synthesis problem with two different initial states that need to be kept private. We consider the specification containing two PCTL properties requiring that both controllers achieve a reward exceeding a minimum threshold λ and a hyperproperty requiring that the achieved values are almost the same (the attacker observes the final rewards). Additionally, we define a set of sensitive states where both controllers have to agree on their actions, encoded using inter-controller constraints $\mathcal{X}_s(\Sigma)$. For both considered variants of the maze (larger-4 and larger-5), there is no pair of memoryless controllers that would satisfy the specification with a non-zero threshold. In Table 5 (left) we report experiments where threshold λ is set to 0, already making the problem unfeasible.

Furthermore, we enhance both MDPs with a 1-bit memory (denoted by suffix +mem). A controller can now properly use the memory to distinguish between otherwise indistinguishable states. With this extension, it is possible to find 2-controllers satisfying probabilistic noninterference with non-zero rewards. In the corresponding experiments in Table 5, the reward thresholds λ are set close (but not equal) to the respective maximal values. This makes the PCTL part of the specification satisfiable, but the overall specification unfeasible. This allows us to inspect how the hyperproperty is leveraged to explore the entire search space. The results show that adding the 1-bit memory enormously enlarges the family size; however, the AR loop is still able to fully explore the family in a reasonable time. Exploring the family of 2-bit controllers is not tractable within the 1-hour timeout.

Opacity is a hyperproperty, used in non-probabilistic robot planning, specifying that for a given state there exist two different paths that both give identical observations to the attacker and reach the planning goal. We adapt opacity to our setting in the following way. For a given initial state, we seek for two controllers that i) achieve an expected reward above a threshold (performance requirement), ii) achieve almost the same reward visible to the attacker (anonymity requirement) and iii) take as many mutually different choices as possible. If the two controllers are indistinguishable for the attacker and yet do not agree on any choice, then anonymisation is fully guaranteed. We encode the last objective as an optimisation over the intra-controller characteristic, further denoted as *distance*, where we maximise the number of different choices.

In order to effectively search for the *optimal* n-controller with respect to such a specification, we need to completely explore the family $\mathcal{R}$. We extend the AR loop to use the distance of the current optimum to improve the pruning strategy: if the current sub-family does not include improving solutions with respect to currently optimal distance, it is promptly pruned without analysing the MDPs with respect to performance and anonymity properties. This way, we extend the deductive reasoning over the families using an additional structural constraint that iteratively changes during the synthesis process.

Table 5 (right) reports the results for memoryless and 1-bit controllers, respectively. The last column reports the obtained maximal distance and the number of choices that need to be synthesised per controller. Note that, due to partial observability, there are fewer choices than states. We see again that adding memory makes the synthesis problem much harder. On the other hand, the memory not only allows us to significantly increase the threshold on the reward, but also improves the achieved distance, i.e. the degree of anonymisation. For the variant larger-4+mem, we synthesized a pair of controllers that differ in all 32 choices, thus ensuring full anonymisation.

6 Conclusions

We proposed a deductive controller synthesis procedure for MDPs with respect to probabilistic hyperproperties expressed in HyperPCTL enriched with additional structural constraints on the resulting controllers. Our approach builds on a symbolic representation of the family of the controllers and an abstraction refinement strategy allowing for an effective exploration of the families. A detailed experimental evaluation demonstrates that our approach considerably outperforms HYPERPROB [19], a state-of-the-art tool based on SMT reasoning. In particular, our approach scales to more complicated synthesis problems where HYPERPROB times out, and, for the first time, effectively supports in the synthesis procedure structural constraints on the resulting controllers. As future work, we plan to improve our approach by investigating counterexamples for probabilistic hyperproperties.

References

1. Ábrahám, E., Bartocci, E., Bonakdarpour, B., Dobe, O.: Parameter synthesis for probabilistic hyperproperties. In: LPAR 2020: 23rd International Conference on Logic for Programming, Artificial Intelligence and Reasoning. EPiC Series in Computing, vol. 73, pp. 12–31. EasyChair (2020). https://doi.org/10.29007/37lf
2. Ábrahám, E., Bartocci, E., Bonakdarpour, B., Dobe, O.: Probabilistic hyperproperties with nondeterminism. In: Hung, D.V., Sokolsky, O. (eds.) ATVA 2020. LNCS, vol. 12302, pp. 518–534. Springer, Cham (2020). https://doi.org/10.1007/978-3-030-59152-6_29
3. Ábrahám, E., Bonakdarpour, B.: HyperPCTL: a temporal logic for probabilistic hyperproperties. In: McIver, A., Horvath, A. (eds.) QEST 2018. LNCS, vol. 11024, pp. 20–35. Springer, Cham (2018). https://doi.org/10.1007/978-3-319-99154-2_2
4. Andriushchenko, R., Bartocci, E., Ceska, M., Pontiggia, F., Sallinger, S.: Deductive controller synthesis for probabilistic hyperproperties. arXiv preprint arXiv:2307.04503 (2023)
5. Andriushchenko, R., Češka, M., Junges, S., Katoen, J.-P.: Inductive synthesis for probabilistic programs reaches new horizons. In: TACAS 2021. LNCS, vol. 12651, pp. 191–209. Springer, Cham (2021). https://doi.org/10.1007/978-3-030-72016-2_11

6. Andriushchenko, R., Ceska, M., Junges, S., Katoen, J.: Inductive synthesis of finite-state controllers for POMDPs. In: Proceedings of UAI 2022: the Thirty-Eighth Conference on Uncertainty in Artificial Intelligence. Proceedings of Machine Learning Research, vol. 180, pp. 85–95. PMLR (2022)

7. Andriushchenko, R., Češka, M., Junges, S., Katoen, J.-P., Stupinský, Š: PAYNT: a tool for inductive synthesis of probabilistic programs. In: Silva, A., Leino, K.R.M. (eds.) CAV 2021. LNCS, vol. 12759, pp. 856–869. Springer, Cham (2021). https://doi.org/10.1007/978-3-030-81685-8_40

8. Baier, C., de Alfaro, L., Forejt, V., Kwiatkowska, M.: Model Checking Probabilistic Systems. In: Clarke, E., Henzinger, T., Veith, H., Bloem, R. (eds.) Handbook of Model Checking, pp. 963–999. Springer, Cham (2018). https://doi.org/10.1007/978-3-319-10575-8_28

9. Baier, C., Funke, F., Piribauer, J., Ziemek, R.: On probability-raising causality in Markov decision processes. In: Bouyer, P., Schröder, L. (eds.) FoSSaCS 2022. LNCS, vol. 13242, pp. 40–60. Springer, Cham (2022). https://doi.org/10.1007/978-3-030-99253-8_3

10. Baier, C., Größer, M., Leucker, M., Bollig, B., Ciesinski, F.: Controller synthesis for probabilistic systems (extended abstract). In: Levy, J.-J., Mayr, E.W., Mitchell, J.C. (eds.) TCS 2004. IIFIP, vol. 155, pp. 493–506. Springer, Boston, MA (2004). https://doi.org/10.1007/1-4020-8141-3_38

11. Baier, C., Katoen, J.: Principles of Model Checking. MIT Press, Cambridge (2008)

12. Barthe, G., Hsu, J.: Probabilistic Couplings from Program Logics, pp. 145–184. Cambridge University Press, Cambridge (2020). https://doi.org/10.1017/9781108770750.006

13. Beutner, R., Finkbeiner, B.: Software verification of hyperproperties beyond k-safety. In: Shoham, S., Vizel, Y. (eds.) CAV 2022. LNCS, vol. 13371, pp. 341–362. Springer, Cham (2022). https://doi.org/10.1007/978-3-031-13185-1_17

14. Bonakdarpour, B., Finkbeiner, B.: Controller synthesis for hyperproperties. In: 33rd IEEE Computer Security Foundations Symposium, pp. 366–379. IEEE (2020). https://doi.org/10.1109/CSF49147.2020.00033

15. Češka, M., Jansen, N., Junges, S., Katoen, J.-P.: Shepherding hordes of Markov chains. In: Vojnar, T., Zhang, L. (eds.) TACAS 2019. LNCS, vol. 11428, pp. 172–190. Springer, Cham (2019). https://doi.org/10.1007/978-3-030-17465-1_10

16. Clarkson, M.R., Finkbeiner, B., Koleini, M., Micinski, K.K., Rabe, M.N., Sánchez, C.: Temporal logics for hyperproperties. In: Abadi, M., Kremer, S. (eds.) POST 2014. LNCS, vol. 8414, pp. 265–284. Springer, Heidelberg (2014). https://doi.org/10.1007/978-3-642-54792-8_15

17. Clarkson, M.R., Schneider, F.B.: Hyperproperties. J. Comput. Secur. **18**(6), 1157–1210 (2010). https://doi.org/10.3233/JCS-2009-0393

18. Dimitrova, R., Finkbeiner, B., Torfah, H.: Probabilistic hyperproperties of Markov decision processes. In: Hung, D.V., Sokolsky, O. (eds.) ATVA 2020. LNCS, vol. 12302, pp. 484–500. Springer, Cham (2020). https://doi.org/10.1007/978-3-030-59152-6_27

19. Dobe, O., Ábrahám, E., Bartocci, E., Bonakdarpour, B.: HYPERPROB: a model checker for probabilistic hyperproperties. In: Huisman, M., Păsăreanu, C., Zhan, N. (eds.) FM 2021. LNCS, vol. 13047, pp. 657–666. Springer, Cham (2021). https://doi.org/10.1007/978-3-030-90870-6_35

20. Dobe, O., Wilke, L., Ábrahám, E., Bartocci, E.: Probabilistic hyperproperties with rewards. In: Deshmukh, J.V., Havelund, K., Perez, I. (eds.) NFM 2022. LNCS, vol. 8837, pp. 146–162. Springer, Cham (2014). https://doi.org/10.1007/978-3-031-06773-0_35

21. Dobe, O., Ábrahám, E., Bartocci, E., Bonakdarpour, B.: Model checking hyperproperties for Markov decision processes. Inf. Comput. **289**, 104978 (2022). https://doi.org/10.1016/j.ic.2022.104978

22. Dräger, K., Forejt, V., Kwiatkowska, M., Parker, D., Ujma, M.: Permissive controller synthesis for probabilistic systems. In: Ábrahám, E., Havelund, K. (eds.) TACAS 2014. LNCS, vol. 8413, pp. 531–546. Springer, Heidelberg (2014). https://doi.org/10.1007/978-3-642-54862-8_44

23. Dwork, C.: Differential privacy. In: Bugliesi, M., Preneel, B., Sassone, V., Wegener, I. (eds.) ICALP 2006. LNCS, vol. 4052, pp. 1–12. Springer, Heidelberg (2006). https://doi.org/10.1007/11787006_1

24. Finkbeiner, B., Hahn, C., Lukert, P., Stenger, M., Tentrup, L.: Synthesizing reactive systems from hyperproperties. In: Chockler, H., Weissenbacher, G. (eds.) CAV 2018. LNCS, vol. 10981, pp. 289–306. Springer, Cham (2018). https://doi.org/10.1007/978-3-319-96145-3_16

25. Finkbeiner, B., Hahn, C., Lukert, P., Stenger, M., Tentrup, L.: Synthesis from hyperproperties. Acta Informatica **57**(1-2), 137–163 (2020). https://doi.org/10.1007/s00236-019-00358-2

26. Goguen, J.A., Meseguer, J.: Security policies and security models. In: IEEE Symposium on Security and Privacy, pp. 11–20. IEEE Computer Society (1982)

27. Gray, J.W., Syverson, P.F.: A logical approach to multilevel security of probabilistic systems. Distrib. Comput. **11**(2), 73–90 (1998). https://doi.org/10.1007/s004460050043

28. Hansson, H., Jonsson, B.: A logic for reasoning about time and reliability. Formal Aspects Comput. **6**(5), 512–535 (1994). https://doi.org/10.1007/BF01211866

29. Hensel, C., Junges, S., Katoen, J., Quatmann, T., Volk, M.: The probabilistic model checker Storm. Int. J. Softw. Tools Technol. Transf. **24**(4), 589–610 (2022). https://doi.org/10.1007/s10009-021-00633-z

30. Hsu, T., Bonakdarpour, B., Kang, E., Tripakis, S.: Mapping synthesis for hyperproperties. In: 35th IEEE Computer Security Foundations Symposium, CSF 2022, Haifa, Israel, 7–10 August 2022, pp. 486–500. IEEE (2022). https://doi.org/10.1109/CSF54842.2022.9919679

31. Jansen, N., Könighofer, B., Junges, S., Serban, A., Bloem, R.: Safe reinforcement learning using probabilistic shields (invited paper). In: Proceedings of CONCUR 2020: International Conference on Concurrency Theory. Leibniz International Proceedings in Informatics (LIPIcs), vol. 171, pp. 1–16 (2020). https://doi.org/10.4230/LIPIcs.CONCUR.2020.3

32. Köpf, B., Basin, D.A.: An information-theoretic model for adaptive side-channel attacks. In: Proceedings of the 2007 ACM Conference on Computer and Communications Security (CCS), pp. 286–296. ACM (2007). https://doi.org/10.1145/1315245.1315282

33. Kwiatkowska, M., Norman, G., Parker, D.: PRISM 4.0: verification of probabilistic real-time systems. In: Gopalakrishnan, G., Qadeer, S. (eds.) CAV 2011. LNCS, vol. 6806, pp. 585–591. Springer, Heidelberg (2011). https://doi.org/10.1007/978-3-642-22110-1_47

34. Norman, G., Parker, D., Zou, X.: Verification and control of partially observable probabilistic real-time systems. In: Sankaranarayanan, S., Vicario, E. (eds.) FORMATS 2015. LNCS, vol. 9268, pp. 240–255. Springer, Cham (2015). https://doi.org/10.1007/978-3-319-22975-1_16

35. O'Neill, K.R., Clarkson, M.R., Chong, S.: Information-flow security for interactive programs. In: Proceedings of CSFW: the 19th IEEE Computer Security Foundations Workshop, pp. 190–201. IEEE Computer Society (2006). https://doi.org/10.1109/CSFW.2006.16
36. Wang, Y., Nalluri, S., Pajic, M.: Hyperproperties for robotics: planning via hyperLTL. In: Proceedings of ICRA: the 2020 IEEE International Conference on Robotics and Automation, pp. 8462–8468. IEEE (2020). https://doi.org/10.1109/ICRA40945.2020.9196874
37. Zaman, E., Ciardo, G., Ábrahám, E., Bonakdarpour, B.: HyperPCTL model checking by probabilistic decomposition. In: ter Beek, M.H., Monahan, R. (eds.) IFM 2022. LNCS, vol. 13274, pp. 209–226. Springer, Cham (2022). https://doi.org/10.1007/978-3-031-07727-2_12

Model Abstraction and Conditional Sampling with Score-Based Diffusion Models

Luca Bortolussi, Francesca Cairoli, Francesco Giacomarra,
and Davide Scassola[✉]

AIlab, University of Trieste, Trieste, Italy
lbortolussi@units.it , davide.scassola@phd.units.it

Abstract. We propose an approach to build and sample surrogate stochastic models leveraging state-of-the-art score-based diffusion approaches, either abstracting a known stochastic process or learning directly the model from data. In particular, we propose a method for efficient conditional sampling from such surrogate models, enforcing logical and consistency constraints on generated samples in a soft fashion. As a preliminary case study, we consider a surrogate SIR model, in both its ergodic and non-ergodic formulations. Using the aforementioned method, we are able to sample trajectories from such models that exhibit desirable features having low probability in the unconstrained models, allowing us to explore epidemiologically relevant scenarios. Although the proposed approach is still a work-in-progress, it has significant potential for applications in epidemiology and other fields. The method is also efficient in the sense that retraining is not needed to generate samples satisfying different constraints.

Keywords: Model abstraction · Score-based diffusion model · Conditional generative model

1 Introduction

Stochastic processes are are a fundamental tool for modelling and simulating complex systems with inherent uncertainty, such as epidemics. However, sampling trajectories from these processes can be computationally demanding, posing scalability issues in practice. State-of-the-art generative models, such as diffusion models [4], could be used to learn the surrogate probability distribution of trajectories directly in the trajectory space. Following this rationale, Score-Based Diffusion models [6] have been efficiently used for conditional time series imputation. However, since the training of a generative model aims at minimizing a certain measure of error, such a model may struggle to learn and reproduce properties of the underlying process, such as the existence of absorbing states. The simulations it produces can thus be inconsistent from a biological or physical perspective. To tackle these issues, we propose a method to incorporate

© The Author(s), under exclusive license to Springer Nature Switzerland AG 2023
N. Jansen and M. Tribastone (Eds.): QEST 2023, LNCS 14287, pp. 307–310, 2023.
https://doi.org/10.1007/978-3-031-43835-6_21

differentiable constraints directly during sampling from a Score-Based Diffusion model [1]. As a case study, we used our approach to generate trajectories from a SIR model in both its ergodic and non-ergodic formulation. In our preliminary results, we show that the samples generated with our procedure are consistent with respect to the constraints imposed at sampling. Moreover, since our method allows us to impose different constraints without retraining the model, it can be used to sample efficiently various regions of the trajectory space, thus exploring relevant scenarios. The organization of this working paper is as follows: Sect. 2 outlines our proposed approach, while Sect. 3 presents the results obtained using our formulation on surrogate SIR models. In the conclusions, we discuss future directions of work.

2   Proposed Approach

In this paper we consider diffusion probabilistic models [4], the state of the art approach to learn models of generative data distributions from data. In particular, we focus on the variant presented in [1], called denoising diffusion probabilistic models (DDPMs). DDPMs are equivalent, during the training phase, to score-based models at multiple noise levels with Langevin dynamics [5]. We based our implementation on the work by Ermon $et\ al.$ [6] in which they build on the idea of DDPMs to develop a model for multivariate time series imputation.

Let $\mathbf{x}_0$ be the available data, in our case a set of trajectories over a d-dimensional space, which is a sample from the space of trajectories $\mathcal{X}$, in principle according to an unknown distribution. Our goal is to learn $p_\theta(\mathbf{x}_0)$, a model distribution which approximates the real data distribution $q(\mathbf{x}_0)$. Consider now a sequence of latent variables $\mathbf{x}_1, \mathbf{x}_2, \ldots, \mathbf{x}_K \in \mathcal{X}$ that represent copies of $\mathbf{x}_0$ perturbed according to a diffusion process applied to the datapoints $\mathbf{x}_0$. In DDPM this process is a Markov chain with the following form:

$$q\left(\mathbf{x}_i \mid \mathbf{x}_{i-1}\right) := \mathcal{N}\left(\sqrt{1 - \beta_i}\mathbf{x}_{i-1}, \beta_i \mathbf{I}\right), \tag{1}$$

where β_i represents the noise level at step i. The reverse process, which denoises $\mathbf{x}_i$ to retrieve $\mathbf{x}_0$, is defined by another Markov chain:

$$p_\theta\left(\mathbf{x}_{i-1} \mid \mathbf{x}_i\right) := \mathcal{N}\left(\mathbf{x}_{i-1}; \boldsymbol{\mu}_\theta\left(\mathbf{x}_i, i\right), \sigma_\theta\left(\mathbf{x}_i, i\right)\mathbf{I}\right) \quad \mathbf{x}_K \sim \mathcal{N}(\mathbf{0}, \mathbf{I}). \tag{2}$$

In diffusion models, the backward process has to be learned during training. In the formulation given in [1] $\boldsymbol{\mu}_\theta$ is a function of the score $\epsilon_\theta\left(\mathbf{x}_i, i\right)$ ($i.e.$, the gradient of the log of the perturbed data distribution at step i) and σ_θ is a function of β_i. In [6] $\epsilon_\theta\left(\mathbf{x}_i, i\right)$ is a neural network and it is learned by score matching [2], that reduces to learning a function that predicts the noise added to $\mathbf{x}_0$ in order to obtain $\mathbf{x}_i$.

We then define the target conditional distribution in the following way:

$$\tilde{q}(\mathbf{x}_0) \propto q(\mathbf{x}_0)e^{c(\mathbf{x}_0)}$$

where $c(\mathbf{x}_0)$ is a differentiable function expressing the satisfaction level of the constraints. It is possible to choose $c(\mathbf{x}_0)$ such that the density of samples that do not satisfy the constraints is minimized. It is easy to show that the following equality holds:

$$\nabla_{\mathbf{x}_0} \ln \tilde{q}(\mathbf{x}_0) = \nabla_{\mathbf{x}_0} \ln q(\mathbf{x}_0) + \nabla_{\mathbf{x}_0} c(\mathbf{x}_0). \tag{3}$$

Then the score of the target constrained distribution can be obtained by summing the score of the unconditional distribution with the gradient of the constraint. In practice, it is necessary to know the score $\forall i \in \{0, 1, ..., K\}$. However, to the best of our knowledge, we have an analytical form only for $i = 0$, which is (3), and for $i = K$, where we can assume $c(\mathbf{x}_K) = 0$ and $\tilde{q}(\mathbf{x}_K)$ is still equal to the prior distribution. We thus propose the following approximation of the noise-dependent score of the target distribution:

$$\tilde{\epsilon}_\theta (\mathbf{x}_i, i) = \epsilon_\theta (\mathbf{x}_i, i) + r(i)\nabla_x c(x_i), \tag{4}$$

where $r(i)$ is a function that modulates the contribution of the constraint, defined such that $r(0) = 1$ and $r(K) = 0$ (in our experiments we mostly use a linear function $r(i) = 1 - \frac{i}{K}$). Sampling from the conditional distribution then reduces to substituting the score of the unconditional model with the modified score $\tilde{\epsilon}_\theta (\mathbf{x}_i, i)$. Notice that since the constraint is applied only during sampling, this method do not require any re-training of the model.

3   Preliminary Results

As a case study, we apply our proposed method for conditional generation of SIR [3] and ergodic SIRS (eSIRS) trajectories. The SIR and eSIRS are widely used to model the spreading of a disease in a population. The SIR model assumes a fixed population of variable size N composed of Susceptible (S), Infected (I), and Recovered (R) individuals. When $I = 0$, the associated state is absorbing. On the other hand, the eSIRS model consider a fixed and constant population and it does not have absorbing states. For both fomulations we consider trajectories with H discretized timesteps, thus for the SIR model we have that the sample space is $\mathcal{X}_{\mathbf{SIR}} := \left(\mathbb{N}_0{}^3\right)^H$, whereas for the eSIRS model the sample space is $\mathcal{X}_{\mathbf{eSIRS}} := \left(\mathbb{N}_0{}^2\right)^H$. To generate the starting data, we use the CSDI method proposed in [6] with given initial states.

In Fig. 1a we use the model trained on SIR data to generate trajectories that satisfy the aforementioned consistency constraints and additionally imposing that the disease vanish at time $t = 5$. In Fig. 1b we show the results obtained for the eSIRS model. Here we imposed non-negative populations as a consistency constraint and the number of infected to be less than 20 for the whole trajectory to represent a limited infection. As can be seen, for both the SIR and eSIRS the generated trajectories mostly satisfy the desired properties while also being consistent with the dynamics of the observed data. This method only allows us to impose soft constraints, so the satisfaction of the target properties is not guaranteed for all the samples, but the failure rate is limited (e.g., in the eSIRS model, around 15%), making rejection sampling a viable approach to strongly enforce constraints.

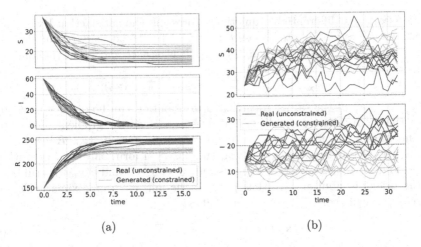

(a)   (b)

Fig. 1. Real and generated trajectories for the constrained SIR (1a) and eSIRS (1b) models. Consistency constraints are almost always satisfied for all trajectories. For SIR model the constraint $I(t) = 0$ for $t \geq 5$ is always satisfied, for eSIRS the constraint $I(t) \leq 20 \; \forall t$ is satisfied in $\approx 85\%$ of samples.

4   Conclusion and Future Works

Being able to incorporate various constraint to explore the sample space efficiently without retraining could be useful in various applications where efficient simulation of *what-if* scenarios is required, and the preliminary results we have obtained are promising. Moreover, the formulation in (4) allows to consider as a constraint any differentiable function, thus all the *Signal Temporal Logic* formulae can be employed as constraints, widening the range of possible applications of our method. Future works include studying new possible candidates for $r(i)$, testing our approach on score-based non-diffusive models and better validating the generated samples of our method via exploration of the sampling space $\mathcal{X}$.

References

1. Ho, J., Jain, A., Abbeel, P.: Denoising diffusion probabilistic models (2020)
2. Hyvärinen, A., Dayan, P.: Estimation of non-normalized statistical models by score matching. J. Mach. Learn. Res. **6**(4) (2005)
3. Kermack, W.O., McKendrick, A.G., Walker, G.T.: A contribution to the mathematical theory of epidemics (1927)
4. Sohl-Dickstein, J., Weiss, E.A., Maheswaranathan, N., Ganguli, S.: Deep unsupervised learning using nonequilibrium thermodynamics (2015)
5. Song, Y., Ermon, S.: Generative modeling by estimating gradients of the data distribution (2020)
6. Tashiro, Y., Song, J., Song, Y., Ermon, S.: CSDI: conditional score-based diffusion models for probabilistic time series imputation (2021)

Probabilistic Counterexample Guidance for Safer Reinforcement Learning

Xiaotong Ji[✉] and Antonio Filieri[✉]

Department of Computing, Imperial College London,
London SW7 2AZ, UK
{xiaotong.ji16,a.filieri}@imperial.ac.uk

Abstract. Safe exploration aims at addressing the limitations of Reinforcement Learning (RL) in safety-critical scenarios, where failures during trial-and-error learning may incur high costs. Several methods exist to incorporate external knowledge or to use proximal sensor data to limit the exploration of unsafe states. However, reducing exploration risks in unknown environments, where an agent must discover safety threats during exploration, remains challenging.

In this paper, we target the problem of safe exploration by guiding the training with counterexamples of the safety requirement. Our method abstracts both continuous and discrete state-space systems into compact abstract models representing the safety-relevant knowledge acquired by the agent during exploration. We then exploit probabilistic counterexample generation to construct minimal simulation submodels eliciting safety requirement violations, where the agent can efficiently train offline to refine its policy towards minimising the risk of safety violations during the subsequent online exploration.

We demonstrate our method's effectiveness in reducing safety violations during online exploration in preliminary experiments by an average of 40.3% compared with QL and DQN standard algorithms and 29.1% compared with previous related work, while achieving comparable cumulative rewards with respect to unrestricted exploration and alternative approaches.

Keywords: Safe reinforcement learning · Probabilistic model checking · Counterexample guidance

1 Introduction

A critical limitation of applying Reinforcement Learning (RL) in real-world control systems is its lack of guarantees of avoiding unsafe behaviours. At its core, RL is a trial-and-error process, where the learning agent explores the decision space and receives rewards for the outcome of its decisions. However, in safety-critical scenarios, failing trials may result in high costs or unsafe situations and should be avoided as much as possible.

N. Jansen and M. Tribastone (Eds.): QEST 2023, LNCS 14287, pp. 311–328, 2023.
https://doi.org/10.1007/978-3-031-43835-6_22

Several learning methods try to incorporate the advantages of model-driven and data-driven methods to encourage safety during learning [16,27]. One natural approach for encouraging safer learning is to analyse the kinematic model of the learning system with specific safety requirements and to design safe exploration [2,15,38] or safe optimisation [1,43,45] strategies that avoid unsafe states or minimise the expected occurrence of unsafe events during training. However, this approach is not applicable for most control systems with partially-known or unknown dynamics, where not enough information is available to characterise unsafe states or events *a priori*.

To increase the safety of learning in environments with (entirely or partially) unknown dynamics, we propose an online-offline learning scheme where online execution traces collected during exploration are used to construct an abstract representation of the visited state-action space. If during exploration the agent violates a safety requirement with unacceptable frequency, a probabilistic model checker is used to produce from the abstract representation minimal counter-example sub-models, i.e., a minimal subset of the abstract state-action space within which the agent is expected to violate its safety requirement with a probability larger than tolerable. These counterexamples are then used to synthesise small-size offline simulation environments within which the agent's policy can be conveniently reinforced to reduce the probability of reiterating safety-violating behaviours during subsequent online exploration. As new evidence from online exploration is gathered the abstract representation is incrementally updated and additional offline phases can be enforced when necessary, until an acceptable safety exploration rate is achieved. Overall, our strategy aims at migrating most trial-and-error risks to the offline training phases, while discouraging the repeated exploration of risky behaviours during online learning. As new evidence is collected during online exploration, the abstract representation is incrementally updated and the current *value* the agent expects from each action is used to prioritise the synthesis of more relevant counterexample-guided simulations.

Our main conceptual contribution in this paper is the use of probabilistic counterexamples to automatically synthesise small-scale simulation submodels where the agent can refine its policy to reduce the risk of violating a safety requirement during learning. In particular, we 1) propose a conservative geometric abstraction model representing safety-relevant experience collected by the agent at any time during online exploration, with theoretical convergence and accuracy guarantees, suitable for the representation of both discrete and continuous state spaces and finite action spaces, 2) adapt minimal label set probabilistic counterexample generation [49] to generate small-scale submodels for the synthesis of offline agent training environments aimed at reducing the likelihood of violating safety requirements during online exploration, and 3) a preliminary evaluation of our method to enhance Q-Learning [48] and DQN [35] agents on problems from literature and the OpenAI Gym, demonstrating how it achieves comparable cumulative rewards while increasing the exploration safety rate by an average of 40.3% compared with QL/DQN, and of 29.1% compared with previous related work [20,23].

2   Background

2.1   Problem Framework

Definition 1. Markov Decision Process (MDP). *An MDP [4] is a tuple* (S, A, s_0, P, R, L), *where S is a set of states, A is a finite set of actions, s_0 is the initial state, $P : S \times A \times S \rightarrow [0, 1]$ is the probability of transitioning from a state $s \in S$ to $s' \in S$ with action $a \in A$, $R : S \times A \rightarrow \mathbb{R}$ is a reward function and $L : S \rightarrow 2^{AP}$ is a labelling function that assigns atomic propositions (AP) to each state.*

A state in S is typically represented by a vector of finite length $n_S \geq 1$. A state space is *discrete* if the elements of the vector are countable, where we assume $S \subseteq \mathcal{Z}^{n_S})$, continuous if $S \subseteq \mathbb{R}^{n_S}$, or hybrid if some elements are discrete and others are continuous. When possible, we omit the cardinality n_S for readability.

Definition 2. Trace. *A finite trace (also called path or trajectory) through an MDP is a sequence $\sigma = s_0, a_0, s_1, a_1, \ldots s_i, a_i, \ldots s_n$, where s_0 is the initial state and $P(s_i, a_i, s_{i+1}) > 0$.*

Definition 3. Policy. *A (deterministic) policy $\pi : S \rightarrow A$ selects in every state s the action a to be taken by the agent.*

Q-learning (QL) [48] is a reinforcement learning algorithm where an agent aims at finding an optimal policy π^* for an MDP that maximises the expected cumulative reward. Given a learning rate $\alpha \in (0, 1]$ and a discount factor $\gamma \in (0, 1]$, such that rewards received after n transitions are discounted by the factor γ^n, the agent learns a value function Q based on the following update rule:

$$Q_t(s, a) = (1 - \alpha)Q_{t-1}(s, a) + \alpha(R(s, a) + \gamma \max_{a' \in A} Q_{t-1}(s', a')) \tag{1}$$

The optimal Q-function Q^* satisfies the Bellman optimality equation:

$$Q^*(s, a) = \mathbb{E}[R(s, a) + \gamma \max_{a' \in A} Q^*(s', a')|s, a] \tag{2}$$

For finite-state and finite-action spaces, QL converges to an optimal policy as long as every state action pair is visited infinitely often [48], but it is not suitable for learning in continuous state spaces. For continuous state spaces instead, the Deep Q-Learning method [35] parameterises Q-values with weights θ as a Q-network and the learning process is adapted to minimising a sequence of loss function L_i at each iteration i (cf. Algorithm 1 in [35]):

$$L_i(\theta_i) = \mathbb{E}[(y_i - Q(s, a; \theta_i))^2], \tag{3}$$

where $y_i = \mathbb{E}[(R(s, a) + \gamma \max_{a' \in A} Q(s', a'; \theta_{i-1})|s, a]$.

During learning, the agent selects the next action among those available in the current state at random with probability $\epsilon_{QL} > 0$ while with probability $1 - \epsilon_{QL}$ it will select an action a yielding $Q^*(s, a)$.

Definition 4. Optimal Policy. *An optimal policy π^*, in the context of Q learning and DQN, is given by $\pi^* = \arg\max_{a \in A} Q^*(s, a), \forall s \in S$.*

2.2 Probabilistic Model Checking and Counterexamples

Probabilistic model checking is an automated verification method that, given a stochastic model – an MDP in our case – and a property expressed in a suitable probabilistic temporal logic, can verify whether the model complies with the property or not [3]. In this work, we use Probabilistic Computational Temporal Logic (PCTL) [19] to specify probabilistic requirements for the safety of the agent. The syntax of PCTL is recursively defined as:

$$\Phi := true \mid \alpha \mid \Phi \wedge \Phi \mid \neg\Phi \mid P_{\bowtie p}\varphi \qquad (4)$$

$$\varphi := X\Phi \mid \Phi U \Phi \qquad (5)$$

A PCTL property is defined by a state formula Φ, whose satisfaction can be determined in each state of the model. $true$ is a tautology satisfied in every state, $\alpha \in AP$ is satisfied in any state whose labels include ($\alpha \in L(s)$), and $\wedge$ and $\neg$ are the Boolean conjunction and negation operators. The modal operator $P_{\bowtie p}\varphi$, with $\bowtie \in \{<, \leq, \geq, >\}$ and $p \in [0, 1]$, holds in a state s if the cumulative probability of all the paths originating in s and satisfying the path formula φ is $\bowtie p$ *under any possible policy*. The Next operator $X\Phi$ is satisfied by any path originating in s such that the next state satisfies Φ. The Until operator $\Phi_1 U \Phi_2$ is satisfied by any path originating in s such that a state s' satisfying Φ_2 is eventually encountered along the path, and all the states between s and s' (if any) satisfy Φ_1. The formula $true\ U\ \Phi$ is commonly abbreviated as $F\Phi$ and satisfied by any path that eventually reaches a state satisfying Φ. A model M satisfies a PCTL property Φ if Φ holds in the initial state s_0 of M [3].

PCTL allows specifying a variety of safety requirements. For simplicity, in this work, we focus on safety requirements specified as upper-bounds on the probability of eventually reaching a state labelled as `unsafe`:

Definition 5. *Safety Requirement.* *Given a threshold $\lambda \in (0, 1]$, the safety requirement for a learning agent is formalised by the PCTL property $P_{\leq\lambda}$ [F `unsafe`], i.e., the maximum probability of reaching a $s \in S$ such that `unsafe` $\in L(s)$ must be less than or equal to λ.*

Definition 6. *Counterexamples in Probabilistic Model Checking.* *A counterexample is a minimal possible sub-model $M_{cex} = (S_{cex}, A_{cex}, s_0, P_{cex})$ derived from the model M, where S_{cex}, A_{cex} are of subsets S and A, containing violating behaviours of a PCTL property from the initial state s_0 in M.*

When a model M does not satisfy a PCTL property, a counterexample can be computed as evidence of the violation [18]. In this work, we adapt the *minimal critical label set* counterexample generation method of [49].

The computation of a minimal possible sub-model requires the solution of a mixed-integer linear optimisation problem that selects the smallest number of transitions from the state-action space original model that allows the construction of violations. An extensive description of the counterexample generation algorithm, including a heuristic to bias the counterexample generation towards including actions that a tabular Q-learning agent is more likely to select is included in [26].

Generating Multiple Counterexamples. For an MDP violating the safety requirement, there can exist, in general, multiple counterexamples (both with minimal or non-minimal sizes), each potentially highlighting different policies that lead to requirement violations [10].

In this work, we use counterexamples to guide the generation of offline training environments where the agent learns to reduce the value of actions that may eventually lead to the violation of safety requirements. We therefore aim at generating multiple, *diverse counterexamples* (if they exist), while keeping each of them at a small size for faster training. Given a counterexample, a different one can be obtained by adding a blocking clause to the minimisation problem, i.e., forcing the optimiser to exclude one or more previously selected action pairs (by imposing the corresponding selector variables $x_\ell = 0$ in the optimisation problem given in [26]). Hence, we can systematically add (an increasing number of) blocking clauses to obtain multiple diverse counterexamples that jointly provide a more comprehensive representation of the different violating behaviors the agent explored at any time.

3   Counterexample-Guided Reinforcement Learning

We assume that the agent does not have prior knowledge about the environment. In particular, it will only discover unsafe states upon visiting them during exploration. During the learning process the agent will iteratively interact with either the actual environment (*online* exploration) or with a counterexample-guided *offline* simulation. The online phases aim at exploring the actual environment and improving the agent's policy expected reward, while acquiring information to build and continuously refine an abstract, compact representation of the control problem. The offline phases expose the agent to simulated, small-size, environments, within which the agent can revise its policy to penalise decisions that may lead to safety violations during the subsequent online phases. In the remaining of the section, we first show how to construct and update an abstract finite MDP that compactly represents safety-relevant aspects of the (parts of) environment explored by the agent at any time (Sect. 3.1). Then, in Sect. 3.2, we introduce the main learning algorithm with the online-offline alternation scheme, and discuss the main challenges of the offline learning phases.

3.1   Safety-Relevant State-Space Abstraction

For simplicity, let us assume the agent has no information about the topology of the state space S at the beginning of the exploration. Each online interaction with the environment (*episode*) can be described by the trace of the states visited by the agent and the actions it took. Besides the reward associated with each state-action pair, we assume states in the trace can be assigned a set of labels. Labels represent properties of the state related specific to the learning problem, e.g., a goal has been reached. We assume a special label unsafe labels the occurrence of unsafe situations the agent should aim to avoid. W.l.o.g., we assume an episode terminates when the agent enters an unsafe state. We will refer to the states of the online environment as *concrete* states.

In this section, we propose an abstraction procedure to construct a finite, abstract MDP that retains sufficient information about the explored concrete environment to enable the synthesis of abstract counterexamples to the safety requirement. Each counterexample will therefore be an abstract representative of a set of possible safety violating behaviors that can happen in the concrete environment. To maintain the size of the abstract MDP tractable – especially in the presence of continuous concrete state spaces – the abstraction will retain only (approximate) safety-relevant information.

We assume that any state not labeled as unsafe is safe to explore and that the unsafe label is time-invariant. Furthermore, the abstraction must preserve at any time a **safety invariant**: *every explored unsafe concrete state should be mapped to an unsafe abstract state.* The finite abstract state space must therefore separate safe and unsafe regions of the concrete space, with only safe concrete states possibly misclassified as unsafe but not vice versa.

Inspired by the idea of casting the learning of geometric concepts as a set cover problem in [8,40], we frame the separation task as a minimal red-blue set cover problem to abstract the explored concrete state space as a finite set of disjoint boxes or polyhedra, each expressed as a set of logical constraints and defined as the intersection of a finite set of hyperplanes.

To formalise the construction of the abstract state-space, we first introduce the notion of coverage of a concrete state by a polyhedra predicate.

Definition 7. *Coverage of a polyhedra predicate Let $\bar{S} \subseteq S = \{s_0, s_1, \dots, s_n\}$ be the set of all explored concrete states, a particular state $s \in \bar{S}$ is covered by a (polyhedra) predicate C_i if $s \in C_i$, where $C_i = \{s \mid \boldsymbol{\omega} s + \boldsymbol{b} \leq 0\}$, in which $\boldsymbol{\omega}$ represents the vector of slopes and $\boldsymbol{b}$ represents the vector of biases corresponding to half-spaces enclosing the predicate.*

Remark 1. The general affine form of the predicate C_i accounts for a variety of common numerical abstract domains, including, e.g., boxes (intervals), octagons, zonotopes, or polyhedra. In this work, we fix $\omega = 1$, i.e., restrict to hyper-boxes. We allow the user to specify a minimum size $d > 0$, which ensures that no dimension of the box will be reduced to a length smaller than d. This coarsening of the abstract domain struck a convenient trade-off between computational cost and accuracy of the abstraction in our preliminary experiments (see Appendix in [26] for additional discussion); the restriction can be lifted for applications requiring different trade-offs [42].

The identification of a finite set of predicates that allow separating the concrete state space preserving the safety invariant can thus be reduced to the following:

Minimal Red-Blue Set Cover Problem. Let $\bar{S} \subseteq S = \{s_0, s_1, \dots, s_n\}$ be the set of all explored concrete states and $U = \{\boldsymbol{u}_0, \boldsymbol{u}_1, \dots, \boldsymbol{u}_m\}$ be the set of explored states assigned the unsafe label, find the minimal set $C = \{\cup_{i=1} C_i\}$ s.t. every element $\boldsymbol{u} \in U$ is covered by some predicate C_i, with an overall false positive rate fpr $\leq f \in (0, 1]$ for safe concrete states $(\boldsymbol{s} \in \bar{S} \setminus U)$ covered by C.

Remark 2. In general, f cannot be zero, since the concrete state space, whether discrete or continuous, may not be perfectly partitioned by a finite set of polyhedra predicates, with smaller values of f possibly resulting in a larger number of predicates $|C|$.

To solve this optimisation problem, we employ a branch and bound method [30] to systematically enumerate possible combinations of predicates. The solution set guarantees all the unsafe concrete states are covered by a Boolean combination of predicates in C, while safe concrete states may also be covered by some predicate C_i, with a prescribed maximum tolerable rate f.

Definition 8. *Safety-relevant Abstraction MDP.* *A safety-relevant abstraction MDP M_a is a tuple $(S_a, A_a, s_{a0}, R_a, P_a, L_a)$, where S_a is the abstract state space, which is the partition of the concrete state space S induced by the boundaries of $C_i \in C$ from the solution of the minimal set cover above, A_a is the set of applicable actions, s_{a0} is the initial abstract state, $P_a : S_a \times A_a \times S_a \to [0, 1]$ is a probability transition function, R_a is the abstract reward function (which will be defined later), and $L_a : S \to \{\texttt{safe}, \texttt{unsafe}\}$ is a labelling function.*

M_a is constructed from the concrete traces collected during online learning, with the satisfaction of the predicates C_i determining the abstraction of concrete states and the abstract transition function is estimated accordingly from the frequencies observed in the concrete traces. The abstraction must preserve the *safety invariant*, therefore it may overapproximate explored unsafe regions, but not underapproximate them. Initially, the entire state space is assumed safe to explore. As new traces are collected during online exploration, the abstract model is incrementally updated: when a concrete state is found to be unsafe, the hyperbox containing its numerical vector representation is split to correct the classification (after the concrete state is wrapped around with a hyperbox of minimal size d, if $d > 0$).

The incremental branch-and-bound refinement of the abstraction could lead to excessive fragmentation of the abstract state space, making it intractably large, particularly for the purpose of counterexample generation. To mitigate this issue, we merge adjacent states, i.e., abstract states sharing at least one separating hyperplane, into a single abstract state, adapting the general notion of probabilistic approximate ϵ−simulation [12] as in the following definition:

Definition 9. *Adjacent ϵ-simulation:* *Let S_l be the partitions of S_a induced by the equivalence relation $s \sim s'$ iff $L_a(s) = L_a(s')$. Then, for a given $\epsilon \in [0, 1]$, two adjacent states $s \in S_a$ and $s' \in S_a$ are ϵ-similar if $(\exists s_l \in S_l)(s \in s_l, s' \in s_l)$ and $(\forall s_l \in S_l)(|P_a(s, a, s_l) - P_a(s', a, s_l)| \leq \epsilon, \forall a \in A_a)$.*

The ϵ-simulation in Definition 9 induces a hierarchical merging scheme. Let level l_0 contain the initial abstract states, which partition the explored concrete state space into a finite set of boxes – one per abstract state – which are labeled as either safe or unsafe. ϵ-similar adjacent states from level l_i are merged into a single state at level l_{i+1}, until no further merge is possible. Besides reducing the number of abstract states, and in turn the cost of generating counterexamples, this hierarchical merging scheme brings the indirect benefit of more aggressively merging abstract states corresponding to safe regions of the concrete state space, while preserving a finer-grained approximation of concrete state space regions in proximity of explored unsafe states, as discussed in [26].

Counterexample-Guided Simulation. If a probabilistic safety requirement can be violated, one or more counterexamples M_{cex} can be generated from the abstract model M_a, where each counterexample includes a (near-)minimal subset of the abstract state-action space. We then use each counterexample as a guide to build an offline, simulation environment where the agent can update its Q-values towards avoiding eventually reaching an unsafe state.

Starting from the initial concrete state s_0, the abstract state s_{cex} in M_{cex} corresponding to the current concrete state is computed. By construction, each counterexample selects one action a from state s_{cex}. The abstract transition is randomly simulated according to the transition function P_{cex} from (s_{cex}, a) and an abstract destination state s'_{cex} is identified. Such abstract state is concretised by sampling from the past concrete traces a transition (s, a, s') where $s \in s_{cex}$ and $s' \in s'_{cex}$. If s'_{cex} is an unsafe state, a penalty (negative reward; the impact of its magnitude is further discussed in [26]) is given to the agent, which has the transitive effect of re-weighting also the Q-value of the actions that led the agent to the current state. The simulation traces can be used by both Q-learning and DQN agents. The simulation terminates when an unsafe state is reached (penalty) or when we fail to concretise an abstract transition, which may happen when concrete safe states are misclassified as unsafe in the abstraction, but there is no actual transition to unsafe states from them. In the latter case, the simulation trace is discarded (no reward). While every simulation within the counterexample is designed to eventually reach an unsafe state with probability 1 by construction [49], to avoid excessive length of a simulation, it can be practical to set an arbitrary, large bound on the maximum number of steps per run as additional termination criterion.

Multiple counterexamples, and corresponding simulations, can be generated up to a maximum simulation budget allowed by the user, adding blocking clauses in random order as described previously. Each counterexample is typically of small-size, which results in short simulation traces, thus reducing the overall cost of each offline learning experience.

3.2   Online-Offline Learning with Counterexample Guidance

Algorithm 1 summarises the main steps of our online-offline learning method with counterexample guidance. We initially assume no knowledge about the environment is given to the agent: both the abstract model M_a and the set of explored paths D are empty (line 3). If prior knowledge was available, either in the form of an initial abstraction or of previously explored paths, M_a and D can be initialised accordingly.

Online Learning. The `onlineQLearning` procedure (line 7) lets the agent operate in the concrete environment with either tabular Q-Learning in discrete state space or DQN in continuous state space. We augment the exploration with a sequential Bayesian hypothesis testing (line 6) that monitors the frequency of violations of the safety requirement, by incrementally updating after each online episode a Beta distribution that estimates the probability of violation [13]. If the odds of such probability exceeding λ is larger than a prescribed Bayes factor β, the online learning phase is interrupted. The updated Q-values/Q-network of the agent are stored and the set of explored traces D is updated (line 8).

Offline Learning. If the online learning phases has been interrupted because by the Bayesian test (line 9), an offline learning phase is triggered to reinforce the avoidance of discovered unsafe behaviors in future online exploration.

First, the abstraction M_a is updated with the current set of online traces D (line 10) to support the generation of current counterexamples. While there are theoretically a finite number of counterexample submodels [49], for large M_a it could be computationally too expensive; instead, up to a maximum number N_{cex} of counterexamples is generated at each iteration. The addition of random blocking clauses (as described in Sect. 2.2) will increase the diversity of the counterexamples within and across different offline learning phases.

The offline simulation traces synthesised from each M_{cex} (line 13) as described in the previous section are used by the agent to update its Q-values/Q-network (line 14), thus penalising the selection of eventually unsafe actions before the next online learning phase begins. Notice that the Bayesian hypothesis test is re-initialised before the next online learning phase (line 6) since the agent is expected to behave differently after an offline learning phase.

Discussion. The interleaving of offline and online learning phases aims at reducing the frequency of unsafe events during the exploration of the environment. This goal is pursued by synthesising simulation environments from counterexamples to the safety requirement computed from an abstraction of the state space explored online. Notably, the offline phases never preclude the exploration of an action during the following online phases, rather they reduce the likelihood of selecting actions that may eventually lead to reaching unsafe states by lowering their Q-values. Due to space limitations, we report here the two main results related to the convergence of our abstraction method and of the online-offline learning process, and defer a more extensive discussion to [26].

Counterexample guidance relies on the abstraction constructed from online exploration phases, which classifies every region of the explored state space as either safe or unsafe. While by construction the abstraction preserves the safety invariant (every explored concrete unsafe state is mapped to an abstract unsafe state), the quality of the offline guidance relies also on controlling the misclassification error of safe concrete regions, which may unduly penalise the exploration of safe states.

Proposition 1. *The maximum misclassification error of a concrete safe state into an abstract unsafe state can eventually be reduced below an arbitrary bound $0 < \bar{u} \le 1$ with probability at least $1 - \delta$ ($0 < \delta < 1$) throughout the exploration.*

Further empirical analysis of the convergence of the abstraction and the impact of the abstraction parameters is provided in [26].

Finally, the following proposition states that the introduction of counterexample guidance does not preclude the convergence of the overall learning process to a maximal reward policy that satisfies the safety requirement, if such policy exists.

Proposition 2. *If there exist maximal-reward policies that satisfy the safety requirement, then the online-offline learning process eventually converges to one of them.*

Further discussion of the convergence properties of the offline-online learning process are included in [26], including elaborating on the validity of the two propositions above. In the next section, we instead report on our preliminary experimental evaluation of the performance of our counterexample-guided learning process.

Algorithm 1: CEX-Guided Reinforcement Learning

---

1 **Input:** Safety requirement bound $\lambda > 0$, max number of CEX N_{cex}, num simulations per CEX NS_{cex}, Bayes factor β, penalty for offline learning traces *penalty*
2 **Output:** Synthesised Policy π
3 Initialise $M_a, D \leftarrow \emptyset$;
4 Initialise Q to random values;
5 **while** *not converged* **do**
6      $\delta_{bht} \leftarrow$ BayesianHypothesisTester(λ, β)
7      $Q, D_{online} \leftarrow$ onlineQLearning(Q, δ_{bht}) ;
8      $D \leftarrow D \cup D_{online}$;
9   **if** $\delta_{bht} > \beta$ **then**
           //cex-guided offline learning
10         $M_a \leftarrow$ refine(M_a, D);   // update the abstract MDP
11   **for** *up to N_{cex}* **do**
12             $M_{cex} \leftarrow$ cexGeneration(M_a, λ)
13   simTraces $\leftarrow$ collectSimulationTraces(M_{cex}, Q, NS_{cex});
14             $Q \leftarrow$ Q-value/Q-network update(simTraces, *penalty*);
15   **end**
16   **end**
17 **end**
18 $\pi \leftarrow \arg\max_a Q$;

---

4   Evaluation

In this section, we present a preliminary experimental evaluation of the performance of our method from two perspectives: 1) the improvement in the exploration safety rate, and 2) the impact on the cumulative reward achieved by the agent. Finally, we briefly discuss the overhead of counterexample guidance and make some observations on the policies it synthesises. Additional experimental results and discussion, including on abstraction effectiveness and sensitivity to hyperparameters can be found in [26].

Environments.   We consider four environments: `DiscreteGrid` from the implementation of [23], the slippery `FrozenLake8x8` from OpenAI Gym [6], `HybridGrid` – where we change the state space from discrete to continuous with the same layout in [23], and `MarsRover` [22] (in particular, the exploration of the melas chasma in the Coprates quadrangle [34]). In all the environments, the agent decides a direction of move between `up`, `down`, `left`, and `right`. We define the objective of the agent as finding a policy with maximum Q-value while avoiding unsafe behaviours with intolerably high probability λ during exploration. Specifically, in the `FrozenLake8x8`, the agent aims to find a walkable path in a 8x8 grid environment with slippery actions while avoiding entering states labelled with H. With the slippery setting, the agent will move in the intended direction with a probability of only 1/3 else will move in either perpendicular directions with equal probabilities of 1/3 respectively. In the `DiscreteGrid` and `HybridGrid`, the agent aims to reach the states labelled with `goal1` and then states labelled with `goal2` in a fixed order while avoiding entering any states labelled with `unsafe` along the path, with 15% probability of moving to random directions at every action. In `HybridGrid`, the distance covered in a move is also randomly sampled from a Gaussian $\mathcal{N}(2, 0.5)$, thus choosing the same direction from a state may reach different states. In the `MarsRover` environment, the agent aims to find one of the target states labelled with `goal` while avoiding reaching the unsafe regions, which in this scenario cannot be perfectly abstracted by boxes or other affine abstract domains. Following [21], the distance covered in each move is sampled uniformly in the range $(0, 10)$, with the addition of further uniform noise from $\mathcal{U}(-0.1, 0.5)$.

Baselines.   We compare the learning performance of our method with classical Q-Learning and DQN [35] as the baseline for discrete and continuous MDPs, respectively. For discrete MDPs, we further compare our method with [20] (referred to as QL-LCRL in the following), using the same set of hyper-parameters as provided in their implementation and the associated tool paper [21,23]. Given an automata corresponding to an LTL property, QL-LCRL guides the agent's exploration of an initially unknown MDP by reshaping on-the-fly the reward function to encourage the exploration of behaviors that satisfy such property. In this application QL-LCRL will encourage a safe exploration by discouraging reaching `unsafe` states.

Implementation and Parameters.   We implemented a standard tabular Q-learning in Python and used the DQN implementation from OpenAI Gym

Baselines [37]. We parameterise the abstraction process with a learning rate α and a discount factor γ for the Q-value/Q-Network updates, and ϵ for adjacent ϵ-bisimulation. The agents move within Cartesian planes of sizes $|S|$ and the minimisation of the abstract models reduces the state space to $|S_a|$, where the minimum size of each box is set to 1 and 0.01 for the discrete and the continuous environments, respectively. We set the safety specification parameter λ according to the intrinsic uncertainty in the respective environments. A summary of the parameters used for each environment is reported in the left side of Table 1 (additional parameters are discussed in [26] to ease reproducibility). We require at least 50 samples to be collected by the Bayesian hypothesis test before it can trigger offline training to reduce false positive triggers.

Table 1. Learning results with corresponding hyperparamters in CEX-guided Learning

Environment	$|S|$	$|S_A|$	λ	α	γ	ϵ	fpr	QL/DQN safe.prob	QL/DQN reward	QL-LCRL safe. prob	QL-LCRL reward	Q-CEXsafe. prob	Q-CEX reward
DiscreteGrid	3200	30	0.2	0.9	0.9	0.01	0.05	0.299	0.924	0.387	0.99	0.572	0.99
FrozenLake8x8	64	50	0.35	0.1	0.9	0.01	0.05	0.339	0.55	0.401	0.54	0.443	0.59
HybridGrid	∞	41	0.2	5e−4	0.9	0.01	0.05	0.548	0.753	−	−	0.687	0.78
MarsRover	∞	149	0.2	5e−4	0.9	0.05	0.15	0.681	0.947	−	−	0.775	0.86

Experimental Results. Figure 1 shows the accumulated safety rates (bottom) and the rolling average of accumulated rewards, indicating the real-time learning performance of the agent. The line is the average across 10 runs, while the shaded region around it is the standard deviation. We do not provide any prior information to the agent. The dashed vertical lines in the figure indicate the average episode number where an offline learning phase in QL/DQN-CEX (our method) is triggered, and the solid horizontal line indicates the target safety rate in corresponding safety specifications. The cumulative rewards of different methods converge to similar values, demonstrating that the performance under guidance (QL-LCRL and Q-CEX) achieves cumulative rewards comparable to the baseline QL and DQN method. As expected, providing additional guidance to discourage actions that may lead to reaching unsafe states with QL/DQN-CEX or QL-LCRL improves the safety rate of the exploration, with QL/DQN-CEX achieving on average higher safety rates.

In DiscreteGrid, the safety rate of online learning in our method exceeds the threshold much faster than other methods. This is due to the rapid convergence of the abstraction thanks to the grid layout that can be accurately and efficiently abstracted (and minimised). In turn, no further offline phases were required after the first 1000 episodes.

In FrozenLake8x8, more online exploration is required to support comprehensive counterexample guidance, partly due to the high uncertainty in the outcome of the actions. Another phenomenon due to high uncertainty is that the agent takes a longer time to reach a stable performance, therefore the offline learning phase is triggered more frequently compared with other environments and also more episodes were required to stably satisfy the safety requirement.

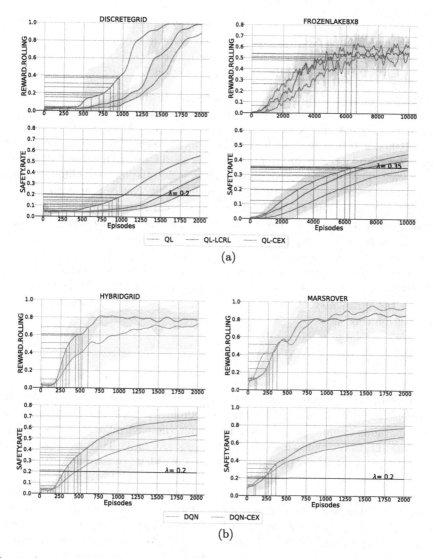

Fig. 1. Average over 10 runs of the accumulated safety rates and rolling rewards in QL/DQN, QL-LCRL [23] and QL/DQN-CEX (our method). The shaded region shows ± standard deviation.

In `HybridGrid` and `MarsRover`, while the baseline eventually achieves a marginally higher cumulative reward in MarsRover, DQN-CEX achieves a higher safety rate – the number of failures experienced by the agent is inversely proportional to the integral of the safety rate curve, which results in significantly fewer failure events. The frequency of offline training phases also decreases over time. This is not surprising due to a safer exploration being possibly less speculative and slower in exploring the optimal policy. QL-LCRL is not applicable to

HybridGrid and MarsRover. Although LCRL [23] applies NFQ-based method for continuous MDPs, NFQ-LCRL trains the agent in a completely offline manner based on randomly sampled data, thus it is not suitable for comparison with our method from the perspective of safe exploration.

Overhead. Counterexamples generation requires solving a MILP problem on the abstract state space. On a Macbook Air with M1 CPU and 8Gb of memory, the average $\pm$ stdev time to solve the optimization problems were 0.71 ± 0.98 s, 0.08 ± 0.06 s, 2.31 ± 3.94 s, and 3.87 ± 4.07 s for FrozenLake8x8, DiscreteGrid, HybridGrid, and MarsRover, respectively. We used Gurobi Optimiser v9.1.0 [17] off-the-shelf. The numbers of counterexamples generated for each offline learning phase were, on average: 11, 6, 19, and 21, respectively. Notice that both counterexample generation and the simulations can be parallelised. While the cost of solving the MILP may be higher, we notice that offline learning is triggered only when the recent online exploration resulted in an unacceptable failure rate. As shown in Fig. 1, thanks to counterexample-guidance, QL/DQN-CEX can achieve the required safety exploration rate much faster, which helps amortizing the initial MILP solution cost. Finally, the optimisation problem might be relaxed to sacrifice the optimality of the solution (i.e., size of the counterexamples) for computation time.

5   Related Work

Safe Exploration. [7,16,27] provide surveys and taxonomies of recent safe reinforcement learning methods with different emphases. Most existing safe exploration methods assume prior knowledge about the environment or the agent's dynamics [11,14,36], known safety constraints [20,25,33,38], or utilise expert guidance [15,24,39,52] to provide guarantees of safety constraints satisfaction during exploration. A different class of methods [32,44,47] utilises surrogate Gaussian process models to characterise the unknown dynamics and optimise the unknown function with a *prior* model structure. There are only a few methods [5,22] tackling safe exploration without known model structure. [22] focuses on solving continuous, unknown MDPs into sub-tasks using an online RL framework under LTL specification with less emphasis on safety rate but blocking actions unsafe according to the specification, and [5] trains a safety layer/critic used for filtering out probabilistic unsafe actions with an offline dataset. While motivated by the same idea of safer exploration without prior knowledge, our method can be initialised with none or any amount of previously explored paths, meanwhile it converged to cumulative rewards comparably or better than baseline methods.

Offline RL. Offline RL can be seen as a data-driven formulation of RL, where the agent collects transitions using the behaviour policy instead of interacting with the environment [31]. The biggest challenge in offline RL is the bootstrapping error: the Q-value is evaluated with little or no prior knowledge and propagated through Bellman equation [28]. [41,50] regularise the behaviour policy while optimising to address this issue. [9,29] alternatively update the Q values in more conservative ways to learn a lower bound of the Q function using uncertainty estimated with the sampled information. From the safe RL perspective, [46] optimises a risk-averse criterion using data previously collected by a *safe* policy offline, and [51] learns a constrained policy maximizing the long-term reward based on offline data without interaction in the concrete environment using a constrained penalised Q-Learning method. These methods have similar motivation of combining offline learning with risk-averse RL as ours, while focusing on continuous control setting instead. Besides utilising offline learning to reduce unsafe online exploration, we alternate online and offline learning to keep the abstract knowledge about the environment up to date and increase the risk aversion of the agent based on the most current evidence it collected online.

6 Conclusion

We presented our investigation of a safer model-free reinforcement learning method using counterexample-guided offline training. We proposed an abstraction strategy to represent the knowledge acquired during online exploration in a succinct, finite MDP model that can consistently and accurately describe safety-relevant dynamics of the explored environment. Counterexample generation methods from probabilistic model checking are then adapted to synthesise small-scale simulation environments capturing scenarios in which the decisions of the agent may lead to the violation of safety requirements. The agent can then train offline within this minimal submodels by replaying concrete transitions recorded during past online exploration consistent with the counterexample, using a reward scheme focused on reducing the likelihood of selecting actions that may eventually lead to visiting again explored unsafe concrete states. The q-values penalized during the offline training phases implicitly reduce the risk of repeating unsafe behaviors during subsequent online exploration, while newly explored paths feedback information to the next offline learning phase.

The alternation of online exploration – and abstraction refinement – and counterexample guided learning can ultimately lead to higher safety rates during exploration, without significant reduction in the achieved cumulative reward, as demonstrated in our preliminary evaluation on problems from previous literature and the OpenAI Gym. While this paper focused on improving Q-Learning (and the related DQN algorithm), the fundamental framework is not specific to Q-Learning, and we plan to explore its impact on other learning algorithms in future work.

Data availability. A prototype Python implementation of our method is available at Github: https://github.com/xtji/CEX-guided-RL.

References

1. Achiam, J., Held, D., Tamar, A., Abbeel, P.: Constrained policy optimization. In: International Conference on Machine Learning, pp. 22–31. PMLR (2017)
2. Alshiekh, M., Bloem, R., Ehlers, R., Könighofer, B., Niekum, S., Topcu, U.: Safe reinforcement learning via shielding. In: Thirty-Second AAAI Conference on Artificial Intelligence (2018)
3. Baier, C., Katoen, J.P.: Principles of Model Checking. MIT press, Cambridge (2008)
4. Bellman, R.: A Markovian decision process. J. Math. Mech., 679–684 (1957)
5. Bharadhwaj, H., Kumar, A., Rhinehart, N., Levine, S., Shkurti, F., Garg, A.: Conservative safety critics for exploration. arXiv preprint arXiv:2010.14497 (2020)
6. Brockman, G., et al.: Openai gym. arXiv preprint arXiv:1606.01540 (2016)
7. Brunke, L., Greeff, M., Hall, A.W., Yuan, Z., Zhou, S., Panerati, J., Schoellig, A.P.: Safe learning in robotics: from learning-based control to safe reinforcement learning. Ann. Rev. Control Rob. Auton. Syst. **5**, 411–444 (2022)
8. Bshouty, N.H., Goldman, S.A., Mathias, H.D., Suri, S., Tamaki, H.: Noise-tolerant distribution-free learning of general geometric concepts. J. ACM (JACM) **45**(5), 863–890 (1998)
9. Buckman, J., Gelada, C., Bellemare, M.G.: The importance of pessimism in fixed-dataset policy optimization. arXiv preprint arXiv:2009.06799 (2020)
10. Češka, M., Hensel, C., Junges, S., Katoen, J.-P.: Counterexample-driven synthesis for probabilistic program sketches. In: ter Beek, M.H., McIver, A., Oliveira, J.N. (eds.) FM 2019. LNCS, vol. 11800, pp. 101–120. Springer, Cham (2019). https://doi.org/10.1007/978-3-030-30942-8_8
11. Dalal, G., Dvijotham, K., Vecerik, M., Hester, T., Paduraru, C., Tassa, Y.: Safe exploration in continuous action spaces. arXiv preprint arXiv:1801.08757 (2018)
12. Desharnais, J., Laviolette, F., Tracol, M.: Approximate analysis of probabilistic processes: logic, simulation and games. In: 2008 Fifth International Conference on Quantitative Evaluation of Systems, pp. 264–273. IEEE (2008)
13. Downey, A.: Think Bayes. O'Reilly Media, Sebastopol (2021)
14. Fulton, N., Platzer, A.: Safe reinforcement learning via formal methods: toward safe control through proof and learning. In: Proceedings of the AAAI Conference on Artificial Intelligence, vol. 32, no. 1 (2018)
15. Garcia, J., Fernández, F.: Safe exploration of state and action spaces in reinforcement learning. J. Artif. Intell. Res. **45**, 515–564 (2012)
16. García, J., Fernández, F.: A comprehensive survey on safe reinforcement learning. J. Mach. Learn. Res. **16**(1), 1437–1480 (2015)
17. Gurobi Optimization, LLC: Gurobi Optimizer Reference Manual (2022). https://www.gurobi.com
18. Han, T., Katoen, J.P., Berteun, D.: Counterexample generation in probabilistic model checking. IEEE Trans. Softw. Eng. **35**(2), 241–257 (2009)
19. Hansson, H., Jonsson, B.: A logic for reasoning about time and reliability. Formal Aspects Comput. **6**(5), 512–535 (1994)
20. Hasanbeig, M., Abate, A., Kroening, D.: Logically-constrained reinforcement learning. arXiv preprint arXiv:1801.08099 (2018)
21. Hasanbeig, M., Kroening, D., Abate, A.: LCRL: Certified policy synthesis via logically-constrained reinforcement learning - implementation. https://github.com/grockious/lcrl

22. Hasanbeig, M., Kroening, D., Abate, A.: Deep reinforcement learning with temporal logics. In: Bertrand, N., Jansen, N. (eds.) FORMATS 2020. LNCS, vol. 12288, pp. 1–22. Springer, Cham (2020). https://doi.org/10.1007/978-3-030-57628-8_1

23. Hasanbeig, M., Kroening, D., Abate, A.: LCRL: certified policy synthesis via logically-constrained reinforcement learning. In: Abraham, E., Paolieri, M. (eds.) Quantitative Evaluation of Systems, QEST. LNCS, vol. 13479, pp. 217–231. Springer, Cham (2022). https://doi.org/10.1007/978-3-031-16336-4_11

24. Huang, J., Wu, F., Precup, D., Cai, Y.: Learning safe policies with expert guidance. In: Proceedings of the 32nd International Conference on Neural Information Processing Systems, NIPS'18, Curran Associates Inc., pp. 9123–9132 (2018)

25. Jansen, N., Könighofer, B., Junges, S., Bloem, R.: Shielded decision-making in mdps. arXiv preprint arXiv:1807.06096 (2018)

26. Ji, X., Filieri, A.: Probabilistic counterexample guidance for safer reinforcement learning (extended version). arXiv preprint arXiv:2307.04927 (2023)

27. Kim, Y., Allmendinger, R., López-Ibáñez, M.: Safe learning and optimization techniques: towards a survey of the state of the art. In: Heintz, F., Milano, M., O'Sullivan, B. (eds.) TAILOR 2020. LNCS (LNAI), vol. 12641, pp. 123–139. Springer, Cham (2021). https://doi.org/10.1007/978-3-030-73959-1_12

28. Kumar, A., Fu, J., Tucker, G., Levine, S.: Stabilizing Off-Policy Q-Learning via Bootstrapping Error Reduction. Curran Associates Inc. (2019)

29. Kumar, A., Zhou, A., Tucker, G., Levine, S.: Conservative q-learning for offline reinforcement learning. Adv. Neural. Inf. Process. Syst. **33**, 1179–1191 (2020)

30. Lawler, E.L., Wood, D.E.: Branch-and-bound methods: a survey. Oper. Res. **14**(4), 699–719 (1966)

31. Levine, S., Kumar, A., Tucker, G., Fu, J.: Offline reinforcement learning: Tutorial, review, and perspectives on open problems. arXiv preprint arXiv:2005.01643 (2020)

32. Liu, A., Shi, G., Chung, S.J., Anandkumar, A., Yue, Y.: Robust regression for safe exploration in control. In: Learning for Dynamics and Control, pp. 608–619. PMLR (2020)

33. Mason, G.R., Calinescu, R.C., Kudenko, D., Banks, A.: Assured reinforcement learning with formally verified abstract policies. In: 9th International Conference on Agents and Artificial Intelligence (ICAART), York (2017)

34. McEwen, A.S., et al.: Recurring slope lineae in equatorial regions of Mars. Nature Geosci. **7**(1), 53–58 (2014)

35. Mnih, V., et al.: Playing atari with deep reinforcement learning. arXiv preprint arXiv:1312.5602 (2013)

36. Moldovan, T.M., Abbeel, P.: Safe exploration in Markov decision processes. arXiv preprint arXiv:1205.4810 (2012)

37. OpenAI: Stable baselines version 3 - dqn. https://stable-baselines3.readthedocs.io/en/master/modules/dqn.html

38. Pham, T.H., De Magistris, G., Tachibana, R.: Optlayer-practical constrained optimization for deep reinforcement learning in the real world. In: 2018 IEEE International Conference on Robotics and Automation (ICRA), pp. 6236–6243. IEEE (2018)

39. Prakash, B., Khatwani, M., Waytowich, N., Mohsenin, T.: Improving safety in reinforcement learning using model-based architectures and human intervention. arXiv preprint arXiv:1903.09328 (2019)

40. Sharma, R., Gupta, S., Hariharan, B., Aiken, A., Nori, A.V.: Verification as learning geometric concepts. In: Logozzo, F., Fähndrich, M. (eds.) SAS 2013. LNCS, vol. 7935, pp. 388–411. Springer, Heidelberg (2013). https://doi.org/10.1007/978-3-642-38856-9_21

41. Siegel, N.Y., et al.: Keep doing what worked: Behavioral modelling priors for offline reinforcement learning. arXiv preprint arXiv:2002.08396 (2020)
42. Singh, G., Püschel, M., Vechev, M.: A practical construction for decomposing numerical abstract domains. Proc. ACM Program. Lang. **2**(POPL) (2017)
43. Stooke, A., Achiam, J., Abbeel, P.: Responsive safety in reinforcement learning by pid Lagrangian methods. In: International Conference on Machine Learning, pp. 9133–9143. PMLR (2020)
44. Sui, Y., Gotovos, A., Burdick, J., Krause, A.: Safe exploration for optimization with gaussian processes. In: International Conference on Machine Learning, pp. 997–1005. PMLR (2015)
45. Tessler, C., Mankowitz, D.J., Mannor, DS.: Reward constrained policy optimization. arXiv preprint arXiv:1805.11074 (2018)
46. Urpí, N.A., Curi, S., Krause, A.: Risk-averse offline reinforcement learning. arXiv preprint arXiv:2102.05371 (2021)
47. Wachi, A., Sui, Y., Yue, Y., Ono, M.: Safe exploration and optimization of constrained mdps using Gaussian processes. In: Proceedings of the AAAI Conference on Artificial Intelligence, vol. 32 (2018)
48. Watkins, C.J., Dayan, P.: Q-learning. Mach. Learn. **8**(3), 279–292 (1992)
49. Wimmer, R., Jansen, N., Vorpahl, A., Ábrahám, E., Katoen, J.-P., Becker, B.: High-level counterexamples for probabilistic automata. In: Joshi, K., Siegle, M., Stoelinga, M., D'Argenio, P.R. (eds.) QEST 2013. LNCS, vol. 8054, pp. 39–54. Springer, Heidelberg (2013). https://doi.org/10.1007/978-3-642-40196-1_4
50. Wu, Y., Tucker, G., Nachum, O.: Behavior regularized offline reinforcement learning. arXiv preprint arXiv:1911.11361 (2019)
51. Xu, H., Zhan, X., Zhu, X.: Constraints penalized q-learning for safe offline reinforcement learning. In: Proceedings of the AAAI Conference on Artificial Intelligence, vol. 36, pp. 8753–8760 (2022)
52. Zhou, W., Li, W.: Safety-aware apprenticeship learning. In: Chockler, H., Weissenbacher, G. (eds.) CAV 2018. LNCS, vol. 10981, pp. 662–680. Springer, Cham (2018). https://doi.org/10.1007/978-3-319-96145-3_38

Symbolic Semantics for Probabilistic Programs

Erik Voogd[1]([⊠])(ID), Einar Broch Johnsen[1](ID), Alexandra Silva[2](ID), Zachary J. Susag[2](ID), and Andrzej Wąsowski[3](ID)

[1] University of Oslo, Oslo, Norway
erikvoogd@live.nl
[2] Cornell University, Ithaca, New York, USA
[3] IT University of Copenhagen, Copenhagen, Denmark

QEST
Artifact
Evaluation
2023

Accepted

Abstract. We present a new symbolic execution semantics of probabilistic programs that include observe statements and sampling from continuous distributions. Building on Kozen's seminal work, this symbolic semantics consists of a countable collection of measurable functions, along with a partition of the state space. We use the new semantics to provide a full correctness proof of symbolic execution for probabilistic programs. We also implement this semantics in the tool symProb, and illustrate its use on examples.

1 Introduction

Probabilistic programming languages are designed to make probabilistic computations easier to express for a broader scientific community. They can be used to model behaviour based on data that carries uncertainty or randomness, as found in, e.g., robotics [30], machine learning [12,23], statistics [11], and cryptography [14]. Besides traditional programming constructs, a key aspect of a probabilistic language is the ability to *sample* random values, in order to represent the uncertainty that occurs in the real world. It is essential that these programming languages have a rigorous foundation, such that correctness and safety properties can be guaranteed when designing and implementing tools for probabilistic program analysis, optimization, and compilation.

Probabilistic semantics was first studied by Kozen [19], using measure theory to relate denotational and operational semantics of imperative probabilistic programs. The *denotational* semantics represents states as probability measures over program variables and programs as measure transformers. It enables one to effectively reason about programs as a whole. The *operational* semantics, on the other hand, is a computation model describing step-by-step computation in terms of a measurable function. A correctness theorem for the operational semantics states that the produced sets of outputs are distributed correctly according to the measure transformer of the denotational semantics.

To enable more detailed reasoning about probabilistic programs, we introduce a new semantics, inspired by symbolic execution techniques. For non-probabilistic programs, symbolic execution [1,18] has been very successful in program analysis techniques such as debugging, test generation, and verification (e.g., [3,5–8,13,15,16]). Its appeal arises from the fact that one symbolic execution is an abstraction of possibly infinitely many executions in the concrete state space, that all share a single execution path through the program. Symbolic execution interprets program variables symbolically, such that assignments update a symbolic substitution and conditional statements produce so-called path conditions that must be true for the program's input variables to execute each path. Thus, symbolic execution generates pairs consisting of a symbolic substitution and a path condition. If the Boolean path condition holds in some initial state, then running the program is the same as applying the corresponding substitution to the initial state.

We use the new semantics to provide a full correctness proof of symbolic execution of probabilistic programs with respect to a denotational semantics á la Kozen. The correctness proof is highly nontrivial: one semantics deals with the program's symbolic execution traces, the other interprets programs as measure transformers. Our approach is to first prove a one-to-one correspondence between traces and the elements of our new symbolic semantics. The latter is equivalent to the operational semantics first introduced by Kozen [19], and hence to the denotational semantics. The full details of this proof are included in the extended version of this paper [32].

We consider a language with *observe* statements: in some interpretations of Bayesian inference, a programmer can express to have observed a value as a sample from some discrete or continuous distribution [28]. In others, including the imperative language that we consider, observe statements are given the semantics of asserting the truth of a provided Boolean formula. Observe statements were not studied in Kozen's work, and an operational semantics for probabilistic programming with observe statements has not been formally defined in that setting. For our correctness proof, therefore, we must extend both denotational and operational semantics with an interpretation of *observe*, and state and prove the corresponding correctness theorem anew.

Intuitively, observe statements in an operational semantics can be modelled by *rejection sampling*. That is, any execution that violates the observed formula is aborted. The probability mass will then reside in execution traces where the observe condition holds, and the probability of the set of traces where it does not hold will be annihilated. We formalize this semantics and prove its correctness with respect to a denotational semantics that interprets observe b as a measure transformer that restricts measures to the set corresponding with the Boolean formula b. This is an interpretation based on Bayes' theorem.

Contributions and Overview. We start the paper by introducing the language using an example program and discussing the challenges in the technical development (Sect. 2). The subsequent technical sections of the paper present the following contributions:

- An in-depth description of symbolic execution of probabilistic programs (Sect. 3), which supports discrete and continuous sampling as well as Bayesian inference through observing Boolean formulas of positive-measure sets. In symbolic execution, sampled values will be represented by symbolic variables.
- As a main contribution we provide a full correctness proof of symbolic execution of probabilistic programs (Sect. 4) by introducing a new *symbolic* semantics for imperative probabilistic programs and proving correspondence with established semantics (which we extend to support *observe* statements).
- To showcase symbolic execution of probabilistic programs, we have developed the tool symProb that performs bounded symbolic execution according to our new sym-

bolic semantics (Sect. 5). At the end of a symbolic execution, symProb reports the path condition, the substitution mapping, and the set of Boolean formulas which correspond to the program's observe statements. We report on running symProb on a series of examples to show how our semantics can be applied.

2 Probabilistic Programming

We introduce the language through an example and then present its grammar.

2.1 Example Program

Consider the probabilistic program on the right, modelling the gender distribution among people taller than 200 centimeters. In the Bernoulli sampling statement bern(0.51) for the variable gender we assume that 51% of the total population is male. Among men, height is normally distributed with mean 175 and variance 72, and among women with mean 161 and variance 50. The last line conditions the distribution on people taller than 200 centimeters.

```
gender ~ bern(0.51);
if (gender = 1) {
  height ~ norm(175,72);
} else {
  height ~ norm(161,50);
}
observe (height >= 200);
```

Symbolic execution has been very effective in analysis of non-probabilistic programs. The technique builds variable substitutions by analyzing assignments, and resolves conditional branching and iteration by use of nondeterminism. The conditionals are stored under variable substitution as a Boolean formula – called the *path condition* – that needs to be satisfied by the initial state for the corresponding substitution to be a representation of the program. Some problems arise when programs have discrete and continuous sample and observe statements, such as the above.

Consider for example a Bernoulli sampling statement $x_i \sim$ bern(e). One could introduce additional sets of symbolic variables for sampling statements, but it is practically infeasible to do so for each possible bias e, especially if one allows parameters to be arithmetic expressions. In our approach, we assume a finite number of *primitive* distributions (that is, unparameterized) and encode *parameterized* distributions using these primitives. A Bernoulli sampling statement $x_i \sim$ bern(e) is then encoded by use of uniform continuous sampling from the unit interval $[0, 1]$, written $x_i \sim$ rnd, as follows:

```
xᵢ ~ rnd; if (xᵢ < t) {xᵢ := 1} else {xᵢ := 0}
```

A denotational semantics can be used to show soundness of such encodings.

For Gaussian distributed samples, the primitive distribution used is the *standard* Gaussian distribution, which has mean zero and variance one. To obtain a sample from a Gaussian distribution with mean e_1 and variance e_2 – both arithmetic expressions – a standard Gaussian sample is scaled (multiplied by $\sqrt{e_2}$) and translated (add e_1).

With these encodings, one trace through the example program above has final substitution σ and path condition ϕ, given by

$$\sigma = \{\text{gender} \mapsto 1, \text{height} \mapsto z_0\sqrt{72} + 175\} \qquad \phi \equiv y_0 < 0.51 \wedge 1 = 1$$

Here, y_0 is a symbolic variable that is uniformly distributed over the interval $[0, 1]$. As it turns out, the probability of the path condition as measured by the input measure is the *prior* probability 0.51 of the trace.

The obtained Boolean formula from *observe* statements in this symbolic execution is

$$\psi \equiv z_0 \cdot \sqrt{72} + 175 \geq 200,$$

where z_0 is a symbolic variable representing a standard Gaussian sample. Measuring the set represented by this Boolean formula with the input measure yields exactly the *likelihood* of the model we have in mind. To keep the prior and the likelihood separate, we collect Boolean conditions under *observe* statements in what we coin the *path observations*.

This paper formally describes the procedure above for general probabilistic programs, and proves correctness with respect to a denotational semantics. Building on work by Kozen [19], we choose to interpret substitutions from symbolic execution as measurable functions on the value space, and path conditions and path observations as measurable sets of values for the program variables. These can then be compared against Kozen's denotational semantics, where program states are subprobability measures over the domain of values for the program variables, and a program p is a transformer of input to output measures $[\![p]\!] : \mathcal{M}(\mathbb{R}^n) \to \mathcal{M}(\mathbb{R}^n)$. The use of measure theory is unavoidable in order to handle continuous random variables. For example the semantics of `height ~ norm(175,72)`, given an input $\mu \in \mathcal{M}(\mathbb{R})$, is the measure:

$$[\![\texttt{height} \sim \texttt{norm(175,72)}]\!](\mu) : A \mapsto \gamma_{175,72}(A)$$

Here, $\gamma_{175,72}$ denotes the Gaussian measure with expected value 175 and variance 72.

After defining the language formally in Sect. 2.2, we provide a detailed description of symbolic execution of probabilistic programs in Sect. 3. There, towards a correctness proof, we also introduce the *symbolic semantics*. This is a *big-step* semantics for symbolic execution—this is stated in Theorem 4. Section 4 is aimed at proving correctness with respect to a measure-transforming semantics. The correctness statement (Theorem 5) says that the sum of probabilities of all symbolic execution paths, as measured by the input measure, expressed using the final substitutions, path conditions, and path observations, is the same as the probability under the denotational semantics. The technical contribution is concluded with a proof of concept tool that implements symbolic execution and we perform some experiments with it (Sect. 5).

2.2 Language

To express programs like the above example, we consider a language that contains basic imperative constructs (assignments, sequential composition, conditionals, and loops) together with constructs to manipulate random variables (sampling) and observe statements that allow conditioning on an observation defined by a Boolean condition:

$$
\begin{array}{lll}
\mathbb{E} \ni e ::= q \in \mathbb{Q} & \mathbb{B} \ni b ::= \texttt{False} & \mathbb{P} \ni p ::= \texttt{skip} \\
\quad\mid \ \texttt{x}_i & \quad\mid \ \texttt{True} & \quad\mid \ \texttt{x}_i := e \\
\quad\mid \ \texttt{op}(e_1,\dots,e_{\sharp \texttt{op}}) & \quad\mid \ e \, \Diamond \, e & \quad\mid \ \texttt{x}_i \sim \texttt{rnd} \\
& \quad\mid \ b \parallel b & \quad\mid \ \texttt{observe } b \\
& \quad\mid \ b \, \&\& \, b & \quad\mid \ p \, \mathbin{\raise.3ex\hbox{$_\circ$}} \, p \\
\texttt{x}_i \in \mathcal{X} & \quad\mid \ !b & \quad\mid \ \texttt{if } b \, p \texttt{ else } p \\
\Diamond \in \{<, \leq, =, != , \geq, >\} & & \quad\mid \ \texttt{while } b \, p
\end{array}
$$

It defines *expressions* $e \in \mathbb{E}$ over program variables $\texttt{x}_i \in \mathcal{X}$ (where $i \in \mathbb{N}$) and constants $q \in \mathbb{Q}$ using operators op of arity $\sharp\texttt{op}$. Expressions $e \in \mathbb{E}$ are interpreted as functions $\overline{e} : \mathbb{R}^n \to \mathbb{R}$ that compute the value of e given a valuation $v : \{0, 1, \dots, n - 1\} \to \mathbb{R}$ of the program variables. Formally, op is a function $\overline{\texttt{op}} : \mathbb{R}^m \to \mathbb{R}$ where $m = \sharp\texttt{op}$, and

$$
\overline{q}(v) = q, \quad \overline{\texttt{x}_i}(v) = v(i), \quad \overline{\texttt{op}(e_1,\dots,e_{\sharp\texttt{op}})}(v) = \overline{\texttt{op}}(\overline{e_1}(v),\dots,\overline{e_{\sharp\texttt{op}}}(v))
$$

The grammar also defines *Boolean expressions* $b \in \mathbb{B}$ that can relate expressions and apply the standard Boolean operators. We slightly overload notation and interpret Boolean expressions b as subsets $\overline{b}$ of $\mathbb{R}^n$ where the formula holds. For example, $\overline{\texttt{True}} = \mathbb{R}^n$ and if $b = e_1 \, \Diamond \, e_2$ then

$$
\overline{b} = \{v \in \mathbb{R}^n \mid \overline{e_1}(v) \, \Diamond \, \overline{e_2}(v)\}.
$$

Boolean conjunction, disjunction, and negation are respectively interpreted as set intersection, union, and complement in $\mathbb{R}^n$.

The Boolean expressions are used in the if and while conditions and in observe statements of *program statements* $p \in \mathbb{P}$. Whenever some program $p \in \mathbb{P}$ is fixed, there is also a fixed amount of n variables.

Sampling statements $\texttt{x}_i \sim \texttt{rnd}$ draw uniform random samples from the unit interval $[0, 1]$. For presentation purposes, we sample from only one primitive distribution in the formal grammar, and we do so at the level of statements rather than in expressions. Expressions can be made probabilistic, however, by first sampling and then using the variable in a (Boolean) expression. We also stress again that the language can be extended to support sampling from a multitude of other primitive distributions, both discrete and continuous, without affecting any of the results in this paper.

With the extensions described in Sect. 2.1, the probabilistic program presented there is in the language generated by our grammar.

3   Symbolic Execution

The inductive rules that define a transition system implementing symbolic execution of probabilistic programs are presented in Fig. 1. The *symbolic states* in this transition system are quintuples consisting of the following data: (i) a *program* p that is to be executed; (ii) a *substitution* σ of program variables to symbolic expressions (defined shortly), which captures past program behavior; (iii) a *sampling index* [1] $k \in \mathbb{N}$; (iv) the

---

[1] One for each primitive distribution: we use one in this paper for presentation purposes. Two sampling indices are used in the tool symProb.

path condition as a precondition for this symbolic trace; and (v) the *path observation* as a means to bookkeep which states will be accepted by *observe* statements throughout execution. Progression of the system depends only on the program syntax (i). Below we provide a detailed description of the components (ii)-(v). The substitutions (ii) and path conditions (iv) follow established methods from symbolic execution [3]; the notion of path observation (v) is novel.

The program skip cannot make a transition and represents the *terminated* program. The system is nondeterministic due to the pairs of rules for *if* and *while* statements. The system has infinite symbolic traces due to Rule iter-T. Customarily, $\xrightarrow{*}$ denotes the reflexive-transitive closure of $\longrightarrow$.

$$\frac{}{(x_i := e, \sigma, k, \phi, \psi) \longrightarrow (\text{Skip}, \sigma[x_i \mapsto \sigma e], k, \phi, \psi)} \text{ asgn}$$

$$\frac{}{(x_i \sim \text{rnd}, \sigma, k, \phi, \psi) \longrightarrow (\text{Skip}, \sigma[x_i \mapsto y_k], k+1, \phi, \psi)} \text{ smpl}$$

$$\frac{}{(\text{observe } b, \sigma, k, \phi, \psi) \longrightarrow (\text{Skip}, \sigma, k, \phi, \psi \wedge \sigma b)} \text{ obs}$$

$$\frac{}{(\text{Skip} \, \S \, p, \sigma, k, \phi, \psi) \longrightarrow (p, \sigma, k, \phi, \psi)} \text{ seq-0}$$

$$\frac{(p, \sigma, k, \phi, \psi) \longrightarrow (p', \sigma', k', \phi', \psi')}{(p \, \S \, q, \sigma, k, \phi, \psi) \longrightarrow (p' \, \S \, q, \sigma', k', \phi', \psi')} \text{ seq-n}$$

$$\frac{}{(\text{if } b \, p_1 \text{ else } p_2, \sigma, k, \phi, \psi) \longrightarrow (p_1, \sigma, k, \phi \wedge \sigma b, \psi)} \text{ if-T}$$

$$\frac{}{(\text{if } b \, p_1 \text{ else } p_2, \sigma, k, \phi, \psi) \longrightarrow (p_2, \sigma, k, \phi \wedge \sigma! b, \psi)} \text{ if-F}$$

$$\frac{}{(\text{while } b \, p, \sigma, k, \phi, \psi) \longrightarrow (\text{Skip}, \sigma, k, \phi \wedge \sigma! b, \psi)} \text{ iter-F}$$

$$\frac{}{(\text{while } b \, p, \sigma, k, \phi, \psi) \longrightarrow (p \, \S \, \text{while } b \, p, \sigma, k, \phi \wedge \sigma b, \psi)} \text{ iter-T}$$

Fig. 1. Inductive transition rules for symbolic execution

3.1   Symbolic Substitutions

To capture the behavior of a symbolic trace, a *substitution* assigns to every program variable a *symbolic expression*. Symbolic expressions are generated by the following grammar (recall that $\sharp$op denotes the arity of op):

$$\mathbb{SE} \ni \mathsf{E} ::= q \in \mathbb{Q} \mid x_i \in \mathcal{X} \mid y_k \in \mathcal{Y} \mid \text{op}(\mathsf{E}, \ldots, \mathsf{E}_{\sharp \text{op}})$$

$\mathbb{SE}$ extends $\mathbb{E}$ with the set $\mathcal{Y} = \{y_0, y_1, \dots\}$ as base cases, representing the random samples that may be drawn during execution. When the language samples from more than one primitive distribution, a set of symbolic variables $\mathcal{Y}_{\mathcal{D}}$ is used for every primitive distribution $\mathcal{D}$. For each such distribution $\mathcal{D}$, one will need a sampling index $k_{\mathcal{D}}$ in the transition system. We use $\mathcal{Z} = \{z_0, z_1, \dots\}$ for symbolic variables representing standard normal samples in our examples and in the tool.

Thus, formally, substitutions are maps $\sigma : \mathcal{X} \to \mathbb{SE}$. We sometimes write $\sigma(i)$ in lieu of $\sigma(x_i)$. The *updated* substitution $\sigma[i \mapsto \mathsf{E}]$ denotes the substitution σ' such that $\sigma'(j) = \sigma(j)$ for $j \neq i$ and $\sigma'(i) = \mathsf{E}$. Any substitution σ inductively extends to expressions $e \in \mathbb{E}$ by $\sigma(\mathsf{op}(e_1, \dots, e_{\sharp\mathsf{op}})) := \mathsf{op}(\sigma(e_1), \dots, \sigma(e_{\sharp\mathsf{op}}))$.

If symbolic execution is to conform with a denotational semantics, the symbolic substitutions have to be interpreted as concrete state transformers. For this, first, symbolic *expressions* are interpreted as functions $\mathbb{R}^{n+\omega} \to \mathbb{R}$, where the first n arguments are the values of the program variables, and the rest is an infinite stream of samples to be drawn. Here, one may use extra streams of sample spaces for any additional primitive distributions in the language. The mapping $|\cdot| : \mathbb{R}^{n+\omega} \to \mathbb{R}$ equips symbolic expressions with a formal interpretation as follows:

$$|q| : \rho \mapsto q, \qquad |x_i| : \rho \mapsto \rho_i, \qquad |y_k| : \rho \mapsto \rho_{n+k},$$
$$|\mathsf{op}(\mathsf{E}_1, \dots, \mathsf{E}_{\sharp\mathsf{op}})| : \rho \mapsto \overline{\mathsf{op}}(|\mathsf{E}_1|(\rho), \dots, |\mathsf{E}_{\sharp\mathsf{op}}|(\rho))$$

The symbols y_k thus pick the k-th sample available in $\mathbb{R}^\omega$. Symbolic expressions free of y_k are just program expressions, and their interpretations agree in the following way:

Lemma 1 (Substitution lemma for expressions). *For all $e \in \mathbb{E}$, $|e|(\rho) = \overline{e}(\rho|_n)$.*

This interpretation of $\mathsf{E} \in \mathbb{SE}$ as a function $\mathbb{R}^{n+\omega} \to \mathbb{R}$ extends naturally to symbolic substitutions $\sigma \in \mathbb{SE}^n$ as functions $\mathbb{R}^{n+\omega} \to \mathbb{R}^n$, by mere point-wise application after substitution. However, since we want to compose substitutions (their interpretations) due to sequencing, the codomain must be extended from $\mathbb{R}^n$ to $\mathbb{R}^{n+\omega}$:

Definition 2 (Interpretation of symbolic substitutions). *Let $\sigma : \mathcal{X} \to \mathbb{SE}$ be a substitution and $k \in \mathbb{N}$ a sampling index. The interpretation of σ at sampling index k, denoted $|\sigma|_k$, is the function $\mathbb{R}^{n+\omega} \to \mathbb{R}^{n+\omega}$ defined by*

$$|\sigma|_k : \rho \mapsto (|\sigma(x_0)|(\rho), \dots, |\sigma(x_{n-1})|(\rho), \rho_{n+k}, \rho_{n+k+1}, \rho_{n+k+2}, \dots)$$

The first n values of $|\sigma|_k(\rho)$ are just pointwise applications of $|\cdot|$. The remaining elements of $|\sigma|_k(\rho)$ are the same as those of ρ, but left-shifted k positions. This formalizes the idea that $|\sigma|_k$ has already drawn k samples.

3.2 Path Conditions and Path Observations

The Boolean formulas for the condition in *branching* and *iteration* statements are aggregated under substitution to build the *path condition* of a symbolic trace. The path condition represents the unique part of the input space that triggers the symbolic trace. Similarly, *observed* Boolean formulas under substitution make up the *path observation*, and represent the part of the input space that will lead to acceptance of *observe* statements in this trace.

Path conditions and observations are expressed as *symbolic Boolean expressions*:

$$\mathbb{SB} \ni \mathsf{B} ::= \bot \mid \top \mid \mathsf{E} \Diamond \mathsf{E} \mid \mathsf{B} \vee \mathsf{B} \mid \mathsf{B} \wedge \mathsf{B} \mid \neg \mathsf{B}$$

Subtitutions $\sigma : \mathcal{X} \to \mathbb{SE}$ extend through $\mathbb{E} \to \mathbb{SE}$ further to Booleans, as in $\mathbb{B} \to \mathbb{SB}$, in a completely trivial way. For example, $\sigma(e_1 \Diamond e_2 \,\&\&\, \texttt{True}) = \sigma(e_1) \Diamond \sigma(e_2) \wedge \top$.

Path conditions and path observations – their symbolic Boolean expressions – are interpreted as subsets of $\mathbb{R}^{n+\omega}$ where the formula is satisfied. The mapping $|\cdot|$ equips symbolic Boolean expressions with a formal interpretation as follows:

$$|\bot| = \emptyset, \qquad\qquad\qquad |\mathsf{B}_1 \vee \mathsf{B}_2| = |\mathsf{B}_1| \cup |\mathsf{B}_2|,$$

$$|\top| = \mathbb{R}^{n+\omega}, \qquad\qquad |\mathsf{B}_1 \wedge \mathsf{B}_2| = |\mathsf{B}_1| \cap |\mathsf{B}_2|,$$

$$|\neg\mathsf{B}| = \mathbb{R}^{n+\omega} \setminus |\mathsf{B}|, \qquad |\mathsf{E}_1 \Diamond \mathsf{E}_2| = \{\rho \in \mathbb{R}^{n+\omega} \mid |\mathsf{E}_1|(\rho) \Diamond |\mathsf{E}_2|(\rho)\}.$$

Here we overload notation of $|\cdot|$ to use it for both expressions and Boolean expressions.

3.3 Final Configurations

Let σ_0 denote the *initial* substitution $\{x_i \mapsto x_i\}_{x_i \in \mathcal{X}}$. We call the quadruple (σ, k, ϕ, ψ) the (symbolic) *configuration* of a state $(p, \sigma, k, \phi, \psi)$ and $(\sigma_0, 0, \top, \top)$ is the *initial configuration*. The configurations we are mostly interested in result from a finite symbolic execution trace starting from the initial configuration:

$$\Gamma_p := \{(\sigma, k, \phi, \psi) \mid (p, \sigma_0, 0, \top, \top) \xrightarrow{\ *\ } (\texttt{skip}, \sigma, k, \phi, \psi)\}$$

is called the set of *final configurations* of p.

For a program $p \in \mathbb{P}$ and a configuration $(\sigma, k, \phi, \psi) \in \Gamma_p$, p transforms inputs $\rho \in |\phi| \cap |\psi|$ to the output $|\sigma|_k(\rho)$. That is, p behaves like $|\sigma|_k$ on $|\phi| \cap |\psi|$. Execution of p on inputs from $|\phi| \setminus |\psi|$ leads to an unsatisfied *observe* statement.

Example 3. Consider the example program in Sect. 2.1 (call it p) that models the gender distribution among people taller than 200 cm. Due to Bernoulli sampling, it contains an additional *if* statement hidden in the encoding. The program has thus *four* symbolic traces; their respective final configurations $\gamma_1, \gamma_2, \gamma_3, \gamma_4 \in \Gamma_p$ are:

Final	Substitution σ	k_y	k_z	Path condition ϕ	Path observation ψ
γ_1	$\{g \mapsto 1, h \mapsto z_0\sqrt{72} + 175\}$	1	1	$y_0 < 0.51 \wedge 1 = 1$	$z_0\sqrt{72} + 175 \geq 200$
γ_2	$\{g \mapsto 1, h \mapsto z_0\sqrt{50} + 161\}$	1	1	$y_0 < 0.51 \wedge 1 \neq 1$	$z_0\sqrt{50} + 161 \geq 200$
γ_3	$\{g \mapsto 0, h \mapsto z_0\sqrt{72} + 175\}$	1	1	$y_0 \geq 0.51 \wedge 0 = 1$	$z_0\sqrt{72} + 175 \geq 200$
γ_4	$\{g \mapsto 0, h \mapsto z_0\sqrt{50} + 161\}$	1	1	$y_0 \geq 0.51 \wedge 0 \neq 1$	$z_0\sqrt{50} + 161 \geq 200$

The final configurations γ_2 and γ_3 have unsatisfiable path conditions. Path conditions represent the *priors* in the model and path observations represent *likelihoods*.

3.4 Symbolic Semantics

Now we introduce symbolic semantics for probabilistic programs with the main purpose of proving correctness of symbolic execution, which we just described. The new semantics is a set of functions, and can in fact be considered a denotational semantics for symbolic execution of probabilistic programs. Defined directly on the syntax of programs, the semantics consists of the interpretations of all final symbolic configurations—this is Theorem 4 below.

For a triple $(F, B, \mathcal{O})$ in the definition below, F is the final substitution of a trace, B is the path condition, and $\mathcal{O}$ is the path observation. Recall that $\mathcal{B}(X)$ is the Borel σ-algebra of X and let $\mathcal{B}(X, Y)$ denote the space of Borel measurable functions $X \to Y$. Then, for programs $p \in \mathbb{P}$ in n variables, the sets

$$\mathcal{F}_p \subset \mathcal{B}(\mathbb{R}^{n+\omega}, \mathbb{R}^{n+\omega}) \times \mathcal{B}(\mathbb{R}^{n+\omega}) \times \mathcal{B}(\mathbb{R}^{n+\omega})$$

are defined inductively on the structure of p as follows:

- For *inaction* skip, the state remains unaltered and there is no restriction on the path condition or the path observation:

$$\mathcal{F}_{\text{skip}} := \{(\rho \mapsto \rho, \mathbb{R}^{n+\omega}, \mathbb{R}^{n+\omega})\}$$

- An *assignment* has no restriction on the precondition, but the state is updated according to the assignment:

$$\mathcal{F}_{x_i := e} := \{(\rho \mapsto \rho[i \mapsto \bar{e}(\rho|_n)], \mathbb{R}^{n+\omega}, \mathbb{R}^{n+\omega})\}$$

Only the first n values $\rho|_n$ of the state ρ are needed to evaluate e, and $\rho[i \mapsto a]$ denotes the state ρ' where $\rho'(j) = \rho(i)$ if $j \neq i$ and $\rho'(i) = a$.
- When *sampling* we also merely perform an appropriate state update:

$$\mathcal{F}_{x_i \sim \text{rnd}} := \{(\rho \mapsto \text{sample}_i(\rho), \mathbb{R}^{n+\omega}, \mathbb{R}^{n+\omega})\}$$

Here, $\text{sample}_i(\rho)$ is an updated stream ρ' that has drawn one sample:

$$\text{sample}_i(\rho) = (\rho_0, \ldots, \rho_{i-1}, \rho_n, \rho_{i+1}, \ldots, \rho_{n-1}, \ \rho_{n+1}, \rho_{n+2}, \ldots)$$

- *Observing* a Boolean formula only updates the path observation:

$$\mathcal{F}_{\text{observe } b} := \{(\rho \mapsto \rho, \mathbb{R}^{n+\omega}, \bar{b} \times \mathbb{R}^{\omega})\}$$

- When *sequencing* two programs p_1 and p_2, range over all pairs of executions $(F_1, B_1, \mathcal{O}_1) \in \mathcal{F}_{p_1}$ and $(F_2, B_2, \mathcal{O}_2) \in \mathcal{F}_{p_2}$ and compose them. The first path condition B_1 should be satisfied and, after executing the first component F_1, the second path condition B_2 should be satisfied (and sim. for the path observation):

$$\mathcal{F}_{p_1 \, \S \, p_2} := (F_2 \circ F_1, B_1 \cap F_1^{-1}[B_2], \mathcal{O}_1 \cap F_1^{-1}[\mathcal{O}_2]) \mid (F_i, B_i, \mathcal{O}_i) \in \mathcal{F}_{p_i}, i = 1, 2\}$$

- The two branches of an *if* statement are put together in a binary union of sets. The path conditions are updated accordingly ($-^{C}$ denotes complement):

$$\mathcal{F}_{\text{if } b \ p \text{ else } q} := \{(F, B \cap (\overline{b} \times \mathbb{R}^{\omega}), \mathcal{O}) \mid (F, B, \mathcal{O}) \in \mathcal{F}_p\}$$
$$\cup \{(F, B \cap (\overline{b}^{C} \times \mathbb{R}^{\omega}), \mathcal{O}) \mid (F, B, \mathcal{O}) \in \mathcal{F}_q\}$$

- In a *while* statement, the union is over every possible number of iterations m. For $m = 0$, the behavior is that of skip and the precondition is the negation of the Boolean formula. Every next number $m + 1$ of loop iterations takes all possible executions of m iterations, pre-composes all possible additional iterations, and updates the preconditions accordingly:

$$\mathcal{F}_{\text{while } b \ p} := \bigcup_{m=0}^{\infty} \mathbb{F}_{b,p}^m \{(v \mapsto v, \overline{b}^{C}, \mathbb{R}^{n+\omega})\},$$

where $\mathbb{F}_{b,p}^m$ denotes m applications of the mapping $\mathbb{F}_{b,p}$ from $\mathcal{B}(\mathbb{R}^{n+\omega}, \mathbb{R}^{n+\omega}) \times \mathcal{B}(\mathbb{R}^{n+\omega}) \times \mathcal{B}(\mathbb{R}^{n+\omega})$ to itself that pre-composes an additional iteration of the loop:

$$\mathbb{F}_{b,q} : \mathcal{F} \mapsto \{ (F \circ F_q, (\overline{b} \times \mathbb{R}^{\omega}) \cap B_q \cap F_q^{-1}[B], \mathcal{O}_q \cap F_q^{-1}[\mathcal{O}])$$
$$\mid (F, B, \mathcal{O}) \in \mathcal{F}, (F_q, B_q, \mathcal{O}_q) \in \mathcal{F}_q\}$$

The sets $\mathcal{F}_p$ form a big-step semantics for symbolic execution of probabilistic programs:

Theorem 4. *For any program $p \in \mathbb{P}$, there is a one-to-one correspondence between final configurations $(\sigma, k, \phi, \psi) \in \Gamma_p$ and triples $(F, B, \mathcal{O}) \in \mathcal{F}_p$ such that $F = |\sigma|_k$, $B = |\phi|$, and $\mathcal{O} = |\psi|$.*

In this correspondence, we consider all final configurations $(\sigma, k, \phi, \psi) \in \Gamma_p$ with unsatisfiable path condition, i.e., $|\phi| = \emptyset$, equivalent. Similarly, all triples $(F, B, \mathcal{O})$ for which $B = \emptyset$ are considered equivalent.

4   Correctness

The symbolic execution engine described in the previous section is now proven correct with respect to a denotational semantics. Following Kozen [19], probabilistic programs are mappings of measures on the Borel measurable space of $\mathbb{R}^n$, where n is the number of program variables. Observe statements were not considered in his work. They *have* been studied [4] in this context of measure transformer semantics, but in the absence of unbounded loops.

To be fully precise, we need to recall some definitions from measure theory. A *measurable space* (X, Σ) is a set X equipped with a *σ-algebra* Σ, i.e., a set $\Sigma \subseteq \mathcal{P}(X)$ that (i) contains the emptyset, (ii) is closed under complements in X, and (iii) is closed under countable union. Elements of Σ are called *measurable sets*. The *Borel σ-algebra* $\mathcal{B}(X)$ is the one generated by the open sets of X. Whenever we say that functions or sets are measurable, we mean with respect to the Borel σ-algebras. A *(sub)probability measure* on (X, Σ) is a function $\mu : \Sigma \to [0, 1]$ such that $\mu(\emptyset) = 0$ and $\mu(\bigcup_{i \in I} A_i) = \sum_{i \in I} \mu(A_i)$ for any countable disjoint family of sets $(A_i)_{i \in I} \subseteq \Sigma$. We let $\mathcal{M}(X)$ denote the set of measures on the Borel σ-algebra of X; this makes $\mathcal{M}(\mathbb{R}^n)$ the state space in the denotational semantics.

The denotational semantics of a program p is a function $[\![p]\!] : \mathcal{M}(\mathbb{R}^n) \to \mathcal{M}(\mathbb{R}^n)$, pushing a measure corresponding to the current program state forward, according to the statement being interpreted. The inductive definition is as follows:

- *Skip* does not change the given measure: $[\![\texttt{skip}]\!](\mu) = \mu$.
- For an *assignment*, let α_e^i be the function that updates the i-th value appropriately:

$$\alpha_e^i : \mathbb{R}^n \to \mathbb{R}^n, \quad (x_1, \ldots, x_n) \mapsto (x_1, \ldots, x_{i-1}, \overline{e}(x_1, \ldots, x_n), x_{i+1}, \ldots, x_n)$$

This function is measurable, so its preimages can be measured by μ. Thus, the semantics of assignments as *pushforward* measures $[\![\mathsf{x}_i := e]\!](\mu) := \mu \circ (\alpha_e^i)^{-1}$ is well-defined. Intuitively, the measure $[\![\mathsf{x}_i := e]\!](\mu)$ measures with μ the set of values in $\mathbb{R}^n$ that lead to appropriately updated values for the i-th variable.
- *Sampling* updates the measure component of the corresponding variable with the distribution measure λ to be sampled from:

$$[\![\mathsf{x}_i \sim \mathtt{rnd}]\!](\mu) : A_1 \times \cdots \times A_n \mapsto \mu(A_1 \times \cdots \times A_{i-1} \times \mathbb{R} \times A_{i+1} \times \cdots \times A_n) \cdot \lambda(A_i).$$

For λ, we use the Lebesgue measure on the unit interval for uniform continuous sampling. Other measures can be used for other primitive sampling statements. The measure $[\![\mathsf{x}_i \sim \mathtt{rnd}]\!](\mu)$ here is defined on the rectangles of $\mathbb{R}^n$, and by Carathéodory's extension theorem, defines a unique measure on all of $\mathbb{R}^n$.
- When *observing*, we restrict the measure to the observed measurable set. For measurable $B \subseteq \mathbb{R}^n$, let $e_B(\mu)$ denote the subprobability measure $A \mapsto \mu(A \cap B)$. Then $[\![\texttt{observe } b]\!](\mu) := e_{\overline{b}}(\mu)$. Note that this measure is not normalized.
- *Sequencing* is just composition: $[\![p \,\mathbf{;}\, q]\!] = [\![q]\!] \circ [\![p]\!]$.
- *Branching* restricts the measure of one branch (resp. the other) to the measurable subset where the condition is true (resp. false). Formally,

$$[\![\texttt{if } b \; p_1 \texttt{ else } p_2]\!](\mu) = ([\![p_1]\!] \circ e_{\overline{b}})(\mu) + ([\![p_2]\!] \circ e_{\overline{\neg b}})(\mu)$$

This sum of measures is setwise. The measure is first restricted by $e_{\overline{b}}$ (or $e_{\overline{\neg b}}$) to a subprobability measure over the part of the state space where the condition is true (or false) and then passed on to be transformed by $[\![p_1]\!]$ (or $[\![p_2]\!]$).
- Interpreting *iterations* as repeated unfolding of if statements, the infinite sum

$$[\![\texttt{while } b \; p]\!](\mu) = \left(\sum_{m=0}^{\infty} e_{\overline{\neg b}} \circ ([\![p]\!] \circ e_{\overline{b}})^m \right)(\mu)$$

describes the semantics of while loops.

Now $[\![p]\!]$ is a positive operator of norm at most one for all p defined by the grammar in Sect. 2.2. This means that any subprobability measure is mapped to a subprobability measure [2]. Without while and observe statements, this norm would be exactly one.

Intuitively, this is because programs would then always terminate, and no probability mass would ever be lost either by diverging while loops or by conditional probability. The measure semantics is not normalized.

Recall that Γ_p is the set of final configurations of symbolic execution of a program p.

Theorem 5 (Correctness of Symbolic Execution of Probabilistic Programs). *Let $p \in \mathbb{P}$ be a program, $\mu \in \mathcal{M}(\mathbb{R}^n)$ a distribution measure over the input variables, and $A \subseteq \mathbb{R}^n$ a measurable set. Then*

$$[\![p]\!](\mu)(A) = \sum_{(\sigma,k,\phi,\psi) \in \Gamma_p} (\mu \otimes \lambda^\omega)\big(|\sigma|_k^{-1}[A \times \mathbb{R}^\omega] \cap |\phi| \cap |\psi|\big)$$

A friendly reminder that the measure λ was some chosen measure implicitly used in the denotational semantics; λ^ω denotes the product measure of infinitely many copies of it.

Proof (Sketch). For every program p, there is a function f_p such that

- $(\mu \otimes \lambda^\omega)\big(f_p^{-1}[A \times \mathbb{R}^\omega]\big) = [\![p]\!](\mu)(A)$, and
- $(\mu \otimes \lambda^\omega)\big(f_p^{-1}[A \times \mathbb{R}^\omega]\big) = \sum_{(F,B,\mathcal{O}) \in \mathcal{F}_p} (\mu \otimes \lambda^\omega)(F^{-1}[A \times \mathbb{R}^\omega] \cap B \cap \mathcal{O}).$

The function f_p is basically the operational semantics defined in [19], but extended to an observe construct that rejects unsatisfied observed formulas. The theorem is proven by chaining these two equalities and applying Theorem 4. See [32] for the full proof.

Table 1. Performance metrics for symProb on a series of case studies. For the "Random Z2 Walk (i)" cases, i refers to the number of times the main while loop was unrolled.

Case Study	Number of Paths		Samples	Lines	Time (sec.)
	Actual	Discarded			
BurglarAlarm	4	12	4	26	0.31
DieCond	20	0	20	17	2.15
Grass	28	36	6	21	0.80
MurderMystery	2	2	2	12	0.06
NeighborAge	4	4	4	14	0.10
NeighborBothBias	7	5	4	19	0.24
NoisyOr	256	–	8	37	2.33
Piranha	3	1	2	12	0.07
Random Z2 Walk (1)	16	–	4	14	0.19
Random Z2 Walk (2)	52	–	6	14	0.60
Random Z2 Walk (4)	712	–	10	14	8.37
Random Z2 Walk (8)	159,436	–	18	14	2167.90
SecuritySynthesis	1	0	8	14	0.04
TrueSkillFigure9	1	0	9	20	0.03
TwoCoins	3	1	2	6	0.05

5 Implementation and Experiments

We have developed symProb[2] as a prototype implementation of the symbolic execution technique presented in Sect. 3. symProb takes as input a probabilistic program written in the language described in Sect. 2.2 (up to some natural imperative-style syntactic additions such as keywords for *else* and delimiter use of parentheses and braces). symProb performs *bounded* symbolic execution, meaning that all while loops are unrolled a finite number of times, to ensure termination. Finite loops are fully unrolled while all other loops are unrolled a configurable, yet fixed, number of times. symProb reports the final configuration: the final substitution σ, the amount of samples, the path condition ϕ, and the path observation ψ. All symbolic configurations reported by symProb are expressed as uninterpreted symbolic expressions. symProb is written in Rust in around 2,000 lines of code and uses Z3 [22] for real numbers to determine branch satisfiability.

We have executed symProb on a series of examples sourced from PSI [10], R2 [24], "Fun" programs from Infer.NET [21], and Barthe et al. [2]. We summarize our findings in Table 1. All the experiments were done on a machine with 3.3GHz Intel Core i7-5820K and 32 GB of RAM, running Linux 6.3.2-arch1-1.

One feat of symbolic execution of probabilistic programs that symProb implements, is that paths with an unsatisfiable *observe* statement can be filtered out, since they have zero likelihood in the probabilistic model. symProb therefore *discards* such paths.

For each experiment, we report the following metrics: the number of traces explored, the number of traces which were discarded due to a failed observe statement ('–' if there were no observe statements in the program), the maximum number of random samples drawn, the number of lines of code, and the time symProb took to explore all paths.

We make a few general observations about the results. Outside of the DieCond, Grass, NoisyOr, and Random Z2 Walk (4,8) examples, symProb terminates in under a second. While these are relatively small examples, they still showcase a range of path counts and number of random samples drawn. Additionally, of the case studies that had observe statements present, many paths were discarded due to reaching an unsatisfiable observe statement. The Grass case study, for example, which involves six Bernoulli samples about weather conditions and such, had 36 paths discarded. In Sect. 7, we discuss how we might utilize this information in future work to optimize the performance of probabilistic programs with observe statements.

6 Related Work

Symbolic transition systems as described in this work have recently been formalized in a non-probabilistic setting by de Boer and Bonsangue [3]. From that starting point, this work extends to sampling in probabilistic programming by incorporating symbolic random variables $\{y_0, y_1, \dots\}$ in the symbolic substitutions and path conditions. Furthermore, we introduced *path observations* to keep track of observe statements.

Denotational semantics for Bayesian inference is an active research area [9,17,27, 28]. Mostly, the focus has been on *discrete* probabilistic programs [17,25], whereas we

---

[2] Source code for symProb and all experiments are available on Zenodo [31].

support sampling of both discrete *and* continuous distributions, thereby generalizing the probabilistic choice based on discrete sampling used in these works.

Staton [27,28] describes *observe-from* statements in his denotational semantics as an encoding for a *score* construct used to attach a likelihood scoring to an execution trace. The construct observe e from $\mathcal{D}$, where $\mathcal{D}$ is some distribution, is then sugar for score $f_{\mathcal{D}}(e)$, where $f_{\mathcal{D}}$ is the probability density function for the distribution $\mathcal{D}$. The score value in that semantics is therefore akin to the probability measure of our path observation. Staton's work considers language semantics of a functional nature, where the difficulty mainly lies in handling higher-order functions. In an imperative language, this is generally not a major concern. This allows us to consider the more general construct of observing Boolean formulas (of positive measure), rather than observing fixed samples (which may have zero measure). To the best of our knowledge, our work provides the first semantics for Boolean observe statements as a forward state transformer (as proposed by Kozen [19]) for an imperative probabilistic programming language in the presense of unbounded loops.

Sampson et al. [26] used symbolic execution to transform probabilistic programs into a Bayesian network. Their semantics was aimed at verification of certain probabilistic assertions, however, and lacked a formal foundation in measure theory. Luckow et al. combined symbolic execution with model counting to analyze programs with discrete distributions and nondeterministic choice [20]. Their work used schedulers to reduce Markov Decision Processes to Markov Chains, in contrast to our work with continuous distributions and observe statements, founded in measure theory. Susag et al. [29] considered symbolic execution for programs with random sampling to automatically verify quantitative correctness properties over unknown inputs. While they also used symbolic random variables, they were more focused on verifying correctness properties of "real-world" programs (e.g., written in C++) that use random sampling. They solely considered discrete distributions and did not support observe statements.

Gehr et al. [10] developed PSI, *probabilistic symbolic inference*, a tool that enables programmers to perform posterior distribution, expectation, and assertion queries through symbolic inference. Their symbolic reasoning engine works on symbolic representations of probability distributions, whereas we have symbolic terms which are interpreted as *random variables* of which we do not know the distribution in principle. By default, PSI computes a symbolic representation of the joint posterior distribution represented by the given probabilistic program. In contrast, symProb explores all paths through a given probabilistic program and reports on the path condition, substitution map, and path observation of each path. We see these two tools as complementary, as inference is only one aspect of probabilistic programming.

7   Conclusion and Future Work

We have defined new symbolic semantics for imperative probabilistic programs supporting forward execution and conditioning (Bayesian inference), which we proved to be a big-step semantics for symbolic execution of probabilistic programs. To support Bayesian inference, we extended Kozen's denotational semantics to include observe statements of positive-measure events. Significantly, the symbolic semantics thus theoretically supports implementation of a symbolic executor for imperative probabilistic programs with continuous domain variables. We introduced *path observations* to keep track of the part of the initial state space that lead to acceptance of observe statements for each symbolic trace. The symbolic transition system is implemented in our prototype tool symProb. Its effectiveness has been demonstrated on example programs, producing results with path conditions and path observations that correctly represent (prior) path probabilities and likelihoods of the symbolic traces, as expected in light of Theorem 5. Since path conditions represent priors and path observations represent likelihoods, we believe it to be beneficial to conceptually separate the two in a symbolic executor.

Interestingly, our semantics and its accompanying correctness results enable one to prove the correctness of certain program transformations. More generally, we believe the symbolic semantics has potential applications in model checking tools, deductive proof systems, optimizations, and reasoning about termination. For example, one can exploit this theory to commute sample and observe statements under appropriate substitutions. In complex probabilistic models, the placement of observe statements may then become an optimization problem. We plan to explore such program transformations both theoretically and experimentally in the future. Moreover, we intend to investigate the limiting behavior of our correctness result, which contains an infinite sum in the presence of unbounded loops. One naturally wonders how fast the sum approximates the true posterior of the program, and how different structures in a program influence the speed of this approximation.

A shortcoming of our semantics for *observe* statements is that zero-measure events cannot be observed in our language. Denotationally, the measure would then always yield zero; operationally, *almost all* simulations will be aborted. Zero-measure observed Boolean conditions may be included in future work, for example by considering *measure couplings* and *disintegrations* [9].

Data Availability Statement. The source code for symProb and the experiments we performed with it are available on Zenodo [31].

References

1. Baldoni, R., Coppa, E., D'Elia, D.C., Demetrescu, C., Finocchi, I.: A survey of symbolic execution techniques. ACM Comput. Surv. **51**(3), 50:1-50:39 (2018)
2. Barthe, G., Katoen, J.P., Silva, A.: Foundations of Probabilistic Programming. Cambridge University Press, Cambridge (2020)
3. de Boer, F.S., Bonsangue, M.: Symbolic execution formally explained. Formal Aspects Comput., 617–636 (2021). https://doi.org/10.1007/s00165-020-00527-y
4. Borgström, J., Gordon, A.D., Greenberg, M., Margetson, J., Van Gael, J.: Measure transformer semantics for Bayesian machine learning. Log. Methods Comput. Sci. **9**(3) (2013)
5. Cadar, C., Dunbar, D., Engler, D.R.: KLEE: unassisted and automatic generation of high-coverage tests for complex systems programs. In: Draves, R., van Renesse, R. (eds.) Proceedings 8th USENIX Symposium on Operating Systems Design and Implementation (OSDI 2008), pp. 209–224. USENIX Association (2008)
6. Cadar, C., Ganesh, V., Pawlowski, P.M., Dill, D.L., Engler, D.R.: EXE: automatically generating inputs of death. In: Juels, A., Wright, R.N., di Vimercati, S.D.C. (eds.) Proceedings

13th ACM Conference on Computer and Communications Security (CCS 2006), pp. 322–335. ACM (2006)

7. Cadar, C., et al.: Symbolic execution for software testing in practice: preliminary assessment. In: Taylor, R.N., Gall, H.C., Medvidovic, N. (eds.) Proceedings 33rd International Conference on Software Engineering (ICSE 2011), pp. 1066–1071. ACM (2011)

8. Cadar, C., Sen, K.: Symbolic execution for software testing: three decades later. Commun. ACM **56**(2), 82–90 (2013)

9. Dahlqvist, F., Silva, A., Danos, V., Garnier, I.: Borel kernels and their approximation, categorically. In: Staton, S. (ed.) Proceedings of the Thirty-Fourth Conference on the Mathematical Foundations of Programming Semantics (MFPS 2018). Electronic Notes in Theoretical Computer Science, vol. 341, pp. 91–119. Elsevier (2018)

10. Gehr, T., Misailovic, S., Vechev, M.: PSI: exact symbolic inference for probabilistic programs. In: Chaudhuri, S., Farzan, A. (eds.) CAV 2016. LNCS, vol. 9779, pp. 62–83. Springer, Cham (2016). https://doi.org/10.1007/978-3-319-41528-4_4

11. Gelman, A., Carlin, J.B., Stern, H.S., Rubin, D.B.: Bayesian Data Analysis, 2nd edn. Chapman and Hall/CRC, London (2004)

12. Ghahramani, Z.: Probabilistic machine learning and artificial intelligence. Nature **521**(7553), 452–459 (2015)

13. Godefroid, P., Klarlund, N., Sen, K.: DART: directed automated random testing. In: Sarkar, V., Hall, M.W. (eds.) Proceedings of the ACM SIGPLAN Conference on Programming Language Design and Implementation (PLDI 2005), pp. 213–223. ACM (2005)

14. Goldwasser, S., Micali, S.: Probabilistic encryption. J. Comput. Syst. Sci. **28**(2), 270–299 (1984)

15. de Gouw, S., Rot, J., de Boer, F.S., Bubel, R., Hähnle, R.: OpenJDK's Java.utils.Collection.sort() is broken: the good, the bad and the worst case. In: Kroening, D., Păsăreanu, C.S. (eds.) CAV 2015. LNCS, vol. 9206, pp. 273–289. Springer, Cham (2015). https://doi.org/10.1007/978-3-319-21690-4_16

16. Hentschel, M., Bubel, R., Hähnle, R.: The symbolic execution debugger (SED): a platform for interactive symbolic execution, debugging, verification and more. Int. J. Softw. Tools Technol. Transf. **21**(5), 485–513 (2019)

17. Holtzen, S., Van den Broeck, G., Millstein, T.: Scaling exact inference for discrete probabilistic programs. Proc. ACM Program. Lang. **4**(OOPSLA), 1–31 (2020)

18. King, J.C.: Symbolic execution and program testing. Commun. ACM **19**(7), 385–394 (1976)

19. Kozen, D.: Semantics of probabilistic programs. In: 20th Annual Symposium on Foundations of Computer Science (SFCS 1979), pp. 101–114. IEEE (1979)

20. Luckow, K., Păsăreanu, C.S., Dwyer, M.B., Filieri, A., Visser, W.: Exact and approximate probabilistic symbolic execution for nondeterministic programs. In: Proceedings of the 29th ACM/IEEE International Conference on Automated Software Engineering (ASE 2014), pp. 575–586 (2014)

21. Minka, T., Winn, J., Guiver, J., Zaykov, Y., Fabian, D., Bronskill, J.: Infer.net 0.3 (2018)

22. de Moura, L., Bjørner, N.: Z3: an efficient SMT solver. In: Ramakrishnan, C.R., Rehof, J. (eds.) TACAS 2008. LNCS, vol. 4963, pp. 337–340. Springer, Heidelberg (2008). https://doi.org/10.1007/978-3-540-78800-3_24

23. Murphy, K.P.: Probabilistic Machine Learning: An Introduction. MIT Press, Cambridge (2022)

24. Nori, A., Hur, C.K., Rajamani, S., Samuel, S.: R2: An efficient MCMC sampler for probabilistic programs. In: Proceedings of the AAAI Conference on Artificial Intelligence, vol. 28 (2014)

25. Olmedo, F., Gretz, F., Jansen, N., Kaminski, B.L., Katoen, J.P., McIver, A.: Conditioning in probabilistic programming. ACM Trans. Program. Lang. Syst. (TOPLAS) **40**(1), 1–50 (2018)

26. Sampson, A., Panchekha, P., Mytkowicz, T., McKinley, K.S., Grossman, D., Ceze, L.: Expressing and verifying probabilistic assertions. In: O'Boyle, M.F.P., Pingali, K. (eds.) Proceedings of the 35th ACM SIGPLAN Conference on Programming Language Design and Implementation (PLDI 2014), pp. 112–122. ACM (2014)

27. Staton, S.: Commutative semantics for probabilistic programming. In: Yang, H. (ed.) ESOP 2017. LNCS, vol. 10201, pp. 855–879. Springer, Heidelberg (2017). https://doi.org/10.1007/978-3-662-54434-1_32

28. Staton, S., Yang, H., Wood, F.D., Heunen, C., Kammar, O.: Semantics for probabilistic programming: higher-order functions, continuous distributions, and soft constraints. In: Grohe, M., Koskinen, E., Shankar, N. (eds.) Proceedings of the 31st Annual ACM/IEEE Symposium on Logic in Computer Science (LICS 2016), pp. 525–534. ACM (2016)

29. Susag, Z., Lahiri, S., Hsu, J., Roy, S.: Symbolic execution for randomized programs. Proc. ACM Program. Lang. 6(OOPSLA) (2022)

30. Thrun, S.: Probabilistic robotics. Commun. ACM **45**(3), 52–57 (2002)

31. Voogd, E., Johnsen, E.B., Silva, A., Susag, Z.J., Wasowski, A.: Artifact for Symbolic Semantics for Probabilistic Programs (2023). https://doi.org/10.5281/zenodo.8139552

32. Voogd, E., Johnsen, E.B., Silva, A., Susag, Z.J., Wasowski, A.: Symbolic Semantics for Probabilistic Programs (extended version) (2023). https://doi.org/10.48550/arXiv.2307.09951

Verification of Quantum Systems Using Barrier Certificates

Marco Lewis[1]([✉]) [iD], Paolo Zuliani[1,2] [iD],
and Sadegh Soudjani[1,3] [iD]

[1] Newcastle University, Newcastle upon Tyne, UK
m.j.lewis2@newcastle.ac.uk
[2] Università di Roma "La Sapienza", Rome, Italy
[3] Max Planck Institute for Software Systems, Kaiserslautern, Germany

Abstract. Various techniques have been used in recent years for verifying quantum computers, that is, for determining whether a quantum computer/system satisfies a given formal specification of correctness. Barrier certificates are a recent novel concept developed for verifying properties of dynamical systems. In this article, we investigate the usage of barrier certificates as a means for verifying behaviours of quantum systems. To do this, we extend the notion of barrier certificates from real to complex variables. We then develop a computational technique based on linear programming to automatically generate polynomial barrier certificates with complex variables taking real values. Finally, we apply our technique to several simple quantum systems to demonstrate their usage.

Keywords: barrier certificates · dynamical systems · quantum systems

1 Introduction

Quantum computers are powerful devices that allow certain problems to be solved faster than on classical computers. The research area focusing on the formal verification of quantum devices and software has witnessed the extension of verification techniques from classical systems [7,26] to the quantum realm. Classical techniques that have been used include theorem provers [13,17], Binary Decision Diagrams [5,33], SMT solvers [6,29] and other tools [14,30].

Quantum systems evolve according to the Schrödinger equation from some initial state. However, the initial state may not be known completely in advance. One can prepare a quantum system by making observations on the quantum objects, leaving the quantum system in a basis state, but this omits the global phase which is not necessarily known after measurement. Further, the system could be disturbed through some external influence before it begins evolving.

P. Zuliani—Currently at Università di Roma; work predominately done at Newcastle University.

This can slightly change the quantum state from the basis state to a state in superposition or possibly an entangled state. Articles around quantum state preparation and the uncertainty when preparing states include, for example, [8, 21, 23].

By taking into account these uncertain factors, a set of possible initial states from which the system evolves can be constructed. From this initial set, we can see if the system evolves according to some specified behaviour such as reaching or avoiding a particular set of states. As an example, consider a single qubit system that evolves according to a Hamiltonian $\hat{H}$ implementing the controlled-NOT operation. Through measurement and factoring in for noise, we know the system starts close to $|10\rangle$. The controlled-NOT operation keeps the first qubit value the same and so we want to verify that, as the system evolves via $\hat{H}$, the quantum state does not evolve close to $|00\rangle$ or $|01\rangle$.

The main purpose of this work is to study the application of a technique called *barrier certificates*, used for verifying properties of classical dynamical systems, to check properties of quantum systems similar to the one mentioned above. The concept of barrier certificates has been developed and used in Control Theory to study the safety of dynamical systems from a given set of initial states on real domains [24]. This technique can ensure that given a set of initial states from which the system can start and a set of unsafe states, the system will not enter the unsafe set. This is achieved through separating the unsafe set from the initial set by finding a *barrier*.

Barrier certificates can be defined for both deterministic and stochastic systems in discrete and continuous time [3, 16]. The concept has also been used for verification and synthesis against complicated logical requirements beyond safety and reachability [15]. The conditions under which a function is a barrier certificate can be automatically and efficiently checked using SMT solvers [4]. Such functions can also be found automatically using learning techniques even for non-trivial dynamical systems [22].

Dynamical systems are naturally defined on real domains ($\mathbb{R}^n$). To handle dynamical systems in complex domains ($\mathbb{C}^n$), one would need to decompose the system into its real and imaginary parts and use the techniques available for real systems. This has two disadvantages, the first being that this doubles the number of variables being used for the analysis. The second disadvantage is that the analysis may be easier to perform directly with complex variables than their real components. As quantum systems use complex values, it is desirable to have a technique to perform the reachability analysis using complex variables.

In this paper, we explore the problem of safety verification in quantum systems by extending barrier certificates from real to complex domains. Our extension is inspired by a technique developed by Fang and Sun [11], who studied the stability of complex dynamical systems using Lyapunov functions (where the goal is to check if a system eventually stops moving). Further, we provide an algorithm to generate barrier certificates for quantum systems and use the algorithm to generate barriers for several examples.

2   Background

2.1   Safety Analysis

We begin by introducing the problem of safety for dynamical systems with real state variables $x \in \mathbb{R}^n$. More details can be found in [24]. A continuous dynamical system is described by

$$\dot{x} = \frac{\mathrm{d}x}{\mathrm{d}t} = f(x), \quad f : \mathbb{R}^n \to \mathbb{R}^n,$$

where the evolution of the system is restricted to $X \subseteq \mathbb{R}^n$ and f is usually Lipschitz continuous to ensure existence and uniqueness of the differential equation solution. The set $X_0 \subseteq X$ is the set of initial states and the unsafe set $X_u \subseteq X$ is the set of values that the dynamics $x(t)$ should avoid. These sets lead to the idea of safety for real continuous dynamical systems:

Definition 1 (Safety). *A system, $\dot{x} = f(x)$, evolving over $X \subseteq \mathbb{R}^n$ is considered safe if the system cannot reach the unsafe set, $X_u \subseteq X$, from the initial set, $X_0 \subseteq X$. That is for all $t \in \mathbb{R}_+$ and $x(0) \in X_0$, then $x(t) \notin X_u$.*

The safety problem is to determine if a given system is safe or not. Numerous techniques have been developed to solve this problem [12]. Barrier certificates are discussed in Sect. 2.2. Here, we describe two other common techniques.

Abstract Interpretation One way to perform reachability analysis of a system is to give an abstraction [9,10] of the system's evolution. Given an initial abstraction that over-approximates the evolution of the system, the abstraction is improved based on false bugs. False bugs are generated when the current abstraction enters the unsafe space but the actual system does not. This method has been investigated for quantum programs in [32], where the authors can verify programs using up to 300 qubits.

Backward and Forward Reachability A second approach is to start from the unsafe region and reverse the evolution of the system from there. A system is considered unsafe if the reversed evolution enters the initial region. This is backward reachability. Conversely, forward reachability starts from the initial region and is considered safe if the reachable region does not enter the unsafe region. Both backward and forward reachability are discussed in [18,27,28].

2.2   Barrier Certificates

Barrier certificates [24] are another technique used for safety analysis. This technique attempts to divide the reachable region from the unsafe region by putting constraints on the initial and unsafe set, and on how the system evolves. The benefit of barrier certificates over other techniques is that one does not need to compute the system's dynamics at all to guarantee safety, unlike in abstract interpretation and backward (or forward) reachability.

A barrier certificate is a differentiable function, $B : \mathbb{R}^n \to \mathbb{R}$, that determines safety through the properties that B has. Generally, a barrier certificate needs to meet the following conditions:

$$B(x) \leq 0, \forall x \in X_0 \tag{1}$$

$$B(x) > 0, \forall x \in X_u \tag{2}$$

$$x(0) \in X_0 \implies B(x(t)) \leq 0, \forall t \in \mathbb{R}_+. \tag{3}$$

Essentially, these conditions split the evolution space into a (over-approximate) reachable region and an unsafe region, encapsulated by Conditions (1) and (2) respectively. These regions are separated by a "barrier", which is the contour along $B(x) = 0$.

Condition (3) prevents the system evolving into the unreachable region and needs to be satisfied for the system to be safe. However, Condition (3) can be replaced with stronger conditions that are easier to check. For example, the definition of one simple type of barrier certificate is given.

Definition 2 (Convex Barrier Certificate). *For a system $\dot{x} = f(x)$, $X \subseteq \mathbb{R}^n$, $X_0 \subseteq X$ and $X_u \subseteq X$, a function $B : \mathbb{R}^n \to \mathbb{R}$ that obeys the following conditions:*

$$B(x) \leq 0, \forall x \in X_0$$

$$B(x) > 0, \forall x \in X_u$$

$$\frac{\mathrm{d}B}{\mathrm{d}x} f(x) \leq 0, \forall x \in X, \tag{4}$$

is a convex barrier certificate.

Note that in Condition (4): $\frac{\mathrm{d}B}{\mathrm{d}x}\frac{\mathrm{d}x}{\mathrm{d}t} = \frac{\mathrm{d}B}{\mathrm{d}t}$. This condition can be viewed as a constraint on the evolution of the barrier as the system evolves over time.

Now, if a system has a barrier certificate, then the system is safe. We show the safety theorem for convex barrier certificates.

Theorem 1. *If a system, $\dot{x} = f(x)$, has a convex barrier certificate, $B : \mathbb{R}^n \to \mathbb{R}$, then the system is safe [24].*

Proofs of Theorem 1 are standard and can be found in, *e.g.*, [24]. The intuition behind the proof is that since the system starts in the negative region and the barrier can never increase, then the barrier can never enter the positive region. Since the unsafe set is within the positive region of the barrier, this set can therefore never be reached. Thus, the system cannot evolve into the unsafe set and so the system is safe. Figure 1 shows an example of a dynamical system with a barrier based on the convex condition.

Remark 1. The term "convex" is used for these barriers as the set of barrier certificates satisfying the conditions in Definition 2 is convex. In other words, if B_1 and B_2 are barrier certificates for a system, the function $\lambda B_1 + (1 - \lambda)B_2$ is also a barrier certificate for any $\lambda \in [0, 1]$. See [24] for similar details.

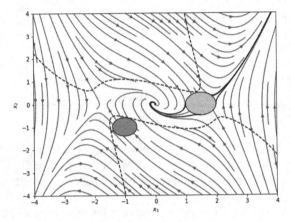

Fig. 1. Example adapted from Section V-A in [24]. The initial region is the green circle centred at $(1.5, 0)$ and the system evolves according to the dynamical system given by differential equations $\dot{x} = [x_2, -x_1 + \frac{1}{3}x_1^3 - x_2]$. The unsafe region is the red circle centred at $(-1, -1)$ and is separated from the initial region by a barrier, the dashed purple line defined by $B(x) = 0$ where $B(x) = -13 + 7x_1^2 + 16x_2^2 - 6x_1^2x_2^2 - \frac{7}{6}x_1^4 - 3x_1x_2^3 + 12x_1x_2 - \frac{12}{3}x_1^3x_2$.

There are a variety of different barrier certificates to choose from with different benefits, *e.g.*, the convex condition given is simple but may not work for complicated or nonlinear systems. In comparison, the non-convex condition given in [24] changes Condition (4) such that $\frac{dB}{dx}f(x) \leq 0; \forall x \in X, B(x) = 0$ (instead of $\forall x \in X$). This is a weaker condition allowing for more functions to be a suitable barrier certificate. However, a different computational method is required because the set of such barrier certificates is non-convex. Each barrier certificate requires a different proof that if the system has a satisfying barrier certificate, then the system is safe. It should be noted that Theorem 1 only has a one way implication, a system does not necessarily have a barrier certificate even if it is safe. In [31], the authors showed the converse holds for systems defined on a compact manifold and using convex barrier certificates.

3   Complex-Valued Barrier Certificates

Now we wish to extend the use of barrier certificates into a complex space ($\mathbb{C}^n$). We use $i = \sqrt{-1}$ as the imaginary unit in the rest of the paper. The complex dynamical systems considered are of the form

$$\dot{z} = \frac{dz}{dt} = f(z), \quad f : \mathbb{C}^n \to \mathbb{C}^n,$$

which evolves in $Z \subseteq \mathbb{C}^n$. The initial and unsafe sets are defined in the usual way except now we have $Z_0 \subseteq Z$ and $Z_u \subseteq Z$, respectively. The notion of safety for this system is similar to Definition 1.

Definition 3 (Safety). *A complex system, $\dot{z} = f(z)$, with $Z \subseteq \mathbb{C}^n$, $Z_0 \subseteq Z$ and $Z_u \subseteq Z$, is considered safe if for any $z(0) \in Z_0$, then $\forall t \in \mathbb{R}^+, z(t) \notin Z_u$.*

Whilst it is easy to extend the safety problem and required definitions into the complex plane, extending the notion of barrier certificates requires particular attention. Conditions (1), (2) and (3) are changed respectively to

$$B(z) \leq 0, \forall z \in Z_0; \tag{5}$$
$$B(z) > 0, \forall z \in Z_u; \tag{6}$$
$$z(0) \in Z_0 \implies B(z(t)) \leq 0, \forall t \in \mathbb{R}_+. \tag{7}$$

Many barrier certificates use differential equations to achieve Condition (7), which restricts the class of complex functions that can be used. Differentiable complex functions must satisfy the Cauchy-Riemann equations. For $z_j = x_j + iy_j$, $g(z) : \mathbb{C}^n \to \mathbb{C}$ and $g(z) = g(x, y) = u(x, y) + iv(x, y)$, the Cauchy-Riemann equations (for several variables) are

$$\frac{\partial u}{\partial x_j} = \frac{\partial v}{\partial y_j} \qquad \frac{\partial u}{\partial y_j} = -\frac{\partial v}{\partial x_j}.$$

We say g is holomorphic if all its partial derivatives, $\frac{\partial g(z)}{\partial z_j}$, satisfy the Cauchy-Riemann equations and therefore $\frac{dg(z)}{dz}$ exists. When we consider our barrier certificates, $B(z)$, we want them to be holomorphic on z to make use of the already available theory.

Using an adapted technique developed by Fang and Sun [11] allows us to reason about barrier certificates in the complex plane. We begin by introducing a family of complex functions that are key to our technique.

Definition 4 (Conjugate-flattening function). *A function, $b : \mathbb{C}^n \times \mathbb{C}^n \to \mathbb{C}^n$, is conjugate-flattening if $\forall z \in \mathbb{C}^n, b(z, \bar{z}) \in \mathbb{R}$.*

Definition 5 (Complex-valued barrier function). *A function, $B : \mathbb{C}^n \to \mathbb{R}$, is a complex-valued barrier function if $B(z) = b(z, \bar{z})$ where $b : \mathbb{C}^n \times \mathbb{C}^n \to \mathbb{C}^n$ is a conjugate-flattening, holomorphic function.*

Suppose now that we have a system that evolves over time, $z(t)$. To use the complex-valued barrier function, $B(z(t))$, for barrier certificates we require the differential of B with respect to t. The chain rule reveals that

$$\frac{dB(z(t))}{dt} = \frac{db(z(t), \overline{z(t)})}{dt} = \frac{db(z, u)}{dz}\bigg|_{u=\bar{z}} \frac{dz(t)}{dt} + \frac{db(z, u)}{du}\bigg|_{u=\bar{z}} \frac{du(t)}{dt}\bigg|_{u=\bar{z}},$$

where $\frac{db(z,u)}{dz} = \left[\frac{\partial b(z,u)}{\partial z_1}, \frac{\partial b(z,u)}{\partial z_2}, \ldots, \frac{\partial b(z,u)}{\partial z_n}\right]$ is the gradient of $b(z, u)$ with respect to z and the gradient is defined with respect to u in a similar way. Now noting that $\frac{du(t)}{dt}\big|_{u=\bar{z}} = \text{Re}\{f\} - i\,\text{Im}\{f\} = \overline{\frac{dz(t)}{dt}}$, we get that

$$\frac{dB(z(t))}{dt} = \frac{db(z, u)}{dz}\bigg|_{u=\bar{z}} f(z) + \frac{db(z, u)}{du}\bigg|_{u=\bar{z}} \overline{f(z)}. \tag{8}$$

Given Eq. (8), barrier certificates that include a differential condition can be extended into the complex domain quite naturally. For example, the convex barrier certificate is extended to the complex domain.

Definition 6 (Complex-valued Convex Barrier Certificate). *For a system $\dot{z} = f(z)$, $Z \subseteq \mathbb{C}^n$, $Z_0 \subseteq Z$ and $Z_u \subseteq Z$; a complex-valued barrier function $B : \mathbb{C}^n \to \mathbb{R}$, $B(z) = b(z, \bar{z})$, that obeys the following conditions,*

$$b(z, \bar{z}) \leq 0, \forall z \in Z_0 \tag{9}$$

$$b(z, \bar{z}) > 0, \forall z \in Z_u \tag{10}$$

$$\left.\frac{db(z, u)}{dz}\right|_{u=\bar{z}} f(z) + \left.\frac{db(z, u)}{du}\right|_{u=\bar{z}} \overline{f(z)} \leq 0, \forall z \in Z, \tag{11}$$

is a complex-valued convex barrier certificate.

The reason for defining a barrier certificate from complex variables rather than their real and imaginary parts is so that we avoid doubling the number of variables used in the analysis. Additionally, this provides us with a framework for defining other barrier certificates based on complex variables. With Definition 6, we can ensure the safety of complex dynamical systems:

Theorem 2. *If a complex system, $\dot{z} = f(z)$, has a complex-valued convex barrier certificate, $B : \mathbb{C}^n \to \mathbb{R}$, then the system is safe.*

Proposition 1. *The set of complex-valued barrier certificates satisfying the conditions of Definition 2 is convex.*

4   Generating Satisfiable Barrier Certificates for Quantum Systems

We now describe how to compute a complex-valued barrier function. Throughout, let $\dot{z} = f(z)$, $Z \subseteq \mathbb{C}^n$, $Z_0 \subseteq Z$ and $Z_u \subseteq Z$ be defined as before. We introduce a general family of functions that will be used as "templates" for complex barrier certificates.

Definition 7. *A k-degree polynomial function is a complex function, $b : \mathbb{C}^n \to \mathbb{C}$, such that*

$$b(z_1, \ldots, z_n) = \sum_{\alpha \in A_{n,k}} a_\alpha z^\alpha \tag{12}$$

where $A_{n,k} := \{\boldsymbol{\alpha} = (\alpha_1, \ldots, \alpha_n) \subseteq \mathbb{N}_0^n : \sum_{j=1}^n \alpha_j \leq k\}$, $a_\alpha \in \mathbb{C}$, and $z^\alpha = \prod_{j=1}^n z_j^{\alpha_j}$.

The family of k-degree polynomials are polynomial functions where no individual term of the polynomial can have a degree higher than k. Note that k-degree polynomial functions are holomorphic. Further, some k-degree polynomials are conjugate-flattening. For example, the 2-degree polynomial $b(z_1, u_1) =$

$z_1 u_1$ is conjugate-flattening since $z\bar{z} = |z|^2$, whereas the 1-degree polynomial $b(z_1, u_1) = z_1$ is not. Thus, a subset of this family of functions are suitable to be used for barrier certificates as complex-valued barrier functions.

The partial derivative of the polynomials in Eq. (12) is required for ensuring the function meets Condition (11). The partial derivative of the function is

$$\frac{\partial b}{\partial z_j} = \sum_{\alpha \in A_{n,k}} a_\alpha \alpha_j z_j^{-1} z^\alpha. \tag{13}$$

We write

$$B(a, z) := b(a, z, \bar{z}) := \sum_{\substack{(\alpha,\beta) \in A_{2n,k} \\ \alpha = (\alpha_1,...,\alpha_n) \\ \beta = (\alpha_{n+1},...\alpha_{2n})}} a_{\alpha,\beta} z^\alpha \bar{z}^\beta,$$

where $a = (a_{\alpha,\beta}) \in \mathbb{R}^{|A_{2n,k}|}$ is a vector of real coefficients to be found and $\bar{z}^\beta = \prod_{j=1}^n \bar{z}_j^{\alpha_{n+j}}$.

The following (polynomial) inequalities find the coefficient vector:

$$\textbf{find } a^T$$
$$\textbf{subject to } B(a, z) \leq 0, \forall z \in Z_0$$
$$B(a, z) > 0, \forall z \in Z_u$$
$$\frac{dB(a, z)}{dt} \leq 0, \forall z \in Z \tag{14}$$
$$B(a, z) \in \mathbb{R}$$
$$-1 \leq a_{\alpha,\beta} \leq 1.$$

The coefficients, $a_{\alpha,\beta} \in \mathbb{R}$, are restricted to the range $[-1, 1]$ since any barrier certificate $B(a, z)$, can be normalised by dividing B by the coefficient of greatest weight, $m = \max|a_{\alpha,\beta}|$. The resulting function $\frac{1}{m}B(a, z)$ is still a barrier certificate. A barrier certificate generated from these polynomial inequalities can then freely be scaled up by multiplying it by a constant.

4.1   An Algorithmic Solution

One approach of solving the inequalities in (14) is to convert the system to real numbers and solve using sum of squares (SOS) optimisation [24]; another method is to use SMT solvers to find a satisfiable set of coefficients; or it is possible to use neural network based approaches to find possible barriers [1,22].

For our problem, time-independent quantum systems exhibit periodic behaviour [19]; that is for all $t \in \mathbb{R}^+$, $z(t) = z(t + T)$ for some T. By interpreting the barrier as a function over time and considering a periodic system $z(t)$, we get that $B(t) = B(z(t)) = B(z(t+T)) = B(t+T), \forall t \in \mathbb{R}^+$. Therefore, if a given system is periodic, then the barrier is periodic as well. To satisfy this property within the program in (14), we further require that $\frac{dB(a,z)}{dt} = 0$ rather than $\frac{dB(a,z)}{dt} \leq 0$, which allows the problem to be turned into a linear program.

Whilst there are other properties that ensure a function is periodic, these would involve non-polynomial terms such as trigonometric functions. Further, linear programs tend to be solved faster than SOS methods. This is because SOS programs are solved through semidefinite programming techniques, which are extensions of linear programs and therefore harder to solve.

We begin by transforming the differential constraint, $\frac{dB(a,z)}{dt} = 0$. To obey the third condition for the complex-valued convex barrier certificate, we can substitute terms in Eq. (8) with the partial derivatives from Eq. (13). Essentially one will end up with an equation of the form

$$(\mathbf{A}a)^{\top}\zeta = 0,$$

where ζ is a vector of all possible polynomial terms of $z_j, \overline{z}_j$ with degree less than k,[1] and $\mathbf{A}$ is a matrix of constant values. By setting $\mathbf{A}a = \mathbf{0}$ the constraint is satisfied. Therefore, each row of the resultant vector, $(\mathbf{A}a)_j = 0$, is added as a constraint to a linear program.

To transform the real constraint $(B(a,z) \in \mathbb{R})$ note that if $x \in \mathbb{C}$, then $x \in \mathbb{R}$ if and only if $x = \overline{x}$. Therefore, $B(a,z) - \overline{B(a,z)} = 0$ and we have

$$B(a,z) - \overline{B(a,z)} = \sum_{\substack{(\alpha_j) \in A_{2n,k} \\ \alpha = \{\alpha_1,...,\alpha_n\} \\ \beta = \{\alpha_{n+1},...\alpha_{2n}\}}} a_{\alpha,\beta} z^{\alpha} \overline{z}^{\beta} - \sum_{\substack{(\alpha_j) \in A_{2n,k} \\ \alpha' = \{\alpha_1,...,\alpha_n\} \\ \beta' = \{\alpha_{n+1},...\alpha_{2n}\}}} \overline{a}_{\alpha',\beta'} z^{\beta'} \overline{z}^{\alpha'}$$

$$= \sum_{\substack{(\alpha_j) \in A_{2n,k} \\ \alpha = \{\alpha_1,...,\alpha_n\} \\ \beta = \{\alpha_{n+1},...\alpha_{2n}\}}} (a_{\alpha,\beta} - \overline{a}_{\beta,\alpha}) z^{\alpha} \overline{z}^{\beta},$$

where the final summation is achieved by grouping terms where $\alpha' = \beta$ and $\beta' = \alpha$.

The whole polynomial is equal to 0 if all coefficients are 0. Thus, taking the coefficients and noting that $a_j \in \mathbb{R}$ gives the transformed constraints $a_{\alpha,\beta} = a_{\beta,\alpha}$ for $\alpha = (\alpha_j)_{j=1}^{n}, \beta = (\alpha_j)_{j=n+1}^{2n}, (\alpha_j) \in A_{2n,k}$. These constraints to the coefficients are then also added to the linear program.

The final constraints we need to transform are the constraints on the initial and unsafe set: $B(a,z) \leq 0$ for $z \in Z_0$ and $B(a,z) > 0$ for $z \in Z_u$, respectively. We begin by noting that $B(a,z) = c + b(a,z,\overline{z})$ where $b(a,z,\overline{z})$ is a k-degree polynomial (with coefficients a) and $c \in \mathbb{R}$ is a constant. When considering the differential and real constraint steps, c is not involved in these equations since c does not appear in the differential term and c is cancelled out in the real constraint $(c - \overline{c} = c - c = 0)$.

Considering the initial and unsafe constraints, we require that

$$\forall z \in Z_0, \ c + b(a,z,\overline{z}) \leq 0, \text{ and}$$
$$\forall z \in Z_u, \ c + b(a,z,\overline{z}) > 0.$$

---

[1] e.g., for $k = 2$ acceptable terms include $z_j^a, z_j z_l, z_j \overline{z}_l, \overline{z}_j^a, \overline{z}_j \overline{z}_l$ for $0 \leq a \leq 2$.

Algorithm 1 Computing the barrier certificate using linear programming

1: Solve the linear program

$$\textbf{find } a^T$$
$$\textbf{subject to } \mathbf{A}a = 0$$
$$a_{\alpha,\beta} = a_{\beta,\alpha} \qquad\qquad \text{for } \alpha = \{\alpha_j\}_{j=1}^n, \beta = \{\alpha_j\}_{j=n+1}^{2n},$$
$$-1 \leq a_j \leq 1. \qquad\qquad \text{and } \{\alpha_j\}_{j=1}^{2n} \in A_{2n,k}$$

2: $c \leftarrow \min_{z \in Z_0} -b(a, z, \overline{z})$
3: **if** $c > \max_{z \in Z_u} -b(a, z, \overline{z})$ **then return** $B(a, z) = c + b(a, z, \overline{z})$
4: **else   fail**

Therefore, c is bounded by

$$\max_{z \in Z_u} -b(a, z, \overline{z}) < c \leq \min_{z \in Z_0} -b(a, z, \overline{z}).$$

Finding $c = \min_{z \in Z_0} -b(a, z, \overline{z})$ and then checking $\max_{z \in Z_u} -b(a, z, \overline{z}) < c$ will ensure the initial and unsafe constraints are met for the barrier. The final computation is given in Algorithm 1.

Note that the algorithm can fail since the function b may divide the state space in such a way that a section of Z_0 may lie on the same contour as a section of Z_u. This means that either the function b is unsuitable or the system is inherently unsafe.

5   Application to Quantum Systems

We consider quantum systems that evolve within Hilbert spaces $\mathcal{H}^n = \mathbb{C}^{2^n}$ for $n \in \mathbb{N}$. We use the computational basis states $|j\rangle \in \mathcal{H}^n$, for $0 \leq j < 2^n$, as an orthonormal basis within the space, where $(|j\rangle)_l = \delta_{jl}$.[2] General quantum states, $|\phi\rangle \in \mathcal{H}^n$, can then be written in the form

$$|\phi\rangle = \sum_{j=0}^{2^n-1} z_j |j\rangle,$$

where $z_j \in \mathbb{C}$ and $\sum_{j=0}^{2^n-1} |z_j|^2 = 1$.[3] Quantum states reside within the unit circle of $\mathbb{C}^{2^n}$. For simplicity, we consider quantum systems that evolve according to the Schrödinger equation

$$\frac{d|\phi\rangle}{dt} = -i\hat{H}|\phi\rangle, \qquad\qquad (15)$$

where $\hat{H}$ is a Hamiltonian and $|\phi\rangle$ is a quantum state.[4] A Hamiltonian is a Hermitian matrix, which is a complex matrix $\hat{H}$ such that $\hat{H}^\dagger = \overline{\hat{H}^\top}$.

---

[2] δ_{jl} is the Kronecker delta, which is 1 if $j = l$ and 0 otherwise.
[3] For readers familiar with the Dirac notation, $z_j = \langle j|\phi\rangle$ and $\overline{z_j} = \langle\phi|j\rangle$.
[4] We set the Planck constant $\hbar = 1$ in the Schrödinger equation.

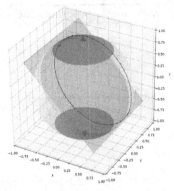

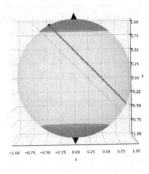

(a) Isometric view of system   (b) Top down view of system

Fig. 2. System evolution on a Bloch sphere. The initial state of the system is $\sqrt{0.9}|0\rangle +$ $i\sqrt{0.1}|1\rangle$ (the black dot) and evolves according to the black line (in an anti-clockwise rotation with a period of $t = \pi$). The green surface around the north pole ($|0\rangle$) is the initial region, Z_0, and the red surface around the south pole ($|1\rangle$) is the unsafe region, Z_u. The blue surface is the plane of the barrier function when $B(z) = 0$, with $x < -z$ being the unsafe region.

Note that the solution to the system is $z(t) = e^{i\hat{H}t}z(0) = U(t)z(0)$ where $U(t)$ is a unitary matrix ($U^{-1} = U^{\dagger}$). Throughout, we write $|\phi\rangle = z$ and so the Schrödinger equation becomes $\frac{dz}{dt} = -i\hat{H}z = \dot{z}$.

Remark 2. The Schrödinger equation in (15) can be translated into a linear time-invariant system over reals by separating the real and imaginary parts of the variables. Although reachability analysis of such linear systems has been studied extensively through set propagation with multiple techniques available in the literature [2], these techniques are mainly focused on finite-time reachability. In our work, we study infinite-time reachability thus utilising the concept of barrier certificate without doubling the number of variables.

Again, we can make use of Algorithm 1 since time-independent Schrödinger equations are periodic [19]. In the rest of this section, we make use of Algorithm 1 in order to find suitable barrier certificates for operations that are commonly used in quantum computers. The example of Hamiltonians we consider are of these operations [20,25].

5.1   Hadamard Operation Example

The evolution of the Hadamard operation, $H = \frac{1}{\sqrt{2}} \begin{pmatrix} 1 & 1 \\ 1 & -1 \end{pmatrix}$, is given by $\hat{H}_H = \begin{pmatrix} 1 & 1 \\ 1 & -1 \end{pmatrix}$ and $|\phi\rangle$ is one qubit, $z_0|0\rangle + z_1|1\rangle$. We have $z(t) = \begin{pmatrix} z_0(t) \\ z_1(t) \end{pmatrix}$ and

$$\dot{z} = -i\hat{H}_H z = -i \begin{pmatrix} z_0 + z_1 \\ z_0 - z_1 \end{pmatrix}.$$

The system evolves over the surface of the unit sphere, $Z = \{(z_0, z_1) \in \mathbb{C}^2 : |z_0|^2 + |z_1|^2 = 1\}$. The initial set is defined as $Z_0 = \{(z_0, z_1) \in Z : |z_0|^2 \geq 0.9\}$ and the unsafe set as $Z_u = \{(z_0, z_1) \in Z : |z_0|^2 \leq 0.1\}$. Note that the definitions of Z_0 and Z_u are restricted by Z, therefore $|z_1|^2 \leq 0.1$ and $|z_1|^2 \geq 0.9$ for Z_0 and Z_u respectively. A barrier function computed by Algorithm 1 is

$$B(z) = \frac{11}{5} - 3z_0\overline{z_0} - z_0\overline{z_1} - \overline{z_0}z_1 - z_1\overline{z_1}.$$

By rearranging and using properties of the complex conjugate, we find that

$$B(z) = 2(\frac{1}{10} - |z_0|^2 + \frac{1}{2} - \text{Re}\{z_0\overline{z_1}\}). \tag{16}$$

The first term of the barrier ($\frac{1}{10} - |z_0|^2$) acts as a restriction on how close to $|0\rangle$ as $|\phi\rangle$ evolves, whereas the second term ($\frac{1}{2} - \text{Re}\{z_0\overline{z_1}\}$) is a restriction on the phase of the quantum state. Next, we double check that B is indeed a barrier certificate.

Proposition 2. *The system evolving according to Eq. (5.1), initial set Z_0 and unsafe set Z_u is safe.*

A visualisation on a Bloch sphere representation of the example system and its associate barrier are given in Fig. 2.

5.2 Phase Operation Example

The evolution of the phase operation $S = \begin{pmatrix} 1 & 0 \\ 0 & i \end{pmatrix}$ is given by the Hamiltonian

$\hat{H}_S = \begin{pmatrix} 1 & 0 \\ 0 & -1 \end{pmatrix}$ for a single qubit $z_0|0\rangle + z_1|1\rangle$. Thus, the evolution of the system for $z(t) = \begin{pmatrix} z_0(t) \\ z_1(t) \end{pmatrix}$ is

$$\dot{z} = -i\begin{pmatrix} z_0 \\ -z_1 \end{pmatrix}. \tag{17}$$

Again, Z represents the unit sphere as described previously. Two pairs of safe and unsafe regions are given. The first pair $Z_1 = (Z_0^1, Z_u^1)$ is given by

$$Z_0^1 = \{(z_0, z_1) \in Z : |z_0|^2 \geq 0.9\}, \quad Z_u^1 = \{(z_0, z_1) \in Z : |z_1|^2 > 0.11\};$$

and the second pair $Z_2 = (Z_0^2, Z_u^2)$ is given by

$$Z_0^2 = \{(z_0, z_1) \in Z : |z_1|^2 \geq 0.9\}, \quad Z_u^2 = \{(z_0, z_1) \in Z : |z_0|^2 > 0.11\}.$$

The pair Z^1 starts with a system that is close to the $|0\rangle$ state and ensures that the system cannot evolve towards the $|1\rangle$ state. The pair Z^2 has similar behaviour with respective states $|1\rangle$ and $|0\rangle$. The system for each pair of constraints is considered safe by the following barriers computed by Algorithm 1:

$$B_1(z) = 0.9 - z_0\overline{z_0}, \quad B_2(z) = 0.9 - z_1\overline{z_1},$$

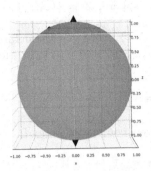

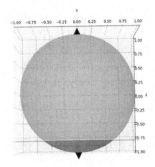

(a) Evolution with initial and unsafe states Z^1. The barrier at $B_1(z) = 0$ is a flat plane that borders Z_0^1.

(b) Evolution with initial and unsafe states Z^2. Similarly, $B_2(z) = 0$ is a flat plane that borders Z_0^2.

Fig. 3. State evolution of (17) demonstrated on a Bloch sphere.

where B_1 is the barrier for Z^1 and B_2 is the barrier for Z^2.[5] The system with different pairs of regions can be seen on Bloch spheres in Fig. 3. Again, both functions B_1 and B_2 are valid barrier certificates.

Proposition 3. *The system given by Eq. 17 with the set of initial states Z_0^1 and the unsafe set Z_u^1 is safe.*

Proposition 4. *The system given by Eq. 17 with the set of initial states Z_0^2 and the unsafe set Z_u^2 is safe.*

These barriers give bounds on how the system evolves, *i.e.*, the system must only change the phase of the system and not the amplitude. This can be applied in general by combining barriers to show how a (disturbed) system is restricted in its evolution.

5.3 Controlled-NOT Operation Example

The final example we consider is the controlled-NOT (CNOT) operation acting on two qubits; a control qubit, $|\phi_c\rangle$, and a target qubit, $|\phi_t\rangle$, with the full quantum state being $|\phi_c\phi_t\rangle$. The CNOT operation performs the NOT operation on a target qubit ($|0\rangle \rightarrow |1\rangle$ and $|1\rangle \rightarrow |0\rangle$) if the control qubit is set to $|1\rangle$ and does nothing if the control qubit is set to $|0\rangle$. The CNOT operation and its associated Hamiltonian are given by

$$\text{CNOT} = \begin{pmatrix} 1 & 0 & 0 & 0 \\ 0 & 1 & 0 & 0 \\ 0 & 0 & 0 & 1 \\ 0 & 0 & 1 & 0 \end{pmatrix}, \quad \hat{H}_{\text{CNOT}} = \begin{pmatrix} 0 & 0 & 0 & 0 \\ 0 & 0 & 0 & 0 \\ 0 & 0 & 1 & -1 \\ 0 & 0 & -1 & 1 \end{pmatrix}.$$

---

[5] These barriers can similarly be written using the Dirac notation.

The system $z(t) = (z_j(t))_{j=0,\dots,3}$ evolves according to

$$\dot{z} = -\mathrm{i} \begin{pmatrix} 0 \\ 0 \\ z_2 - z_3 \\ -z_2 + z_3 \end{pmatrix}.$$

This system evolves over $Z = \{(z_0, \dots, z_3) \in \mathbb{C}^4 : \sum_{j=0}^{3} |z_j|^2 = 1\}$. Using this as our system, various initial and unsafe regions can be set up to reason about the behaviour of the CNOT operation.

Control in $|0\rangle$ Here we consider the following initial and unsafe regions

$$Z_0 = \{(z_j)_{j=0}^3 \in \mathbb{C}^4 : |z_0|^2 \geq 0.9\},$$
$$Z_u = \{(z_j)_{j=0}^3 \in \mathbb{C}^4 : |z_1|^2 + |z_2|^2 + |z_3|^2 \geq 0.11\}.$$

The initial set, Z_0, encapsulates the quantum states that start in the $|00\rangle$ state with high probability and Z_u captures the states that are not in the initial region with probability greater than 0.11. These regions capture the behaviour that the quantum state should not change much when the control qubit is in the $|0\rangle$ state. Using Algorithm 1, the barrier $B(z) = 0.9 - z_0\overline{z_0}$ can be generated to show that the system is safe.

A similar example can be considered where the initial state $|00\rangle$ is replaced with $|01\rangle$ instead (swap z_0 and z_1 in Z_0 and Z_u). The behaviour that the state of the system should not change much is still desired; the function $B(z) = 0.9 - z_1\overline{z_1}$ is computed as a barrier to show this behaviour is met.

Control in $|1\rangle$ Now consider when the initial region has the control qubit near the state $|1\rangle$. The following regions are considered:

$$Z_0 = \{(z_j)_{j=0}^3 \in \mathbb{C}^4 : |z_2|^2 \geq 0.9\},$$
$$Z_u = \{(z_j)_{j=0}^3 \in \mathbb{C}^4 : |z_1|^2 + |z_2|^2 \geq 0.11\}.$$

This system starts close to the $|10\rangle$ state and the evolution should do nothing to the control qubit. Note that the specified behaviour does not captures the NOT behaviour on the target qubit. Our Algorithm 1 considers this system safe by outputting the barrier certificate $B(z) = 0.9 - z_2\overline{z_2} - z_3\overline{z_3}$. This is also the barrier if the system were to start in the $|11\rangle$ state instead.

6   Conclusions

In this paper, we extended the theory of barrier certificates to handle complex variables and demonstrated that barrier certificates can be extended to use complex variables. We then showed how one can automatically generate simple complex-valued barrier certificates using polynomial functions and linear programming techniques. Finally, we explored the application of the developed techniques by investigating properties of time-independent quantum systems.

There are numerous directions for this research to take. In particular, one can consider (quantum) systems that are time-dependent, have a control component or are discrete-time, *i.e.*, quantum circuits. Data-driven approaches for generating barrier certificates based on measurements of a quantum system can also be considered. A final challenge to consider is how to verify large quantum systems. Techniques, such as Trotterization, allow Hamiltonians to be simulated either by simpler Hamiltonians of the same size or of lower dimension. How barrier certificates can ensure safety of such systems is a route to explore.

Acknowledgements. M. Lewis is supported by the UK EPSRC (project reference EP/T517914/1). The work of S. Soudjani is supported by the following grants: EPSRC EP/V043676/1, EIC 101070802, and ERC 101089047.

Data Availability Statement. An artifact with an implementation of the algorithm from Sect. 4 and case studies from Sect. 5 was submitted to the QEST 2023 Artifact Evaluation. The public repository is available on GitHub: https://github.com/marco-lewis/quantum-barrier-certificates

Proof Availability. Proofs for Theorem 2; Proposition 1 and 2; and the derivation of (16) in Sect. 5.1 are given in the appendix of the preprint version of this paper, available at https://arxiv.org/abs/2307.07307.

References

1. Abate, A., et al.: FOSSIL: a software tool for the formal synthesis of Lyapunov functions and barrier certificates using neural networks. In: Proceedings of the 24th International Conference on Hybrid Systems: Computation and Control. ACM (2021). https://doi.org/10.1145/3447928.3456646
2. Althoff, M., Frehse, G., Girard, A.: Set propagation techniques for reachability analysis. Annu. Rev. Control Robot. Auton. Syst. **4**, 369–395 (2021)
3. Ames, A.D., et al.: Control barrier functions: theory and applications. In: 18th European control conference (ECC), pp. 3420–3431. IEEE (2019)
4. Bak, S.: t-barrier certificates: a continuous analogy to k-induction. In: 6th IFAC Conference on Analysis and Design of Hybrid Systems, pp. 145–150 (2018). https://doi.org/10.1016/j.ifacol.2018.08.025
5. Burgholzer, L., Wille, R.: Advanced equivalence checking for quantum circuits. IEEE Trans. Comput. Aided Des. Integr. Circuits Syst. **40**, 1810–1824 (2021). https://doi.org/10.1109/TCAD.2020.3032630
6. Chareton, C., Bardin, S., Bobot, F., Perrelle, V., Valiron, B.: An automated deductive verification framework for circuit-building quantum programs. In: ESOP 2021. LNCS, vol. 12648, pp. 148–177. Springer, Cham (2021). https://doi.org/10.1007/978-3-030-72019-3_6
7. Clarke, E.M., et al.: Model Checking, 2nd edn. MIT Press, Cambridge (2018)
8. Combes, J., Wiseman, H.M.: Quantum feedback for rapid state preparation in the presence of control imperfections. J. Phys. B. At. Mol. Opt. Phys. **44**(15), 154008 (2011). https://doi.org/10.1088/0953-4075/44/15/154008
9. Cousot, P.: Abstract interpretation based formal methods and future challenges. In: Wilhelm, R. (ed.) Informatics. LNCS, vol. 2000, pp. 138–156. Springer, Heidelberg (2001). https://doi.org/10.1007/3-540-44577-3_10

10. Cousot, P., Cousot, R.: Abstract interpretation: a unified lattice model for static analysis of programs by construction or approximation of fixpoints. In: Proceedings of the 4th ACM SIGACT-SIGPLAN Symposium on Principles of Programming Languages, pp. 238–252 (1977). https://doi.org/10.1145/512950.512973
11. Fang, T., Sun, J.: Stability analysis of complex-valued nonlinear differential system. J. Appl. Math. **2013**, 621957 (2013). https://doi.org/10.1155/2013/621957
12. Fränzle, M., Chen, M., Kröger, P.: In memory of Oded Maler: automatic reachability analysis of hybrid-state automata. ACM SIGLOG News **6**(1), 19–39 (2019). https://doi.org/10.1145/3313909.3313913
13. Hietala, K., et al.: Proving quantum programs correct. In: 12th International Conference on Interactive Theorem Proving, pp. 21:1–21:19. Leibniz International Proceedings in Informatics (LIPIcs) (2021). https://doi.org/10.4230/LIPIcs.ITP.2021.21
14. Honarvar, S., Mousavi, M.R., Nagarajan, R.: Property-based testing of quantum programs in Q#. In: Proceedings of the IEEE/ACM 42nd International Conference on Software Engineering Workshops, pp. 430–435 (2020). https://doi.org/10.1145/3387940.3391459
15. Jagtap, P., Soudjani, S., Zamani, M.: Formal synthesis of stochastic systems via control barrier certificates. IEEE Trans. Autom. Control **66**(7), 3097–3110 (2021). https://doi.org/10.1109/TAC.2020.3013916
16. Lavaei, A., Soudjani, S., Abate, A., Zamani, M.: Automated verification and synthesis of stochastic hybrid systems: a survey. arXiv preprint arXiv:2101.07491 (2021)
17. Liu, J., et al.: Formal verification of quantum algorithms using quantum Hoare logic. In: Dillig, I., Tasiran, S. (eds.) CAV 2019. LNCS, vol. 11562, pp. 187–207. Springer, Cham (2019). https://doi.org/10.1007/978-3-030-25543-5_12
18. Mitchell, I.M.: Comparing forward and backward reachability as tools for safety analysis. In: Bemporad, A., Bicchi, A., Buttazzo, G. (eds.) HSCC 2007. LNCS, vol. 4416, pp. 428–443. Springer, Heidelberg (2007). https://doi.org/10.1007/978-3-540-71493-4_34
19. Mizuta, K., Fujii, K.: Optimal Hamiltonian simulation for time-periodic systems. Quantum **7**, 962 (2023). https://doi.org/10.22331/q-2023-03-28-962
20. Mozyrsky, D., Privman, V., Hotaling, S.P.: Design of gates for quantum computation: the NOT gate. Int. J. Mod. Phys. B **11**(18), 2207–2215 (1997). https://doi.org/10.1142/S0217979297001143
21. Murta, B., Cruz, P.M.Q., Fernández-Rossier, J.: Preparing valence-bond-solid states on noisy intermediate-scale quantum computers. Phys. Rev. Res. **5**, 013190 (2023). https://doi.org/10.1103/PhysRevResearch.5.013190
22. Peruffo, A., Ahmed, D., Abate, A.: Automated and formal synthesis of neural barrier certificates for dynamical models. In: TACAS 2021. LNCS, vol. 12651, pp. 370–388. Springer, Cham (2021). https://doi.org/10.1007/978-3-030-72016-2_20
23. Plesch, M., Brukner, Č: Quantum-state preparation with universal gate decompositions. Phys. Rev. A **83**, 032302 (2011). https://doi.org/10.1103/PhysRevA.83.032302
24. Prajna, S., Jadbabaie, A., Pappas, G.J.: A framework for worst-case and stochastic safety verification using barrier certificates. IEEE Trans. Autom. Control **52**, 1415–1428 (2007). https://doi.org/10.1109/TAC.2007.902736
25. Santos, A.C.: Quantum gates by inverse engineering of a Hamiltonian. J. Phys. B Atom. Mol. Opt. Phys. **51**(1), 015501 (2017). https://doi.org/10.1088/1361-6455/aa987c

26. Seligman, E., Schubert, T., Kumar, M.V.A.K.: Formal Verification: An Essential Toolkit for Modern VLSI Design. Morgan Kaufmann Publishers Inc., Burlington (2015)
27. Esmaeil Zadeh Soudjani, S., Abate, A.: Precise approximations of the probability distribution of a Markov process in time: an application to probabilistic invariance. In: Ábrahám, E., Havelund, K. (eds.) TACAS 2014. LNCS, vol. 8413, pp. 547–561. Springer, Heidelberg (2014). https://doi.org/10.1007/978-3-642-54862-8_45
28. Soudjani, S., Abate, A.: Quantitative approximation of the probability distribution of a Markov process by formal abstractions. Logical Meth. Comput. Sci. **11** (2015). https://doi.org/10.2168/LMCS-11(3:8)2015
29. Tao, R., et al.: Giallar: Push-button verification for the Qiskit quantum compiler. In: Proceedings of the 43rd ACM SIGPLAN International Conference on Programming Language Design and Implementation, pp. 641–656 (2022). https://doi.org/10.1145/3519939.3523431
30. van de Wetering, J.: ZX-calculus for the working quantum computer scientist. arXiv preprint arXiv:2012.13966 (2020)
31. Wisniewski, R., Sloth, C.: Converse barrier certificate theorem. In: 52nd IEEE Conference on Decision and Control, pp. 4713–4718 (2013). https://doi.org/10.1109/CDC.2013.6760627
32. Yu, N., Palsberg, J.: Quantum abstract interpretation. In: Proceedings of the 42nd ACM SIGPLAN International Conference on Programming Language Design and Implementation, pp. 542–558 (2021). https://doi.org/10.1145/3453483.3454061
33. Zulehner, A., Wille, R.: Advanced simulation of quantum computations. IEEE Trans. Comput. Aided Des. Integr. Circ. Syst. **38**, 848–859 (2017). https://doi.org/10.1109/TCAD.2018.2834427

Author Index

N. Jansen and M. Tribastone (Eds.): QEST 2023, LNCS 14287, pp. 363–364, 2023.
https://doi.org/10.1007/978-3-031-43835-6

Printed in the United States
by Baker & Taylor Publisher Services